AF477573

Electronic Power Control

VOLUME 1 : **POWER ELECTRONICS**

JEAN POLLEFLIET

ACADEMIA PRESS

©   The Author: Jean.Pollefliet@telenet.be

Eekhout 2
B - 9000 Gent (Belgium)
Tel. 00 32 9 233 80 88   Fax 00 32 9 233 14 09
Info@academiapress.be   www.academiapress.be

Distribution BENELUX
J.Story-Scientia Scientific Booksellers
P. Van Duyseplein 8
B – 9000 Gent (Belgium)
info@story.be
www.story.be

Jean Pollefliet

Electronic Power Control. Volume 1: Power Electronics

Gent, Academia Press, 2011, 504 pages.

Dutch: First edition 1986 , Seventh edition 2011

Seventh edition (English version): 2011

Illustrations and lay-out: Jean Pollefliet

Translation: Paul Fogarty, Rotterdam University

Illustrations cover: © Schott Solar

                    © Vestas Windtechnology A/S

ISBN 978 90 382 1791 8

D/2011/4804/196

U 1665

To my wife Gilberte

PREFACE

This book first appeared in 1986 and after 25 years has reached the seventh edition. With this seventh edition the book is now available in English.

Every edition saw continuous updating rearranging as well as addition of material and chapters. At the same time attention was also paid to the didactic aspects. This is not just important for students but also for the large group of people who use the book for self study.

With the sixth edition the book was divided into two parts, one on power electronics and the other on electronic motor control. This allows both volumes may be used independently of each other.

In this edition we continue to use the tradition of white and green pages. The green pages contain the mathematical derivations which in the first case are not necessary for studying the electronics. Once a sufficiently high level and the desire for specialist knowledge the reader can choose to make use of the green pages without disturbing the continuity of the study.

To mention a few numerical details, this book contains more than seven hundred figures, a hundred photos and more than fifty fully worked problems.

The purpose of the book is to explain the principles and applications of power electronics. Electronic switches and converters are studied in volume 1 and drive technology and positioning systems are dealt with in volume 2. A new chapter on "e-mobility" has been added to volume 2.

The largest part of this book is distilled from more than 40 years of lessons, talks and projects. The most important source of information is my students, especially the few hundred of whom I was the mentor I guided during their thesis for Master of Applied Engineering Sciences.

These I quided in wich I remain thankful and indebted to them.

To my publisher Peter Laroy of Academia Press I wish to express my thanks for many years of pleasant cooperation.

Our thanks also goes out to Paul Fogarty from Rotterdam University for the accurate English translation.

I would also like to thank Prof. dr. ir. Bernard Baeyens of the Ibague University (Colombia) for correcting and improving the Spanish technical vocabulary.

Last but not least, we have to thank the advertisers. As a result of their support, we have been able tot minimize the book shop price of our textbook.

In conclusion we wish the readers of this book a fruitful study.

Blankenberge, Belgium, June 2011

Jean.Pollefliet@telenet.be

With thanks to the advertisers:

Hitachi	p. 3.48-49
LEM	p. 3-38-39
Semikron	p. 3.16-17
Siemens	p. 3.12-13

TABLE OF CONTENTS

PART 1 - SEMICONDUCTOR SWITCHES

1. PHILOSOPHY OF POWER CONTROL

2. POWER DIODES

3. TRANSISTOR POWER SWITCHES

4. THYRISTORS

5. NOTES

PART 2 - POWER CONVERTERS

7. UNCONTROLLED RECTIFIERS WITH INDUCTIVE LOAD

8. LINE-FREQUENCY PHASE-CONTROLLED RECTIFIERS

9. AC-CONTROLLERS

B. SINGLE-PHASE INVERTER

15. APPLICATIONS OF POWER ELECTRONICS

VOLUME 2:   ELECTRONIC MOTOR CONTROL

PRINCIPAL SYMBOLS

α	*transistor current gain*
α	*firing angle thyristor* (rad)
β	*conduction angle thyristor* (rad)
B	*magnetic flux density* ($T = Wb/m^2$)
AC	*alternating current*
DC	*direct current*
δ	*duty ratio* (%)
e	*instantaneous e.m.f.* (V)
E	*RMS-value elektromotive force (e.m.f.)* (V) / *DC-e.m.f.* (V)
E	*electric field intensity* (V/m)
E_{on}, E_{off}	*energy dissipation during transistor switching "on" and "off" respectively* (J)
f	*frequency* (Hz)
ϕ	*flux per pole DC- machine / rotating air gap flux induction motor* (Wb)
ϕ_{S1}	*flux one stator winding of an induction motor* (Wb)
g_{fs}	*transconductance* (Siemens / mho)
g_m	*transconductance coefficient* (Siemens / mho)
H	*magnetic field intensity* (A/m)
h_{FE}	*current gain common emitter connection*
$i\,/\,\hat{i}$	*instantaneous current* (A) / *peak value of sinusoidal current* (A)
$i_0\,/\,I_0$	*output current of a circuit* (A)
i_μ	*magnetizing current* (A)
$\hat{i}_\mu$	*peak value of magnetizing current* (A)
I_{AV}	*average value of a semiconductor current* (A)
$I_{RMS}\,/\,I$	*r.m.s. value of current* (A) / *DC-current* (A)
J	*(polar) moment of inertia* (kgm^2)
L_b	*load self inductance*
L_0	*magnetizing inductance (transformer / induction motor)* (H)
$\mathcal{L}$	*Laplace transform*
μ_0	*permeability of free space* ($4.\pi.10^{-7} H/m$)
μ_r	*relative permeability*
M	*momentum (of torque)* (Nm)
M_{em}	*electromagnetic momentum(of torque)* (Nm)
M_J	*accelerating or decelerating momentum (of torque) due to inertia* (Nm)
M_k	*peak value momentum (of torque) induction motor* (Nm)
M_t	***total** momentum of load torque (mechanical load M_L + static friction torque M_F + windage torques M_W ...)* (Nm)

$\mathscr{R}$	reluctance (A/Wb)
$\mathscr{F}$	magnetomotive force (m.m.f.) (Aw)
N_{Se}	equivalent sinusoidal (stator) winding induction motor
n	motor speed (r.p.m. or rad/s)
n_S	synchronous speed (rotating stator field) induction motor (r.p.m.)
n_R	speed rotating rotor field induction motor (r.p.m.)
η	efficiency of operation (%)
P	DC-power (W) / average power (W)
P_e	eddy current loss density (W/m³)
P_h	hysteresis loss density (W/m³)
p	number of pole pairs DC-machine
p	number of pole pairs stator winding induction motor
$\sigma_R.L_0$	leakage inductance rotor induction motor
$\sigma_S.L_0$	leakage inductance stator induction motor
R_b	load resistance
s	Laplace operator
T	time period (s)
T	temperature (°C ; K)
t_{on}	time to switch on a power semiconductor (switch) (μs; ns)
t_{off}	time to switch off a power semiconductor (switch) (μs; ns)
t_{ON}	time that the power semiconductor is conducting (ON-state) (μs ; ms)
t_{OFF}	time that the power semiconductor is blocking (OFF-state) (μs ; ms)
t_d	delay time (to switch a transistor on) (μs ; ns)
t_f	fall time during switching off transistor (μs ; ns)
t_r	rise time during switching on transistor (μs ; ns)
t_s	storage time (to build off the space charge in a BJT) (μs ; ns)
τ	time constant (s)
v	instantaneous voltage (V)
v_0 / V_0	output voltage of a circuit (V)
v_s / V_s	supply voltage (V)
$\hat{v}$	peak value sinusoidal voltage (V)
V	voltage (DC, average, ...) (V)
V_L / V_F	line voltage / phase voltage in a three-phase system (V)
V_{RMS}	root mean square voltage (V)
V_{di}	**(dc-)** average voltage for **ideal** rectifier (V)
$V_{di\alpha}$	**(dc-)** average voltage for **ideal** controlled rectifier with **firing angle** α (V)
v	speed (m/s)
W	energy (J)
ω	angular frequency (rad/s)

PART 1

SEMICONDUCTOR SWITCHES

1 THE PHILOSOPHY OF POWER CONTROL

CONTENTS

1. Controlling electrical energy using switches
2. Switching matrix
3. Controllable semiconductors
4. Properties of switches
5. Commutation
6. Power converters
7. Power frequency domain
8. Evaluation

In the first half of the twentieth century electronics was synonymous with telecommunications. At that time this included telephony, radio and TV technology. During the Second World War (W.W.II) radar technology was developed and also the first electrical servo systems were built. Directly after W.W.II these servo systems were directly responsible for the birth of industrial automation. A new field was started: industrial electronics, also known as power electronics. This enabled electrical power to be controlled using electronic technology. Examples of this are control of motors (speed and torque), temperature control of ovens and buildings, lighting levels of lamps.

Electronic power control is a marriage of typical electronic technology and applications from the field of power engineering. It is a difficult marriage because power engineers don't easily think in terms of electronic components and micro second time scales and on the other hand electronic engineers have little concept of motors in the megawatt range. So there is a need for engineers that have a knowledge of both fields.

The control of electrical power is extremely important for the following reasons:
1. Environmental reasons: preserving the environment requires cleaner electrical power
2. Reliable applications: increase of speed and accuracy
3. Energy saving: efficiency is very important and the accurate control of electrical power is a priority.

Only electronics can realize this level of control.

1. CONTROL OF ELECTRICAL ENERGY USING SWITCHES

In power electronics to control the flow of electrical energy with maximum efficiency **switches** are always used. To illustrate this:

Imagine the resistive element of an oven with $R_v = 5\ \Omega$ connected to a voltage of 500V (fig. 1-1). To control the electrical energy in the oven a variable resistor R_1 is required to be connected between the supply and the load R_v. For various values of R_1 we calculate the energy delivered by the supply and the energy consumed by the oven. Additionally we calculate the efficiency of this circuit: $\eta = \dfrac{P_{oven}}{P_{source}}$. The results of the calculation are presented in table 1-1. The efficiency as a function of R_1 is shown in fig. 1-2.

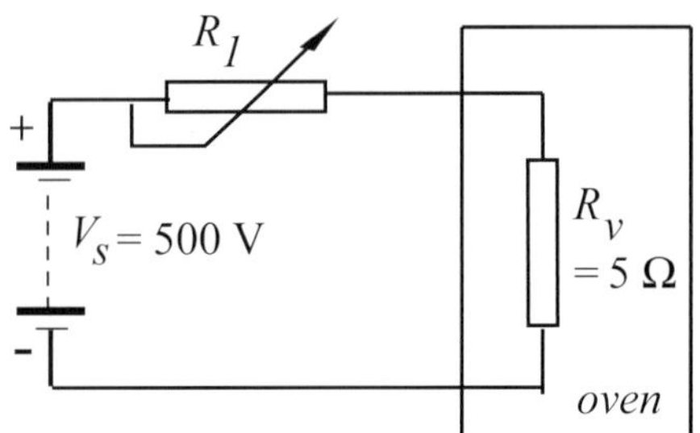

Fig. 1-1: Power control using a resistor

Fig. 1-2: Efficiency of circuit in fig. 1-1

Table 1-1

$R_1\ (\Omega)$	$P_{source}\ (kW)$	$P_{oven}\ (kW)$	$\eta\ (\%)$
0	50	50	100
5/3	37.5	28.125	75
5	25	12.5	50
15	12.5	3.125	25
∞	0	0	—

Next we examine the circuit in fig 1-3 in which the series resistance is replaced by a switch S. This switch is periodically opened and closed. We call t_{ON} the time that the switch is closed and t_{OFF} the time that the switch is open. The time taken to open (t_{on}) and close (t_{off}) the switch is negligible.

We determine the duty cycle of the circuit using :
$$\delta = \frac{t_{ON}}{t_{ON} + t_{OFF}} = \frac{t_{ON}}{T} \qquad (1\text{-}1)$$

T is the period time. Multiplying by 100 gives the δ in %. Table 1-2 shows the calculated values of P_{source}, P_{oven} and the efficiency η for a number of values of δ. The efficiency as a function of the duty cycle is shown in fig. 1-4 .

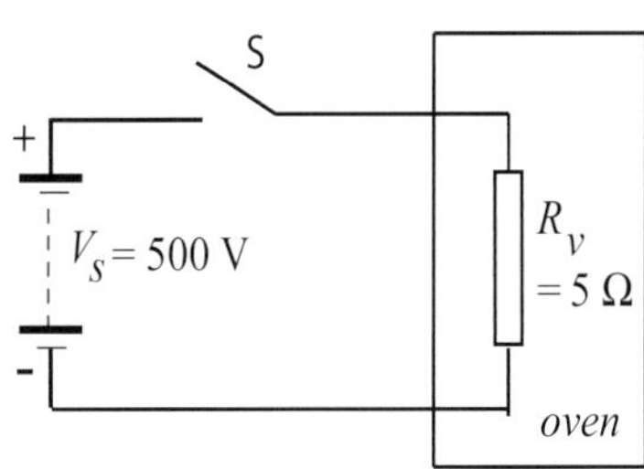

Fig. 1-3: Power control using a switch

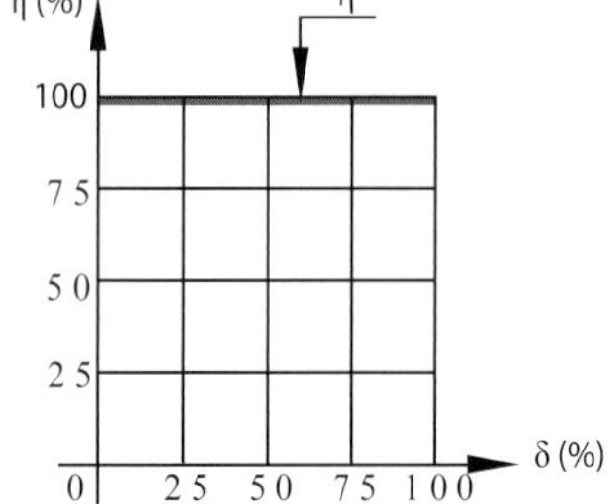

Fig. 1-4: Efficiency of configuration in fig 1-3

Table 1-2

δ (%)	P_{source} (kW)	P_{oven} (kW)	η (%)
0	0	0	---
25	12.5	12.5	100
50	25	25	100
75	37.5	37.5	100
100	50	50	100

Comparing the graphics of fig.1-2 with fig. 1-4 shows that efficiency of power control is the reason switches are used instead of continuously varying elements. If the power flux is passed through the switches and the **switches** are power semiconductors, then we use the term **power electronics**.

2. SWITCHING MATRIX

Every energy transformation can be derived to a switching matrix (fig.1-5). The voltage between "b" output lines is formed by selected connections with "a" input lines during a specific part of the working cycle. Via suitable control signals the individual switches are operated. The output voltage is composed of segments of the input voltage.

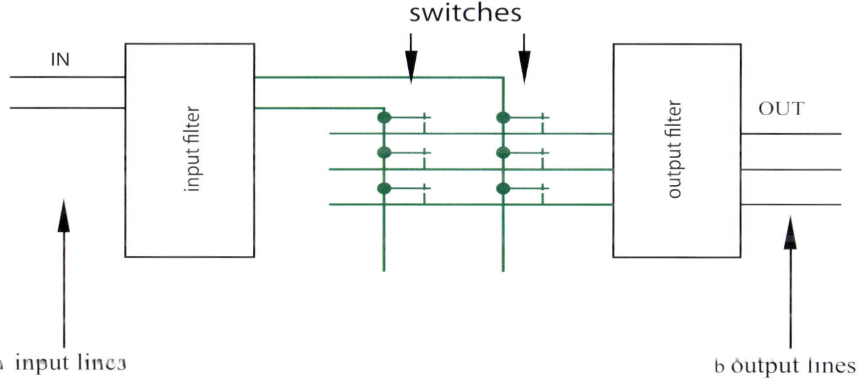

Fig. 1-5: General switching matrix

To connect "a" lines with "b" lines a x b switches are required (fig. 1-5). In the configuration of fig. 1-6 one line is common to input and output and therefore no switch is required for this line. Here b = 1 and a = 1 and (a x b) one switch is sufficient. With the exception of the triac semiconductors are unidirectional, in other words they conduct in only one direction. To make the circuit bidirectional, that is with a reversible output and input function, 2 x a x b switches are required. Example: A dc-motor with (b =) 2 lines bidirectionally connected with (a =) 3 lines from an AC grid requires 12 switches.

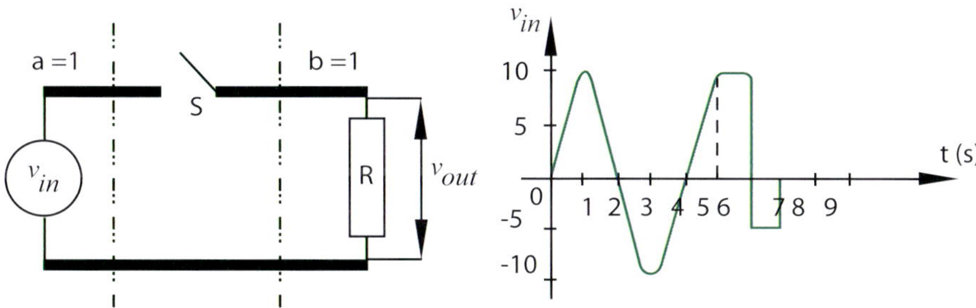

Fig. 1-6: Simple switching matrix

Filters

As a result of switching action undesired frequencies are generated in the supply lines which may necessitate an input filter. On the other hand the switching of the output produces a waveform that is only an approximation of the desired waveform. An output filter helps to approximate the ideal waveform by blocking undesirable ripple voltages and currents from reaching the load.

Filters significantly increase the cost price, weight and volume of the installation. It should be determined if filters are necessary for a particular application and if so, how much should be filtered.

3. CONTROLLABLE SEMICONDUCTORS

As far as controllability is concerned, three types of semiconductors can be distinguished:

TYPE 1 = NON CONTROLLABLE SWITCHES = DIODE:
        conducts automatically when the anode is positive in relation to the cathode and reverse biases before the current goes through zero.

TYPE 2 = CONTROLLED ON SWITCHES = THYRISTOR:
        conducts in response to a control signal and reverse biases before the current goes to zero.

TYPE 3 = CONTROLLED ON OFF SWITCHES = TRANSISTORS, GTO, MCT, IGCT:
        via control signals the switches conducts or reverse biases.

Note that a higher type number corresponds to more possibilities. A type -2 switch can be used as a type-1 switch but not the other way around.

4. PROPERTIES OF SWITCHES

A switch is characterised by the voltage ($V_{nominal}$) which it can continually handle in the open position and the current ($I_{nominal}$) which it can conduct in the closed position. These static properties can be seen in fig 1-7, in which the difference between an ideal switch and a real switch is highlighted.

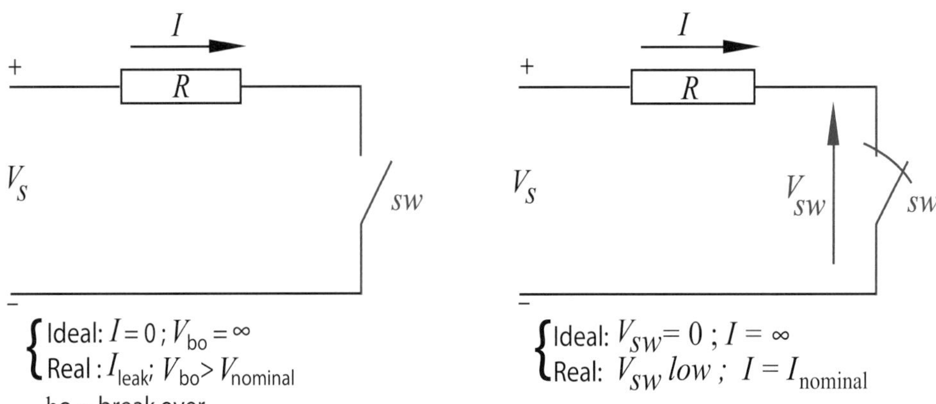

Fig. 1-7: Static properties of ideal and real switches

A switch is also characterized by its dynamic properties, especially the time to close (t_{on}) and open (t_{off}). This is illustrated in fig. 1-8. Once again the distinction between an ideal and real switch is clear.

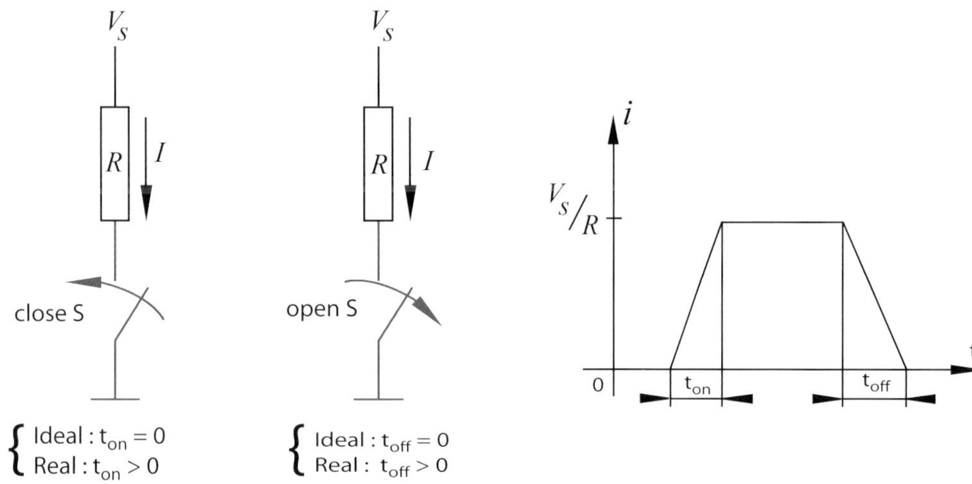

Fig. 1-8: Dynamic properties of a switch

In fig. 1-9 we can also now determine the operating points A and B for a switch in a I-V graph. This has been done for both the ideal and real switch of fig 1-7.

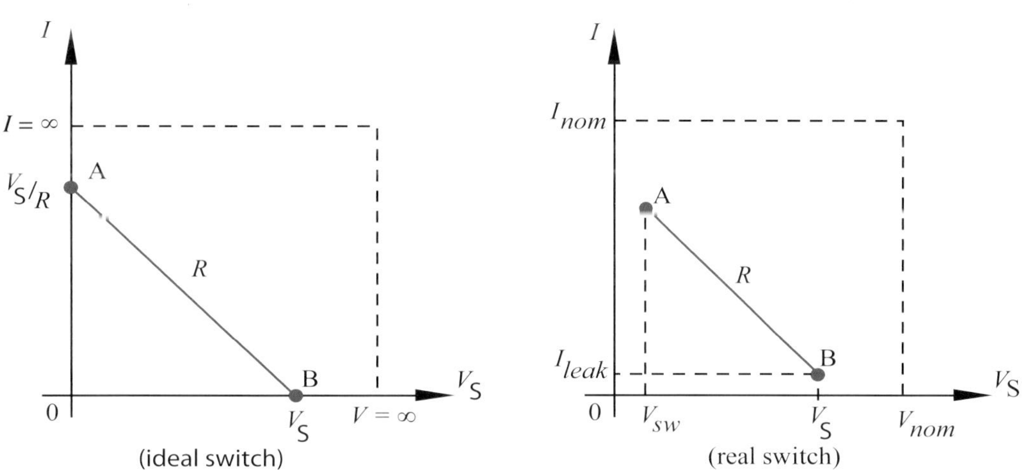

Fig. 1-9: Operating points in a V-I graph

5. COMMUTATION

Commutation is the process whereby the current through a switch goes through zero. If the supply is AC then it can be **natural** or **net commutation** (line commutation) because:

1. The current through the switch goes to zero because of the zero crossover of the grid, or

2. The circuit is switched via another switch to a higher potential.

With a type 3 switch commutation can be "artificial" at any random time. This is referred to as forced commutation but in fact all commutation is forced so to distinguish it the term **artificial commutation** is preferred. It is also possible that the load characteristic is such that the switch is forced into commutation, this is **load commutation**.

6. POWER CONVERTERS

In the industrial world electrical energy is available in two forms:

1. As alternating current (AC) via a distribution network

2. As direct current (DC) via a battery system or via an overhead cable or rail for traction systems.

In addition to the two distribution forms (DC and AC) of electrical energy we can also distinguish two types of consumers:

1. Direct current consumers

2. Alternating current consumers

To regulate the power flow between current source (DC or AC) and consumer (DC or AC) four types of power converters are used (fig. 1-10):

1. **DC controller**: An AC voltage from the grid is converted into a controllable DC voltage

2. **AC controller**: An AC voltage from the grid is converted into a variable AC voltage with the same frequency

3. **Chopper**: A constant DC voltage is converted to another constant or variable DC voltage

4. **Inverter**: A DC voltage is converted into an AC voltage which may be controlled or not

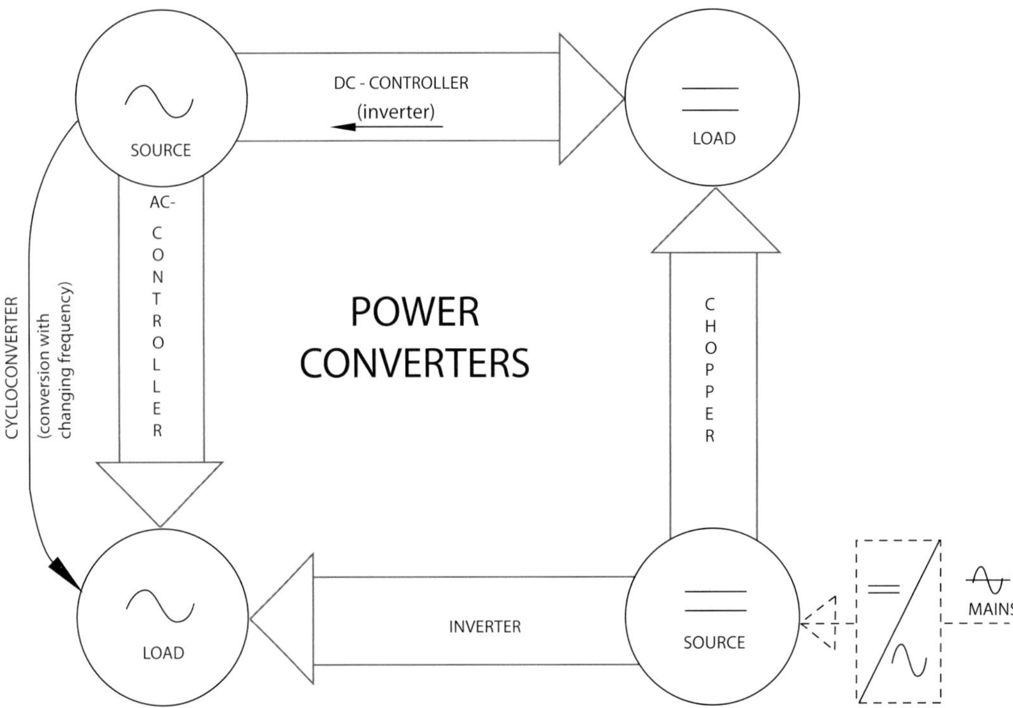

Fig 1-10: Power converters

DC and AC controllers make use of natural commutation since the supply is AC. Type 2 switches are used. In certain circumstances a DC controller can work in reverse as an inverter. Choppers and inverters make use of artificial commutation and use type 3 switches. A special category is the frequency converter. We distinguish between single stage and double stage converters. With a so called cycloconverter the frequency conversion takes place in one stage. In the two stage frequency converter the AC supply is first rectified and the DC voltage is converted back to AC via an inverter. The amplitude and frequency of the AC voltage may be varied.

7. POWER FREQUENCY DOMAIN

Fig. 1.1 illustrates the state of the art in power switches. Currently we see that with the exception of an SCR of 8.5 kV that thyristors are limited to 4 kV - 4 kA. This is a maximum of 16 MW per switch. This is considered the mid power domain.

These days an SCR is also known as Phase Controlled Thyristor (PCT).

There are for example single IGBT's of 1200V - 3600A, 1700 V - 2400 A and 3300V - 1500A.

In a half bridge using IGBT's a typical rating would be 1700V -1000A.

State of the art components use 125 mm Si wafers.

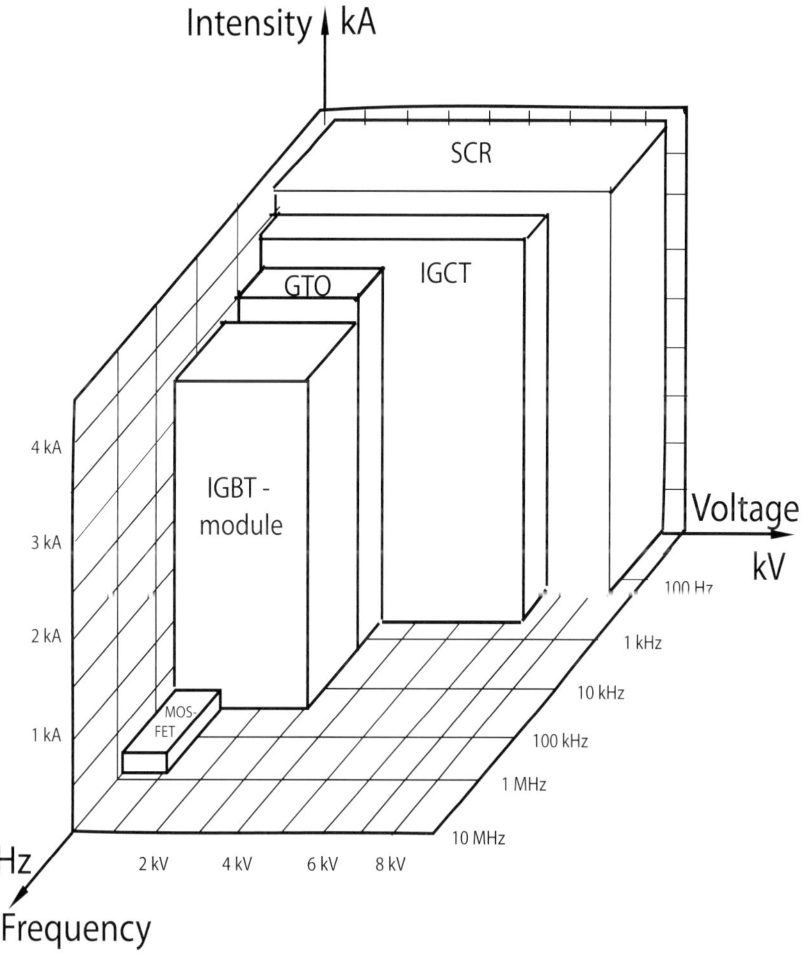

Fig 1-11: Properties of power switches

Transport of electrical energy over long distances can be more economical using DC transmission. For example when the distance is greater than 250 km then transmission of 400kV-1200MW is more economical using HVDC (high voltage direct current). With submarine cables the distance is even shorter. In China and Brazil where large hydro electric power stations are more than a 1000 km from the big cities HVDC is used.

Example: the Ultra High Voltage DC (UHVDC) of 800kV between Xiangjiala and Shanghai. The firm ABB developed a 6-inch 8.5kV thyristor for this application.

Another application of HVDC is the connection between AC - grids of different frequencies. An example is the Garabi back to back station which connects the 60 Hz grid of Brazil and the 50 Hz grid of Argentina.

In table 1-3 an overview is provided of a number of properties of type 3 switches. The indicated maximum values of voltage and current will not be attainable by one switch simultaneously. We find for example mosfets for 1000V - 6A or for 100V - 225A.

Table 1-3

Type - 3 switch	GTO (IGCT)	Transistors		
		bipolar	powermosfet	IGBT
Maximum ratings				
Maximum blocking voltage (V)	5000 (10000)	1200	1200	3300
Maximum current (A)	4000 (2000)	500	225	3600
Switch on time (μs)	5	2	0.1	0.2
Switch off time (μs)	25 (7)	25	0.5	1
Max.switching frequency (kHz)	5	50	1 MHz	200
Gate power	large	average	very small	very small
Si-surface area of the switch	small	large	very large	very large

8. EVALUATION

1.1 What is state of the art?

1.2 What type of power converters can be used between a DC source and DC consumer?

1.3 The switching frequency of a switch is 10 kHz and the duty cycle is 10%. How long is the switch open during one work cycle?

1.4 When did industrial electronics begin?

1.5 What is the definition of power electronics?

1.6 What is a typical switching frequency for an IGBT? What do the letters IGBT stand for?

1.7 What is considered high current in power electronics?

1.8 Name three methods by which a power electronic switch can commutate?

1.9 In fig. 1-3 if $T = 50\,\mu s$ en $\delta = 0.25$, sketch $I_{oven} = f(\text{time})$

1.10 Sketch the output voltage of the circuit shown in fig. 1-6, if the switch is closed between 0 and 3 s and between 5.5 and 6.5 s.

1.11 Re-evaluate question 1-9 with the following data

    1. $\delta = 0.25$ and $t_{on} = t_{off} = 1\,\mu s$; 2. $\delta = 0.2$ and $t_{on} = t_{off} = 12.5\,\mu s$

1.12 What do the following terms mean: AC, DC, GTO, SCR, bipolar, mosfet, IGCT?

2 POWER DIODES

CONTENTS

1. SEMICONDUCTORS

In electrical engineering in addition to conductors and insulators there are also semiconductors. What are the characteristics of such a material? The most important property of a semiconductor is that its electrical conductance in one direction is much better than in the other direction. This property is crucial for a diode in a rectifier for example.

Another property of a semiconductor is that at an average ambient temperature the specific resistance lies between that of metals and an insulator.

To conduct semiconductors need energy in the form of heat, light or a strong electric field.

At absolute zero (-273° C) a semiconductor is an insulator. A photo sensitive semiconductor in complete darkness adopts the properties of an insulator.

The most commonly used semiconductor at the moment is without doubt silicon (Si).

Siliconcarbide (SiC) is making inroads in power applications as a diode. Presently a SiC diode costs considerably more than a Si diode but the cost difference is narrowing all the time.

Germanium and selenium are also used as semiconductors. Semiconductor technology also makes use of chemical connections. Examples are lead sulphide (PbS), gallium arsenide (GaAs), cadmium sulphide (Cds), indium phosphide (InP).

Arsenides, sulfides and selenides are primarily photosensitive semiconductors.

Thermistors, that is temperature sensitive resistors, are generally made of semiconductor oxides.

In addition to referring to the **material** the term **semiconductor** is frequently used to refer to the components: diodes, transistors, thyristors etc. that are made from semiconductor material.

2. *I-V* CHARACTERISTIC OF A JUNCTION DIODE.

The contact area of a PN-junction determines in large measure the admissible current.
For example there are diodes (IN4001, IN4003,…) that allow a current of 1A, while (large) power diodes allow 250A or more.

Fig 2-1 shows the *I-V* curve of a junction diode.

Here we observe:

1. a low voltage drop over the forward conducting diode
2. the reverse current is low
3. beyond a certain voltage (zener voltage V_Z) the reverse conduction increases sharply.
   For a rectifier diode we need to remain below V_Z

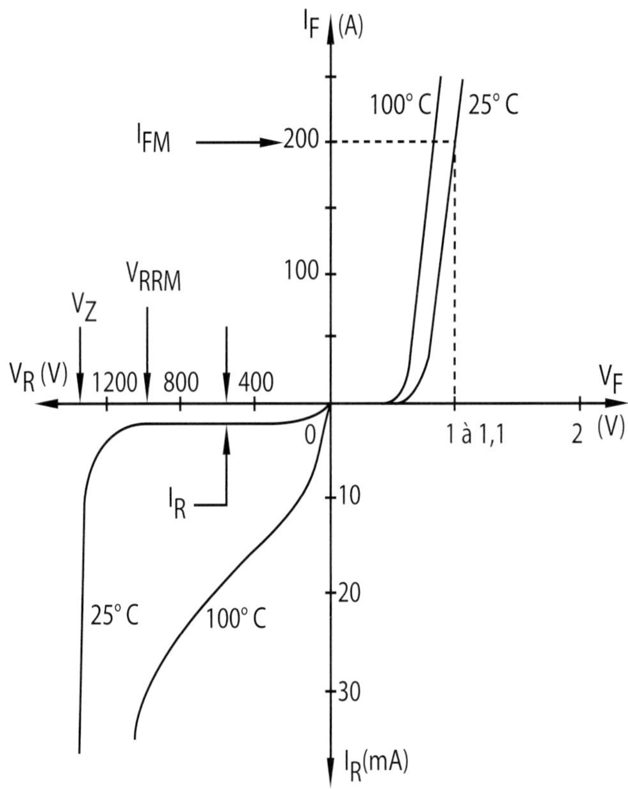

Fig. 2-1: *I-V* characteristic of a junction diode.

4. V_{RRM} is the maximum allowable reverse repetitive peak voltage

5. the dynamic resistance $r_T = \Delta V / \Delta I$ (2-1)   is small (mΩ) in the forward direction

   and large (MΩ) in the reverse direction

6. the temperature influences the *I-V* curve, especially I_R
7. I_{FM} is the maximum allowable peak forward current
8. I_{FM} and V_{RRM} are two important specifications of a rectifier diode
9. Fig. 1 on p. 2.11 shows the I_F-V_F curve of an 1N4001. Take note of the logarithmic I_F- axis.

3. POWER DIODES

It may be necessary that a diode handle large currents (e.g. 100 to 1000A) at high reverse voltages (e.g. 1 to 5kV) and with good dynamic behavior. In addition power diodes need to be able to tolerate high junction temperatures. In order to allow the reverse voltage of a PN junction to be as large as possible we need to maintain the field strength in the barrier region as low as possible. This can be achieved by using small concentrations of majority carriers. In practice though the resistance of the diode material would be so large at the required current that the voltage drop would be tens of volts resulting in excessive power dissipation in the diode. The three layer diode can provide a solution to this problem. In the nearby future SiC diodes (see p. 3.34) will probably become the diode of choice, especially in converters working at higher frequencies (16/20/50kHz…).

3.1 Three layer structure

A three layer diode is constructed of zones of varied doping levels (fig. 2.2a). The outer zones (P and N) are heavily doped. The s-zone has a thickness of 50 to 400 μm and may be a P- or an N-zone but in any case with a very low doping level. Since the depletion zone near a junction primarily expands in the region of lowest doping (in this case the s region) we can reduce the field strength (E) as a consequence of the reverse voltage (V_R) by increasing the width x of this zone.

$$E = \frac{V_R}{x} \qquad (\text{V/m}) \qquad (2\text{-}2)$$

By means of this technique we obtain diodes with a reverse voltage of several thousand volts and voltage drops of 1 to 2 V with a current of for example 2000A. The maximum junction temperature is then 200° C. The SPT-diode (Soft Punch Through) with its soft recover characteristic is a further development of this three layer structure. They are constructed to handle reverse voltages of 3.3/4.5/6.5kV. Fig. 2-2b shows the cross section of an SPT diode.

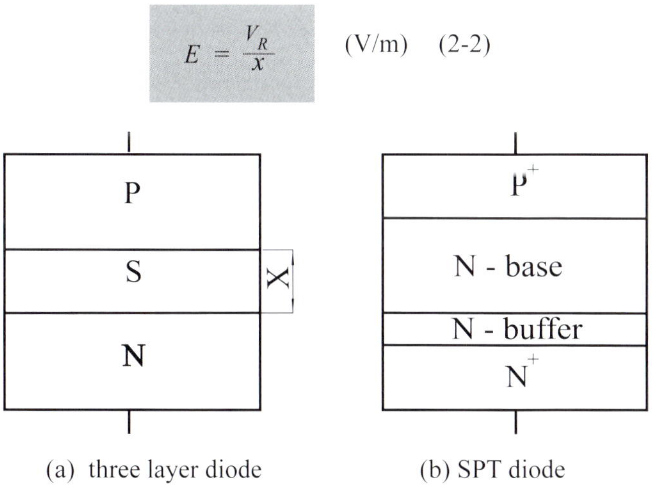

Fig. 2-2 Structure of Si power diodes

3.2 Dynamic behavior of a power diode

In power electronics diodes are not only used in low frequency grid rectifiers but also in high frequency applications (kHz to tens of kHz). Applications include freewheel diodes and snubber diodes in choppers and inverters or rectifier diodes in switch mode power supplies. In these switching applications the dynamic diode properties are very important. The turn off properties are not only important for determining the switching losses of the diode but also the switching losses of the associated transistor switch.

A. Turn-on

When a diode starts to conduct (turn-on) a certain amount of time is needed to fill the central layer with charge carriers. During the forward recovery time t_{FR} (fig. 2-3) the forward voltage drop across the diode is greater than the nominal voltage drop of 1,1V. The thickness of the diodes central layer is proportional to the blocking voltage which means that a "high voltage diode" will have a greater V_{FP} than a low voltage diode. This turn-on overvoltage V_{FP} increases with temperature (+ 0.8% /°C) and the turn on time increases by approximately 0.4%/°C.

The turn on behavior of the diode depends upon:
- the applied voltage V
- the forward current I_F.
- the pitch dI_F/dt
- the junction temperature T_J
- technology used to construct diode

B. Turn-off

During conduction there is a surplus of minority charge carriers in each layer of the diode (holes in the N-layer and electrons in the P-layer). The total space charge Q_1 of a P^+N-diode is for the most part made up of holes in the N-layer. At turn-off this surplus of charge carriers in the central layer will not immediately disappear. As a result the diode remains conducting for a time t_{rr}. This is shown in fig. 2-4. A part of the charge Q_0 of Q_1 disappears through recombination and the rest Q_{rr} is responsible for the reverse current under the influence of the applied reverse voltage.

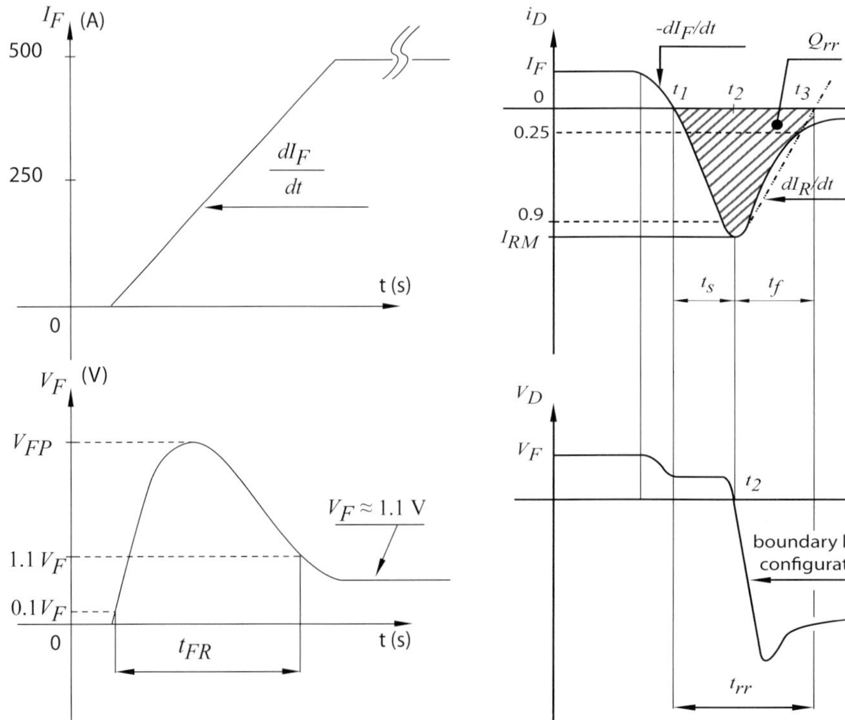

Fig. 2-3: Voltage drop diode during switch on   Fig. 2-4: Reverse recovery diode

Due to the presence of the space charge the diode for a time $(t_2 - t_1) = t_s$ remains polarized in the forward direction, but at time t_2 the charge carriers in the vicinity of the diode junction have gone and under the influence of the reverse voltage the barrier layer builds up quickly. The reverse current finally reaches its steady state value I_R .

The time $t_3 - t_1 = t_{rr}$ is the reverse recovery time. To determine t_3 a line is drawn through $0.9 . I_{RM}$ and $0.25 . I_{RM}$. We set $t_3 - t_2 = t_f$.
The reverse recovery time t_{rr} (100 ns to 10μs) depends upon
- the current I_F flowing through the diode at that instant
- the reverse voltage V_R that is applied
- the value of the reverse voltage
- the rate at which the current decreases $(- dI_F/dt)$
- the junction temperature T_J

The reverse recovery time t_{rr} is often used by circuit designers but does not correspond to any practical situation. Therefore t_{rr} is often determined under test conditions:

$$I_F = 1A; \quad \frac{dI_F}{dt} = -15A/\mu s \text{ and } V_R = -30V; \quad T_J = 25°C$$

In order to collect and have meaningful data available the firm Thompson CSF defines the values I_{RM} and $t_{I_{RM}}$ under the following conditions:

$$I_F = 2 . I_{AV}; \quad \frac{dI_F}{dt} = -4.I_{AV}/\mu s; \quad T_J = 100°C$$

In a modern switching diode $\frac{dI_R}{dt}$ is limited and this is referred to as a "soft recovery" diode as opposed to the " snap off recovery" diodes which often resulted in high frequency oscillation.

Remark

The dissipation at turn-on is small compared to turn- off. At the grid frequency (50 Hz) the total dissipation due to turn-on and turn-off is negligible. The diode losses can become significant in circuits which operate at higher frequencies (kHz) such as solar converters, etc.

The turn-off can change somewhat if there is a self inductance present in series with the diode (e.g. as in fig. 2-5). This can result in large overvoltage V_{RM}.
Suppose diode D is conducting (I_F). When the switch S is closed the current I_F decays, according to $dI_F / dt = V_R / L$

During reverse recovery the current dI_R / dt will result in an induced voltage across the

inductor L (fig.2-5) : $E_L = L . dI_R / dt = V_R \dfrac{dI_R / dt}{dI_F / dt}$. This voltage E_L together with V_R is in

reverse across the diode so that $V_{RM} = V_R + E_L = V_R (1 + \dfrac{dI_R/dt}{dI_F/dt})$.

Here we see that dI_F/dt is determined by the circuit and dI_R/dt by the diode.

A small conversion gives :

$$\frac{V_{RM}}{V_R} = 1 + \frac{dI_R/dt}{dI_F/dt}$$

(2-3)

Fig. 2-5 shows how this reverse voltage can look.

Conclusions :

. the turn-on properties are mostly insignificant. They can play a role in low voltage applications such as for example in the base circuit of a transistor.

. the turn-off properties are important:

1. soft recovery prevents oscillations and high reverse voltage.
2. a small I_{RM} limits the switching losses in the diode and accompanying transistor switches as we shall see later in fig. 3-43 c and d.

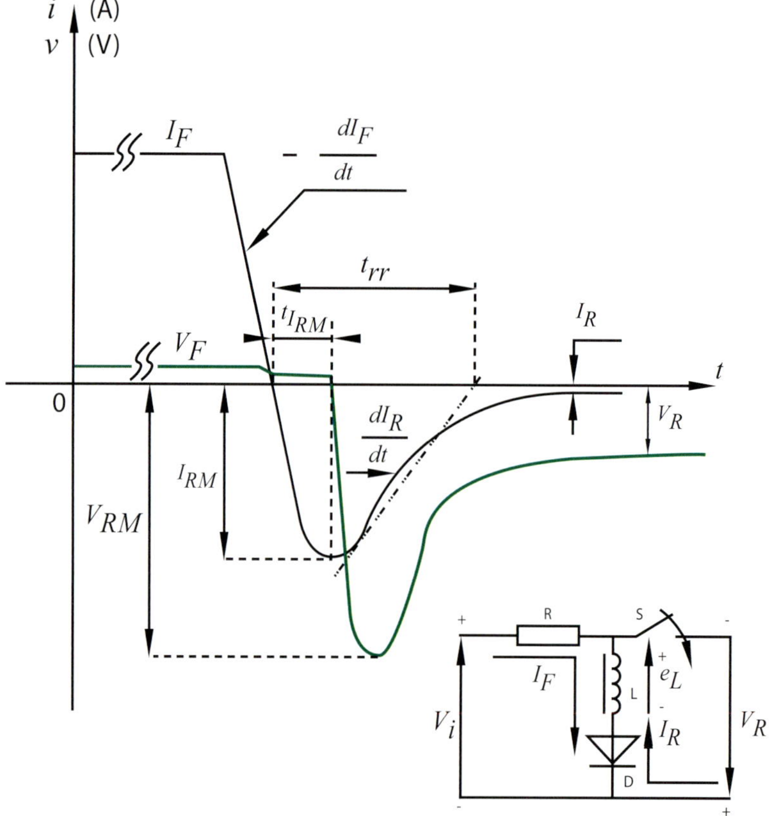

Fig. 2-5: Diode turn-off when a self inductance is present in series with the diode

3.3 Schottkydiodes

Walter Schottky, a German physicist (1886-1959) and inventor of the screen grid vacuum tube (1915) proposed in the nineteen thirties the theory of the potential energy in the barrier region of a metal rectifier. The diode is named after him since the operating principle is based on a metal-semiconductor transition. The most important property of a Schottky diode is the very fast switching time (small recovery time!) and the low forward voltage drop (0.6 to 0.8 V) at nominal current.

For other rectifier diodes this voltage drop is approximately 1.1V. Schottky diodes are normally constructed up to 100V. In 2001 Advanced Power Technology presented a Schottky diode rated for 200V-100A. In the same year the firm General Semiconductor constructed surface mount Schottky rectifiers with a voltage drop of 0.385 V with a current of 0.5A. Above 100 V the pin diode is used when a fast diode is required. In the literature the Schottky diode is sometimes referred to as a "hot carrier diode". The Schottky diode is important as a power diode especially in switched mode power supplies.

Remark

With the recently developed siliciumcarbide (SiC) technology (see p. 3-34) use is made of SiC-Schottky diodes with a blocking voltage of 1700V and currents up to 25A.

4. DATA OF A POWER DIODE

4.1 Forward voltage drop

The forward voltage drop (V_F) depends upon the chosen semiconductor material and for Si diodes is typically 1.1V at $T_J = 25°C$

4.2 Power dissipation

In connection with the cooling of (power)diodes it is important to focus on the power dissipation of a diode.

From the I_F - V_F characteristic of a diode (fig. 2-6) we can derive: $v_F = V_{T0} + r_T \cdot i_F$

The instantaneous losses are: $p_F = v_F \cdot i_F$

Average:
$$P_F = \frac{1}{T} \int_0^T v_F \cdot i_F \cdot dt$$

$$P_F = \frac{1}{T} \int_0^T v_{T0} \cdot i_F \cdot dt + \frac{1}{T} \int_0^T r_T \cdot i_F^2 \cdot dt$$

$$P_F = V_{T0} \cdot I_{FAV} + r_T \cdot I_{F(RMS)}^2$$

$$P_F = V_{T0} \cdot I_{FAV} + r_T \cdot a^2 \cdot I_{FAV}^2 \qquad (2\text{-}4)$$

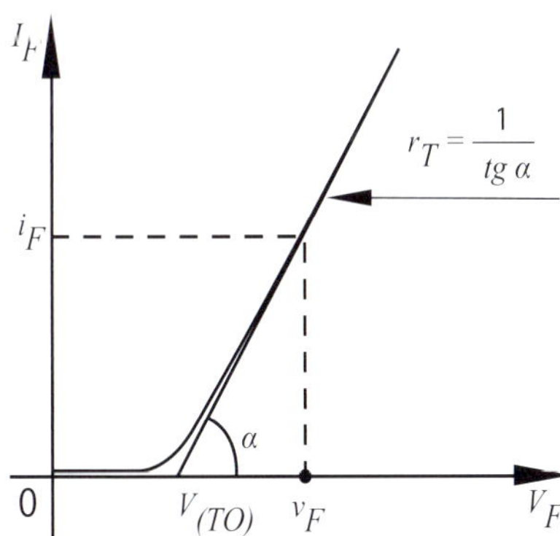

Fig. 2-6: Forward V-I characteristic diode

Here in :

P_F	= dissipated power (W)
I_{FAV}	= average value of the diode current (A)
$I_{F(RMS)}$	= effective value of the diode current (A)
r_T	= dynamic resistance (Ω) to be determined from the diode characteristic (see fig. 2-6)
V_{T0}	= intersection of the line r_T and the V_F - axis (see fig. 2-6)
a	= form factor (see tables 7-3 and 7-4 on pages 7.21 and 7.22)

4.3 Limits (ratings) of VOLTAGE and CURRENT

1. REVERSE VOLTAGE

In fig. 2-7 we see the limit of the reverse voltage of a diode

V_{RWM} = max. of the periodic reverse voltage
V_{RRM} = max. repetitive peak reverse voltage
V_{RSM} = max. non repetitive peak inverse voltage
For the explanation of the subscripts see p. 2.9

2. FORWARD CURRENT

$I_{F(AV)max}$ = maximum average forward current calculated over a full period in the case of a resistively loaded half-wave rectifier (a = 1.57)

$I_{F(RMS)\ max}$ = maximum RMS forward current ($I_{F(RMS)}$ = 1.57 . $I_{F(AV)}$)

I_{FSM} = surge forward current

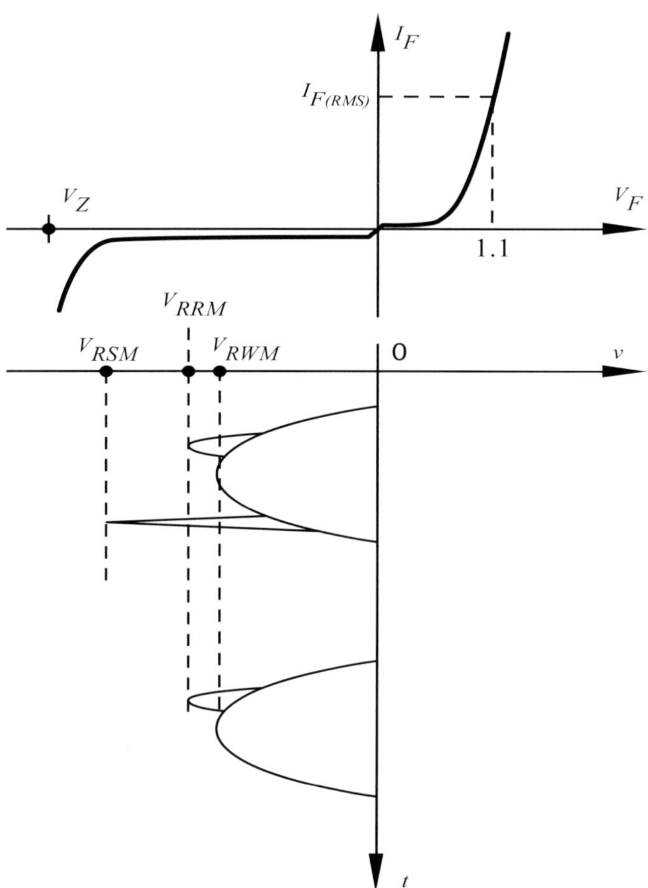

Fig. 2-7: Reverse voltage of a diode

4.4 Subscripts

F	= forward
AV	= average
RMS	= root mean square
R	= reverse (when it is the first letter)
	= repetitive (when it is the second letter)

V_R = reverse; V_{FRM} = forward repetitive maximum

V_{RRM} = reverse repetitive maximum

W	= working = sinusoidal regime
S	= surge (= amplitude of the peak) = one off peak
M	= maximum = the largest allowed value.

4.5 Cooling

With the help of expression (2-4) we can determine the dissipated power in a diode.   Consider an SKN400 (Semikron). On p.2.12 we find: V_{T0} = 0.9V and r_T = 0.5mΩ . In the case of a half wave single phase rectifier we see from table 7-3 that a = 1.57. With for example I_{FAV} = 300A it follows from (2-4): P = 381W. Since the manufacturer also provides the thermal resistance of the diode we can also determine the required heat sink. Normally the manufacturer of (large) power diodes provides graphs from which the power dissipation and case temperature can be derived.

In fig.2-10 such graphs are provide for a Semikron SKN400 diode. With I_{FAV} = 300A we see for a single phase rectifier (sin 180°) that $P \approx 385$W.

This agrees with the calculation we just made. In addition we see in fig. 2-10 that when $P = 385$W the case temperature $T_{case} \approx 118°$C.

We will return to this subject on p 5.26 (5.5: heat sinks for large power semiconductors) in which the right hand side of fig. 2-10 will be considered.

4.6 I-squared t ($= I^2 \cdot t$)

For a well specified component (fuse, semiconductor etc) the $I^2 . R . t$ value represents the energy which will heat the resistance and cause it to blow. As the resistance of the component is known, $I^2 \cdot t$ is a measure of the fusing energy. I is the RMS value of the current (I_{RMS})

From the maximum fault current over one period we can calculate $I^2 . t$ for a rectifier diode.

For an SKN400 I_{FSMmax} = 7500A ($T_{Jmax.}$)

In fig. 5-17 we find $I_{RMS} = \dfrac{\hat{i}_\mu}{2}$ so, $I_{FS(RMS)max}$ = 37500A, and over one full period :

$I^2 . t = 3750^2 . 20 . 10^{-3} \approx 280000$ A^2 s. This value is also provided in the datasheets of the diode.

The $I^2 .t$ value of the fuse must be less than 280000 A^2 s.

5. EXCERPTS FROM DATA BOOKS

5.1 Details of 1N4001…1N4007 excerpts from the MOTOROLA data book

Maximum ratings

RATING		1N4001	1N4002	1N4003	1N4004	1N4005	1N4006	1N4007	unit
Peak Repetitive Reverse Voltage Working Peak Reverse Voltage DC Blocking Voltage	V_{RRM} V_{RWM} V_R	50	100	200	400	600	800	1000	Volts
Non-Repetitive Peak Reverse Voltage (halfwave, single phase, 60 Hz).	V_{RSM}	60	120	240	480	720	1000	1200	Volts
RMS Reverse Voltage	$V_{R(RMS)}$	35	70	140	280	420	560	700	Volts
Average Rectified Forward Current resistive load (T_s=75°C)	I_0				1				Amp.
Non-Repetitive Peak Surge Current	I_{FSM}				30 (for 1 cycle)				Amp.
Operating and Storage Junction Temperature	T_j , T_{stg}				-65 to +175				°C

Electrical characteristics

		typ.	max.	unit	
Maximum instantaneous Forward Voltage Drop	V_F	0,93	1,1	Volts	
Maximum full-cycle Average Forward Voltage Drop (I_0=1A)	$V_{F(AV)}$	-	0,8	Volts	
Maximum Reverse Current (rated DC voltage) T_j=25°C/100°C	I_R	0,05/1	10/50	μA	
Maximum Full-Cycle Average Reverse Current (I_0=1A; T_{lead}=75°C)	$I_{R(AV)}$	-	30	μA	

Excerpts from data books (continued)

Excerpts from the Motorola data book (1N4001 ...1N4007)

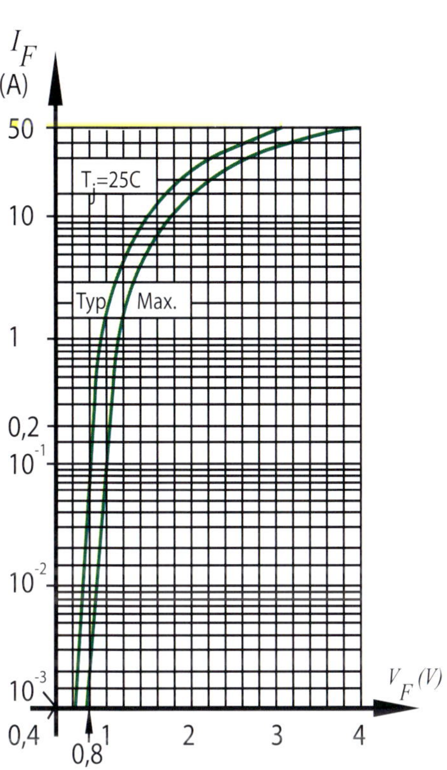

Fig. 1: V-I curve

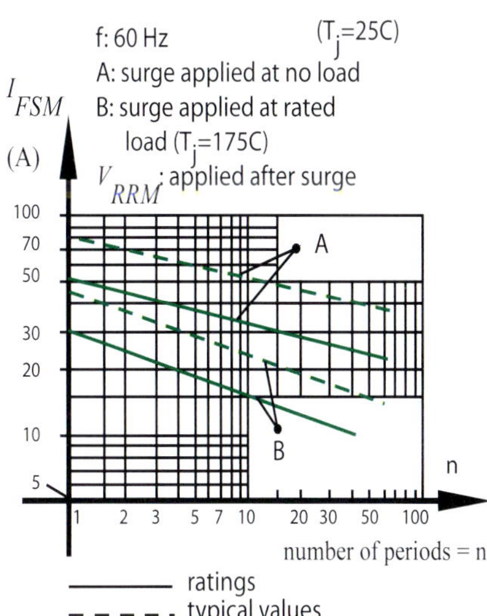

Fig. 2: Surge current

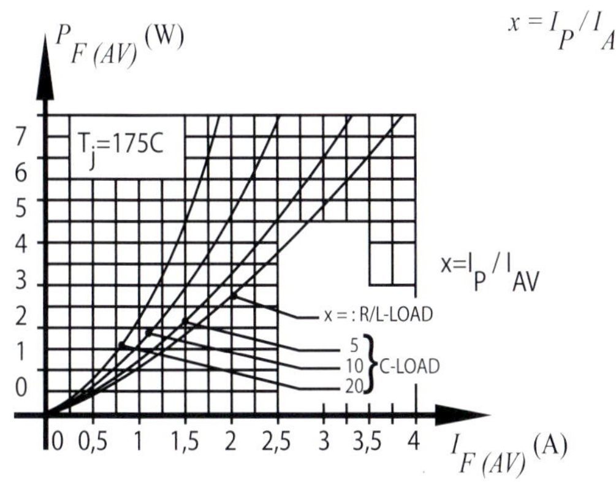

Fig. 3: Power dissipation

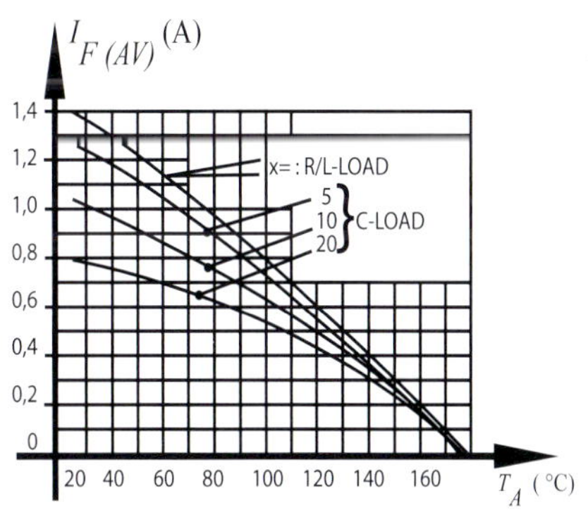

T_A = ambient temperature

Fig. 4: Maximum average diode current
versus temperature

$(R_{thj-a}= 85° C)$

Fig. 2-8: Graphs of 1N4001 ... 1N4007

5.2 SEMIKRON: data of SKN400

SKN 400

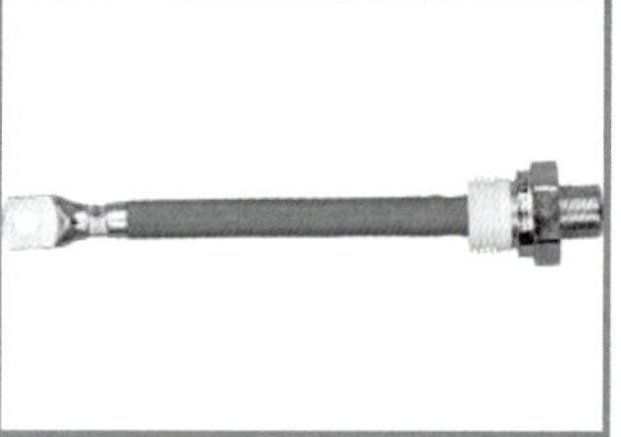

Stud Diode

Rectifier Diode

SKN 400

Features

- Reverse voltages up to 3000 V
- Hermetic metal case with ceramic insulator with extra long creepage distances
- Threaded stud ISO M24 x 1,5
- SKN: anode to stud

Typical Applications*

- High voltage rectifier diode, especially for traction applications
- Cooling via heatsinks
- Non-controllable and half-controllable rectifiers
- Free-wheeling diodes
- Recommended snubber network: RC: 1 µF, 20 Ω (P_R = 2 W), R_p = 25 kΩ (P_R = 20 W)

V_{RSM}	V_{RRM}	I_{FRMS} = 700 A (maximum value for continuous operation)		
V	V	I_{FAV} = 400 A (sin. 180; T_c = 100 °C)		
1800	1800	SKN 400/18		
2400	2400	SKN 400/24		
2700	2700	SKN 400/27		
3000	3000	SKN 400/30		

Symbol	Conditions	Values	Units
I_{FAV}	sin. 180; T_c = 85 (100) °C	445 (400)	A
I_D	K 0,55; T_a = 45 °C; B2 / B6	310 / 450	A
	K 0,55F; T_a = 35 °C; B2 / B6	700 / 1000	A
I_{FSM}	T_{vj} = 25 °C; 10 ms	9000	A
	T_{vj} = 160 °C; 10 ms	7500	A
i²t	T_{vj} = 25 °C; 8,3 ... 10 ms	400000	A²s
	T_{vj} = 160 °C; 8,3 ... 10 ms	280000	A²s
V_F	T_{vj} = 25 °C; I_F = 1200 A	max. 1,45	V
$V_{(TO)}$	T_{vj} = 160 °C	max. 0,9	V
r_T	T_{vj} = 160 °C	max. 0,5	mΩ
I_{RD}	T_{vj} = 160 °C; V_{RD} = V_{RRM}	max. 60	mA
Q_{rr}	T_{vj} = 160 °C; - di_F/dt = 10 A/µs	400	µC
$R_{th(j-c)}$		0,11	K/W
$R_{th(c-s)}$		0,01	K/W
T_{vj}		- 40 ... + 160	°C
T_{stg}		- 55 ... + 160	°C
V_{isol}		-	V~
M_s	to heatsink	60	Nm
a		5 * 9,81	m/s²
m	approx.	500	g
Case		E 17	

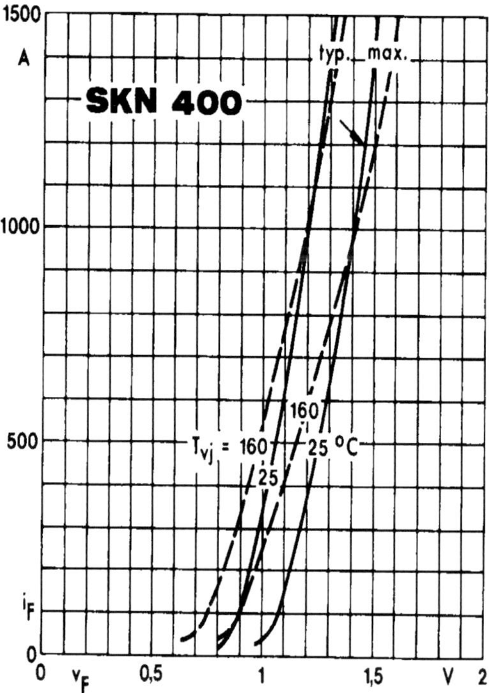

Fig. 2-9: Forward I - V characterictic SKN400 (Semikron)

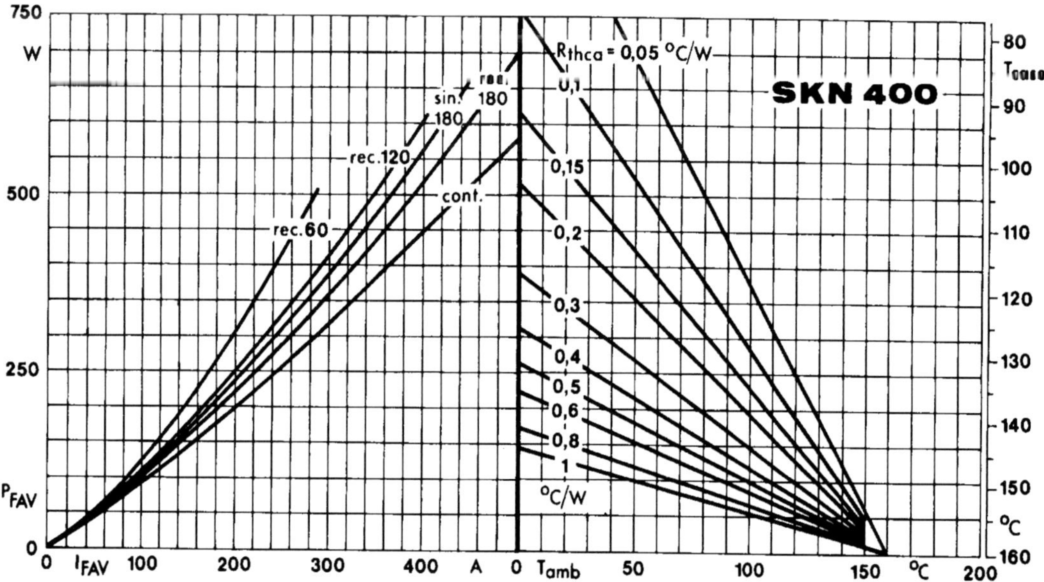

Fig. 2-10: Power dissipation versus current and case temperature (SKN400 - Semikron)

6. EVALUATION

2.1   Name two important properties of a semiconductor.

2.2   Does the arrow in a diode symbol indicate the direction of conventional current flow or the direction of electron flow?

2.3   What are the values of I_{FAV} and V_{RRM} by a 1N4007 diode?

2.4   What is the reverse recovery time of a diode?

2.5   What are the main properties of a Schottky diode?

2.6   How large is the elementary charge of an electron and what is its mass?

2.7   What are the majority charge carriers in P-Si? What is the donor in impure silicium?

2.8   How many electrons need to flow through a cross sectional area for a current of 10pA to be measured?

2.9   Find the value of V_{RRM} of a 1N4003 diode (Motorola), of an SKN400/27 (SEMIKRON)

2.10   Find the maximum I_{FAV} for the diodes mentioned in question 2.10.

2.11   What is the maximum permissible junction temperature of a 1N4005?

2.12   What is the maximum instantaneous forward voltage drop across a 1N4005?

2.13   What are the typical values of the forward voltage drop across an SKN400 when I_F =500A and T_J is respectively 25°C and 160°C.

2.14   What does "dynamic resistance" mean in connection with a diode ?

2.15   What do the following subscripts mean in combination with diode variables : F / S /M ?

Photo Hitachi: Automotive diodes

3   TRANSISTOR POWER SWITCHES

CONTENTS

1 BIPOLAR JUNCTION TRANSISTOR (BJT)

1.1 BJT as a switch

We want to use the NPN transistor shown in fig. 3-1 as a switch. During conduction, that is as a closed switch , V_{CE} needs to be as low as possible. In the case of an ideal switch V_{CE} is zero.

During reverse bias, that is as an open switch, I_C should be zero and $V_{CE} = V_{CC} = 48$V.

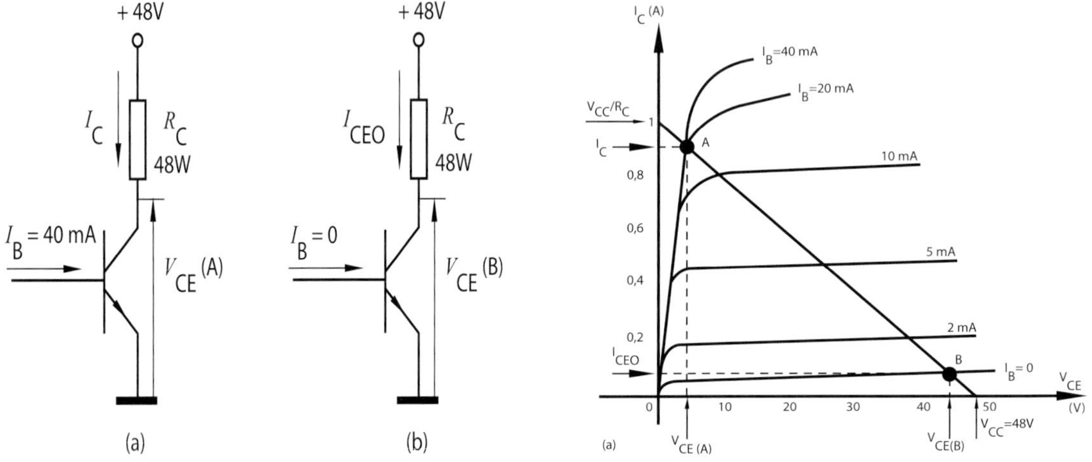

Fig. 3-1: Transistor as switch

Fig 3-2: Q points of transistor switch in an I_C - V_{CE} curve

Fig. 3-2 shows the Q points of a transistor in an I-V curve. The load line corresponds to the data of fig. 3-1. The two switch conditions correspond to the Q points A and B in fig. 3-2.

In this diagram we can read the following information:

	Point A	**Point B**
. base current	40 mA	null
. collector current	almost 1 A	leakage current I_{CE0}
. collector emitter voltage	low	almost 48 V

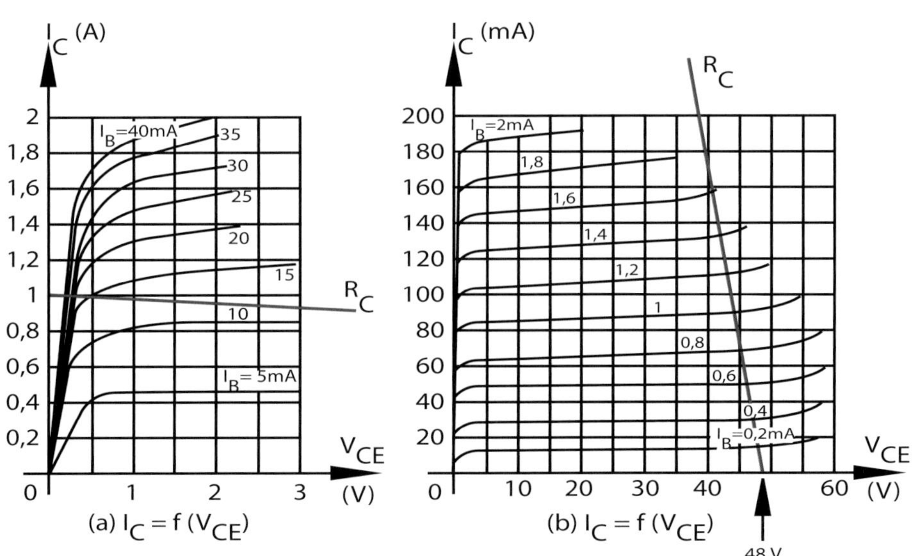

Fig. 3-3: I_C - V_{CE} curves of a transistor switch

To accurately determine the coordinates A and B fig. 3-2 is insufficient. We need extra details such as those of fig. 3-3 and 3-4.

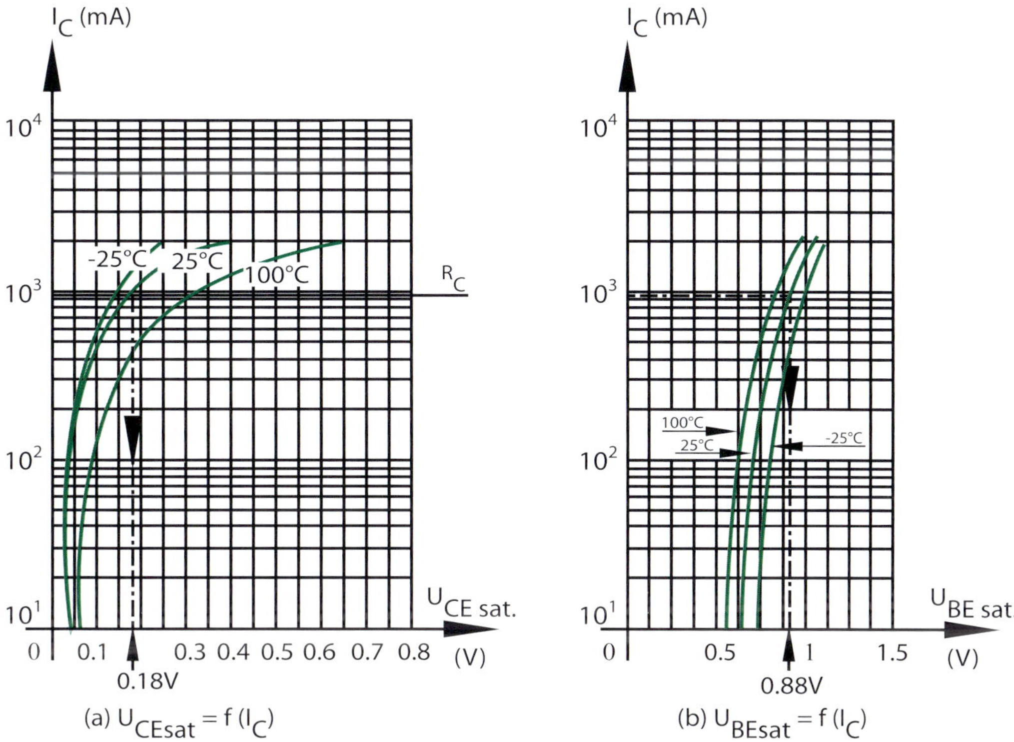

Fig 3-4: I_C - V_{CE} saturation curves of a BJT

Bipolar transistor (e.g. type BSX…or BU..)

Ratings: $V_{CE0} = 60$V; $V_{BE0} = 5$V; $I_C = 2$A; $T_J = 100°$ C; $P_{tot} = 5$W;

$\qquad I_{CE0} = 10$ to 150nA (typical 50nA); $h_{FE} = 40$ to 150 ($I_C = 100$mA)

Dynamic properties: $t_{on} < 0.2$ µs ; $t_{off} < 1$ ms

Characteristics: see fig. 3-3 and 3-4

1.2 Exact coordinates of the operating point

In fig. 3-3a the load line is shown with $I_B = 40$mA and $V_{CE} < 0.25$V. The exact value is easier to read from fig. 3-4a : $V_{CEsat} = 0.18$V (at $T_J = 25°$ C). The subscript "sat" means saturated.

We see clearly in fig 3-3a that even with a base current larger than 40mA the lower limit has been reached for V_{CE}.

The maximum collector current now flows through the transistor (at that specific R_C and V_{CC}): the transistor is saturated.

Calculation shows: $I_C = \dfrac{V_{CC} - V_{CEsat}}{R_C} = \dfrac{48 - 0.18}{48} = 0.966$ A

At saturation we can approximate: $I_C \approx \dfrac{V_{CC}}{R_C} = \dfrac{48}{48} = 1$ A

1.3 Driving the base – Hard saturation

The relationship between the collector and the base current is defined as follows
$h_{FE} = I_C/I_B$. Keeping in mind that $h_{FEmin} < h_{FE} < h_{FEmax}$ we can calculate I_B to obtain
a specific I_C .
The transistor in question has an h_{FE} of 40 to 150 so that $I_{Bmax} = I_C / h_{FEmin} = 966/40 = 24.15\text{mA}$.
Apparently it is not necessary to make $I_B = 40\text{mA}$ as we had predicted earlier since with 25mA the
transistor is in saturation.
In general we can say that to drive a transistor into saturation we drive the transistor to beyond the
knee or turn of the I_C - V_{CE} - curve (see fig. 3-3a).
V_{CEsat} is sometimes called the V_{CE} - knee. In fig. 3-4b we can read that when $I_C = 966\text{mA}$ at
$T_J = 25°$ C, the base-emitter voltage $V_{BEsat} = 0.88\text{V}$.
Take note that $V_{BEsat} > V_{CEsat}$. The base collector diode is forward polarized or in other words:
in heavy saturation the collector-base barrier layer is reduced. The transition from the linear area to
saturation is at $V_{CB} = 0$ V (fig. 3-5).
If we drive more current into the base the transistor will become more and more saturated.
We cannot drive the transistor beyond the slant limit line in the I_C - V_{CE} curve since this is hard
saturation. Between $V_{CB} = 0$ and hard saturation the region is known as quasi-saturation.

1.4 Operating range of a transistor

In fig. 3-5 the operating areas of a transistor are shown:
- **hard saturation:**   transistor as switch (closed)
- **quasi-saturation:**   from this region the transistor can move quickly to the cut off region.
  This would take considerably more time from the hard saturation region.
- **linear region:**   transistor used as an amplifier.
- **cut off region:**   transistor can function as an open switch.

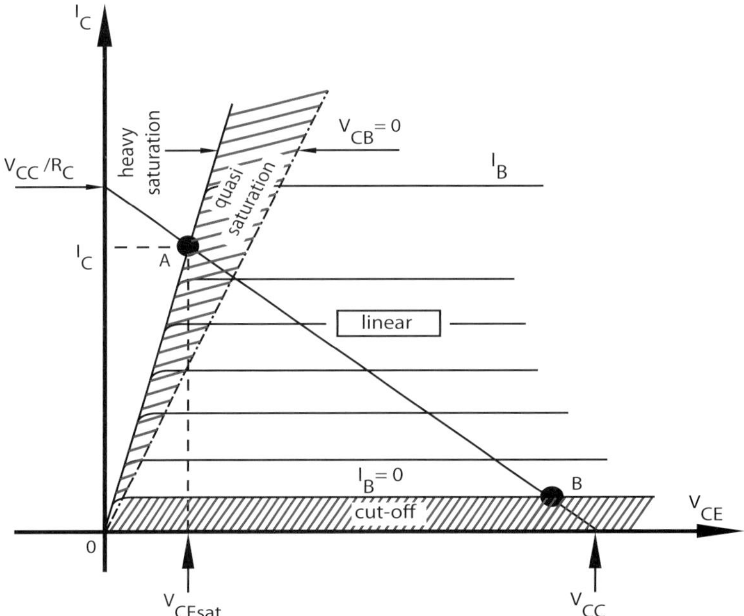

Fig. 3-5: Operating regions of a bipolar transistor

1.5 Comparing a transistor to an ideal switch

In fig. 3-6 we compare a transistor with an ideal switch. The details from the figures 3-1, 3-3 and 3-4 have been used.

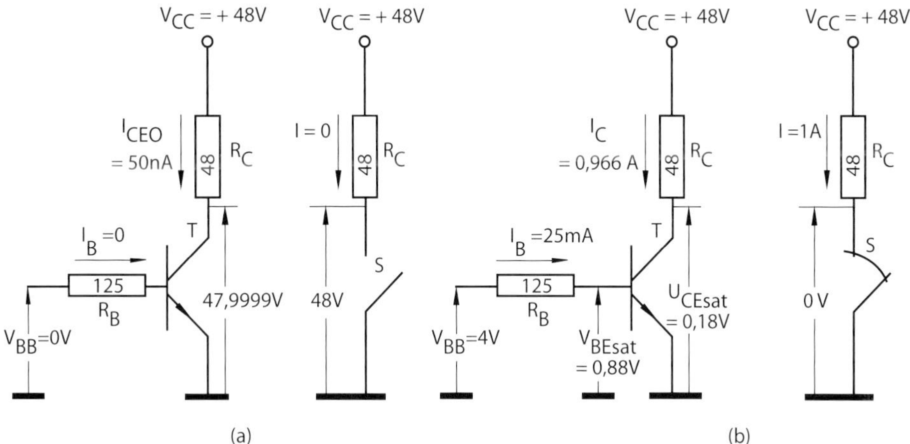

Fig. 3-6 : Non ideal behavior of a transistor as switch

From fig. 3-6 we see the non ideal behavior of a transistor as a switch :

. transistor in cut off: a small leakage current flows $I_{CE0} = 50\text{nA}$

. transistor in saturation: there is a small voltage drop $V_{CEsat} = 0.18\text{V}$ across the terminals of the transistor.

1.6 Determining the base resistor

$$I_{Bmax} = \frac{I_C}{h_{FEmin}} = \frac{V_{CC} - V_{CEsat}}{R_C \cdot h_{FEmin}} \quad \text{and also:} \quad I_{Bmax} = \frac{V_{BB} - V_{BEsat}}{R_B}$$

so that: $\quad R_B = \frac{(V_{BB} - V_{BEsat})}{(V_{CC} - V_{CEsat})} \cdot R_C \cdot h_{FEmin}$

practically: $V_{BEsat} = 0.85$ V and $V_{CEsat} = 0.2$ à 0.3 V

so that:
$$R_B \leq \frac{(V_{BB} - 0.85)}{V_{CC}} \cdot R_C \cdot h_{FEmin} \tag{3-1}$$

Numeric example 3-1:

Given a transistor with a $h_{FE} = 40$ and the data from fig.3-6b. Calculate the value of the base resistor. Repeat the question for a transistor with a $h_{FE} = 200$.
Solution:

$$h_{FE} = 40 : \quad R_B \leq \frac{V_{BB} - V_{BEsat}}{V_{CC} - V_{CEsat}} \cdot R_C \cdot h_{FEmin} \qquad\qquad h_{FE} = 200$$

$$= \frac{4 - 0.88}{48 - 0.18} \cdot 48 \cdot 40 = 125.26\Omega \qquad\qquad R_B \leq \frac{4 - 0.88}{48 - 0.18} \cdot 48 \cdot 200 = 626.35\Omega$$

$$\rightarrow\rightarrow\rightarrow \quad R_B = 120\Omega \qquad\qquad\qquad \rightarrow\rightarrow\rightarrow \quad R_B = 620\Omega$$

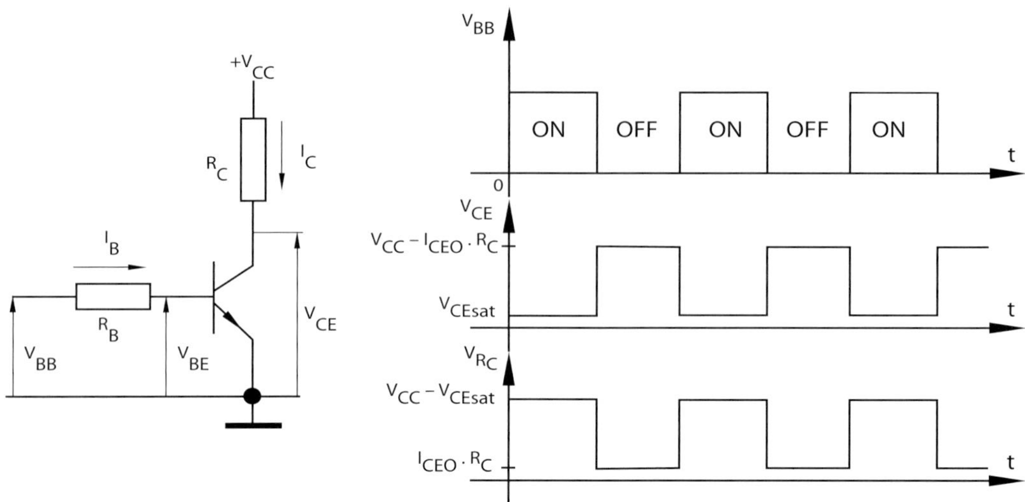

Fig. 3-7: Transistor as a switch with common emitter configuration

1.7 Ratings of a BJT

In the datasheet of a transistor maximum values are stated. These may not be exceeded, otherwise there is a large risk that the transistor will be destroyed.

These ratings are:

. I_{Cmax} :   maximum collector current
. V_{CE0max} :   maximum collector emitter voltage with no base current
. P_{Cmax} :   maximum collector power that can be dissipated
. T_{Jmax} :   maximum allowable junction temperature

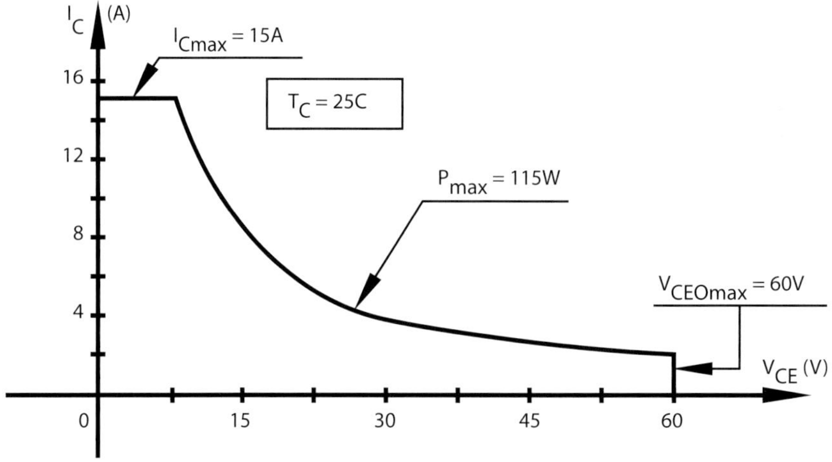

Fig. 3-8: Operating areas of a 2N3055

The maximum V_{CE} - I_C - product (= constant) is shown in an I_C - V_{CE} axis by a hyperbola. The hyperbola is limited by I_{Cmax} and V_{CE0max} as shown in fig. 3-8 which is drawn for a 2N3055.

1.8 Dynamic behaviour with a resistive load

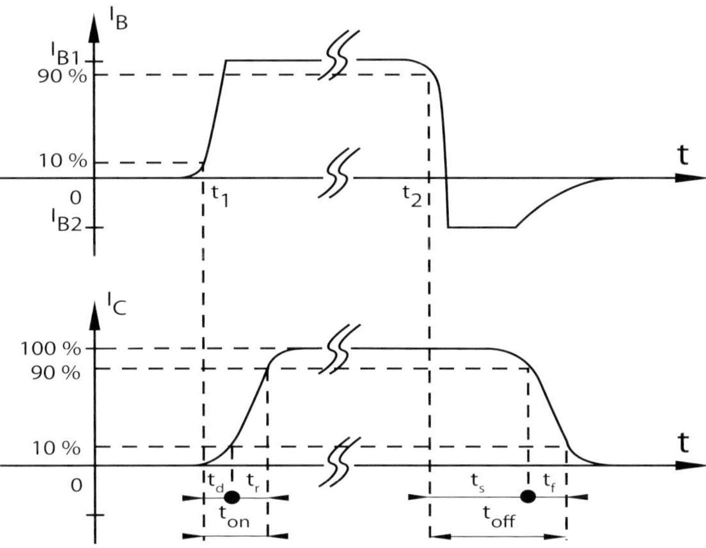

Fig. 3-9: Switch on and off times of a transistor

Fig. 3-9 shows the switch on and switch off times of a transistor switch in the case where it has an resistive load. A time t_{on} is required to forward bias the base-emitter diode (barrier layer decreased) and to bring the transistor to full conduction (injecting enough charge carriers). If we then want to reverse bias the transistor the space charge and the charge carriers need to first decrease, followed by the restoration of the base-collector barrier region. It takes t_{off} before practically no current flows through the transistor.

In fig. 3-9 we see:

$$t_{on} = t_d + t_r = \text{switch on time} \tag{3-2}$$
$$t_{off} = t_s + t_f = \text{switch off time} \tag{3-3}$$

t_d = delay time = time required to reduce the barrier layer between base and emitter

t_r = rise time = time required to inject sufficient charge carriers and to form I_C

t_s = storage time = time required for the saturated storage charge to decrease

t_f = fall time = time required to restore the collector-base barrier

These times coincide with a specific percentage of the collector current and are stated by the manufacturers in accordance with the 10% and 90% values illustrated in fig. 3-9.

In addition note:

. the larger I_{B1} (forward current) the shorter t_d will be

. when the transistor is not saturated : $t_s \approx 0$

. the heavier the saturation the smaller t_{on} will be

. to switch off the transistor the reverse base current (I_{B2}) works as a discharge current

. the switch on time (30 to 300ns) and switch off time (30ns to 1.5µs) become 1 to 2 µs and
   5 to 8µs respectively when large power transistors are used

Take note that when a transistor is not saturated t_{off} is only a few ns.

If fast switching times are needed we can:

. choose a transistor with a small t_{on} and t_{off}

. drive the transistor into saturation since t_{on} becomes small.

. before switch off bring the transistor out of saturation. As a result t_s becomes practically zero
   and the switch off time $t_{off} (\approx t_f)$ very short.

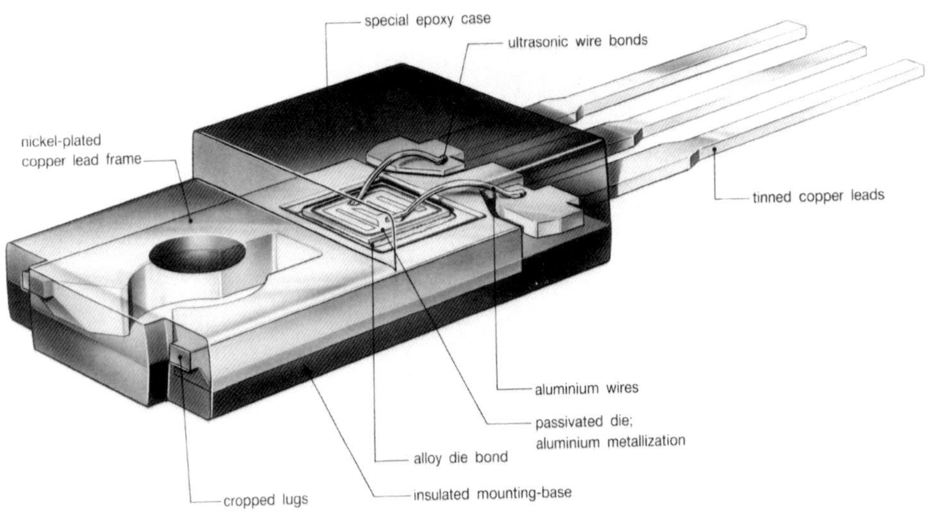

Photo Philips: F (-pack) SOT-186 package

The F-pack provides an integral electrical insulation. This is especially interesting when mounting
a heat sink as no mica or plastic clips are required as was the case with the earlier TO-220 package. The SOT-186 is developed to replace the TO-220. This first F-pack (the SOT-186) from
Philips was introduced in 1986. In addition to transistors diodes and thyristors are available
in F-pack. Normally an F is simply added to the type number (BD645F, BT151F-500R,
BYV72F-200,…).

1.9 Inductive load with a free wheel diode

We supply a base current I_{B1} as in fig. 3-10a. After $5 . \dfrac{L_b}{R_b} = 5 . \dfrac{10^{-1}}{10} = 50$ ms the collector
current reaches its steady state value $I_C = 10$ A. (if we neglect V_{CE} and there is sufficient base
current $I_{B1} = I_C / h_{FE}$ flowing).
By injecting a current I_{B2} instead of I_{B1} the transistor switches off (e.g. in t_{off}=200 ns).

A voltage spike will occur according to $L_b . \dfrac{di_C}{dt} = 10^{-1} . \dfrac{10}{200 . 10^{-9}} = 5$ MV.

Another way of looking at the situation is in terms of energy $W = \dfrac{L_b . I_C^2}{2} = \dfrac{10^{-1} . 10^2}{2} = 5$ J
from the inductor has to dissipate during the switch off of the transistor.

Because of this we can write: $W = E_L . I_{gem} . t_{off}$

With $t_{off} = 200$ ns and $I_{gem} = \dfrac{I_C}{2} = 5$A then this gives: $E_L = \dfrac{W}{I_{gem} . t_{off}} = \dfrac{5}{5.200.10^{-9}} = 5$MV

Reasoning with Lenz's law ($E_L = L_b . \dfrac{di}{dt}$) or with the magnetic energy ($W = \dfrac{L_b I_C^2}{2}$), the result is
the same. The quicker the transistor switches off, the larger the voltage spike across the transistor
will be.

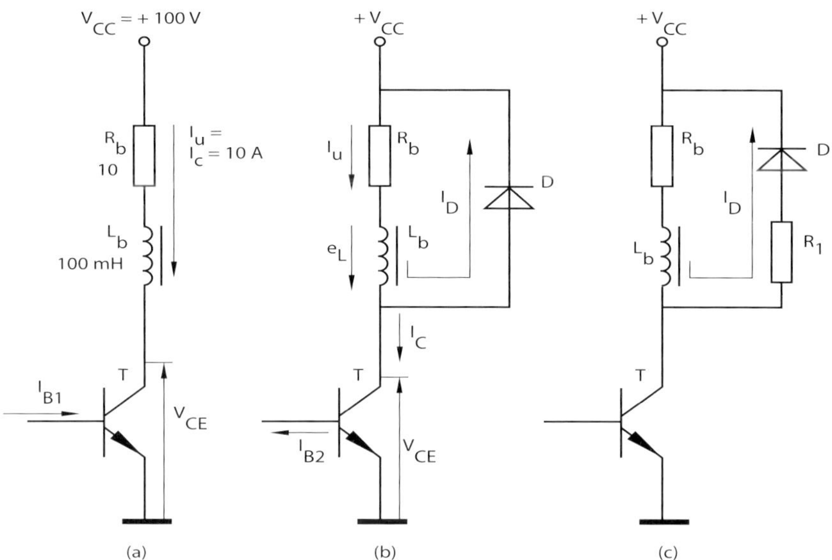

Fig. 3-10: Inductively loaded transistor switch

Looking at it another way : to dissipate the same energy through the transistor with the same average current ($\frac{I_c}{2}$) in a shorter time (t_{off}) is only possible with a higher voltage.

($W = \dfrac{L_b \cdot I_C^2}{2} = V \cdot I_{gem} \cdot t_{off}$). It is clear that the transistor will not survive the over voltage spike. A good solution for this problem is to employ a free wheel diode (fig. 3-10). Other solutions include an RC–series branch in parallel with the $R_b - L_b$ load, or a zener diode parallel over the resistor. The solution using the zener diode is only possible in low power applications.

We will only study the solution using the free wheel diode. When the transistor is switched off the emf e_L ($= L_b \cdot \dfrac{di}{dt}$) will cause a current to flow through the diode. The energy $\dfrac{L_b \cdot I_C^2}{2}$ is discharged via D, and it is referred to as a free wheel diode.

When the diode conducts the voltage $V_{CE} \approx V_{CC}$ is across the transistor. The collector voltage is clamped at V_{CC}. The diode D is also called a clamp diode. To discharge the inductor more rapidly a resistor R_l may be placed in series with D (fig. 3-10c). Indeed the time constant $\dfrac{L_b}{R_b + R_l}$ is less than L_b/R_b.

We can also see that the energy from the inductor is now dissipated in two resistors (R_b and R_l). This will occur quicker than with a single resistor (R_b). The disadvantage of the resistor R_l is a larger collector voltage: $V_{CE} = V_{CC} + I_D \cdot R_l$.

Numeric example 3-2:

1. Assume the transistor in fig. 3-10a switches off in 1µs, how large is the emf in the coil L_b due to the self-inductance?
2. If as in fig. 3-10c a resistor $R_1 = 10\Omega$ is used and the voltage across the diode is 1.1V, determine the collector voltage at the start of switch off of the transistor.
   How much time is required for the energy in the coil to almost completely dissipate?

Solution:

1.
$$e_L = L \cdot \frac{di}{dt} = 10^{-1} \cdot \frac{10}{10^{-6}} = 1 \text{ MV}$$

2.   a) $V_{CE} = 100 + 1.1 + R_1 \cdot I_C = 100 + 1.1 + 10.10 = 201.1\text{V}$

     b) $t = 5 \cdot \dfrac{L_b}{R_b + R_1} = 5 \cdot \dfrac{10^{-1}}{10 + 10} = 25 \text{ ms}$

1.10 Snubber network

To protect a transistor during switch off a snubber network is often used. As a result the switch off losses in the transistor are limited to within the safe operating area of the transistor.
A classic snubber network is composed of a resistor R, a capacitor C and a diode D.
This is the so called RCD snubber (fig. 3-11).

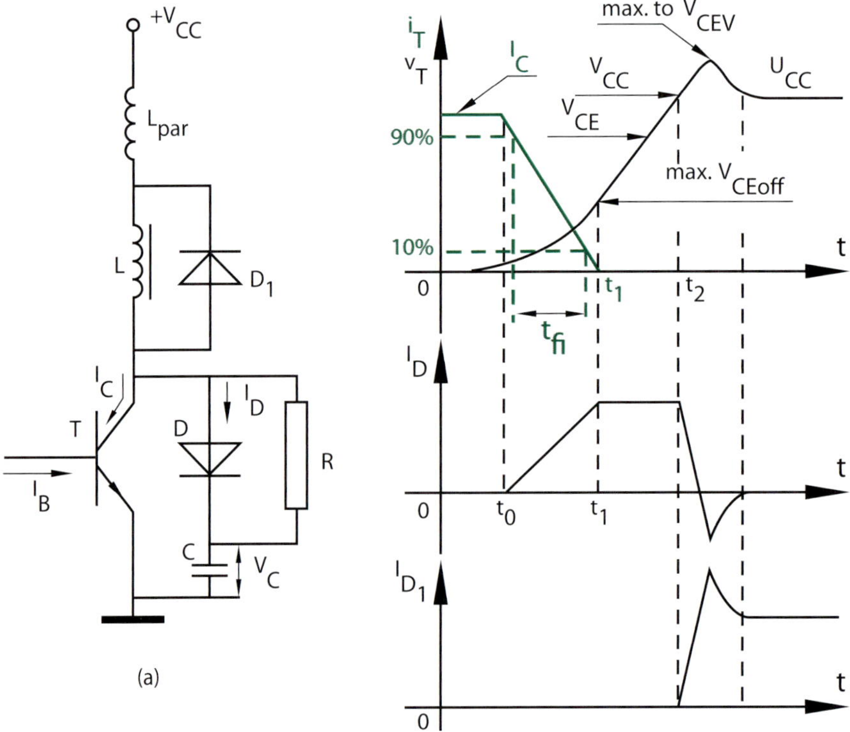

Fig. 3-11: Inductively loaded transistor with an RCD snubber

At switch off of the transistor the current I_C goes to zero in the time t_{fi}. The transistor current now flows via diode D and the capacitor charges to a value V_{CEoff} (less the diode voltage drop of about 1.1V). The coil functions for some time as a constant current source (internal energy $= \frac{1}{2} . L . I_C^2$) until the moment t_2 that C is charged at $+ V_C$.
At t_2 the collector voltage is V_{CC} and the freewheel diode now provides the path for the current. The remaining energy in the coil is discharged via D_1 . The reverse recovery current of D also flows in D_1 (see the peak in the current diagram of fig. 3-11).

If transistor T is again switched on then C will exponentially discharge via R and T with a starting current $\approx \dfrac{V_{CC}}{R}$. The resistor serves to limit the current through the transistor.

A. Capacitor C

During the time t_{fi} an average current $\dfrac{I_C}{2}$ flows as charging current for C. Except for the diode voltage drop $V_C = V_{CE}$.

After t_{fi} we find : $V_{CEoff} \approx V_C = \dfrac{Q}{C} = \dfrac{I_C . t_{fi}}{2 . C}$ or $: C = \dfrac{I_C . t_{fi}}{2 . V_{CEoff}}$.

The larger the value of V_{CEoff} we allow the smaller the value of C we need. If for example we take the maximum value of $V_{CEoff} = V_{CC}$ then V_{CE} will quickly rise .

In the most unfavourable case where C is charged quickly (fig. 3-12) it seems we need to calculate with t_C instead of t_{fi} .

It is clear that the turn off energy of the transistor is given by :

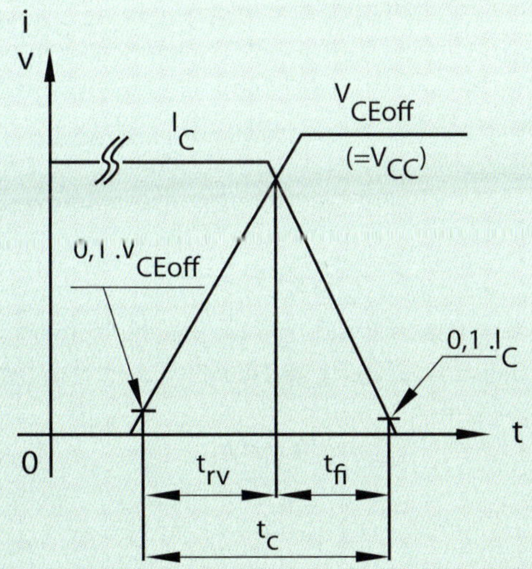

Fig. 3-12: Transistor turn off in the case of hard switching

$$W = I_C . \frac{V_{CEoff}}{2} . t_{rv} + \frac{I_C}{2} . V_{CEoff} . t_{fi}$$

$$= \frac{I_C . V_{CEoff}}{2.} (t_{rv} + t_{fi}) \quad = \frac{I_C . V_{CC} . t_c}{2}$$

with $V_{CEoff} = V_{CC}$

and $t_{rv} + t_{fi} = t_c$

$t_C =$ turn off cross over time (10% V_{CEM} to 10% I_{CM})

The most unfavourable situation is called "hard switching".

The energy in the capacitor,is:

$$W = \frac{1}{2} . C . V_{CEoff}^2 = \frac{1}{2} . C . V_{CC}^2$$

so that $\dfrac{1}{2} . C . V_{CC}^2 = \dfrac{I_C . V_{CC} . t_c}{2}$ or : $C = \dfrac{I_C . t_c}{V_{CC}}$ (3-4)

The capacitor needs to be able to withstand a high RMS current to be able to withstand a very high

dv/dt gradient. In addition the charging current is quite high ($= I_C$!)

B. Diode D

It is necessary to use a soft recovery diode.

C. Resistor R.

The discharge of the snubber capacitor need to take place during the conduction time of the transistor t_{ON} . After 3 to 4 RC the capacitor is practically discharged so that:

$$R = \frac{t_{ON\ (min)}}{(\ 3\ to\ 4\ )\ .\ C} \tag{3-4}$$

This discharge current may not be too large otherwise the total transistor current would be too large. In practice designers fix the maximum of the discharge current ($\frac{V_{CC}}{R}$) at 20 to 25% of I_C . The energy in the capacitor is $\frac{1}{2}$. C . V_{CC}^2 so that the energy dissipated in the resistor can be determined with : $P = \frac{1}{2}$. C . V_{CC}^2 . f_S . The frequency f_S is the frequency with which the transistor is operated.

D. Snubbers with low losses

As can be seen in the previous formulae the losses in the snubber are proportional with the square of the voltage and with the frequency. Snubbers can dramatically influence the efficiency of a transistor circuit because of the losses in the snubber elements, especially the resistor. Designers strive therefore to create low loss snubber circuits, for example by pumping the stored energy from the snubber capacitor back to the supply.

Numeric example 3-3:

In fig. 3-11 a transistor is used with t_c = 160ns at a current I_C of 8A.
If the supply voltage V_{CC} = 300V, determine the value of R and C of the snubber network.
We are given the switching frequency of the transistor as 2kHz and the duty cycle δ is 50%.

Solution:

$$C = \frac{I_c \cdot t_c}{V_{CC}} = \frac{8 \cdot 160 \cdot 10^{-9}}{300} = 4.27\text{nF} \qquad \rightarrow\rightarrow\rightarrow \qquad C = 4.3\text{nF}$$

$$R = \frac{t_{ON\ (min)}}{4\ C} = \frac{\delta \cdot T}{4 \cdot C} = \frac{0.5 \times 5 \times 10^{-4}}{4 \times 4.3 \times 10^{-9}} = 14.53\text{k}\Omega \qquad \rightarrow\rightarrow\rightarrow \qquad R = 15\text{k}\Omega$$

Photo Siemens: With the SINAMICS S120 cabinet modules, Siemens offers a modular cabinet module system for multi-motor drives. The SINAMICS S120 cabinet modules provide a compact and ready-to-use drive solution scalable up to a rating of 4.5MW, for use particularly in machines and plants in the paper machinery sector, rolling mills, test bays and hoisting gear.

Photo Siemens: The SINAMICS S150 drive converter is suitable for demanding drive applications at high outputs.

Is your drive train more than the sum of its parts?

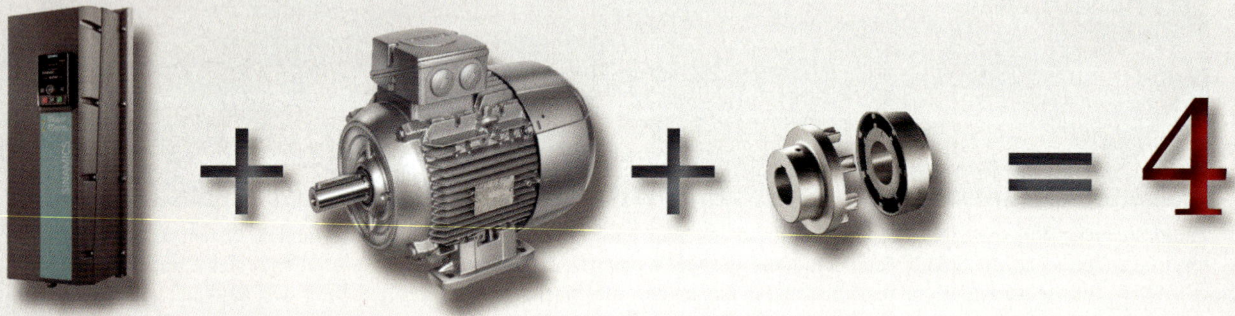

With the Total Drive Train from Siemens it is.

Total Drive Train

Total Drive Train is Siemens' total solution, from power grid to machine axle. This concept encompasses the frequency converter, motor, geared motor, gear unit and coupling. A single specialized Siemens point of contact gives you access to all our knowhow and services. So what's in it for you? A more efficient drive train, keeping you on the right track. **www.siemens.be/drives.**

Answers for Industry.

2. POWER MOSFET

2.1 Mos field effect transistor

Of the existing field effect transistors (JFET and MOSFET) the most commonly used in power electronic applications is the N channel enhancement type. That is the one that will be considered here.

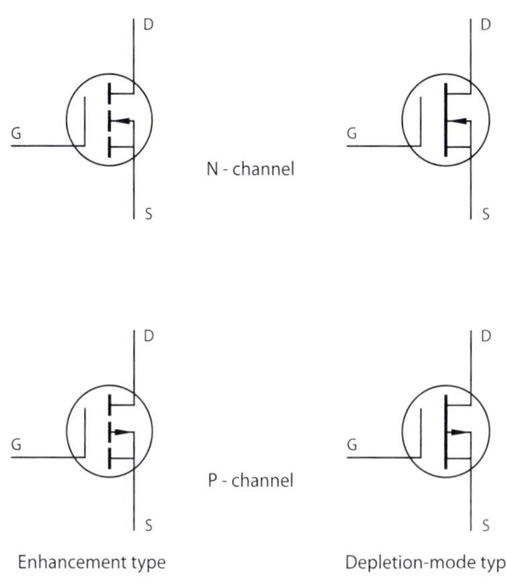

MOS = metal oxide silicon

FET = Field effect transistor

Fig. 3-13 shows the common MOSFET symbols.

Field effect transistors have a source (S), a gate (G) and a drain (D).

MOSFETS are available as N and P channel types.

With the enhancement MOSFET there is no conduction if $V_{GS} = 0$, hence the name enhancement and the broken line between D and S in the MOSFET symbol.

If G and D have the same polarity with reference to S then the MOSFET can conduct.

Fig. 3-13: MOSFET symbols

2.2 Construction

Fig. 3-14 shows the construction of an N channel MOSFET of the enhancement type. In a P type substrate layer two N^+ zones are created 10 to 20 μm from each other.

A dielectric layer of quartz (SiO_2) is placed with a thickness of 0.1 to 0.2 μm.

The electrodes D, G and S are connected using evaporated aluminium. The three layers metal, oxide and silicon are placed one on top of the other, this explains the name: metal oxide silicon which is shortened to MOS.

Some give the term MOS the following meaning : metal oxide semiconductor.

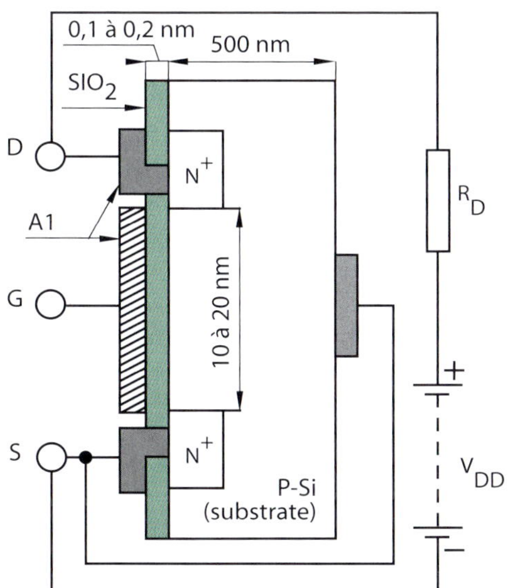

Fig. 3-14: Construction of an N-channel MOSFET of the enhancement type

2.3 Operation

If a positive voltage V_{GS} is applied (fig. 3-15a) then as a result of electrostatic induction an N-zone is created in the P structure between the two existing N^+-zones. Since the N-zone consists of minority charge carriers (from the P substrate) we refer to this as an inversion layer. Source (S) and drain (D) are now connected via the N-channel and with a positive voltage V_{DS} a current I_D will flow from D to S. The larger V_{GS}, the wider the N-channel and the larger the current I_D will be. In fig. 3-15b a schematic representation of the configuration of fig. 3-15a is shown.

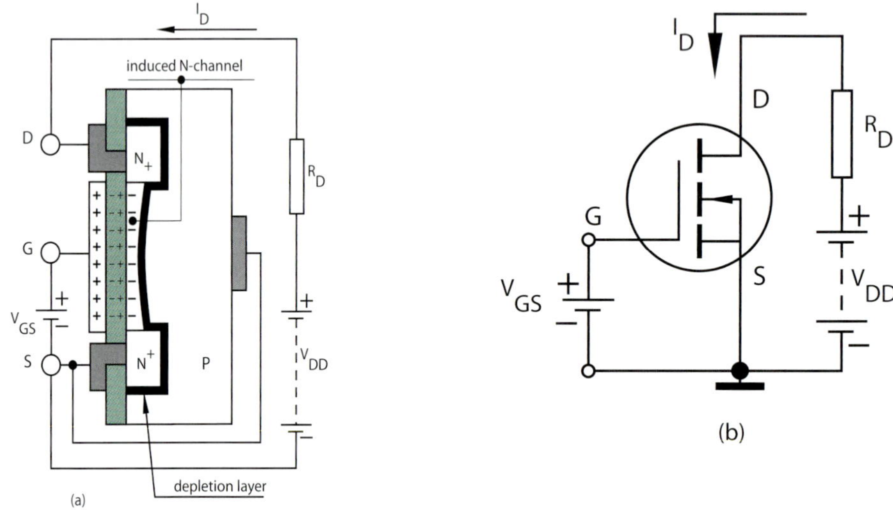

Fig. 3-15: Operation and principal configuration of a N-channel enhancement type MOSFET

2.4 MOSFET characteristics

The characteristic in fig. 3-16a is analogous to the I_C - V_{CE} - characteristic of a bipolar transistor. In the case of a BJT we control the transistor with the base current, while with a MOSFET we work with voltage control. The drain current consists of one type of charge carrier therefore we refer to a MOSFET as a **unipolar transistor** in contrast to the BJT.

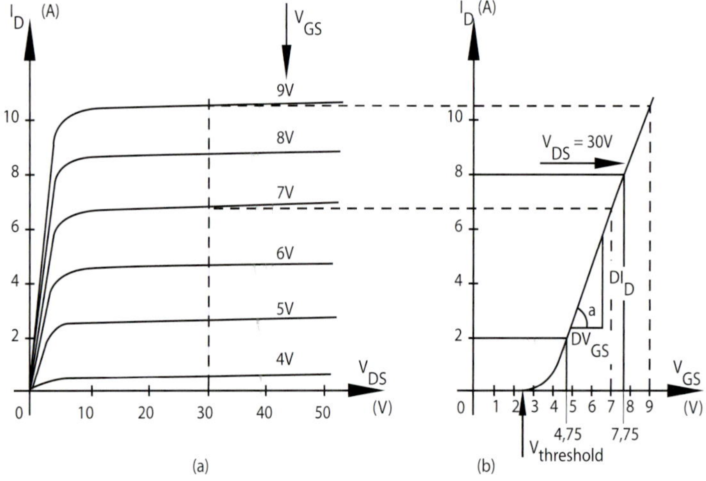

Fig. 3-16 (a and b): Characteristics of an N-channel enhancement MOSFET

From the output characteristic we can derive the transfer characteristic (fig. 3-16b). This transfer provides us with a very important specification , namely the transconductance:

$$g_{fs} = (\frac{\Delta I_D}{\Delta V_{GS}}) \text{ with } V_{DS} = C^{te} \; ; \; g_{fs} = tg\,\alpha \tag{3-6}$$

In fig. 3-16 b we see that the gate voltage must exceed a certain minimum $V_{GS(th)}$ before the MOSFET conducts.

The SiO_2 layer between the gate and source is responsible for the very high input resistance of the MOSFET (10^{10} to 10^{15} Ω). A single elementary unit as shown in fig. 3-14 is only capable of handling a drain current of 100μA. The parallel combination of for example 100,000 similar elementary units results in a MOSFET with a drain current of several amps as shown in fig. 3-16a.

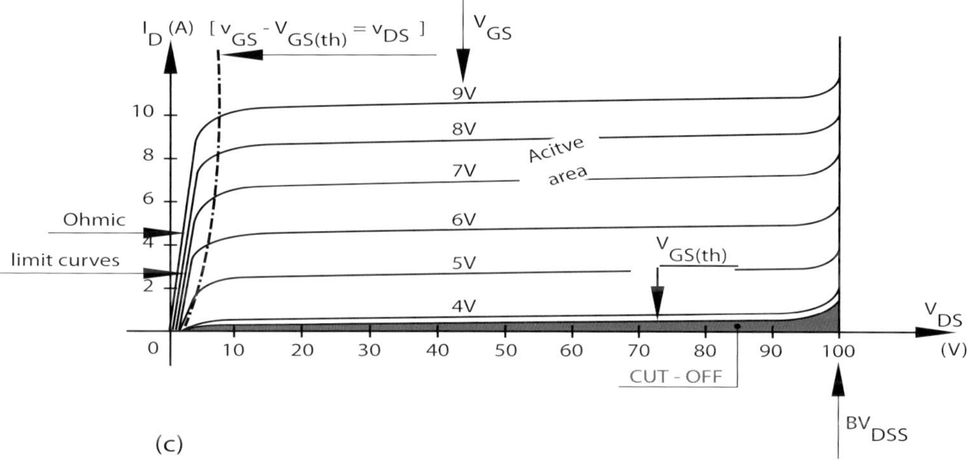

Fig. 3-16c : Operating regions of an N-channel MOSFET

Fig. 3-16c shows the complete I - V characteristic with the three distinct regions:

1. Cut-off: With the gate voltage below the turn-off voltage $V_{GS(th.)}$ there is no current flow and the MOSFET and functions as an open switch. In the characteristic the breakdown voltage BV_{DSS} of the MOSFET is shown. This means that the applied drain source voltage must be less than this BV_{DSS}

BV = Breakdown voltage; DSS = Drain-Source in the common source configuration.

2. Active region: In this region we have an almost linear relationship between drain current and gate source voltage ($i_D = g_{fs} \cdot u_{GS}$)

3. Ohmic area: With a BJT the border between the active area and saturation is given by $V_{CE} = V_{BE}$ or $V_{CB} = 0$, see fig. 3-5. In similar fashion for a MOSFET the border between the active area and saturation is given by: $V_{DG} = 0$ or $V_{GS} - V_{GS(th.)} = V_{DS}$.
When V_{DS} is low, every value of V_{GS} ends up on the border line in the $I_D - V_{DS}$ characteristic. These border lines (fig. 3-16c) all have the same ramp irrespective of the value V_{GS} . On the border line the relationship between I_D and V_{DS} is practically constant, the MOSFET behaves as a resistor R_{DS} . We refer to an OHMIC operating area. In this area or region the MOSFET functions as a closed switch. We refer to the resistance of this "closed" (MOSFET) switch as $R_{DS(ON)}$. The value of $R_{DS(ON)}$ follows from the ramp of the border line in fig. 3-16c, but this is also to be found in the datasheet of the MOSFET.

Numeric Example 3-4:

Determine the transconductance of the MOSFET with the characteristic shown in fig. 3-16, when $V_{DS} = 30V$.

Solution:

In fig. 3-16b we see that when $I_{D1} = 2A$ and $I_{D2} = 8A$ the gate voltage is: $V_{GS1} = 4.75V$ and $V_{GS2} = 7.75V$,

therefore: $g_{fs} = \left(\dfrac{\Delta I_D}{\Delta V_{GS}}\right)_{V_{DS}=30V} = \dfrac{8-2}{7.75-4.75} = 2 \text{ A/V}$

Remarks

1. MOSFET of the depletion type

Fig. 3-17 shows the construction, symbol and characteristic of an N-channel depletion MOSFET.

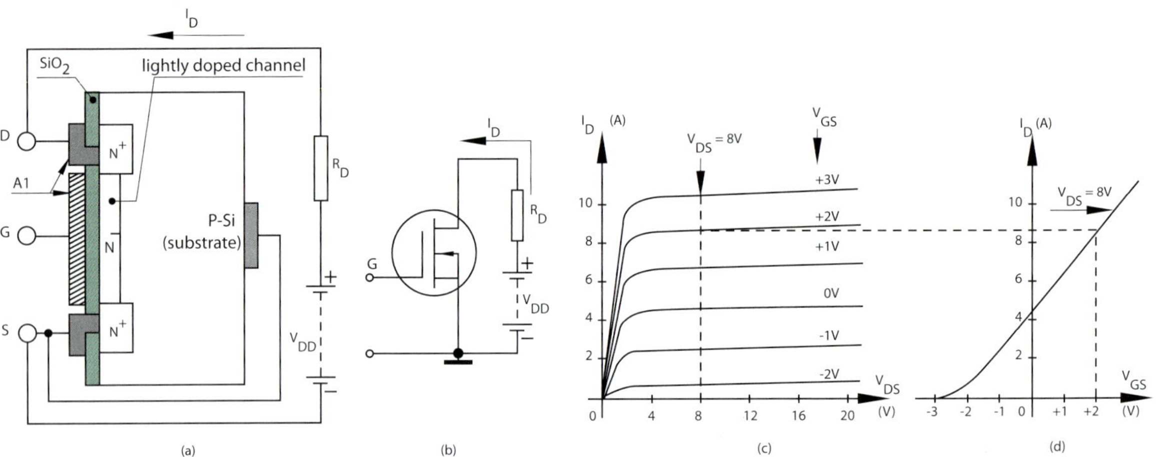

Fig. 3-17: Depletion mode N-channel MOSFET

We note that when $V_{GS} = 0$ the transistor is conducting because an existing N-channel is part of the construction. By making V_{GS} negative the channel narrows and I_D reduces. A positive V_{GS} makes the channel wider and so I_D increases.

2. Static charges

The human body has a capacitance of 100 to 150 pF. This capacitance can charge for example up to $1\,\mu C$. If because of touch this charge transfers to the input capacitance of the MOSFET (e.g. 10pF) then the gate voltage becomes $V_{GS} = \dfrac{Q}{C_{GS}} = \dfrac{10^{-6}}{10 \cdot 10^{-12}} = 100,000V$

IGBT Module

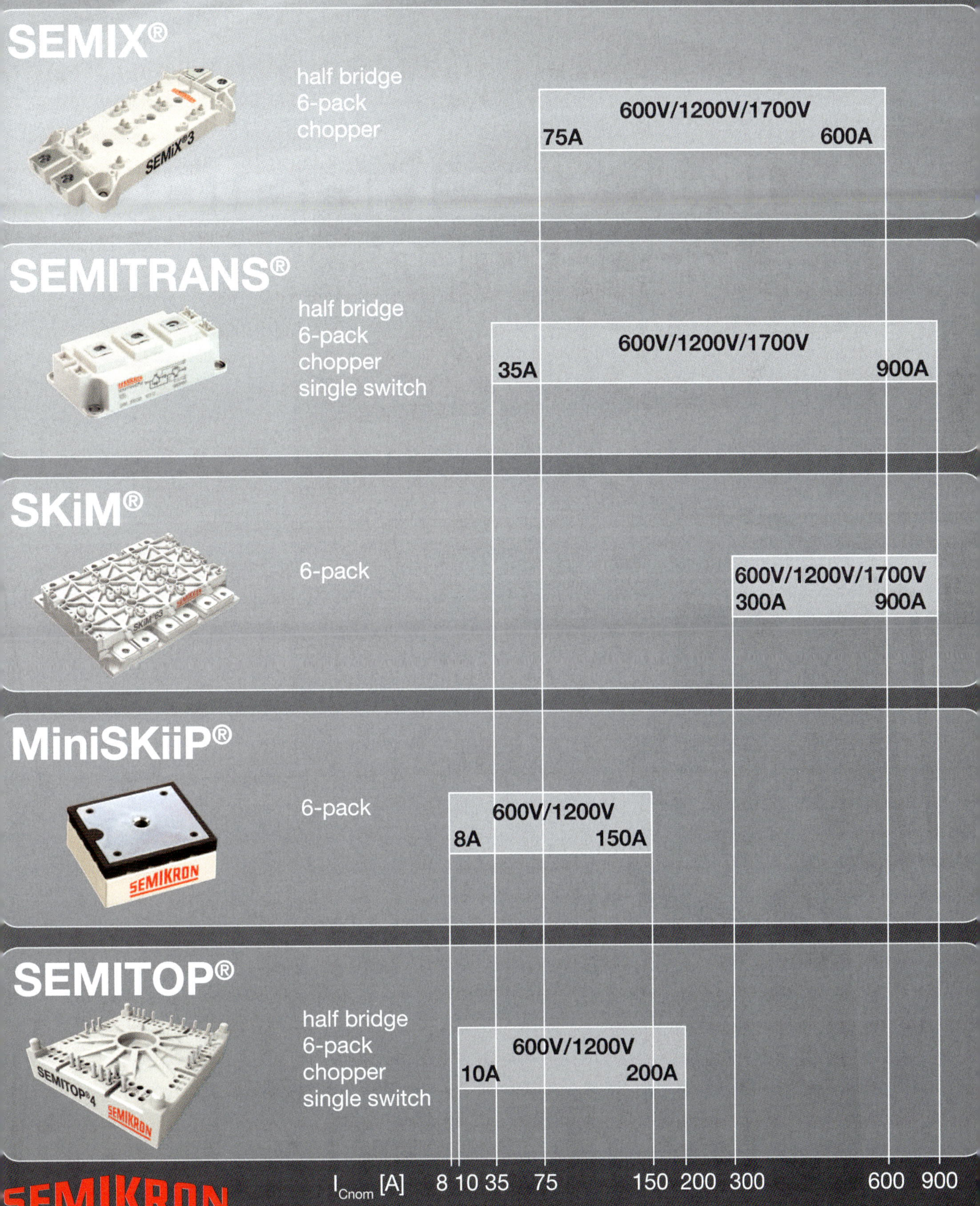
SEMIX®
half bridge
6-pack
chopper
600V/1200V/1700V
75A
600A
SEMITRANS®
half bridge
6-pack
chopper
single switch
600V/1200V/1700V
35A
900A
SKiM®
6-pack
600V/1200V/1700V
300A
900A
MiniSKiiP®
6-pack
600V/1200V
8A
150A
SEMITOP®
half bridge
6-pack
chopper
single switch
600V/1200V
10A
200A
I_{Cnom} [A]
8 10 35 75 150 200 300 600 900

SEMIKRON

SKiN Technology
Wire bond-free

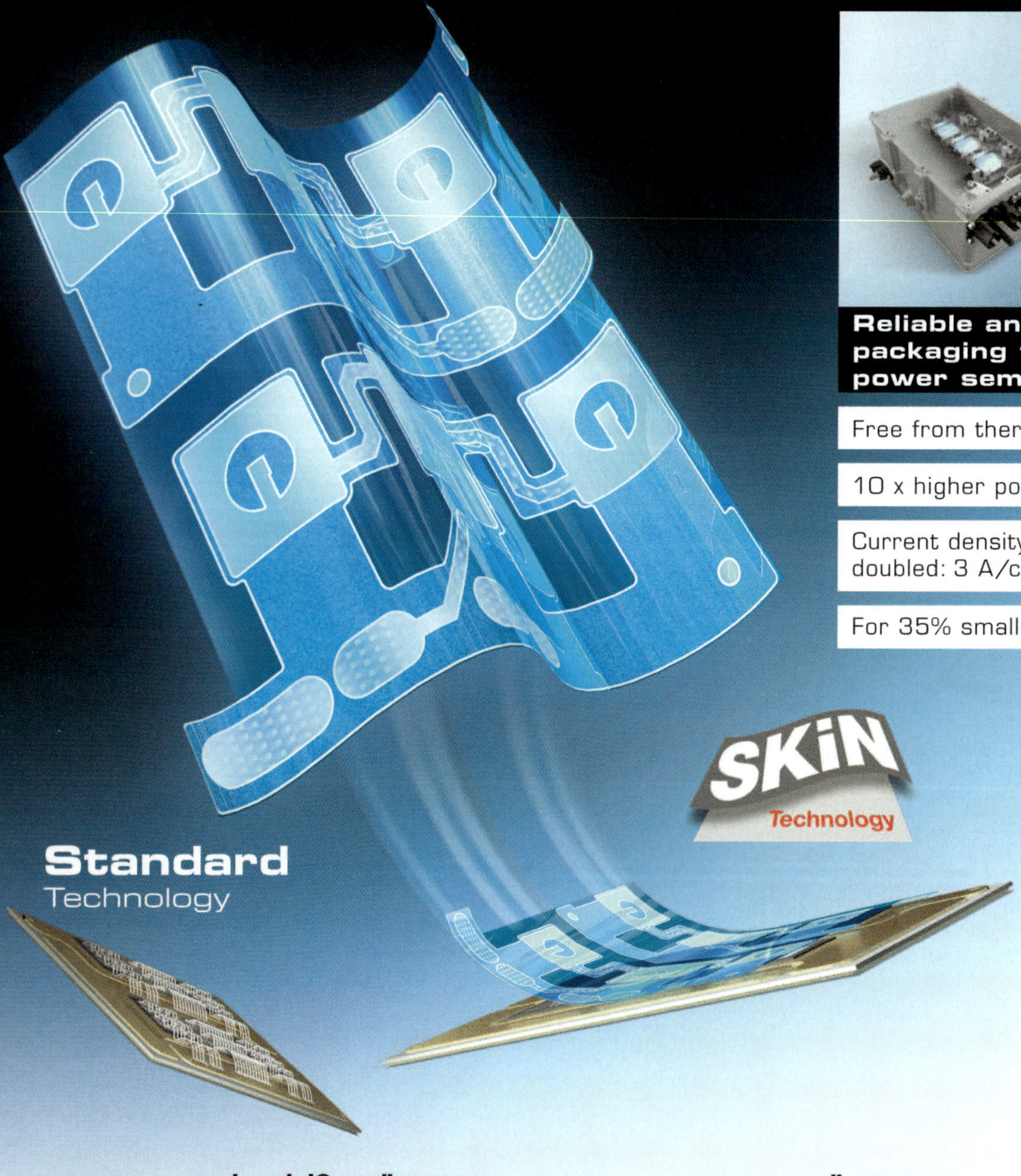

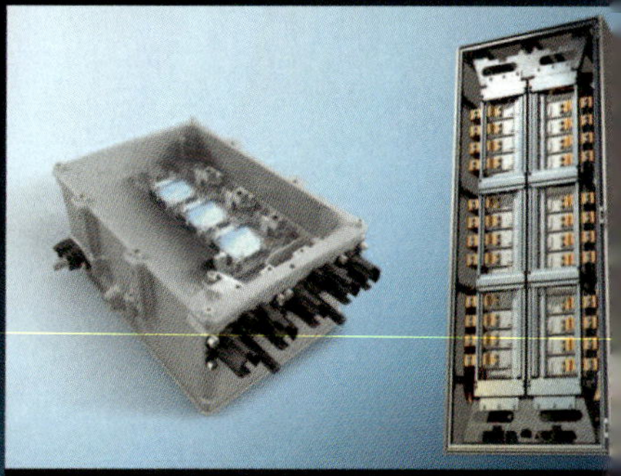

These components are delivered in antistatic packaging and need to be treated with care when they are built into a circuit. The electrodes can be externally short-circuited and it is best to work with earthed soldering equipment. The short circuit is removed after the FET is connected. The input of a MOSFET should never be left "open" in the circuit. In the majority of MOSFETS internal zener diodes protect against ESD (electrostatic discharge).The manufacturer indicates the voltage level of the ESD protection.
The generally accepted level is that of "the human body model" namely 100pF and 1.5kΩ

3. P-channel MOSFET

The *I-V* characteristics of a P-channel MOSFET are the same as an N-channel MOSFET with the exception of the polarity of the current and voltage. For a P-channel the *I-V* characteristics lies in the third quadrant while that of an N-channel lies in the first quadrant.

2.5 The vertical structure of a Power MOSFET

The conventional horizontal structure of a MOSFET as shown in fig. 3-14 is impractical for the construction of a power MOSFET because:
1. The accuracy achieved by the photo masking technology results in relatively long channels and therefore $R_{DS(ON)}$ is too high.
2. Gate-, Drain- and Source metal connections are on the upper surface of the chip which again results in a larger ON resistance.
3 Overlapping of the gate drain and source electrodes results in a large capacitance and limiting of the gain-bandwidth product. These limitations were solved in the beginning of the 1980's by means of a vertical current conduction through the FET.
This was the DMOS (double diffused MOSFET) Currently other techniques are used such as discussed in heading 2.14. This DMOS is referred to as a "classic" or planar MOSFET. Fig. 3-18 shows a cross section of such a structure.

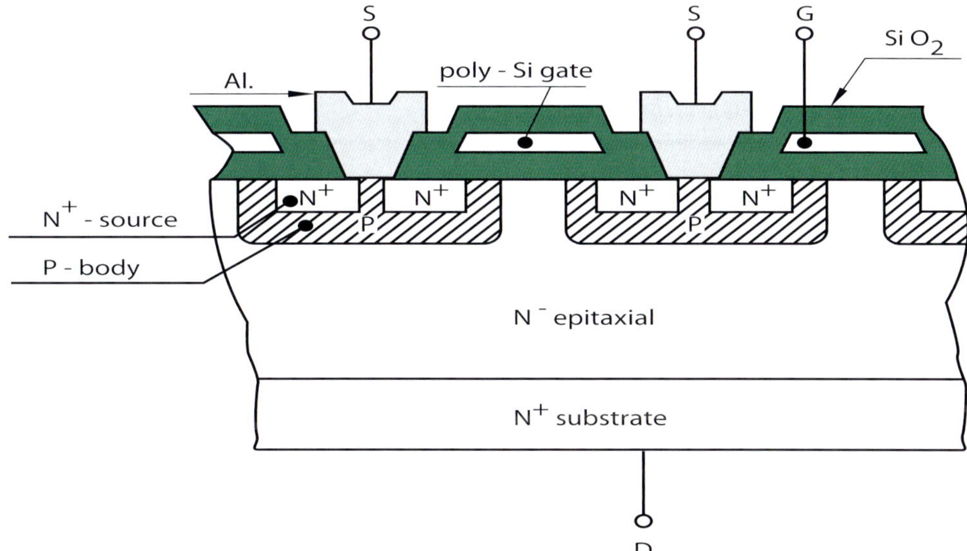

Fig. 3-18: Cross section of a vertical DMOS structure

When a positive voltage is applied between gate and source then the P-layer under the gate changes to an N-area. If V_{GS} is higher than the gate source threshold voltage $V_{GS(th)}$ then a channel is formed whereby the electron flow is horizontal from the source under the gate (fig.3-19) and then vertical down to the drain : therefore the name vertical FET. A channel is formed around the circumference of every source-cell that conducts. All cells are connected in parallel by means of a metal layer that forms the source connection. The drain current leaves the cell in a vertical direction.

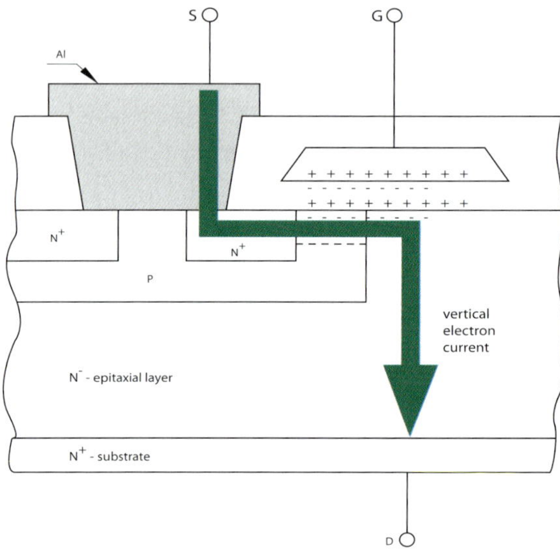

Fig. 3-19: Vertical current flow in a MOSFET

Remarks

Some manufacturers have a registered name for their power MOSFETS, for example HEXFET (International Rectifier),SIMPOS (Infineon Power Mos), ULTRAFET (Harris), etc.
In the construction of power MOSFETS the cell density has increased from 370,000 cells/cm² in 1988 to more than 1,000,000 cells/cm² in 2002 for modern 60V - MOSFETS. The chip surface area is between 0.3 and 1.5 cm² .Trench gate MOSFETS have been built in 2009 with a cell density up to 28 million cells/cm² .

2.6 Equivalent diagram of a power MOSFET

When the MOSFET is in cut-off we note from fig. 3-20 that a parasitic NPN transistor exists. We redraw fig. 3-20 in fig. 3-21 together with the capacitances C_{GS}, C_{GD} and C_{DS} and the material resistances R_K, R_1, R_{DS} R_{BE}. R_K is the resistance of the conducting channel.
In fig. 3-20 the base-collector diode (D_1) is clearly shown. With a good short circuit between base and emitter of the NPN transistor we get the situation of fig. 3-22. We note that diode D_1 formed between source (P –layer) and drain (N⁻ - layer) is in fact an inverse diode.
The resistance $R_{DS(ON)}$ of the conducting MOSFET consists of the sum of R_K and R_{DS} .
With power MOSFETS with a cut-off voltage below $V_{DS} = 100V$ R_{DS} is mostly the result of R_K .
MOSFETS with a $V_{DS} \geq 100V$ have a thicker N⁻ epi-layer whereby especially the resistance R_{DS} of this N⁻ epi-layer is important.

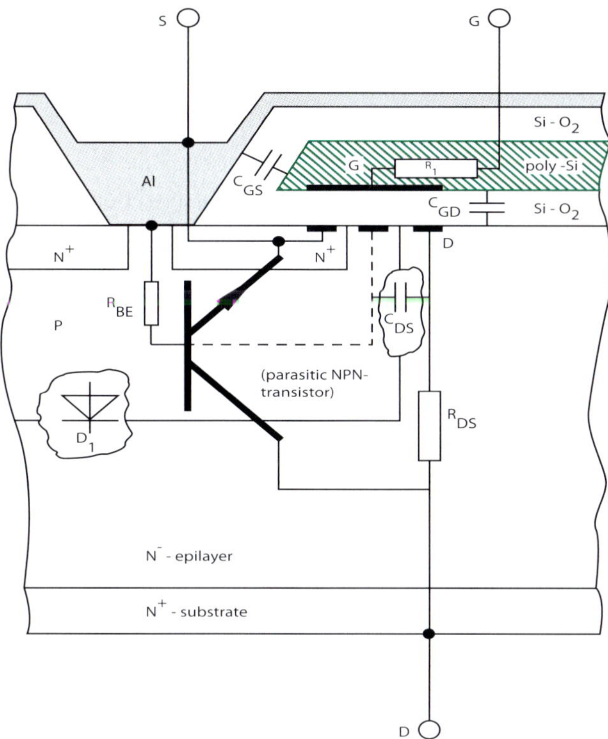

Fig. 3-20: N-channel MOSFET in cut-off mode

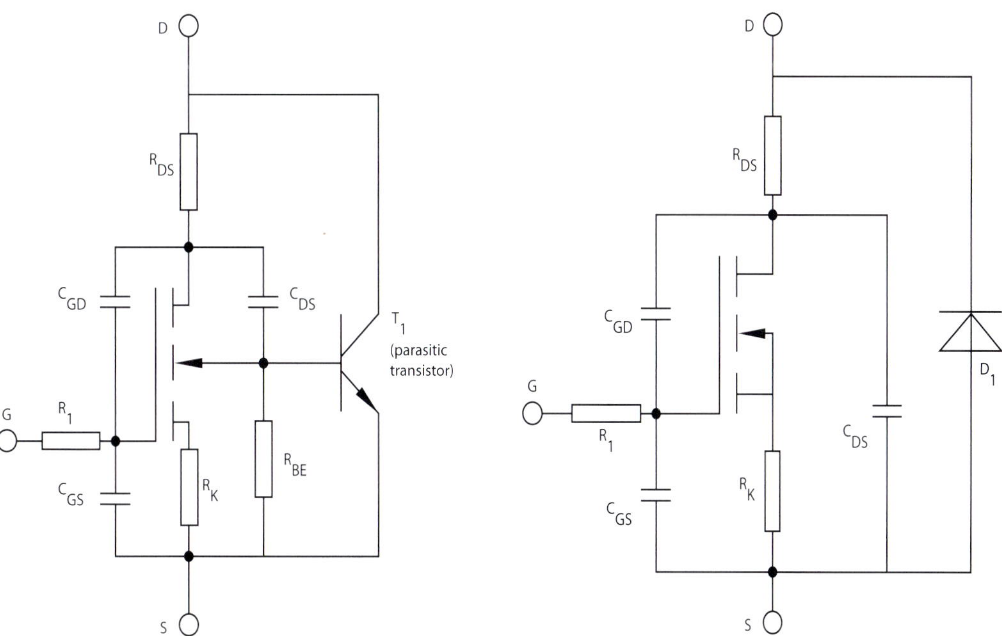

Fig. 3-21: Equivalent diagram N-channel MOSFET   Fig. 3-22: MOSFET with internal freewheel diode

Remarks

1. The internal parasitic diode (inverse) is a normally slow, in other words it has a large recovery time. In most cases an external freewheel diode still needs to be included.
2. Infineon has developed a process whereby the inverse diode of certain MOSFET types have good dynamic properties, in other words it is a Fast Recovery Epitaxial Diode (FRED), hence the name FREDFET. This internal diode can function as a free wheel diode in the case where the MOSFET is required to switch an external inductive load.

2.7 Power MOSFET as switch - $R_{DS(ON)}$

Analogous to the collector load resistance of a junction transistor (fig. 3-1) we place a drain resistor with a MOSFET. Fig. 3-23 shows a so called common source configuration. Just as we did with the transistor we can now also draw a load line in the output characteristic of the MOSFET. (fig. 3-16a).

As you can see in fig. 3-24 the output characteristic of a MOSFET is presented in log-log axes. The advantage of this logarithmic representation is that the active and ohms operating area are clearly readable. It is just as easy to study the area of V_{DS} = 1V as around 20V.

A problem with a log-log axes is that the load line is no longer an easily constructed straight line. To determine the operating point of the circuit shown in fig. 3-23 we replace the graphical construction of the load line with some simple calculations. Before that though we show one more advantage of the log-log axes with the aid of fig.3-25. The graphical representation of a resistor follows from Ohms law:

$V = I \cdot R \longrightarrow\longrightarrow \log V = \log I + \log R$. If I =1A then $\log |U| = \log |R|$. The value of $|R|$ is then equal to the value of the voltage $|V|$.

If we place the load line through the point I (=1A) as in fig. 3-25 then we see for a few examples ($R = 0.1\Omega$ / 0.2Ω / 1Ω / 10Ω) that the value of the resistance can be read from the V-axis.

Instead of constructing the load line in fig. 3-24 we can read directly (from the V-axis) that in the ohmic operating area an IRF530N has as ON $-$ resistor : $R_{DS(ON)}$ = 0.1Ω (if V_{GS} = 7V). With V_{GS} = 10V and $I < I_D < $ 20A we find in fig. 3-24 (after extrapolating left of the point 0.1V on the V_{DS} $-$ axis) that $R_{DS(ON)}$ = 70mΩ .

We reconsider now fig. 3-23 and calculate the operating point (with T_J =25° C):

MOSFET in cut-off :	**MOSFET fully conducting:**
V_{GS} = 0V	V_{GS} = 10V
Leakage current I_{DSS} = 25µA	I_D = 40/4.07 = 9.828A
V_{DS} = 40 $-$ 25 . 10^{-6} . 4 ≈ 40V	V_{DS} = 40 $-$ 4 . 9.828 = 0.688V
V_{R_D} = 25 . 10^{-6} . 4 = 100µV	V_{R_D} = 4 . 9.828 = 39.312V

Under normal circumstances a MOSFET will operate as a conducting switch with a gate voltage $V_{GS} \geq$ 8V.

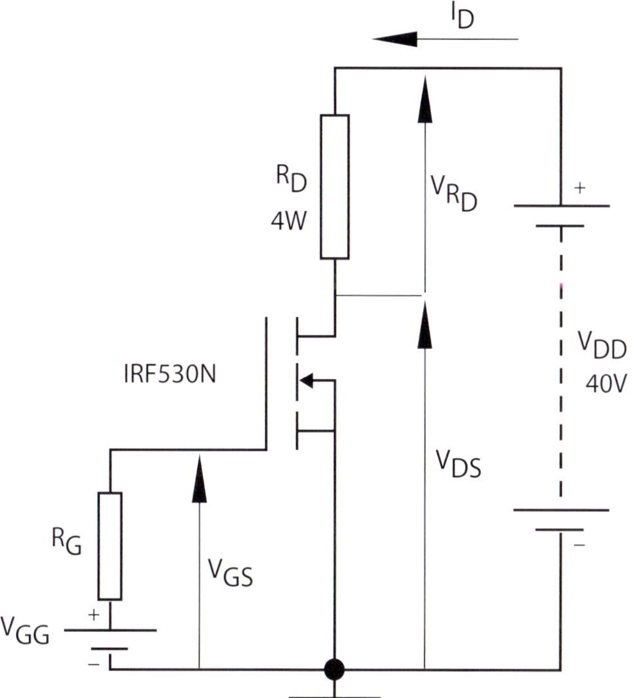

Fig. 3-23: Common source configuration

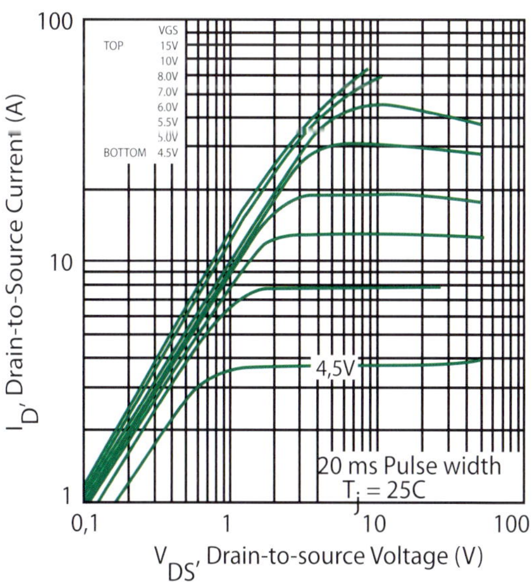

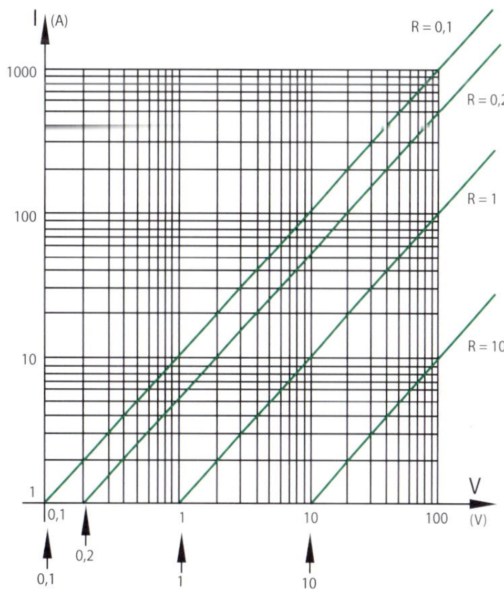

Fig. 3-24: Output characteristic of an IRF530N
(International Rectifier)

Fig. 3-25: Resistance curves in log-log axes

Due to the distribution during construction the manufacturer provides the most unfavourable $R_{DS(ON)}$ value. For an IRF530N this is 110mΩ as can be seen on p. 3-23. Instead of determining $R_{DS(ON)}$ as shown in fig. 3-25 we can take the most unfavourable value of 0.11Ω provided and use this to calculate the operating point.

2.8 Details of an IRF530N(International Rectifier)

HEXFET® Power MOSFET
$V_{DSS} = 100V; R_{DS(on)} = 0.11\Omega; I_D = 17A$
Fast Switching; Fully Avalanche Rated

Description

Fifth Generation HEXFETs from International Rectifier with extremely low on-resistance per silicon area. Extremely efficient and reliable device for use in a wide variety of applications.

The TO-220 package is universally preferred for all commercial industrial applications at power dissipation levels to approximately 50 watts.

TO - 220AB

Absolute Maximum Ratings

	Parameter	Max.	Units
I_D @ $T_C = 25°C$	Continuous Drain Current, V_{GS} @ 10V	17	
I_D @ $T_C = 100°C$	Continuous Drain Current, V_{GS} @ 10V	12	A
I_{DM}	Pulsed Drain Current [a]	60	
P_D @ $T_C = 25°C$	Power Dissipation	79	W
	Linear Derating Factor	0.53	W/°C
V_{GS}	Gate-to-Source Voltage	± 20	V
E_{AS}	Single Pulse Avalanche Energy [b]	150	mJ
I_{AR}	Avalanche Current [a]	9.0	A
E_{AR}	Repetitive Avalanche Energy [a]	7.9	mJ
dv/dt	Peak Diode Recovery dv/dt [c]	5.0	V/ns
T_J	Operating Junction and		
T_{STG}	Storage Temperature Range	- 55 to + 175	°C
	Soldering Temperature, for 10 seconds	300 (1.6 mm from case)	
	Mounting torque, 6-32 or M3 srew	10 Lbf*in (1.1 N•m)	

Thermal Resistance:

$R_{\theta JC} = 1.9\ °C/W; R_{\theta CS} = 0.50\ °C/W; R_{\theta JA} = 62\ °C/W$

Electrical Characteristics @ T_J = 25°C (unless otherwise specified)

	Parameter	Min.	Typ.	Max.	Units	Conditions
$V_{(BR)DSS}$	Drain-to-Source Breakdown Voltage	100	–	–	V	$V_{GS} = 0V, I_D = 250\mu A$
$\Delta V_{(BR)DSS} / \Delta T_J$	Breakdown Voltage Temp. Coefficient	–	0.12	–	V/°C	Reference to 25°C, $I_D = 1mA$
$R_{DS(on)}$	Static Drain-to-Source On-Resistance	–	–	0.11	Ω	$V_{GS} = 10V, I_D = 9{\cdot}0A$ [d]
$V_{GS(th)}$	Gate Threshold Voltage	2.0	–	4.0	V	$V_{DS} = V_{GS}, I_D = 250\mu A$
g_{fs}	Forward Transconductance	6.4	–	–	S	$V_{DS} = 50V, I_D = 9{\cdot}0A$
I_{DSS}	Drain-to-Source Leakage Current	–	–	25	μA	$V_{DS} = 100V, V_{GS} = 0V$
		–	–	250		$V_{DS} = 80V, V_{GS} = 0V, T_J = 150°C$
I_{GSS}	Gate-to-Source Forward Leakage	–	–	100	nA	$V_{GS} = 20V$
	Gate-to-Source Reverse Leakage	–	–	- 100		$V_{GS} = -20V$
Q_g	Total Gate Charge	–	–	44		$I_D = 9.0A$
Q_{gs}	Gate-to-Source Charge	–	–	6.2	nC	$V_{DS} = 80V$
Q_{gd}	Gate-to-Drain ("Miller") Charge	–	–	21		$V_{GS} = 10V$
$t_{d(on)}$	Turn-On Delay Time	–	6.4	–		$V_{DD} = 50V$
t_r	Rise Time	–	27	–		$I_D = 9{\cdot}0A$
$t_{d(off)}$	Turn-Off Delay Time	–	37	–	ns	$R_G = 12\Omega$
t_f	Fall Time	–	25	–		$R_D = 5{\cdot}5\Omega$, See Fig. 10 [d]
L_D	Internal Drain Inductance	–	4.5	–		Between lead, 6mm (O.25in.) from package
L_S	Internal Source Inductance	–	7.5	–	nH	and center of die contact
C_{iss}	Input Capacitance	–	640	–		$V_{GS} = 0V$
C_{oss}	Output Capacitance	–	160	–	pF	$V_{DS} = 25V$
C_{rss}	Reverse Transfer Capacitance	–	88	–		$f = 1.0MHz$, See Fig. 5

Source-Drain Ratings and Characteristics

	Parameter	Min.	Typ.	Max.	Units	Conditions
I_S	Continuous Source Current (Body Diode)	–	–	17	A	
I_{SM}	Pulsed Source Current (Body Diode) [a]	–	–	60		
V_{SD}	Diode Forward Voltage	–	–	1.3	V	$T_J = 25°C$, $I_S = 9.0A$, $V_{GS} = 0V$ [d]
t_{rr}	Reverse Recovery Time	–	130	190	ns	$T_J = 25°C$, $I_F = 9.0A$
Q_{rr}	Reverse Recovery Charge	–	650	970	nC	di/dt = 100A/μs [d]

Notes:

[a] Repetitive rating; pulse width limited by max. junction temperature. (See fig. 11)

[b] $V_{DD} = 25V$, starting $T_J = 25°C$, $L = 3.1mH$, $R_G = 25\Omega$, $I_{AS} = 9.0A$. (See fig. 12)

[c] $I_{SD} \leq 9.0A$, di/dt $\leq 520A/\mu s$, $V_{DD} \leq V_{(BR)DSS}$, $T_J \leq 175°C$

[d] Pulse width $\leq 300\mu s$; duty cycle $\leq 2\%$.

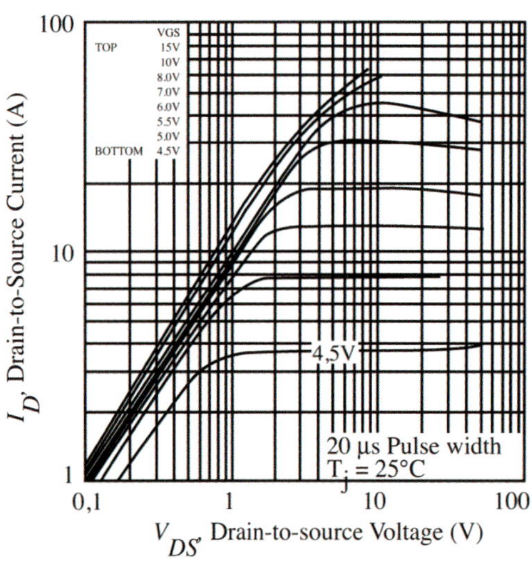

Fig 1. Typical Output Characteristics

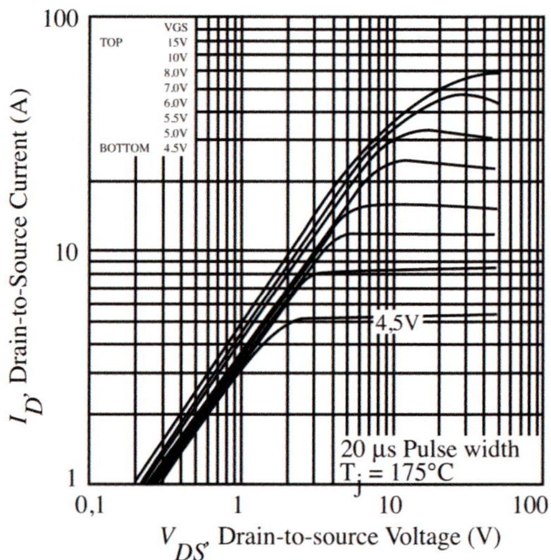

Fig 2. Typical Output Characteristics

Fig. 3-26: Curves of an IRF530N (I.R.)

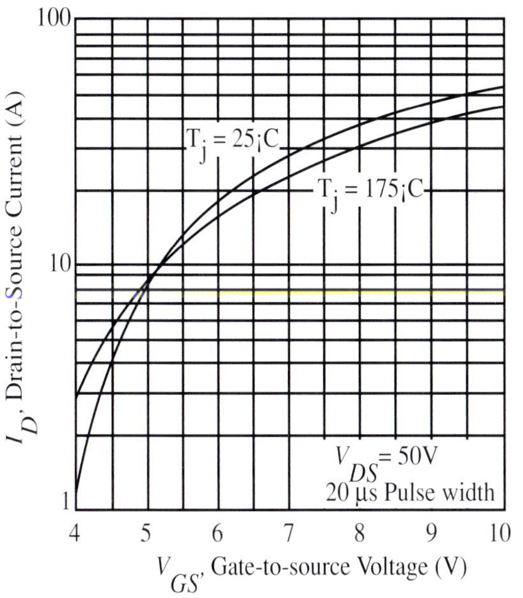

Fig 3. Typical Transfer Characteristics

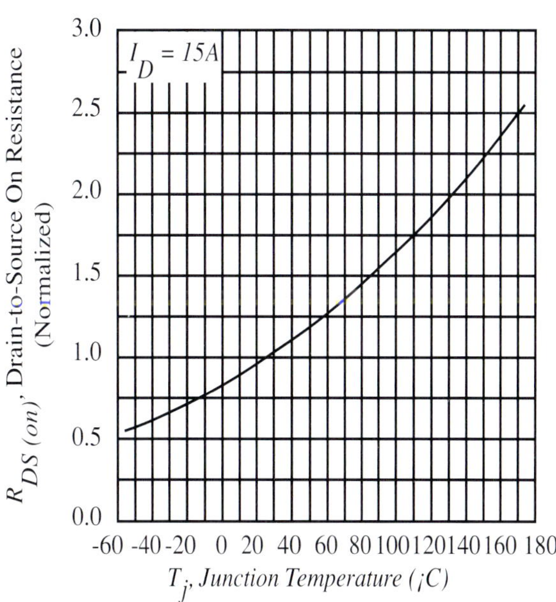

Fig 4. Normalized on-Resistance
Vs. Temperature

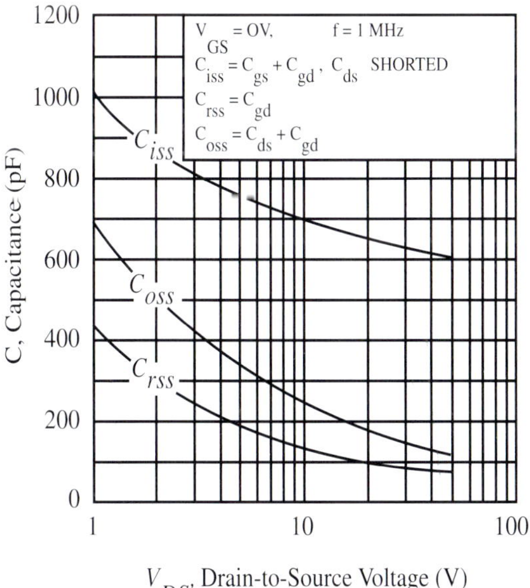

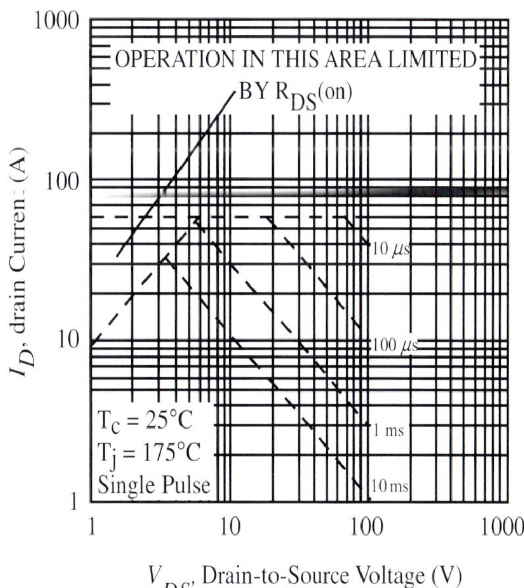

Fig. 3-26: Curves of an IRF530N (I.R.) (continued)

Numeric example 3-5:

What is the power that an IRF530N can switch continuously ?

Solution :

e.g. at $T_C = 100°C \ \rightarrow\rightarrow$ fig.9 (p. 3.26) $\rightarrow\rightarrow I_D = 12A$; $P_{max.} = V_{DD} \times 12$ (W), with $V_{DD} < 100V$

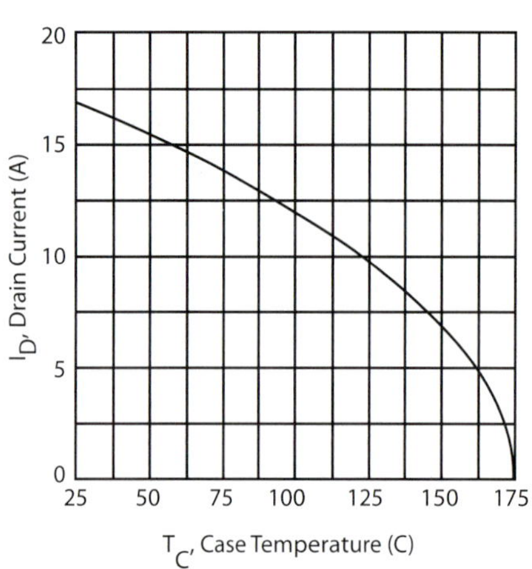

Fig 9. Maximum Drain Current vs.
Case Temperature

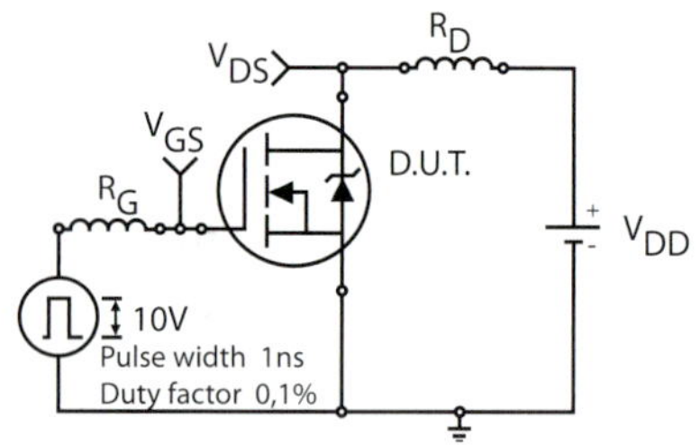

Fig 10. Switching Time Test Circuit

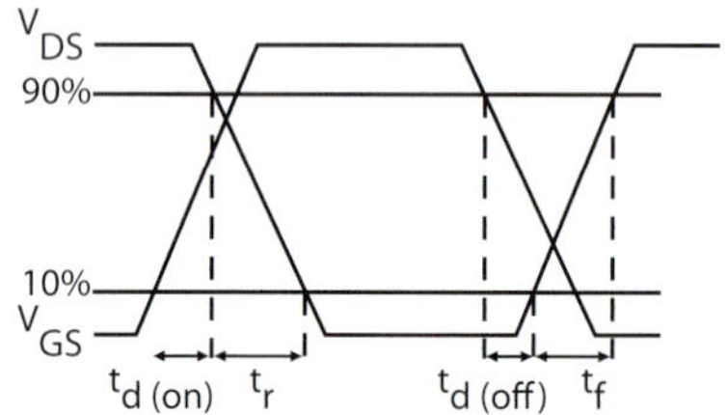

Fig 10b. Switching Time Waveforms

Fig. 3-26: Curves of an IRF530N (I.R.) (continued)

2-9 Comparison of power MOSFET and bipolar transistor

Table 3-1

	BIPOLAR	POWER MOSFET
input impedance	10^3 to 10^5	10^9 to 10^{11}
current gain	10^2 to 10^3	10^5 to 10^6
breakdown voltage	2000 V	1100 V
switch-on time	50ns to 0.5µs	10ns
switch-off time	0.5 to 2µs	10ns
parallel operation	special technology	very easy
ON-resistance	very low $$R_{CEsat} = \frac{V_{CEsat}}{I_C}$$	low to very low $$R_{DS(ON)} = \frac{V_{DS(ON)}}{I_D}$$
control	upto amps (under 0.7 to 1.2V)	Voltage control (nA to mA under 5 to 10V)
charge carriers	minority charge carriers	majority charge carriers

Remarks

The influence of temperature on $R_{DS(ON)}$ is shown in fig. 4 on p. 3-25 for a normalized ON-resistance (1Ω at 25°C). We see clearly that a MOSFET has a positive temperature coefficient.

2.10 Dynamic behaviour of power MOSFETS

2.10.1 Internal capacitance

The response time of a MOSFET is primarily determined by its capacitance and to a much lesser degree by the transit time of the electrons in the very short conduction channel. Fig. 3-27 shows a

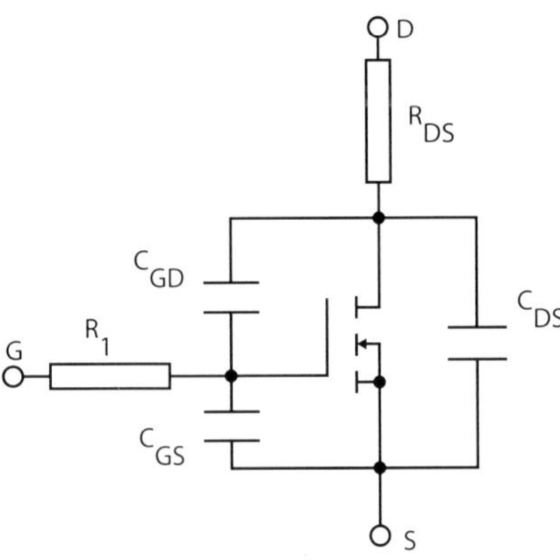

Fig. 3-27: Equivalent circuit of a MOSFET

MOSFET, with the internal resistances and capacitances.

In the datasheets we find three capacitances:

$$C_{iss} = C_{GS} + C_{GD} \ , \ \ C_{DS} \ \text{short circuited}$$
$$C_{oss} = C_{DS} + C_{GD}$$
$$C_{rss} = C_{GD} = C_{MI} \ \ (= \text{Miller capacitance})$$

R_I and R_{DS} are negligible.

The most important delay in switching is given by $C_{iss} . R_G$.

R_G = the external resistance of the control circuit.

2.10.2 Switching behaviour with a resistive load

Fig. 3-28 shows the theoretical pattern and waveform of a MOSFET operating as a switch.
In fig. 3-28 the following symbols mean:

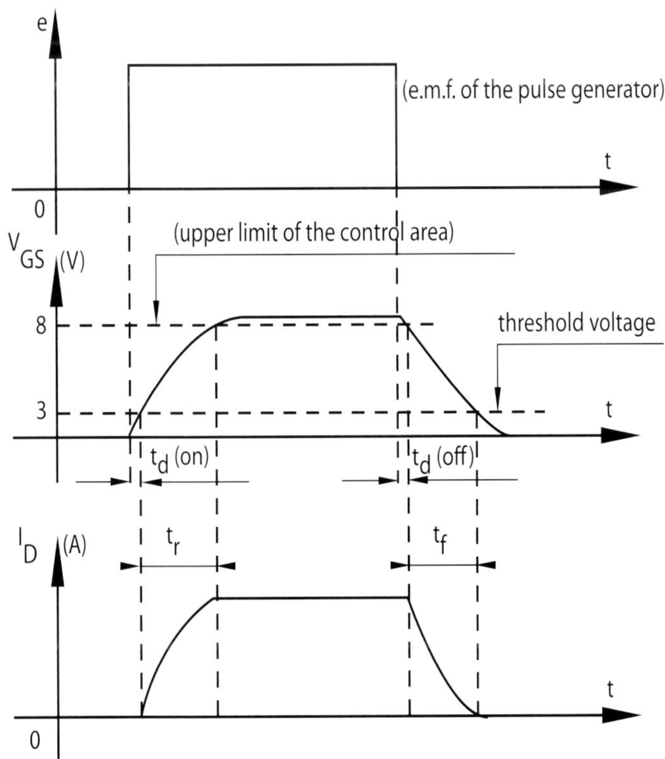

Fig. 3-28: Theoretical waveforms of a MOSFET switch

. $t_{d(on)}$ = delay time at switch-on
= time required to charge C_{iss} to the $V_{GS(th)}$ threshold voltage

. t_r = rise time = time required to charge the input capacitance to V_{GS} that corresponds with full drain current

. $t_{d(off)}$ = delay time at switch-off = time required to discharge the input capacitance from saturation (typically 10V) to the active gate voltage (5 to 8V)

. t_f = fall time = time required to discharge the input capacitance from the active control area to below the threshold voltage.

Numeric Example 3-6:

a) Determine the capacitances of an IRF530N as shown in fig. 3-27 with $V_{GS} = 0V$ and
    $V_{DS} = 10V$
b) Determine the four times shown in fig. 3-28 in the case of the IRF530N, with $V_{DD} = 50V$ and
    $I_D = 9A$

Solution:

a) $C_{iss} = 700pF$; $C_{oss} = 250pF$; $C_{rss} = 140pf$ (fig. 5 on p.3.25)
b) $t_{d(on)} = 6.4ns$; $t_r = 27ns$; $t_{d(off)} = 37ns$; $t_f = 25ns$

2.10.3 Equivalent diagram

The value of C_{GD} depends to a large degree on the voltage V_{DS} , this is because of the changing
thickness of the inversion layer as a result of this voltage. Fig. 3-29 shows this dependence. We
can divide the characteristic from fig. 3-29a into two areas with the border at $V_{GS} = V_{DS}$, in other
words the border between ohmic and active regions. In the active region practically
$i_D = g_{fs} \cdot v_{GS}$ so that fig. 3-29b is an equivalent diagram of the MOSFET in the active region.
When the MOSFET is driven fully closed (in the ohmic area) then fig. 3-29c can be considered
the equivalent circuit diagram

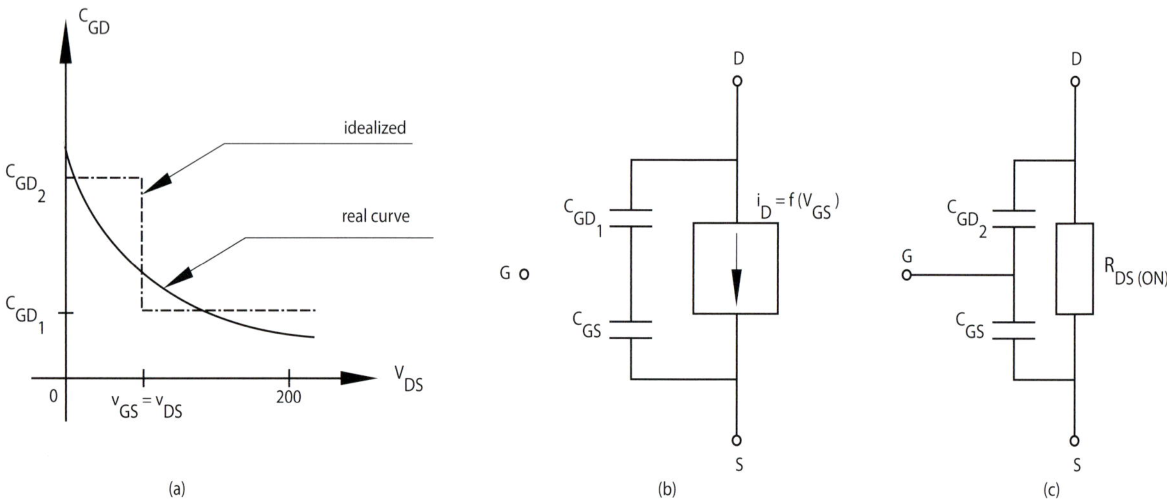

Fig. 3-29: Equivalent diagrams of a power MOSFET

Remarks

1. Inductive loads

    Just like with the BJT, with a MOSFET a free wheel diode D_1 is used to protect the transistor
    against over voltage as a result of switch-off.

2. When voltage V_{DS} is applied to a MOSFET the gate should never be left open or operated
    from a very high impedance. The capacitance between the gate and source (C_{GS}) is then
    charged via the Miller capacitance between the gate and source resulting in a gate voltage
    above the threshold voltage and the MOSFET will remain conducting.
    This can occur when $V_{DS} = 30$ to $40V$.

If a high source voltage is used the MOSFET may be destroyed when the supply is switched on if there is a high impedance between gate and source. This can be prevented if a sufficiently low value resistor is placed between the gate and the source, this prevents the threshold voltage being-reached via de capacitive voltage divider (C_{GD} with C_{GS}) and the applied voltage V_{DS} .

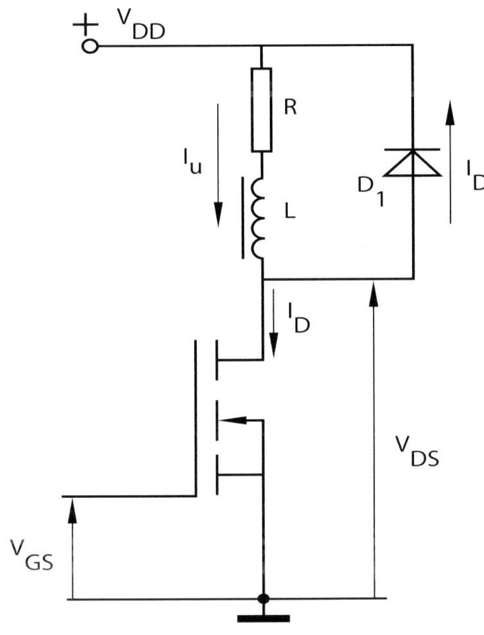

Fig.3-30: Power MOSFET in a chopper circuit

3. Fig. 4 on p.3-25 shows that the MOSFET has a positive temperature coefficient. This has advantages especially when switching MOSFETS in parallel.
   When calculating the heatsink $R_{DS(ON)}$ needs to be taken into account with the actual junction temperature to prevent thermal runaway.

4. Two MOSFETS connected in parallel will each draw half of the current so that per transistor $\left(\dfrac{I_D}{2}\right)^2 . R_{DS(ON)}$ which is a quarter of the value of one MOSFET, assuming that all MOSFETS have the same $R_{DS(ON)}$. One of the consequences is that two parallel MOSFETS together only need half of the heatsink in comparison to the single MOSFET they are replacing. Both MOSFETS need to be connected to the same heatsink to ensure an equal case temperature.

5. The internal wiring, source metallisation,... are measured for continuous (DC) drain current I_D . A constant RMS current may not exceed this I_D value.

That is why in pulsed service , with a duty cycle $\delta = \dfrac{t_{ON}}{T}$ the maximum allowable current I_{peak} is :

$$I_{peak} = \left(\frac{I_{RMS}}{\sqrt{\delta}}\right) = \frac{I_D}{\sqrt{\delta}} \qquad (3\text{-}7)$$

Above this value ensure that $I_{peak} \leq I_{DM}$ (maximum pulsed drain current).

Numeric Example 3-7:

An IRF530N is operated with an f_s = 1000Hz. The MOSFET conducts for 300µs. We are given that V_{DD} = 40V and T_C = 25° C. What is the minimal value of R_D (fig. 3-23) if T_C = 25°C.

Solution:

$$\delta = \frac{t_{ON}}{T} = \frac{300 . 10^{-6}}{10^{-3}} = 0.3.$$

$$(3\text{-}7) \longrightarrow I_{peak} = \frac{I_D}{\sqrt{\delta}} = \frac{17}{\sqrt{0.3}} = 31A$$

By neglecting the voltage drop across the MOSFET this becomes:

$$R_{Dmin} = \frac{V_{DD}}{I_{peak}} = \frac{40}{31} = 1.29\Omega$$

2.11 Ratings

Just as with the BJT, the datasheet of a MOSFET will provide a number of component limits. These " absolute maximum ratings" may under no circumstance be exceeded. The most important ratings are:

V_{DSmax} : maximum drain source voltage ($V_{(BR)DSS}$)
V_{GSmax} : maximum gate source voltage (normally this is $\pm 20V$)
I_{Dmax} : maximum drain DC current
I_{DMmax} : maximum pulsed drain current
P_{Dmax} : maximum power dissipation

2.12 Properties as a power switch

A power MOSFET has a number of interesting properties:

1. Voltage controlled.

The required gate power is low. The control circuit is simpler than that of a bipolar transistor.

2. Large input resistance.

This is responsible for requiring a low control current (nA). Larger currents are only required at switch-on (e.g 150mA) to quickly charge C_{iss} . In comparison a bipolar transistor can require a base current of several amps.

3. High switching frequency.

Switching times of 100ns and less at higher currents. This means high switching frequencies (500kHz and higher).

4. Linear transfer characteristic.

The rate of rise g_{fs} is maximum and almost constant up to high drain currents.
In the case of a bipolar transistor h_{FE} decreases rapidly with increasing collector current.

5. Temperature stability.

Rates of rise and switching time of a power MOSFET are only marginally influenced by temperature change in contrast to the bipolar transistor.

6. No thermal " second breakdown".

As a result of the positive temperature coefficient of the FET there is an internal current division so that the familiar "second breakdown" of the BJT does not exist for a MOSFET. The positive temperature coefficient of MOSFETS is responsible for increasing the resistance (because of heating) in the places where the current density is highest and therefore the current reduces. This mechanism ensures a uniform current density in the MOSFET.

7. Parallel switching.

The positive temperature coefficient of the FET makes it very suitable for parallel switching with other MOSFETS. There is an even current density across the different FETS (explanation, see point 6). In addition the effective ON-resistance decreases with parallel connection.

Remarks

The impact of cosmic radiation on the failure of MOSFETS in aero-space applications has been underestimated as being the same as at sea-level or of what we can call earthly power electronics. Recent research has also shown that cosmic radiation has an influence at sea level on the failure of MOSFETS.

2.13 Control of a MOSFET

2.13.1 Direct control

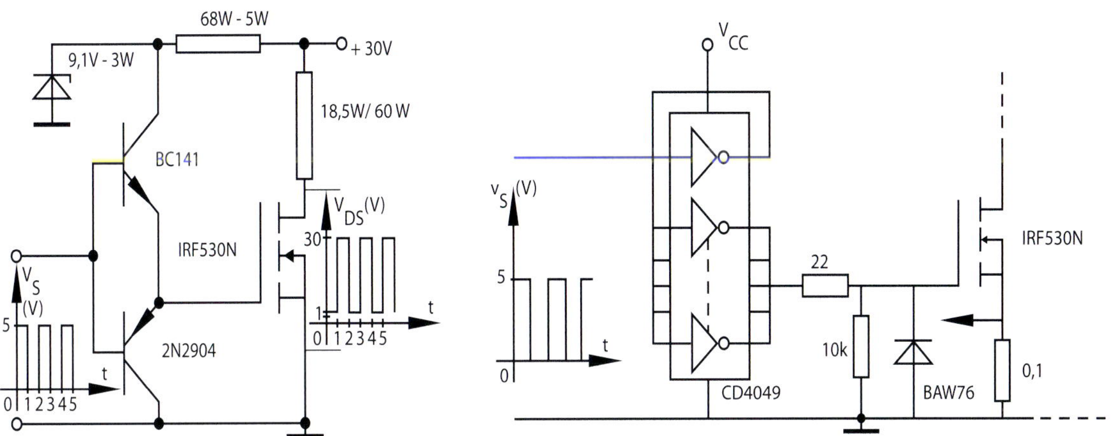

Fig. 3-31: Totem pole MOSFET gate driver   Fig. 3.32: Driving a MOSFET via a hex inverter

Fig. 3-31 shows a totem pole gate driver. This consists of a PNP and an NPN transistor. 30V is too high for the gate of an IRF530N, that is why (9.1V) zener stabilisation is implemented. In fig. 3-32 five inverters from a hex inverter (CD4049) are connected in parallel to ensure sufficient current for quick switch on of the MOSFET. The CD4049 delivers a control pulse with steep flanks. The diode BAW76 serves to protect against an inverse gate voltage. The voltage drop across the inductive free measurement resistor of 0.1Ω in the source circuit can be used for over current detection to protect the MOSFET.

Remarks
Control via a MOSFET driver.
Special IC's exist which function as an interface between TTL and CMOS circuits and the gate of a power MOSFET. An example is the TC1428 from Microchip which is capable of delivering a current of 1.2A (charges a C_{iss} of 1000 pF in 38ns!).

2.13.2 Galvanically isolated control
With power control it is often necessary to galvanically isolate the power circuit and the control circuit, Possibilities include
- transformer coupling
- opto-coupler
- light guides
- special IC's such as the IR2130 from International Rectifier. When studying the three phase inverter we will consider an example with such an IC.

2.14 Manufacture of MOSFETS

New manufacturing processes for MOSFETS are primarily focused on reducing the value of $R_{DS(ON)}$ and if possible increasing the switching frequency and increasing the current density of the drain current. The manufacturers invariably talk about the next "generation" of MOSFETS. By reducing $R_{DS(ON)}$ the voltage drop across the conducting MOSFET is reduced and the switching losses are reduced.

In addition the manufacturers are trying to improve the robustness (over current, over voltage, switch stability), improve the current density, raise the forward blocking voltage (in the case of HV transistors) and reduce the chip surface area.

Some focal points of the present technology:

1. ***Smart power switch***

   Some manufacturers have developed components that in addition to the main goal of switching power also protects itself and have increased functionality. These integrated switches are called **"SMART POWER switches"** or sometimes "intelligent power modules" (**IPM**).

2. ***FOM (figure of merit)***

   With the conventional planar MOSFET we hit a silicon limit, namely that by doubling the blocking voltage $R_{DS(ON)}$ increases by a factor of five. Other technologies (trench gate structure, coolmos super junction structure, hexfet, ...) have succeeded in improving the MOSFET. As mentioned the manufacturers are seeking to reduce $R_{DS(ON)}$ and to increase the switching frequency. The latter is related to the total gate charge Q_g. The lower this charge, the faster it dissipates and the quicker the MOSFET can switch.

   MOSFETS therefore have a sort of quality label : FOM $= R_{DS(ON)} \cdot Q_g$. The smaller the FOM the better. The FOM is also called the R.Q product.

3. ***Trench gate structure***

   In addition to the DMOS structure there are other structures in use with power MOSFET manufacturing. Fig. 3-33 shows the cross section of a Trench-MOS cel.

   Another word for trench is channel. In fig. 3-33 we see that because of the trench construction of the gate that the P-area is vertical. The active Si-surface can therefore be made larger enabling beter control and a smaller resistance. The result is a smaller cell surface area in comparison to a planar MOSFET. This results in a Trench MOSFET with: a smaller $R_{DS(ON)}$, a higher current density, lower forward losses, lower switching losses, a higher blocking voltage.

   Disadvantages of the trench technology in comparison to the planar elements are: inferior short-circuit protection and approximately three times higher gate capacitance.

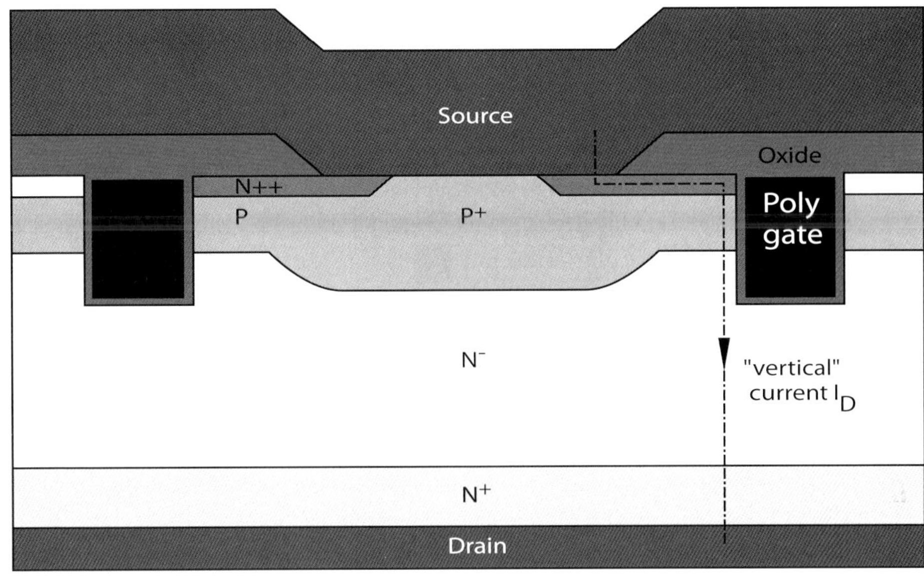

Fig 3-33: Cross section of a Trench-cel

4. *Cool-MOS*.

In 1998 Infineon introduced the Coolmos. With the Coolmos (fig. 3-34) a P area is added to the N^- epilayer. This P area is connected with the P^+ area. As a result the electric field in the forward blocking mode is built up not only vertically but horizontally as well. The result is that the N^- epilayer can be dramatically thinner so that the conduction is much improved. A 600V-Coolmos has one fifth of the forward conducting losses of a classic MOSFET with the same chip surface area. For the same current only one third of the previous chip surface area is required.

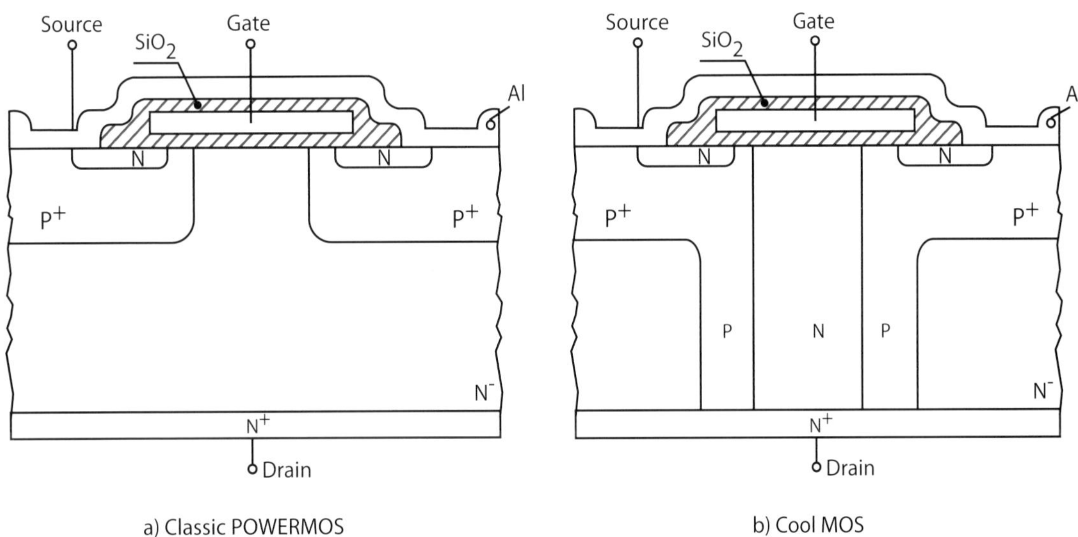

Fig 3-34: Cell structure of a Coolmos

2.15 New developments (SiC and GaN)

Up to now power semiconductors have been made from silicon, a material with a band gap of 1.1 eV. The band gap is the energy difference between the top valence band and the bottom conduction band expressed in electron volts (eV). Presently there is a development to use material with a larger band gap, namely silicon carbide (SiC: 2.86 eV) and gallium nitride (GaN: 3.4 eV).

In contrast to Si SiC has no molten phase and the manufacturers need to work with high pressure and temperatures and simultaneously with gas and a controlled temperature gradient. This makes for a very difficult but not impossible process. SiC Schottky diodes are already widely used. They have a number of interesting properties:

1. Low forward voltage drop and therefore smaller losses
2. High operating temperature (currently more than 225° C)
3. No inverse recovery current as a result of the ultra low Q_{rr} . This causes low losses in the accompanying transistor switch and reduces EMC.
4. No forward recovery voltage as a result the diode can conduct immediately
5. From (3) and (4) a good high frequency capability follows.
6. A small leakage current which remains practically constant by increasing temperature.

Currently 4 inch wafers are in full production and 6 inch wafers are in the starting block. These 150 mm wafers will reduce the cost price of SiC components.
A few examples of SiC Schottky diodes: 1200V-50A and 1700V-25A.

SiC diodes have become standard products and are especially used as free wheel diodes in combination with ultra-fast IGBT's and mosfets in converters at high frequency such as solar inverters, switch mode power supplies etc.

Manufacturers want to make other SiC components such as transistors (MOSFET, JFET, BJT). Recently a manufacturer announced a 1200V-25mΩ SiC FET. We have to wait since it is more than 10 years since the first such announcement. Another material that will make its debut in power electronics is GaN, which has been is used in LED's.
Just as was the case with SiC , GaN can operate at higher frequencies than Si.
An increase of switching frequency of 20 to 100kHz could drastically reduce transformer size.
In addition to reduced volume the weight of the configuration can be reduced with a factor five.
In June 2009 the first GaN transistor was introduced . GaN transistors and diodes possibly threaten the Si world.
GaN diodes have extremely low losses in conduction and are stable at high temperatures.
In many cases air cooling of GaN components is sufficient, this results in a large reduction in weight and volume when compared to equivalent Si and SiC diodes

2.16   MOSFET module SKM111AR (Semitrans)

SKM 111 AR

Absolute Maximum Ratings

Symbol	Conditions [1]	Values	Units
V_{DS}		100	V
V_{DGR}	$R_{GS} = 20\ k\Omega$	100	V
I_D		200	A
I_{DM}		600	A
V_{GS}		$\pm\,20$	V
P_D		700	W
$T_j, (T_{stg})$		$-\,40\ldots+150\ (125)$	°C
V_{isol}	AC, 1 min	2 500	V
humidity	DIN 40 040	Class F	
climate	DIN IEC 68 T.1	40/125/56	
Inverse Diode			
$I_F = -\,I_D$		200	A
$I_{FM} = -\,I_{DM}$		600	A

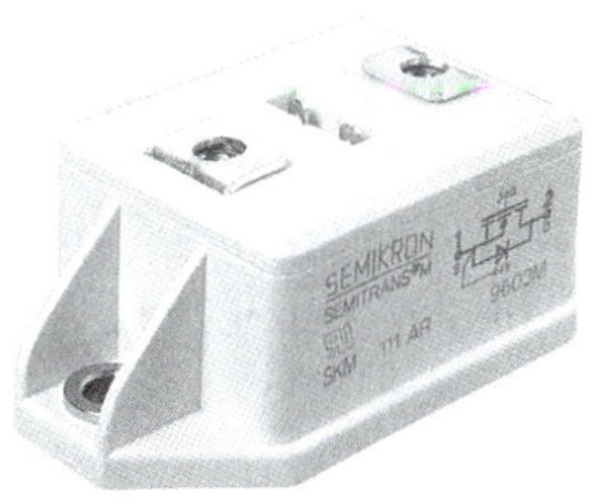

SEMITRANS M1

Characteristics

Symbol	Conditions [1]	min.	typ.	max.	Units
$V_{(BR)DSS}$	$V_{GS} = 0,\ I_D = 0{,}25\ mA$	100	–	–	V
$V_{GS(th)}$	$V_{GS} = V_{DS},\ I_D = 1\ mA$	2,1	3,0	4,0	V
I_{DSS}	$V_{GS} = 0,$ $T_j = 25\ °C$	–	50	250	μA
	$V_{DS} = 100\ V$ $T_j = 125\ °C$	–	300	1000	μA
I_{GSS} [3]	$V_{GS} = 20\ V,\ V_{DS} = 0$	–	10	100	nA
$R_{DS(on)}$	$V_{GS} = 10\ V,\ I_D = 130\ A$	–	7	8,5	mΩ
g_{fs}	$V_{DS} = 25\ V,\ I_D = 130\ A$	60	75	–	S
C_{CHC}		–	–	160	pF
C_{iss}	$V_{GS} = 0$	–	10	13	nF
C_{oss}	$V_{DS} = 25\ V$	–	5	7,5	nF
C_{rss}	$f = 1\ MHz$	–	1,8	2,7	nF
L_{DS}		–	–	20	nH
$t_{d(on)}$	$V_{DD} = 50\ V$	–	60	–	ns
t_r	$I_D = 130\ A$	–	220	–	ns
$t_{d(off)}$	$V_{GS} = 10\ V$	–	270	–	ns
t_f	$R_{GS} = 3,3\ \Omega$	–	200	–	ns
Inverse Diode					
V_{SD}	$I_F = 400\ A,\ V_{GS} = 0$	–	1,25	1,6	V
t_{rr}	$T_j = 25\ °C$ [2]	–	400	–	ns
	$T_j = 150\ °C$ [2]	–	–	–	ns
Q_{rr}	$T_j = 25\ °C$ [2]	–	3,5	–	μC
	$T_j = 150\ °C$ [2]	–	–	–	
Thermal Characteristics					
R_{thjc}		–	–	0,18	°C/W
R_{thch}	M_1, surface 10 μm	–	–	0,05	°C/W

Mechanical Data

M_1	to heatsink, SI Units	4	–	5	Nm
	to heatsink, US Units	35	–	44	lb.in.
M_2	for terminals, SI Units	2,5	–	3,5	Nm
	for terminals, US Units	22	–	24	lb.in.
a		–	–	5x9,81	m/s²
w		–	–	130	g
Case	→ page B 5 – 2		D 15		

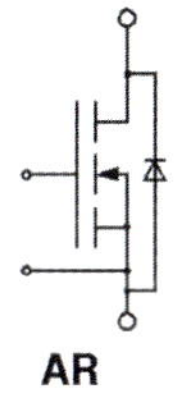

AR

Features

- N Channel, enhancement mode
- Avalanche characteristic
- Short connections and built-in gate resistors to suppress internal oscillations even in critical applications
- Isolated copper baseplate
- All electrical connections on top for easy busbaring
- Large clearances (10 mm) and creepagedistances (13 mm)
- UL recognized, file no. E 63 532

Typical Applications

- Switched mode power supplies
- DC servo and robot drives
- DC choppers
- UPS equipment
- Not suitable for linear amplification

This is an electrostatic discharge sensitive device (ESDS). Please observe the international standard IEC 747-1, Chapter IX.

[1] $T_{case} = 25\ °C$, unless otherwise specified.
[2] $I_F = -\,I_D,\ V_R = 100\ V,\ -\,di_F/dt = 100\ A/μs$

3 IGBT

As we know a bipolar transistor has a low V_{CEsat} even with a large current (also with high voltage transistors). A power MOSFET in contrast requires only low gate power and can operate at high frequency. In the construction of an IGBT (Insulated Gate Bipolar Transistor) the manufacturers sought to combine the benefits of the bipolar transistor and the MOSFET. The first design of an IGBT dates from 1980.

3.1 Construction

Fig 3.35a shows the construction of a single IGBT cell. Here in the N^{+-} substrate from the MOSFET of fig. 3-20 has been replaced with a P-substrate. An IGBT is made up of thousands of such cells which are all connected in parallel. For further study a diagram of a single cell is shown and this is considered as being representative of the complete IGBT. Analogous to fig. 3-20 the internal parasitic components (T_1 , T_2 , etc) are represented in fig. 3-35b.

In fig. 3-35 b and 3-36 note that an N-channel MOSFET is present and a PNP transistor T_1 . The parasitic NPN transistor from fig. 3-20 is also present.

If T_2 and its corresponding R_{BE} are neglected then fig. 3-37a and b are a rudimentary equivalent diagram of fig. 3-36.

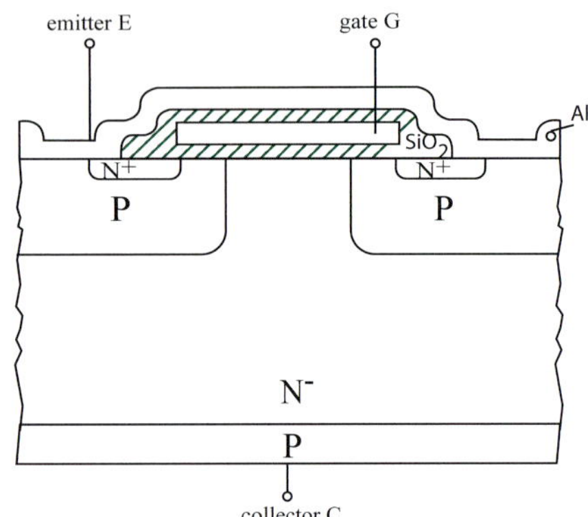

(a) Construction

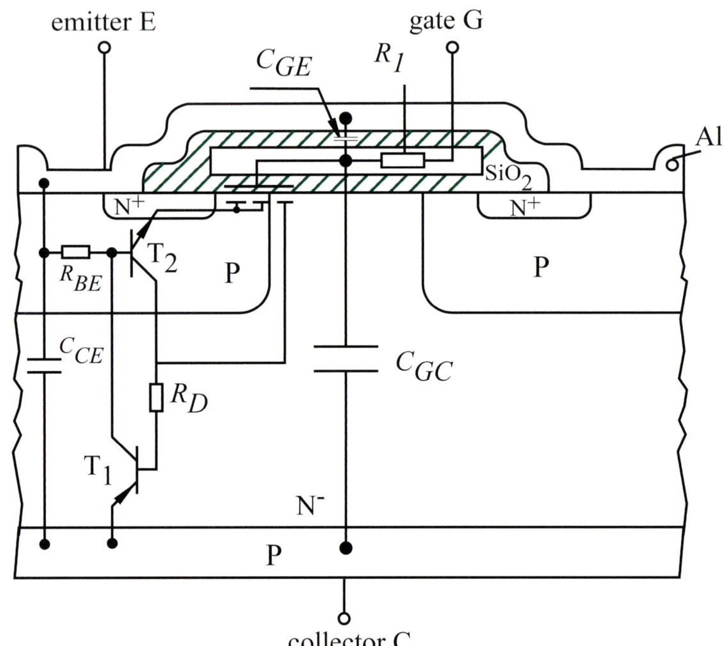

(b) IGBT with internal parasitic elements

Fig. 3-35: Construction of an IGBT

3.2 Operation

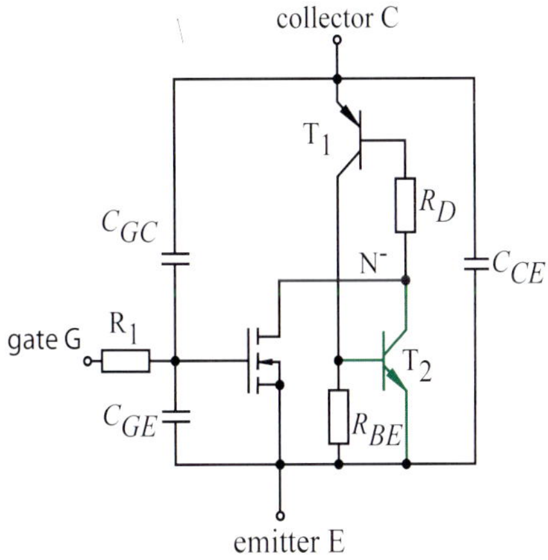

The equivalent circuit as drawn in fig.3-37b consists of a PNP transistor controlled by an N-channel MOSFET in a pseudo-Darlington configuration.

Since there is a MOSFET in the input and a transistor in the output the symbols used in fig.3-37c are easily understood. In the text the bottom symbol of the three will be used.

At the input the IGBT behaves as a MOSFET and at the output approximately the behaviour of a BJT.

Fig. 3-36: Equivalent diagram for fig. 3-35b

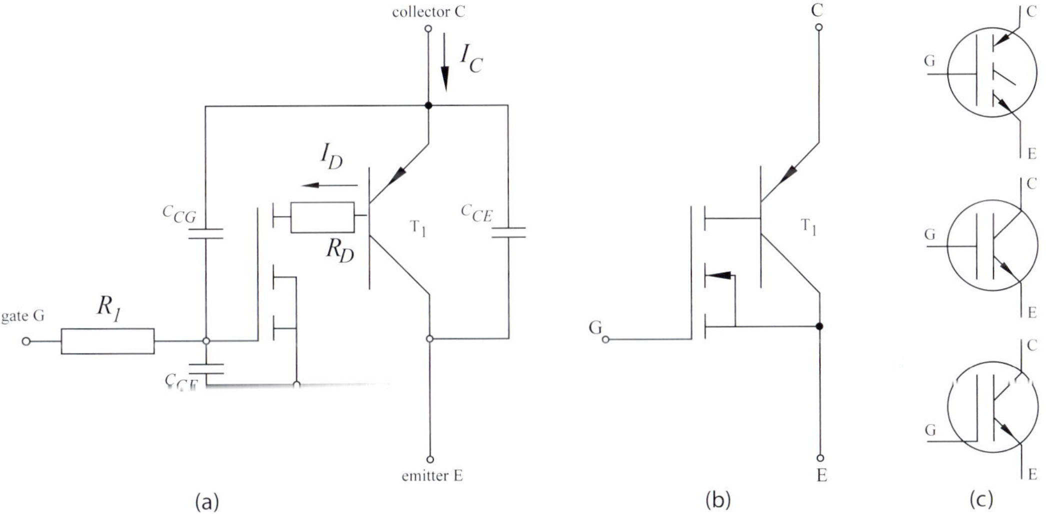

Fig. 3-37: Reduced equivalent diagram of an IGBT and symbols

The physical operation of an IGBT more closely resembles that of a BJT than a power MOSFET. The reason is that the P-substrate of the IGBT is responsible for injecting the minority charge carriers in the N$^-$- layer so that conduction is much better than with a MOSFET that works with majority charge carriers. Since no high voltage exists across the MOSFET part we can for construction purposes use a low voltage type. Low voltage types have a low $R_{DS(ON)}$ as important characteristic. The base of the PNP has no external connection which is expressed in the upper symbol of the IGBT in fig. 3-37c. In the equivalent diagram of fig. 3-37a it is clear that the voltage drop is composed of two parts: on one side the diode voltage drop across the emitter-base junction of T$_1$ and on the other side the voltage drop across R_D and the gating MOSFET. Since it is a low voltage MOSFET the voltage drop across this FET is dependent on the gate control voltage (this is not the case with high voltage MOSFETS !).

We write:
$$V_{CEsat} = V_{BE(T1)} + I_D \cdot (R_D + R_{DS(ON)})$$
(3-8)   with $I_D = \dfrac{I_C}{h_{FE}}$

Remarks

1. We refer once again to a collector and an emitter as in the case of a BJT. Since the current direction through the IGBT corresponds with that of an NPN transistor the emitter arrow is once again shown in the IGBT symbol. Analogous with the MOSFET we call this configuration an N-channel IGBT. We could just as easily have retained the arrow of the T_1 - transistor of fig. 3-37b but that would suggest a PNP-transistor while here we have an N-channel IGBT!

2. The parasitic transistor T_2 together with T_1 forms a thyristor. In the construction it is arranged that this parasitic thyristor cannot latch.

3. The limiting factor in turn-off speed is the lifetime of the minority charge carriers in the N^-- epilayer. This layer forms the base of the PNP. Since the base is not accessible the external circuit has no influence on the switching speed. The charge stored in this base is responsible for the characteristic "tail" current of an IGBT at turn-off (see fig. 3-41) just like a BJT. This tail not only increases the turn-off losses but requires a larger dead time between conduction of two IGBT's in a half bridge circuit.

4. In contrast to MOSFETS, IGBT's have no (integrated) inverse diode. In bridge circuits with inductive loads it is necessary to include separate free wheel diodes. These are obviously fast recovery diodes. These fast recovery diodes have a low charge storage (a few µC) which makes a reverse recovery time of 100 to 200ns possible.
Recently Schottky SiC diodes have appeared as free wheel diodes in bridge converters using IGBT's (and MOSFETS).

Photo Mitsubishi: CM75RX-24S. is an IGBT module with seven IGBT's. Six are used for a three phase inverter and one is used for dynamic braking. This module is suitable for 1200V-75A. Measurements: 122 x 62mm.
CM450DX-24S: half bridge with two IGBT's. Suitable for 1200V-450A. Measurements: 122 x 62mm.

CTSR xx-P and -TP current transducers based on the
Closed Loop Fluxgate Technology for small current measurements

Photo LEM: CTSR current transducer

A new class of Rogowski coil
split-core current transducers: the RT series.

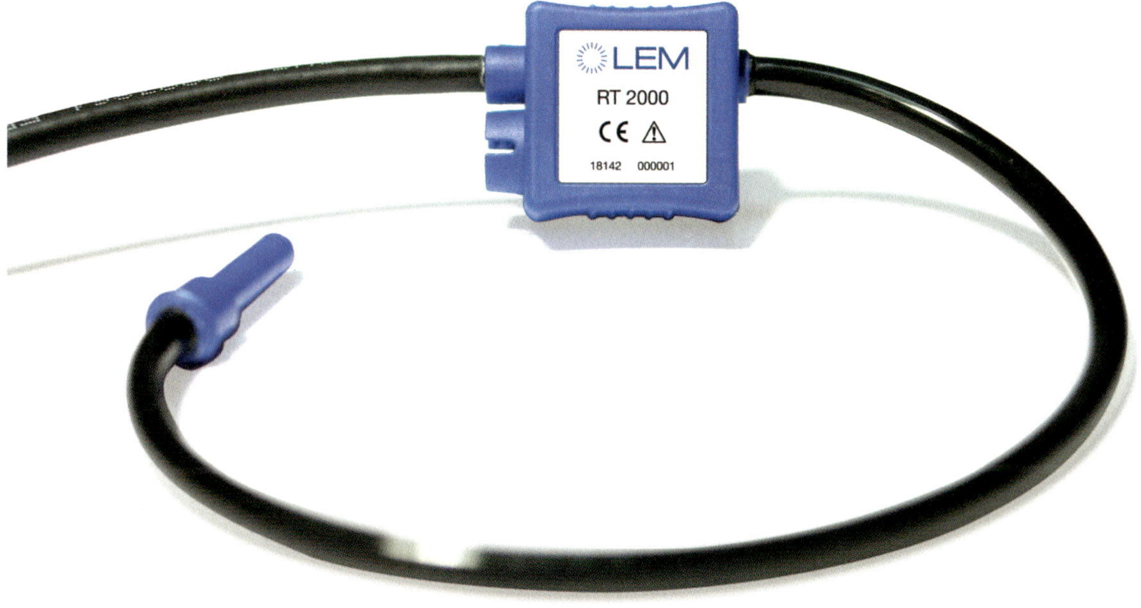

Photo LEM: Rogowski coil

Already part of your vision.
LEM.

Whatever you invent, imagine or develop, LEM's transducers are at the heart of your power electronics applications from the very start. LEM's products, R&D, and people provide knowledge intensive solutions to keep up with your changing industry, allowing your visions to come to life.

www.lem.com   At the heart of power electronics.

3.3 Static properties of an IGBT

3.3.1 Output characteristic

Fig. 3-38 shows the output characteristic of a 15A-1200V IGBT (IGW15T120) from Infineon.
As previously noted with an IGBT we have a series circuit with a MOSFET and a PN diode.
Just past the origin of the V_{CE} - curve of the IGW15T120 we see the break over voltage as a
result of the PN-junction between collector and base. The IGW15T120 is an IGBT with trench and
field stop technology (see paragraph 3.5.3).

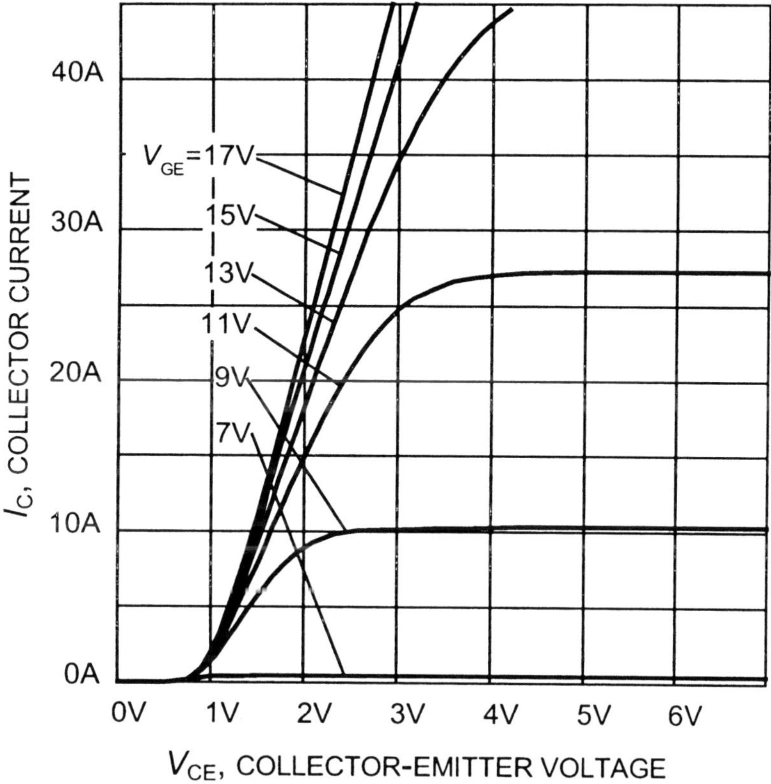

Fig. 3-38: Output characteristics of an IGW15T120

Numeric example 3-8:

a) For an IGP10N60T (Infineon) the following is given: V_{GE} = 15V; I_C = 5A; T_J = 175° C.
   What is the voltage drop across the IGBT?

b) For an SKM400GA12E4 (Semikron) the following is given: V_{GE} = 15V; I_C = 400A;
   T_J = 150° C.
   Determine V_{CEsat} .

Solution:

a) Fig.3-50 (fig.6): V_{CE} = 2.2V

b) Fig.3-51. Typisch: V_{CEsat} = 2.1V

3.3.2 Important properties

a. Transconductance g_{fs}

Just as with the MOSFET the transconductance is an important property of an IGBT.
Similar to the MOSFET:

$$g_{fs} = g_{fe} = \boxed{\; g_{FE} = \left(\frac{\Delta I_C}{\Delta V_{GE}}\right)_{V_{CE} = C^{te}} \;} \tag{3-9}$$

A practical value is for example 6 to 30 Siemens.

b. Breakdown voltage BV_{CES}

The collector emitter breakdown voltage BV_{CES} enables the determination of the supply
voltage for the IGBT. Currently 600V, 1200V and 1700V are the classic BV_{CES} - values.
Some manufacturers have 3300V IGBTs at one's disposal and Infineon even has a
6.5kV-600A type, useful for drive technology in the medium voltage range (2.3 to 4.6kV).

c. Saturation voltage $V_{CE(ON)}$

With an IGBT we no longer refer to $R_{DS(ON)}$ which was the case with the MOSFET but once
again to the saturation voltage $V_{CE(ON)}$ as was the case with a BJT. The output circuit of an
IGBT corresponds with a BJT. As the production processes improve the manufacturers talk
about the "next generation" of IGBTs as was the case with MOSFETS. Characteristic of the
next generation is a low saturation voltage and or a higher switching frequency.
The V_{CEsat} ($= V_{CE(ON)}$) for the recent generation has fallen to 1.8V (600V-types) and
2V (1200V-types). Trench-gate FS types have fallen to a $V_{CEsat} \approx 1.5V$. To maintain a low V_{CEsat}
during conduction we maintain V_{GE} approximately at 15V.

d. Threshold voltage $V_{GE(th.)}$

As the input circuit of an IGBT corresponds to that of a MOSFET the IGBT will also start to
conduct from a certain threshold voltage $V_{GE(th.)}$. This turn-on voltage is approximately 3 to
4V but decreases with increased junction temperature.

e. Collector-current I_C

An IGBT is characterised by a continuous collector current at a specific case temperature
(standard at $T_C = 25°C$ and at $T_C = 100°C$).
Also the maximum acceptable pulse current I_{CM} is given.

f. Maximum gate voltage V_{GE}

The oxide layer on the gate limits the allowable voltage to about ±20V.

g. Clamped inductive load current I_{LM}

This specification indicates the maximum current that can be switched off repetitively with an
inductive load with a free wheel diode as shown in fig. 3-39. This I_{LM} guarantees a square
safe operating area (SOA). This means that an IGBT can withstand a high voltage and a large
current simultaneously. The manufacturer I.R. specifies I_{LM} at $T_J = 150°C$, 80% of $V_{CE(M)}$
and 4 x I_C (at $T_C = 25°C$).

h. SCWT

Short circuit withstand time is the time an IGBT can withstand a short circuit. (10µs is a
considerable value).

i. UIS

Unclamped inductive switching, capability to switch-off an inductive load.

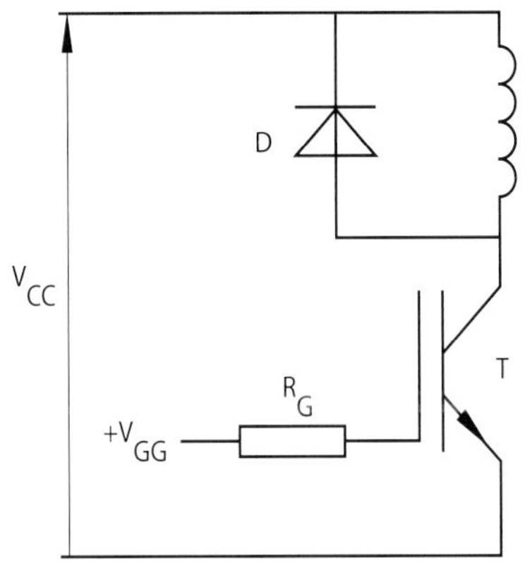

Fig. 3-39: IGBT with inductive load and free wheel diode

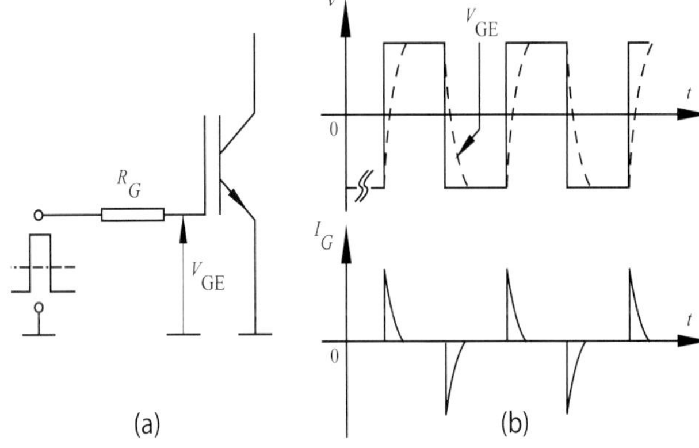

Fig. 3-40: Switch an IGBT on and off.

Numeric example 3-9:

a. The case temperature of an IGP10N60T is 100°C. How large is the maximum collector current?

b. Determine in fig. 7 of fig. 3-50 the transconductance of the IGP10N60T at $T_J = 175$°C and $V_{GE} = 8$V?

Solution:

a. Fig. 3-50 (fig. 4): $I_C = 11.81$A

b. $g_{fs} = \left(\dfrac{\Delta I_C}{\Delta V_{GE}}\right)_{V_{CE} = 20V} = 3.86$ Siemens.

j. Maximum power dissipation P_D

The manufacturer indicates the power that the IGBT can dissipate at a certain case temperature (normally at $T_C = 25$°C and at $T_C = 100$°C).

3.4 IGBT as a switch: dynamic characteristics

Since the equivalent gate circuit of an IGBT is similar to a MOSFET the IGBT is also voltage controlled. This results in an IGBT being faster than a BJT. It is slower than a MOSFET. Just as with the MOSFET it is only with switch-on and switch-off that an appreciable current flows, this is as a result of charging and discharging the input capacitance. Fig. 3-40 shows what this means in practice. The nominal gate current of an IGBT in the on state is in the order of nA in comparison to the mA or A of a BJT. The IGBT family is classified into three categories by International Rectifier: standard, fast and ultra-fast. Standard types can optimally be used in the range between DC and 1 kHz. The fast types can optimally be used in the range between 3 and 10 kHz and the ultra-fast types have minimal switching losses between 10 and 100 kHz. Semikron makes a distinction for its 600V modules between superfast and low saturation types. With their 1700V modules (fourth generation) they speak of high density low saturation. To determine the gating power of an IGBT the following formula may be used: $P_{control\ circuit} = 5 \cdot C_{ies} \cdot V_x^2 \cdot f_{switching}$ (W)

C_{ies} = input capacitance (datasheet: C_{ies} or sometimes C_{iss})

$5 \cdot C_{ies} \approx$ input capacitance at V_{CEsat}

V_x = total rate of rise of the gate voltage

$f_{switching}$ = switching frequency (Hz)

3.4.1 Dynamic behaviour with resistive load

Fig. 3-41 shows an IGBT with a resistive load. The switch-on and switch-off times are also shown.

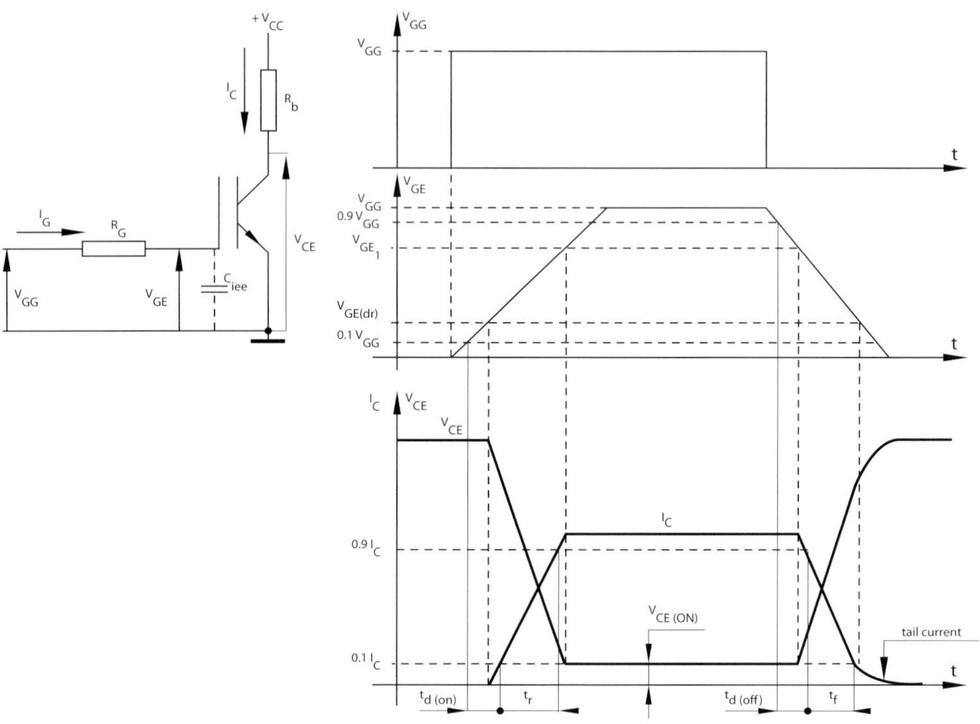

Fig. 3-41: Resistive load. Voltage and current waveforms when switching an IGBT

1. Turn-on

    When a gate voltage is applied V_{GE} increases, dependent upon the series resistor R_G and the input capacitance of the IGBT. After a certain delay time the threshold voltage $V_{GE(th)}$ is reached and collector current begins to flow. This current increases (rise-time) until at a certain value of V_{GE1} it reaches its nominal value.

$$I_C = \frac{(V_{CC} - V_{CE(ON)})}{R_b} \approx \frac{V_{CC}}{R_b} \qquad (3\text{-}10)$$

    While the current I_C increases the collector voltage V_{CE} decreases. The gate voltage increases to the applied voltage V_{GG} but I_C will not increase any further since the nominal value is determined by V_{CC} and R_b . The waveform of I_C is determined by the form of V_{GE} together with the transfer characteristic of the IGBT.

    The mechanism described here is for switch-on of an IGBT.

2. Switch-off

    When the control voltage V_{GG} becomes zero the input capacitance of the IGBT will discharge. The control circuit needs to be able to conduct DC to allow the input capacitance to discharge. When the control circuit voltage is removed it takes a certain time (delay-time) until V_{GE} reaches the value V_{GE1} and from that time on the collector current will begin to decrease (in accordance with the gate voltage V_{GE} and the transfer characteristic of the IGBT). When the threshold voltage has been reached the collector current is theoretically zero.

An IGBT however has a tail current. This is because in the thick N^--epilayer of the PNP output transistor there is still a large number of minority charge carriers present that have yet to recombine. They form the typical tail current that we are familiar with from the high voltage BJT. The time it takes for the current to decrease is called the fall time. The IGBT is now switched off. Switch-on and switch-off are shown in the current and voltage waveforms of fig. 3-41. In reality the delay times at switch-on and switch-off and the rise time and fall time of the current and voltage are not provided exactly as described here but are determined as a function of a percentage value of V_{GE} and I_C. They are provided as follows by the manufacturer:

$t_{d(on)}$: delay time at switch-on: 10% V_{GG} to 10% I_C

t_r : rise time: 10% to 90% I_C

$t_{d(off)}$: delay time at switch-off: 90% V_{GG} to 90 % I_C

t_f : fall time: 90% to 10% I_C

The value of R_G also plays a role in the turn-on and turn-off times so that the value of R_G is also mentioned in the specifications. What was shown in fig. 3-41 as V_{GG} is referred to as V_{GE} in the manufacturers specifications. These switching times are not provided for a resistive load but in the least favourable and most frequently occurring case of an inductive load with a free wheel diode. This will be discussed next.

3.4.2 Switch behaviour with an inductive load with free wheel diode

In most practical circuits an IGBT will be inductively loaded. In fig. 3-42 a part of a chopper circuit is shown. For the operation of a chopper you are referred to the relevant chapter later in this book. The following paragraph can best be understood after studying the chopper. The time constant L_b/R_b of the load is large in relation to the switching period so that the current through the load can be viewed (approximated) as constant. The current with a small ripple is shown in fig. 3-42. The rising part of the current waveform corresponds to the IGBT conducting and the falling part of the waveform to the IGBT blocking (see chopper operation chapter 12). It is assumed the chopper is operating in nominal service and the IGBT is blocking. The output current I_u flows through the diode ($I_D = I_u$) and is decreasing. This situation is similar to the BJT in fig. 3-10b.

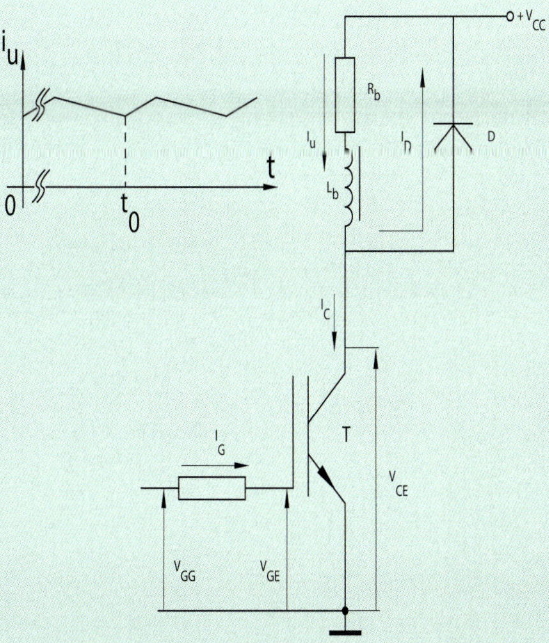

Fig. 3-42: Switching an inductively loaded IGBT

At the instant t_0 (fig. 3-42 and 3-43) the control voltage is applied to the gate of the IGBT. After a delay time the gate voltage will reach the turn-on voltage and the transistor will begin to conduct and will also draw the diode current ($I_D = I_u$). The collector current grows from zero to I_u while the diode current decreases to zero. After this the diode still has an inverse recovery time t_{rr} and the familiar inverse current flows through the free wheel diode. The amplitude of this inverse current is represented by I_{RM}. This current also flows through the IGBT (fig. 3-34c).

When the gate control voltage goes to zero then the IGBT wants to switch of, in other words the voltage V_{CE} increases from $V_{CE(ON)}$ to V_{CC}. During this phase of the switch-off cycle the collector current remains effectively constant (the energy in the inductor will hardly decrease during the ON-state of the transistor) and $V_{GE1} = I_C / g_{fs}$ also remains constant if I_C does not change. At the instant that $V_{CE} = V_{CC}$ the free wheel diode starts to conduct. The free wheel diode now provides a path for the collector current during the fall-time t_f. The collector current decreases while the diode current increases. The values of t_d, t_r en t_f were provided during the study of the resistive load.

Numeric Example 3-10:

a   How large is the dynamic capacitance of an IGP10N60T when $V_{CE} = 25$V and $f = 1$MHz?

b.   With $V_{CC} = 20$V; $T_J = 25°C$ and $V_{GE} = 8$V what is the collector current of an IGP10N60T.

c.   What is the value of the threshold voltage of an SKM400GA12E4? What are the switching times of this IGBT-module?

Solution

a.   p.3.55: $C_{iss} = 551$pF
                            $C_{oss} = 24$pF
                            $C_{rss} = 17$pF

b)   Fig. 3-50 (fig.7) $I_C = 8.4$A

c. $V_{GE(th)} = 5.8$V typical

  $t_{d(on)} = 264$ns; $t_r = 56$ns; $t_{d(off)} = 575$ns; $t_f = 117$ns

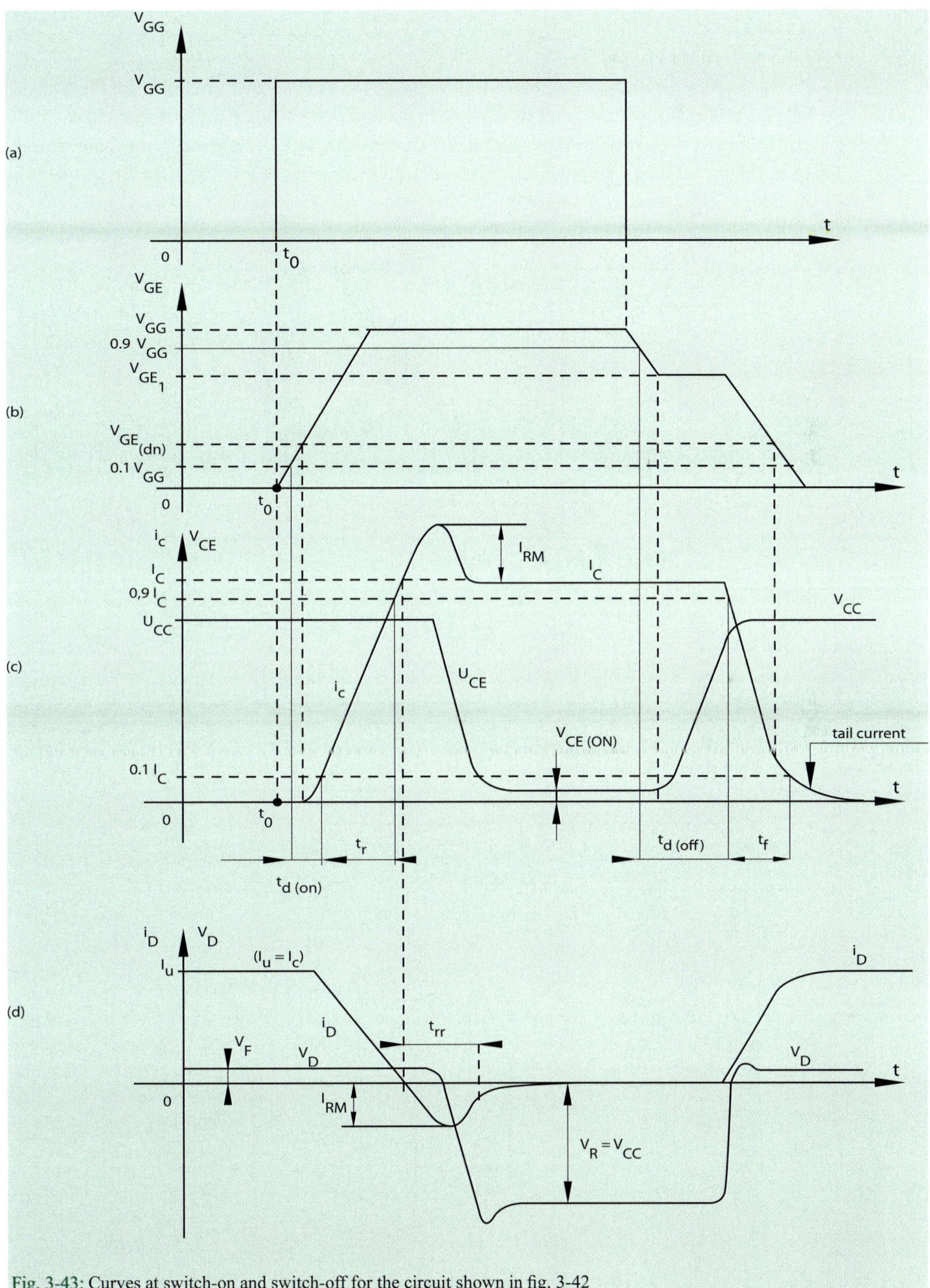

Fig. 3-43: Curves at switch-on and switch-off for the circuit shown in fig. 3-42

3.4.3 Switching losses

Due to the presence of the tail current it is difficult for the manufacturer to specify exactly the fall time of the collector current. The fall time of the collector voltage is also not mentioned. The absence of these two details means that we cannot analytically determine the switching losses. Therefore the manufacturer provides specifications for on and off losses. Fig. 3-44 shows a test circuit from the data book of International Rectifier intended to measure the switching losses.

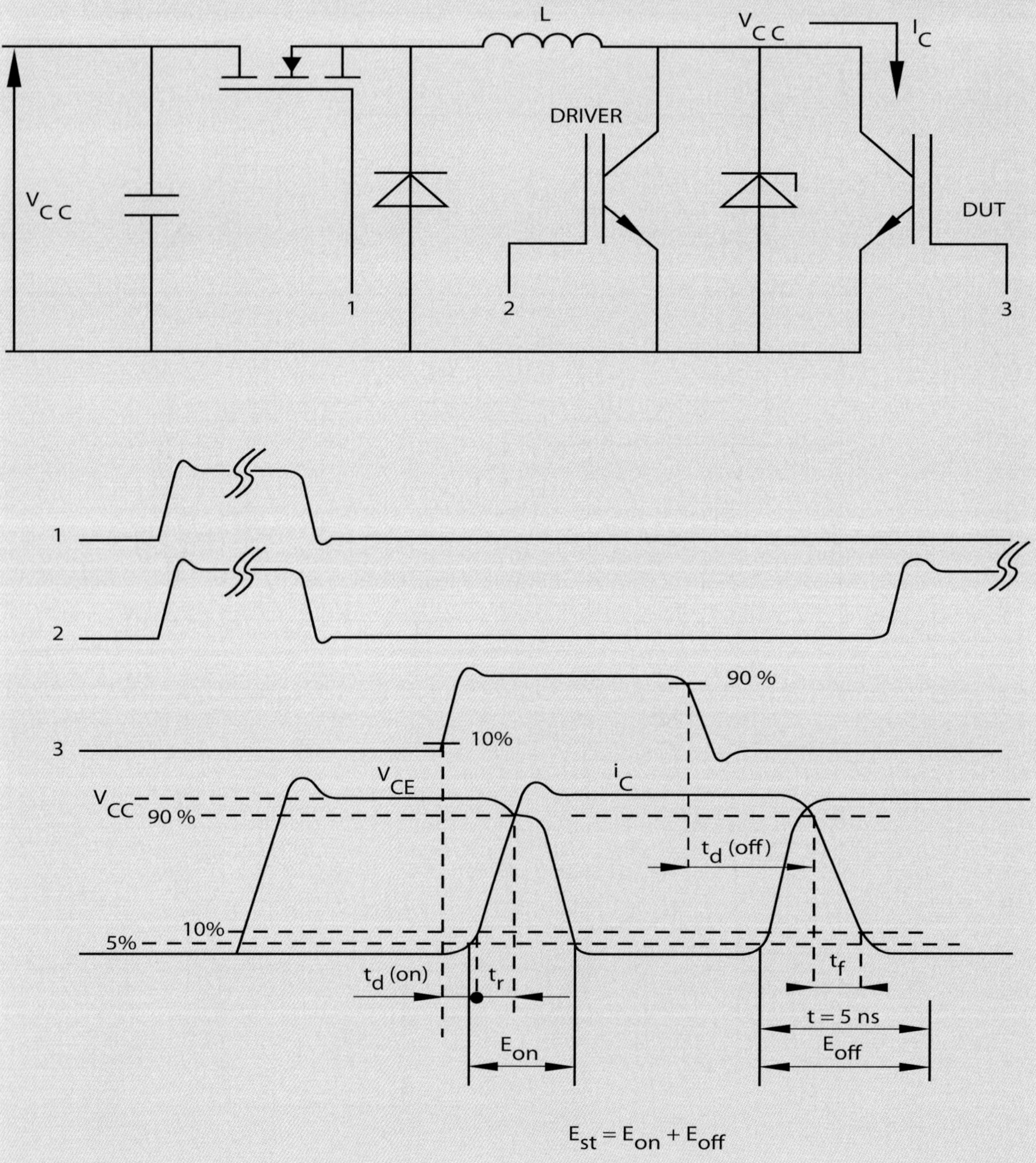

Fig. 3-44: Test circuit (International Rectifier) to determine switching losses of an IGBT

With the driver "on" the current builds up in the coil. At switch-off of the driver the zener diode will conduct and provide a path for the current of the driver. The zener voltage is across the IGBT under test (DUT = device under test). By switching the DUT on and off the accompanying curves are generated. In this way the DUT can be tested under full voltage and full current. The losses in the IGBT are determined as follows:

a. E_{on} =   switch-on losses (mJ) between 5% of the test current and 5% of the test voltage.

b. E_{off} =   switch-off losses (mJ) from 5% of the test voltage during 5µs.
         In this 5µs the start current is included.

c. E_{st} =   total switching losses = $E_{on} + E_{off}$.

d. **Total power dissipation with an inductive load and freewheel diode (clamped inductive collector load):**

By approximation the total energy losses of a chopper circuit such as in fig. 3-42 are given by: $W_{tot.loss} = W_{turn-on} + W_{ON} + W_{turn-off}$.

Here in W_{ON} is the energy which the IGBT converts to heat in the conducting mode:

W_{ON} $= V_{CE(ON)} \cdot I_C \cdot \delta \cdot T$.

$V_{CE(ON)}$ = collector-emitter saturation voltage of the I_C value under consideration.

I_C = collector current

δ = duty cycle chopper ($= t_{ON} / T$)

T = chopper period ($= 1/f_s$ with f_s = chopper frequency)

Here by : $W_{turn-on} + W_{turn-off} = E_{st}$ provided in the IGBT specifications.

The total power loss in the IGBT is given by :

$$P_{tot} = W_{tot.loss} / T = W_{tot.loss} \cdot f_s = E_{st} \cdot f_s + W_{ON} \cdot f_s = E_{st} \cdot f_s + V_{CE(ON)} \cdot I_C \cdot \qquad (3\text{-}11)$$

At a specific duty cycle δ the switching frequency f_s has no influence on the losses when the IGBT is conducting ($V_{CE(ON)} \cdot I_C \cdot \delta$) but has an effect on the switching losses at turn-on and turn-off ($= E_{st} \cdot f_s$)

Numeric example 3-11:

An IGP10N60T works with a resistive load of 40Ω and with a supply voltage of 400V. The IGBT switches at 500Hz and per period the transistor conducts for 1ms with $V_{GE} = 15$V. The junction temperature is 175°C. Determine the power dissipation in this IGBT.

Solution:

$I_{peak} = 400/40 = 10$A ; $T = 1/500 = 2$ms ; $\delta = t_{ON}/T = 1/2 = 0.5$ (50%) ;

$I_C = I_{peak} \cdot \sqrt{\delta} = 10 \cdot \sqrt{0.5} = 7.07$A

$T_J = 175$°C en $I_C = 7.07$A $\longrightarrow$ fig. 3-50 (fig.15) $\longrightarrow E_{st} = 0.625$mJ

Fig. 3-50 (fig.6) : $I_C = 10$A $\longrightarrow V_{CE(ON)} = 1.825$V

$(3\text{-}11)$ $\longrightarrow\longrightarrow P = E_{st} \cdot f_s + V_{CE(ON)} \cdot I_C \cdot \delta$

$$= 0.625 \cdot 10^{-3} \cdot 500 + 1.825 \cdot 10 \cdot 0.5 = 9.4375\text{W}$$

Remarks

1. The losses as a result of the inverse recovery current (of the free wheel diode) which also flow through the IGBT are obviously not included in the switching losses E_{st} provided in the IGBT's datasheet.

2. Just as with the bipolar transistor and the MOSFET the IGBT can switch faster with a negative gate voltage as compared to the situation where the gate voltage is brought to zero (ground) to switch off.

3. For an IGBT and IGBT-modules in the range 30 to 600A we can calculate using 100ns to 1µs for t_{on} and t_{off}. A 1200V-100A IGBT has a typical t_{on} < 200ns.

4. The fall time t_f can lie between 50 and 200ns.

5. Short circuit withstand time (SCWT):

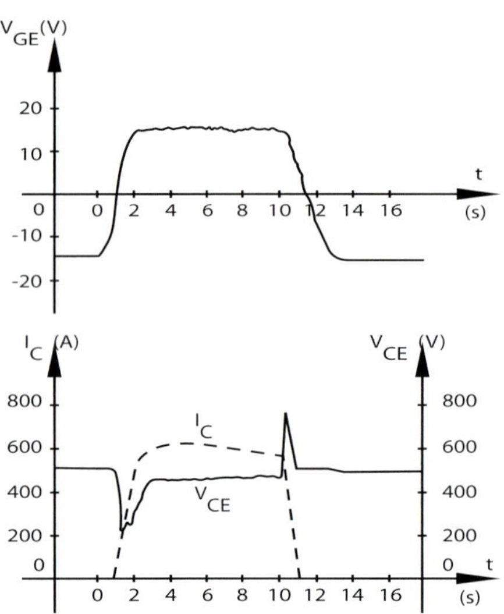

An interesting property of an NPT-IGBT is the fact that it can conduct an overcurrent of approximately ten times its nominal current for tens of microseconds. We are talking about the SCWT (short circuit withstand time). Via a fast overcurrent detection we have the time to switch off the transistor via the gate so that the short-circuit has no negative effect on the IGBT. Fig. 3-45 shows the curves of a short-circuit test from an Infineon 50A-IGBT. Note that at switch off of the IGBT a large voltage spike can result as a result of parasitic self inductance in the collector lead.

Fig. 3-45: Short-circuit test on a 50A-IGBT (Infineon) at T_{Jmax} and 500V

6. Control circuits

The same control circuits can be used as with the MOSFET. To exactly dimension the driver it is necessary to know the gate charge, which means we need to know the input capacitance.

The value of C_{iss} from the datasheet is not directly usable. This is because the C_{iss} is determined by the manufacturer at a voltage below the threshold voltage of the IGBT. As a result the effect of miller capacitance is not taken into account which will be the case in a working circuit. The input capacitance is therefore higher than C_{iss}.

A rule of thumb provides a value of input capacitance of $C_{IN} \approx 5 \cdot C_{iss}$.

With a switching frequency f the control power becomes: $P = f \cdot C_{IN} \cdot V^2$.

The maximum input current (at the start of the ON-pulse) is given by : $I_{Gmax} = \dfrac{V}{R_{Gmin}}$

If f = 5kHz; C_{iss} = 13nF; V_G = 15V and R_G = 6.8Ω , then the (output) power of the control circuit: $P = f \cdot C_{IN} \cdot V^2 = 5 \cdot 10^3 \cdot 65 \cdot 10^{-9} \cdot 15^2 = 73mW$.

The maximum gate current (when the gate pulse is applied) is: $I_{Gmax} = \dfrac{15}{6.8} = 2.2A$.

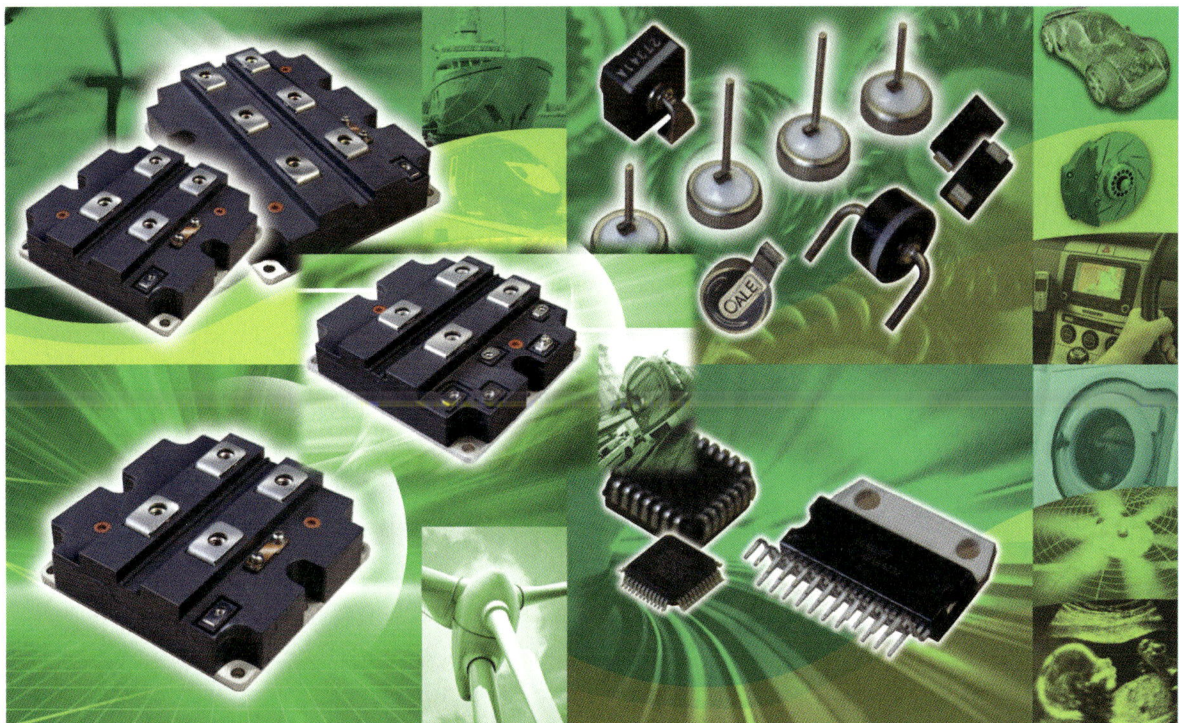

Hitachi Power Semiconductors
www.hitachi.eu/pdd
HITACHI
Inspire the Next

High Voltage IC's
HITACHI
Inspire the Next

Hitachi Power Semiconductors
www.hitachi.eu/pdd
HITACHI
Inspire the Next

3.5 Application specific IGBTs

What items are important in the development of new IGBTs. In the first case the manufacturers seek to improve on existing technology and in the second case they work hard to optimise the casing and control of the IGBT. The market for IGBTs is as switches in the following types of circuits, inductive heating, uninterrupted power supplies (UPS), switch mode power supplies (SMPS), motor control, renewable energy (wind and solar inverters), etc.

The current trend is to develop application specific IGBTs in order to maximise efficiency and performance of the specific circuits.

3.5.1 NPT and PT types of IGBTs

Originally there were two types of IGBT, namely the non-punch-through (NPT) and the punch-through (PT).

The IGBT shown in fig. 3-35 is an NPT-type (fig. 3-46). With the addition of a thick P^+ layer and an N-buffer layer we arrive at a PT-type (fig. 3-47).

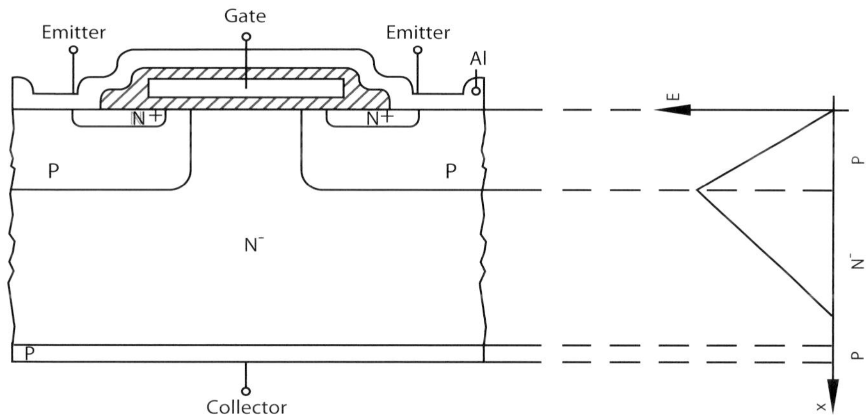

Fig. 3-46: NPT IGBT

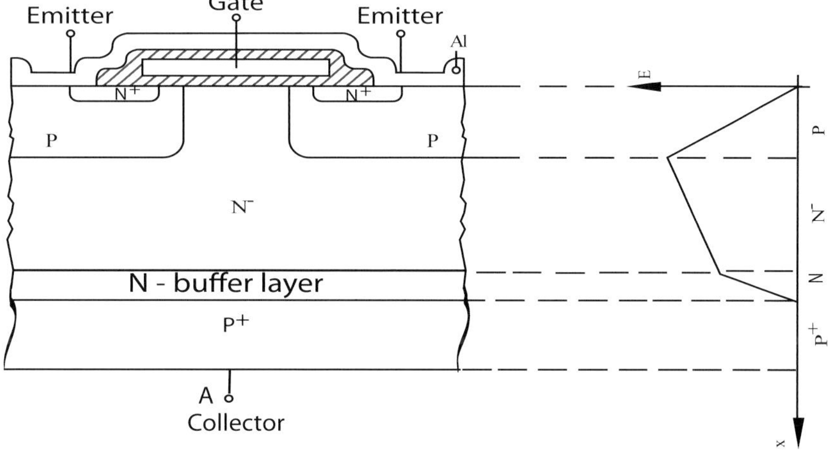

Fig. 3-47: PT IGBT

The PT-type has a thick P^+-substrate and an N-buffer which regulates the injection of holes in the N^--area (base of the transistor T_1 in fig. 3-35/36). The switching speed and the V_{CEsat} are determined by the charge in the N^--buffer layer (doping!) and the life span of the minority charge carriers ($\leq 0.25\mu s$) in the N^--area.

The NPT has no buffer layer but does have a thick N^--layer. Switching properties and V_{CEsat} of this IGBT are determined by the injection possibilities of the P-emitter (transistor T_2 in fig. 3-35b).

Advantages of NPT-types over PT-types:
- positive temperature coefficient. This is interesting from the viewpoint of parallel switching of IGBTs
- smaller (but longer) tail current
- fast switching time and lower switching losses
- lower temperature dependence for switching times, switching losses and tail current
- more robust, and this translates into the SCWT and the UIS (unclamped inductive switching)
- UIS: for the same cell structure an NPT has a better UIS than a PT as a result of the thick N^--layer (and accompanying resistance). The NPT has almost the same UIS as a MOSFET. A PT has a lower UIS, unless specially designed. The quicker the PT-IGBT, the lower its UIS possibilities.

A PT type on the other hand has a much thinner N^--epilayer, which is more economical.
Above 1200V the NPT-type dominates.

Applications of IGBTs in motor control inverters require specific properties:
- low V_{CEsat} to maintain low conduction Losses
- soft turn-on and turn-off to reduce EMC
- SCWT preferably larger than $10\mu s$ since the short circuit in the output of the inverter immediately affects the IGBTs
- non latching operation with a square RBSOA at a minimum of 5 times the nominal current
- positive temperature coefficient for V_{CEsat} with a view to parallel switching of IGBTs
- low Q_{gd} / Q_{gs} ratio to prevent overshoot
- $dv/dt < 7kV/\mu s$ to limit fatigue of the motor windings.

Solar and UPS inverters have the same requirements as mentioned above with the exception of the switching speed which needs to be raised , even at the cost of V_{CEsat} . In the converter section which operates at low frequency (50-60Hz) IGBTs are used with a very low V_{CEsat} .

In the section working at high frequency (16...50kHz) planar IGBTs that are optimised for high frequencies are used. UPS and solar inverters operate at high frequencies to minimise the filter dimensions. The advantage of these filters (see fig. 15-26) in most cases is that the di/dt of the short circuit current is limited and therefore the SCWT does not need to be as large as in motor applications.

3.5.2 Trench NPT-IGBT

The NPT-trench structure was designed to reduce V_{CEsat} and the switch-off losses E_{off} (fig. 3-48). In contrast with a planar NPT-IGBT the trench type has a lower V_{CEsat}, lower switching losses (E_{off}) and a lower thermal resistance R_{th}.

As a result of the trench structure the $R_{DS(ON)}$ is eliminated from the parasitic MOSFET of fig. 3-37a so that (3-8) becomes:

$$V_{CEsat} = V_{BE(T1)} + I_D \cdot R_D \qquad (3\text{-}12)$$

The trench has therefore a lower V_{CEsat} than the NPT shown in fig. 3-46.

3.5.3 Trench-fieldstop (FS) structure

Fig. 3-49 shows the implementation. In comparison to the trench implementation of fig. 3-48 the thickness of the N^--base is much reduced and there is a lightly doped N-buffer added. From the graph of the field strength we can see where the name field stop comes from. Since the N^--base of a trench FS is much thinner than a trench the drift resistance R_D is lower and expression (3-12) shows that V_{CEsat} is reduced in comparison to a trench IGBT.

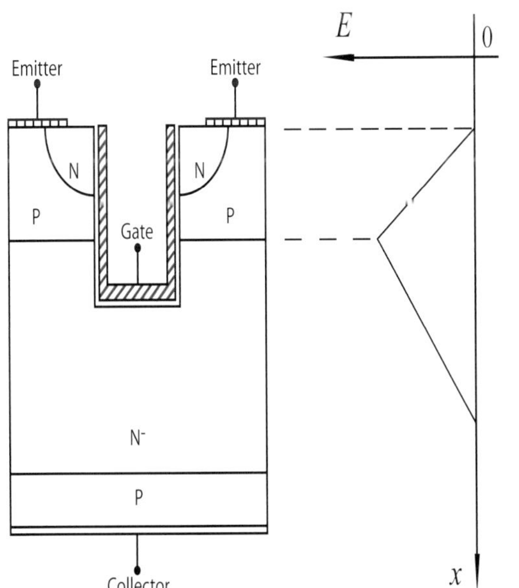

Fig. 3-48: Trench structure of an IGBT

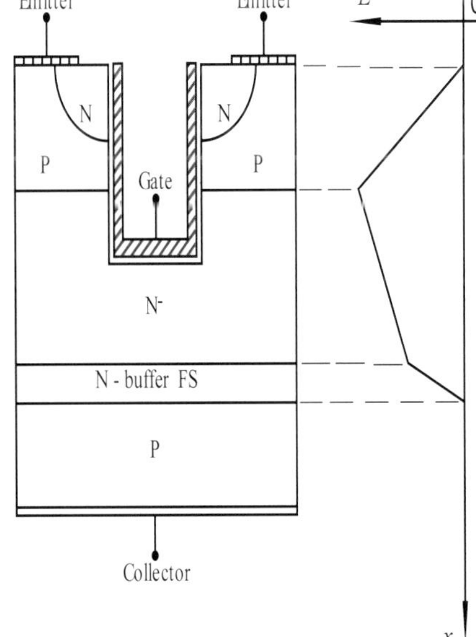

Fig. 3-49: Trench fieldstop IGBT

Photo Hitachi: Superior thermal performance and product lifetime cycling offered by Hitachi's compact grease free direct cooling structures

Photo Hitachi: PDD IGBT 2 is representative for the 4,5kV - 6,5kV family of Hitachi IGBTs rated from 750A - 1200A

3.6 Details of a Hitachi MBN750H65E2

IGBT MODULE

Spec.No.IGBT-SP-09008 R7 P1

MBN750H65E2

Preliminary Specification

Silicon N-channel IGBT 6500V E2 version

FEATURES

* Soft switching behavior & low conduction loss: Soft low-injection punch-through High conductivity IGBT.
* Low driving power due to low input capacitance MOS gate.
* Low noise recovery: Ultra soft fast recovery diode.
* High thermal fatigue durability:
  (delta Tc=70K, N>30,000cycles)
  AlSiC base-plate/AlN substrate

This datasheet is not final version. Changes of this datasheet are reserved.

ABSOLUTE MAXIMUM RATINGS (Tc=25°C)

Item		Symbol	Unit	MBN750H65E2
Collector Emitter Voltage		V_{CES}	V	6,500
Gate Emitter Voltage		V_{GES}	V	±20
Collector Current	DC	I_C	A	750 (Tc=80 ^{o}C)
	1ms	I_{Cp}		1,500
Forward Current	DC	I_F	A	750
	1ms	I_{FM}		1,500
Junction Temperature		T_j	^{o}C	-40 ~ +125
Storage Temperature		T_{stg}	^{o}C	-40 ~ +125
Isolation Voltage		V_{ISO}	V_{RMS}	10,200 (AC 1 minute)
Screw Torque	Terminals (M4/M8)	-	N·m	2/10　　(1)
	Mounting (M6)	-		6　　(2)

Notes: (1) Recommended Value 1.8±0.2/9±1N·m　　(2) Recommended Value 5.5±0.5N·m

ELECTRICAL CHARACTERISTICS

Item		Symbol	Unit	Min.	Typ.	Max.	Test Conditions
Collector Emitter Cut-Off Current		I_{CES}	mA	-	-	25	V_{CE}=6,500V, V_{GE}=0V, Tj=25°C
				-	25	100	V_{CE}=6,500V, V_{GE}=0V, Tj=125°C
Gate Emitter Leakage Current		I_{GES}	nA	-500	-	+500	V_{GE}=±20V, V_{CE}=0V, Tj=25°C
Collector Emitter Saturation Voltage		$V_{CE(sat)}$	V	-	3.2	-	I_C=750A, V_{GE}=15V, Tj=25°C
				3.4	4.3	5.2	I_C=750A, V_{GE}=15V, Tj=125°C
Gate Emitter Threshold Voltage		$V_{GE(TO)}$	V	5.8	6.3	6.8	V_{CE}=10V, I_C=750mA, Tj=25°C
Input Capacitance		C_{ies}	nF	-	130	-	V_{CE}=10V, V_{GE}=0V, f=100kHz, Tj=25°C
Internal Gate Resistance		Rge	Ω	-	0.7	-	V_{CE}=10V, V_{GE}=0V, f=100kHz, Tj=25°C
Switching Times	Rise Time	t_r	μs	2.2	3.2	4.8	V_{CC}=3,600V, Ic=750A
	Turn On Time	t_{on}		2.7	3.9	5.9	Ls=200nH
	Fall Time	t_f		2.2	3.1	4.7	R_G=8.2 Ω　　(3)
	Turn Off Time	t_{off}		4.5	6.4	9.6	V_{GE}=+/-15V, Tj=125°C
Peak Forward Voltage Drop		V_{FM}	V	-	3.6	-	IF=750A, V_{GE}=0V, Tj=25°C
				3.5	3.9	4.4	IF=750A, V_{GE}=0V, Tj=125°C
Reverse Recovery Time		t_{rr}	μs	-	0.8	-	Vcc=3600V, IF=750A, Ls=200nH Tj=125°C
Turn On Loss		$E_{on(10\%)}$	J/p	-	4.9	6.4	Vcc=3600V, Ic= IF=750A, Ls=200nH R_G= 8.2 Ω　　(3) V_{GE}=+/-15V, Tj=125°C
		$E_{on(full)}$		-	5.5	-	
Turn Off Loss		$E_{off(10\%)}$	J/ p	-	3.9	5.1	
		$E_{off(full)}$		-	4.2	-	
Reverse Recovery Loss		$E_{rr(10\%)}$	J/ p	-	2.1	2.7	
		$E_{rr(full)}$		-	2.3	-	

Notes:(3) R_G value is the test condition's value for evaluation of the switching times, not recommended value.
　　Please, determine the suitable R_G value after the measurement of switching waveforms
　　(overshoot voltage, etc.) with appliance mounted.
* Please contact our representatives at order.
* For improvement, specifications are subject to change without notice.
* For actual application, please confirm this spec sheet is the newest revision.

DWN.	K.Yasuda	'10.11.04
CHKD.	Y.Toyoda	'10.11.04
APPD.	Y.Koike	'10.11.04

HITACHI
Inspire the Next

3.7 Details of an Infineon IGP10N60T

IGP10N60T

TrenchStop® Series

Low Loss IGBT in TrenchStop® and Fieldstop technology

- Very low $V_{CE(sat)}$ 1.5 V (typ.)
- Maximum Junction Temperature 175 °C
- Short circuit withstand time – 5µs
- Designed for :
    - Variable Speed Drive for washing machines and air conditioners
    - induction cooking
    - Uninterrupted Power Supply
- TrenchStop® and Fieldstop technology for 600 V applications offers :
    - very tight parameter distribution
    - high ruggedness, temperature stable behaviour
- NPT technology offers easy parallel switching capability due to positive temperature coefficient in $V_{CE(sat)}$
- Low EMI
- Low Gate Charge
- Qualified according to JEDEC[1] for target applications
- Pb-free lead plating; RoHS compliant
- Complete product spectrum and PSpice Models : http://www.infineon.com/igbt/

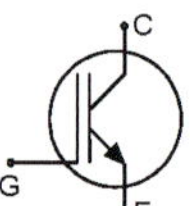

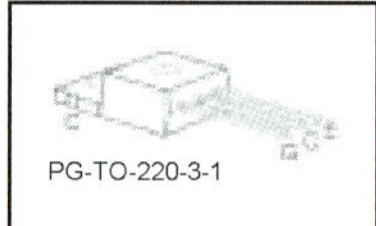

PG-TO-220-3-1

Type	V_{CE}	I_C	$V_{CE(sat), Tj=25°C}$	$T_{j,max}$	Marking Code	Package
IGP10N60T	600V	10A	1.5V	175°C	G10T60	PG-TO-220-3-1

Maximum Ratings

Parameter	Symbol	Value	Unit
Collector-emitter voltage	V_{CE}	600	V
DC collector current, limited by T_{jmax}	I_C		A
$T_C = 25°C$		20	
$T_C = 100°C$		10	
Pulsed collector current, t_p limited by T_{jmax}	I_{Cpuls}	30	
Turn off safe operating area $V_{CE} \leq 600V$, $T_j \leq 175°C$	-	30	
Gate-emitter voltage	V_{GE}	±20	V
Short circuit withstand time[2)] $V_{GE} = 15V$, $V_{CC} \leq 400V$, $T_j \leq 150°C$	t_{SC}	5	µs
Power dissipation $T_C = 25°C$	P_{tot}	110	W
Operating junction temperature	T_j	-40...+175	°C
Storage temperature	T_{stg}	-55...+175	
Soldering temperature, 1.6mm (0.063 in.) from case for 10s		260	

IGP10N60T

TrenchStop® Series

Thermal Resistance

Parameter	Symbol	Conditions	Max. Value	Unit
Characteristic				
IGBT thermal resistance, junction – case	R_{thJC}		1.35	K/W
Thermal resistance, junction – ambient	R_{thJA}		62	

Electrical Characteristic, at T_j = 25 °C, unless otherwise specified

Parameter	Symbol	Conditions	Value min.	Value typ.	Value max.	Unit
Static Characteristic						
Collector-emitter breakdown voltage	$V_{(BR)CES}$	$V_{GE}=0V$, $I_C=0.2mA$	600	-	-	V
Collector-emitter saturation voltage	$V_{CE(sat)}$	$V_{GE} = 15V$, $I_C=10A$				
		$T_j=25°C$	-	1.5	2.05	
		$T_j=175°C$	-	1.8	-	
Gate-emitter threshold voltage	$V_{GE(th)}$	$I_C=0.3mA, V_{CE}=V_{GE}$	4.1	4.6	5.7	
Zero gate voltage collector current	I_{CES}	$V_{CE}=600V$, $V_{GE}=0V$				µA
		$T_j=25°C$	-	-	40	
		$T_j=175°C$	-	-	1000	
Gate-emitter leakage current	I_{GES}	$V_{CE}=0V, V_{GE}=20V$	-	-	100	nA
Transconductance	g_{fs}	$V_{CE}=20V$, $I_C=10A$	-	6	-	S
Integrated gate resistor	R_{Gint}			none		Ω

Dynamic Characteristic

Parameter	Symbol	Conditions	Value min.	Value typ.	Value max.	Unit
Input capacitance	C_{iss}	$V_{CE}=25V$,	-	551	-	pF
Output capacitance	C_{oss}	$V_{GE}=0V$,	-	24	-	
Reverse transfer capacitance	C_{rss}	$f=1MHz$	-	17	-	
Gate charge	Q_{Gate}	$V_{CC}=480V$, $I_C=10A$ $V_{GE}=15V$	-	62	-	nC
Internal emitter inductance measured 5mm (0.197 in.) from case	L_E	TO-220-3-1	-	7	-	nH
Short circuit collector current[1]	$I_{C(SC)}$	$V_{GE}=15V, t_{SC}≤5µs$ $V_{CC} = 400V$, $T_j = 25°C$	-	100	-	A

IGP10N60T

TrenchStop® Series

Switching Characteristic [3)], **Inductive Load**, at T_j=25 °C

Parameter	Symbol	Conditions	Value min.	Value typ.	Value max.	Unit
IGBT Characteristic						
Turn-on delay time	$t_{d(on)}$	T_j=25°C, V_{CC}=400V, I_C=10A, V_{GE}=0/15V, R_G=23Ω, $L_\sigma^{2)}$=60nH, $C_\sigma^{2)}$=40pF Energy losses include "tail" and diode reverse recovery.	-	12	-	ns
Rise time	t_r		-	8	-	
Turn-off delay time	$t_{d(off)}$		-	215	-	
Fall time	t_f		-	38	-	
Turn-on energy	E_{on}		-	0.16	-	mJ
Turn-off energy	E_{off}		-	0.27	-	
Total switching energy	E_{ts}		-	0.43	-	

Switching Characteristic [3)], **Inductive Load**, at T_j=175 °C

Parameter	Symbol	Conditions	Value min.	Value typ.	Value max.	Unit
IGBT Characteristic						
Turn-on delay time	$t_{d(on)}$	T_j=175°C, V_{CC}=400V, I_C=10A, V_{GE}=0/15V, R_G= 23Ω, $L_\sigma^{2)}$=60nH, $C_\sigma^{2)}$=40pF Energy losses include "tail" and diode reverse recovery.	-	10	-	ns
Rise time	t_r		-	11	-	
Turn-off delay time	$t_{d(off)}$		-	233	-	
Fall time	t_f		-	63	-	
Turn-on energy	E_{on}		-	0.26	-	mJ
Turn-off energy	E_{off}		-	0.35	-	
Total switching energy	E_{ts}		-	0.61	-	

IGP10N60T

TrenchStop® Series

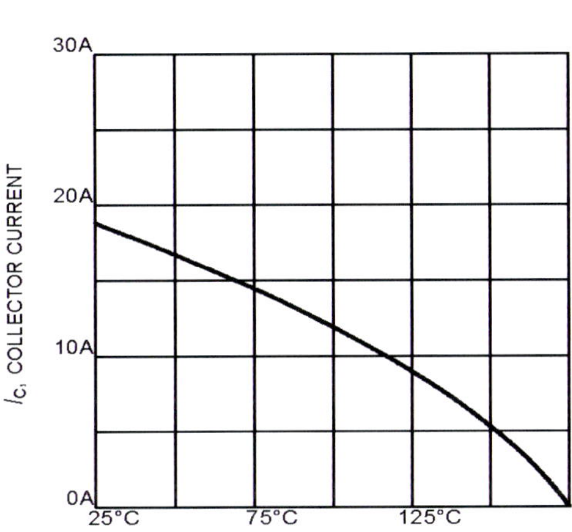

Figure 4. Collector current as a function of case temperature
($V_{GE} \geq 15V$, $T_j \leq 175°C$)

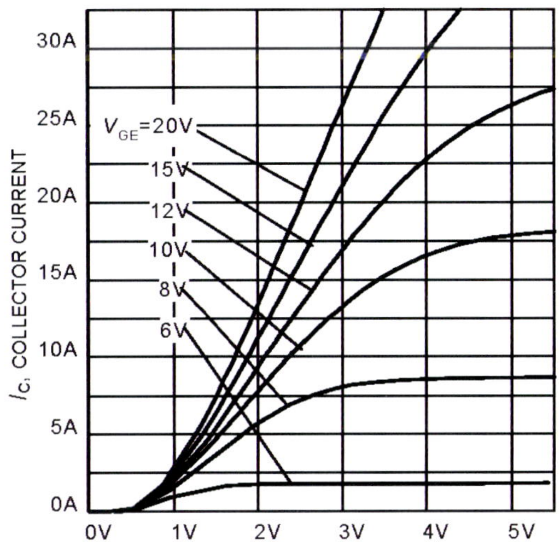

Figure 6. Typical output characteristic
($T_j = 175°C$)

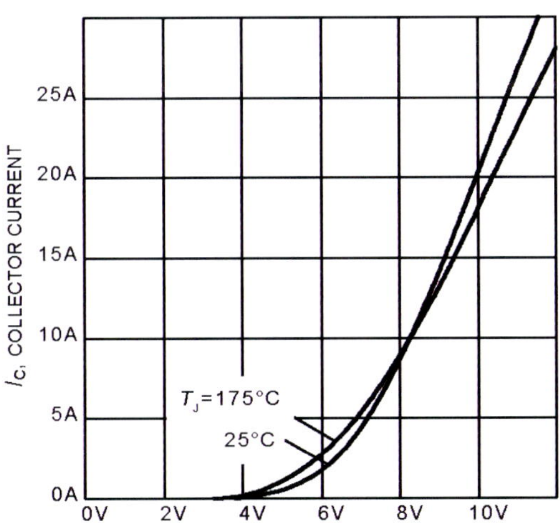

Figure 7. Typical transfer characteristic
(V_{CE}=20V)

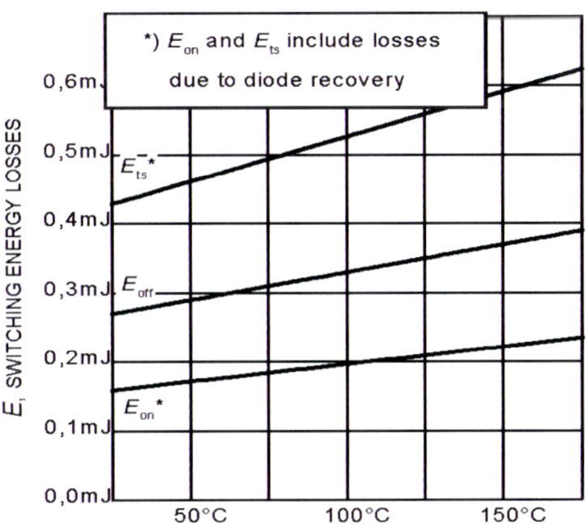

Figure 15. Typical switching energy losses as a function of junction temperature
(inductive load, V_{CE} = 400V, V_{GE} = 0/15V, I_C = 10A, R_G = 23Ω, Dynamic test circuit in Figure E)

Fig. 3-50: Curven van een IGP10N60T van Infineon

3.8 Details of a Semikron SKM400GA12E4

SKM400GA12E4

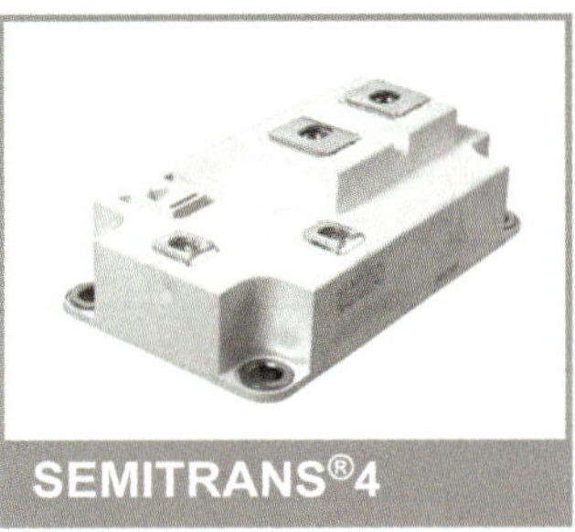

SEMITRANS®4

IGBT4 Modules

SKM400GA12E4

Features
- IGBT4 = 4. Generation (Trench)IGBT
- VCEsat with positive temperature coefficient
- High short circuit capability, self limiting to 6 x I_{CNOM}
- Soft switching 4. Generation CAL diode (CAL4)

Typical Applications
- AC inverter drives
- UPS
- Electronic welders at fsw up to 20 kHz

Remarks
- Case temperature limited to Tc = 125°C max, recomm. Top = -40 ... +150°C, product rel. results valid for Tj = 150°

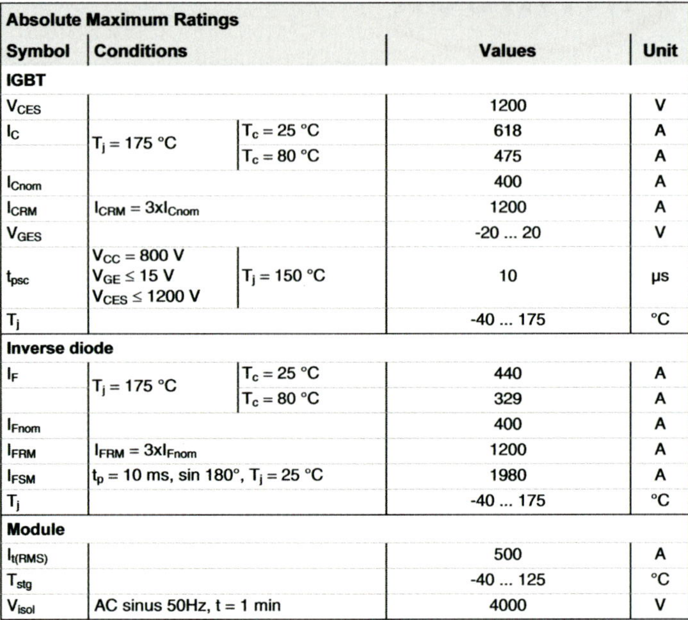

Absolute Maximum Ratings				
Symbol	**Conditions**		**Values**	**Unit**
IGBT				
V_{CES}			1200	V
I_C	$T_j = 175\ °C$	$T_c = 25\ °C$	618	A
		$T_c = 80\ °C$	475	A
I_{Cnom}			400	A
I_{CRM}	$I_{CRM} = 3xI_{Cnom}$		1200	A
V_{GES}			-20 ... 20	V
t_{psc}	$V_{CC} = 800\ V$ $V_{GE} \leq 15\ V$ $V_{CES} \leq 1200\ V$	$T_j = 150\ °C$	10	µs
T_j			-40 ... 175	°C
Inverse diode				
I_F	$T_j = 175\ °C$	$T_c = 25\ °C$	440	A
		$T_c = 80\ °C$	329	A
I_{Fnom}			400	A
I_{FRM}	$I_{FRM} = 3xI_{Fnom}$		1200	A
I_{FSM}	$t_p = 10\ ms,\ sin\ 180°,\ T_j = 25\ °C$		1980	A
T_j			-40 ... 175	°C
Module				
$I_{t(RMS)}$			500	A
T_{stg}			-40 ... 125	°C
V_{isol}	AC sinus 50Hz, t = 1 min		4000	V

Characteristics						
Symbol	**Conditions**		**min.**	**typ.**	**max.**	**Unit**
IGBT						
$V_{CE(sat)}$	$I_C = 400\ A$ $V_{GE} = 15\ V$ chiplevel	$T_j = 25\ °C$		1.8	2.05	V
		$T_j = 150\ °C$		2.2	2.4	V
V_{CE0}		$T_j = 25\ °C$		0.8	0.9	V
		$T_j = 150\ °C$		0.7	0.8	V
r_{CE}	$V_{GE} = 15\ V$	$T_j = 25\ °C$		2.5	2.9	mΩ
		$T_j = 150\ °C$		3.8	4.0	mΩ
$V_{GE(th)}$	$V_{GE}=V_{CE},\ I_C = 15.2\ mA$		5	5.8	6.5	V
I_{CES}	$V_{GE} = 0\ V$ $V_{CE} = 1200\ V$	$T_j = 25\ °C$		0.1	0.3	mA
		$T_j = 150\ °C$				mA
C_{ies}	$V_{CE} = 25\ V$ $V_{GE} = 0\ V$	$f = 1\ MHz$		24.6		nF
C_{oes}		$f = 1\ MHz$		1.62		nF
C_{res}		$f = 1\ MHz$		1.38		nF
Q_G	$V_{GE} = -\ 8\ V...+ 15\ V$			2260		nC
R_{Gint}	$T_j = 25\ °C$			1.9		Ω
$t_{d(on)}$	$V_{CC} = 600\ V$ $I_C = 400\ A$ $V_{GE} = ±15\ V$ $R_{G\ on} = 1\ Ω$ $R_{G\ off} = 1\ Ω$	$T_j = 150\ °C$		264		ns
t_r		$T_j = 150\ °C$		56		ns
E_{on}		$T_j = 150\ °C$		28		mJ
$t_{d(off)}$		$T_j = 150\ °C$		575		ns
t_f	$di/dt_{on} = 7100\ A/µs$	$T_j = 150\ °C$		117		ns
E_{off}	$di/dt_{off} = 3900\ A/µs$	$T_j = 150\ °C$		59		mJ
$R_{th(j-c)}$	per IGBT				0.072	K/W

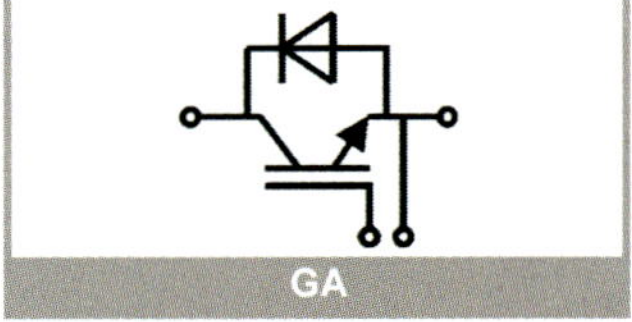

Fig. 3.51: Details of a 1200V-400A IGBT module from Semikron (SKM400GA12E4)

4. EVALUATION

3.1 Why do we use a snubber with a BJT?

3.2 Which value of V_{CB} lies on the border between the linear region and saturation of a BJT?

3.3 From what parts is the switch-on time of a BJT made up? Also for the switch-off time?

3.4 What is a typical value for switch-on time of a heavy power transistor?

3.5 Why do we use a free wheel diode with an inductively loaded BJT? Where does the name "free wheel" come from?

3.6 Why does an inductive collector load discharge quicker if there is a resistor in series with the free wheel diode?

3.7 Between which values of which quantities is the switch-off cross-over time determined for a BJT?

3.8 Redraw fig. 3-8 in a log-log axis.

3.9 What is the meaning of the following terms t_{on}, t_{off}, t_d, t_s, t_f, t_r?

3.10 Name the electrodes of the junction transistor and the field effect transistor.

3.11 What type of MOSFETs can we distinguish in relation to the conduction channel?

3.12 What does the term MOSFET mean?

3.13 What is the expression for the transconductance of a MOSFET?

3.14 Draw a sketch showing the principle of operation of a MOSFET connected in common source.

3.15 Where does the term vertical fet come from?

3.16 What is the threshold voltage of a power MOSFET? Does the existence of a threshold voltage have any advantages?

3.17 Which company produces hexfets? Sipmos? Trenchmos? Ultrafets? Trenchfieldstop?

3.18 From which components is the input capacitance C_{iss} of a MOSFET composed?

3.19 What is $t_{d(off)}$?

3.20 Explain the meaning of I_{DMmax} .

3.21 Which parameters are needed to determine the voltage drop across a fully driven MOSFET (switch)?

3.22 Will a MOSFET switch on if we put 10 mA into the gate? What is the difference if instead we use 100mA?

3.23 What is ESD? How do we protect a MOSFET from ESD?

3.24 What does "human body model" mean?

3.25 Why can MOSFETS be so easily switched in parallel?

3.26 What is a FREDFET?

3.27 What are the maximum ratings (voltage, current) of a Hitachi MBN750H65E2? Is this a PT or an SPT IGBT?

3.28 What do we understand by the ohmic operating area of a MOSFET?

3.29 What is the difference between the I-V curves of a P-channel MOSFET and an N-channel MOSFET?

3.30 What is a "SMART POWER" switch?

3.31 What is the maximum $R_{DS(ON)}$ of an IRF530N?

3.32 What does "Miller capacitance" mean?

3.33 Construct a $P_D = 79W$ curve for $T_J = 25°C$ in fig. 3-26 (fig. 1)

3.34 What is second breakdown of a power transistor?

3.35   Explain the operation of the NPN/PNP totem pole of fig. 3-31?

3.36   Why in fig. 3-32 is one of the six inverters of the CD 4049 separated from the other five?

3.37   What do we mean when we refer to the "short circuit withstand time" of an IGBT?

3.38   Why is there a tail current at switch-off of an IGBT? How long can such a tail current last?

3.39   What does IGBT mean?

3.40   What is the level of saturation voltage of a recent 1200V type IGBT?

3.41   How large is the permissible gate voltage of an IGBT?

3.42   What is meant by "clamped inductive load"?

3.43   What is a Darlington configuration?

3.44   Give a practical value for the transconductance of an IGBT?

3.45   How are the fall time and rise time of an IGBT defined?

3.46   What is meant by DUT?

3.47   How are the switch-on losses E_{on} determined by the manufacturer .
        In which engineering unit is E_{on} expressed?

3.48   Does the switching frequency influence the power loss in a conducting IGBT?
        Explain your answer.

3.49   What is a typical fall time for an IGBT?

4 THYRISTORS

CONTENTS

1. Shockley diode
2. Unidirectional thyristor (SCR)
3. Diac
4. Two directional thyristor (triac)
5. Gate turn-off thyristor (GTO)
6. The MOS controlled thyristor (MCT)
7. Integrated gate commutated thyristor (IGCT)
8. Evaluation

According to the I.E.C. definition a thyristor is a bistable semiconductor with three or more junctions and that can be switched between a conducting and blocking state.

A Shockley diode, SCR, triac, GTO, MCT and IGCT all comply with this definition.

The SCR appeared in 1957 and was used until recently as a power switch in the control of DC and AC motors.

Triacs are used to control AC loads from a few watts up to 10 or 20 kW.

GTO's are predominately used in traction applications and industrial variable frequency drives for the control of AC motors in the MW range.

The MCT arrived on the market in 1992 and remains a switch with many promising properties.

The newest thyristor type, the IGCT from ABB (Asea Brown Boveri) , came on the market in 1996. The IGCT is suitable for use in motor drives in the medium voltage range (2.3/3.3/4.16/6.9 kV).

1. SHOCKLEY DIODE

1.1 Construction

As we see in fig. 4-1 a Shockley diode is made up of four layers of silicon.
This PNPN-configuration is referred to as a four layer diode.
Construction, symbol and elementary equivalent circuit are shown in fig. 4-1.

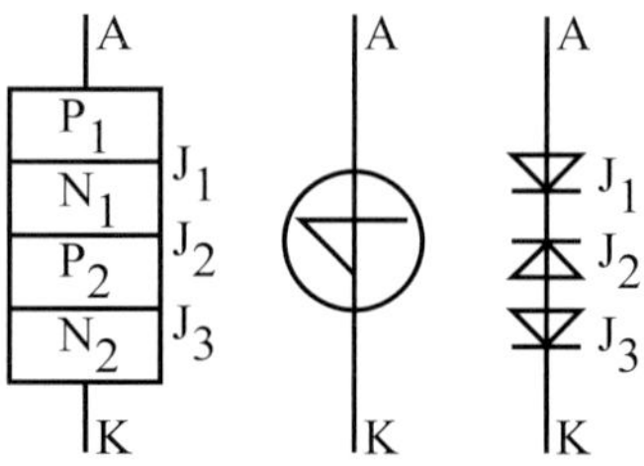

Fig. 4-1: Four layer diode

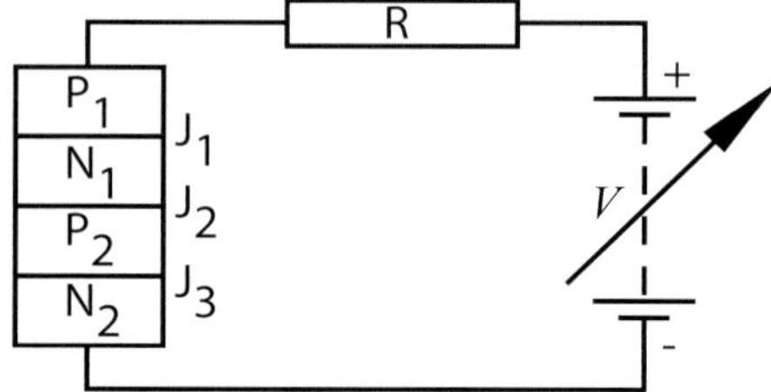

Fig. 4-2: Basic test configuration of a Shockley diode

1.2 Operation

In a test configuration as shown in fig. 4-2 J_1 and J_3 are forward polarised and J_2 is reverse polarised. In this manner practically the full voltage V is across J_2. To understand what happens when the voltage V is increased examine fig. 4-3.
In fig. 4-3a the four layer diode has been replaced by two complementary transistors T_1 and T_2.
In fig. 4-3b the configuration of fig. 4-2 has been redrawn showing these transistors.

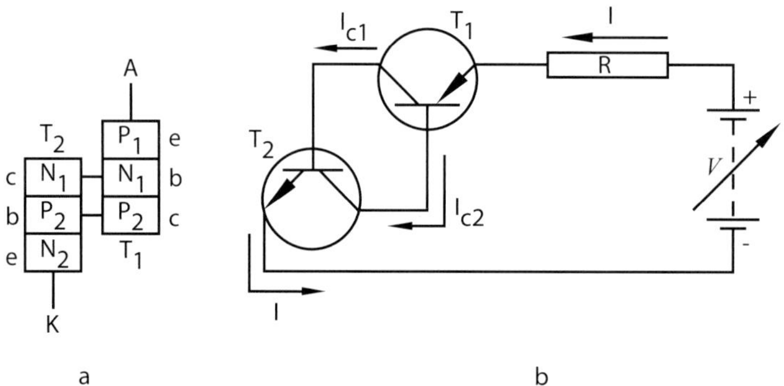

Fig. 4-3: Equivalent circuit of fig. 4-2

Applying Kirchoff's current law to transistor T_1 gives : $I = I_{C1} + I_{C2}$.

In the case of a bipolar transistor we can write : $I_{C1} = \alpha_1 \cdot I + I_{CB01}$ (4-1)

$$I_{C2} = \alpha_2 \cdot I + I_{CB02} \qquad (4\text{-}2)$$

We find therefore : $I = I_{C1} + I_{C2} = \alpha_1 \cdot I + I_{CB01} + \alpha_2 \cdot I + I_{CB02} = (\alpha_1 + \alpha_2) \cdot I + I_{CB0}$

Assume $I_{CB01} + I_{CB02} = I_{CB0}$.

A little rearranging gives:

$$I = \frac{I_{CB0}}{1 - (\alpha_1 + \alpha_2)} \qquad (4\text{-}3)$$

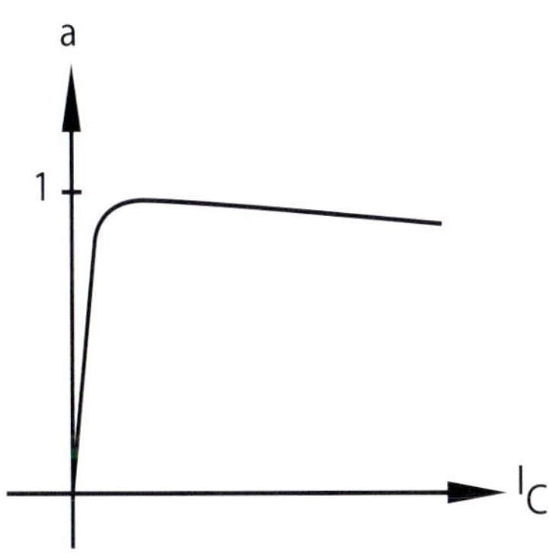

Fig. 4-4: Curve of α of a BJT

With small values of V the leakage current I_{CB0} is very small and from fig. 4-4 it follows that α_1 and α_2 are very low, so that: $I \approx I_{CB0}$. By increasing the voltage V the field strength across J_2 increases, the leakage current increases and α_1 and α_2 increase. Fig. 4-3b reveals that the collector current of the one transistor is exactly the base current of the other transistor. With sufficient increase of α_1 and α_2 the transistors will drive each other into saturation.

At the instant that $\alpha_1 + \alpha_2 = 1$ the current is infinitely large according to expression (4-3).

This does not happen because:

1. I is limited to $I_{max} = \dfrac{V}{R}$.

2. The expressions (4-1) and (4-2) are not valid for saturated transistors.

As a result expression (4-3) is not valid for a conducting Shockley diode. Since junction J_2 is now heavily conducting (fig. 4-2) there will only be a small voltage of 1 to 2 V (J_1 and J_2 !). Practically the entire voltage V is across the load resistor (fig. 4-2 and fig. 4-5).

The voltage at which the diode conducts heavily is called V_{BO} (BO = break over).

1.3 I_a - V_a characteristic

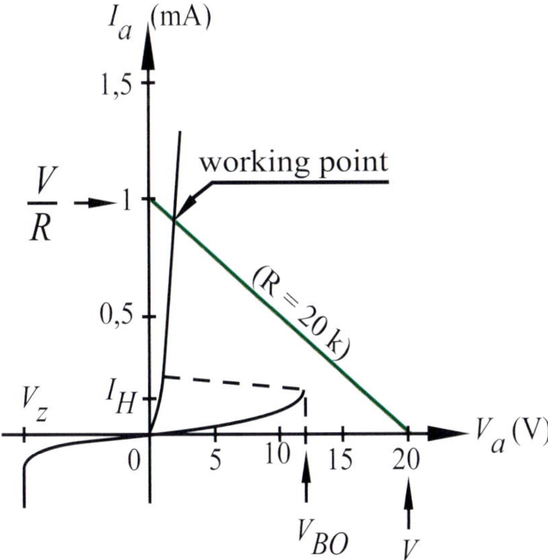

Fig. 4-5: I_a - V_a curves of a Shockley diode

We see in fig. 4-5:

Quadrant 1:

- As the voltage V_a increases from zero to V_{BO} a small current I_a flows.
- At $V > V_{BO}$ the diode conducts heavily (switches on), we obtain the I_a - V_a curve of a conducting diode.
- to switch off the diode ensure $I_a < I_H$. (I_H = holding current).

Quadrant 3:

J_1 and J_3 are reverse polarised: we have the I_a - V_a curve of two reverse biased diodes in series with a zener voltage V_Z which may not be exceeded .

Photo ABB (KWx): New 6 inch thyristors: 5200V / 6500 V - 4100A / 3400A

Photo Semikron: Different types of thyristors

2. UNIDIRECTIONAL THYRISTOR (SCR)

The unidirectional thyristor or SCR (silicon controlled rectifier) is referred as a thyristor in Europe. From the IEC definition the triac, GTO and IGCT may also be called thyristors. Therefore we reserve the name SCR for the unidirectional thyristor.

2.1 Construction of an SCR

The SCR in fig. 4-6 is a PNPN configuration with three connections: anode A, gate G and cathode K. By applying a suitable voltage between gate and cathode the gate-cathode junction will conduct. If there is a positive voltage between anode and cathode a current will flow from anode to cathode. As will be noted while using only a little control power, the thyristor can operate as a switch, switching large powers. Here are some examples of anode current and operating voltage of some thyristors: 25A-25V; 10A-800V; 1900A-2500V; 2500A-1600V.

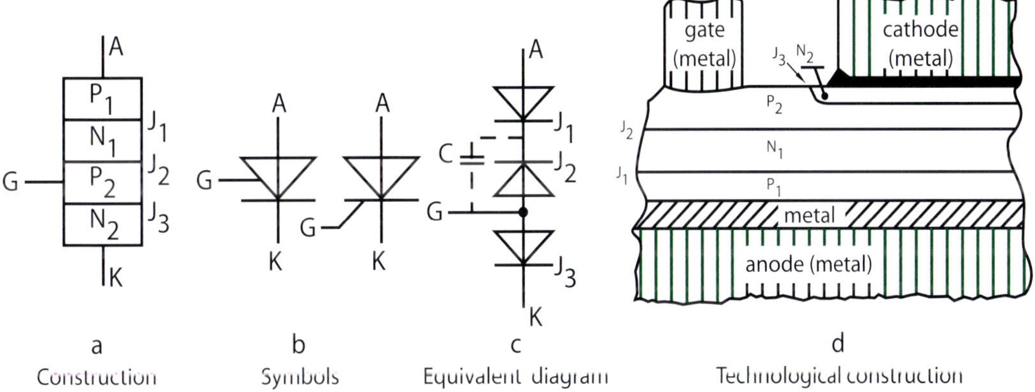

Fig. 4-6: Unidirectional thyristor.

Remarks

1. To switch an SCR on a voltage of 3 to 5 V is needed between gate and cathode with a gate current of 20 mA (for a 10A - SCR) to 500mA (for a 600A - SCR).
   Using a power of approximately 1W we can switch powers of many kVA's!
2. Heavy duty SCR's sometimes have four connections : anode, gate and two cathode-connections. The thinnest cathode connection is connected to the control circuit. By braiding this thin cathode lead with the gate connection any externally generated interference is not only applied to the gate but also inversely in the cathode so that they cancel each other and prevent spurious firing of the SCR.
3. Fig. 4-6c shows how to test an SCR with a DMM (digital multimeter). There is a PN-junction between gate and cathode and by measuring with the DMM in diode mode we can measure the junction voltage of the unconnected SCR. Between anode and cathode a large resistance should be measured in both directions. If an infinite resistance is measured this may indicate a defect or break in the SCR.

2.2 Operation of an SCR

Fig. 4-7a shows the fundamental circuit with

- the power circuit with V, R_b and I_D.
- the control circuit with V_{GK} and I_G.

In fig. 4-7b a two transistor equivalent circuit is shown similar to the Shockley diode. The barrier layer capacitance of J_2 is also shown. We will return to this capacitance later.

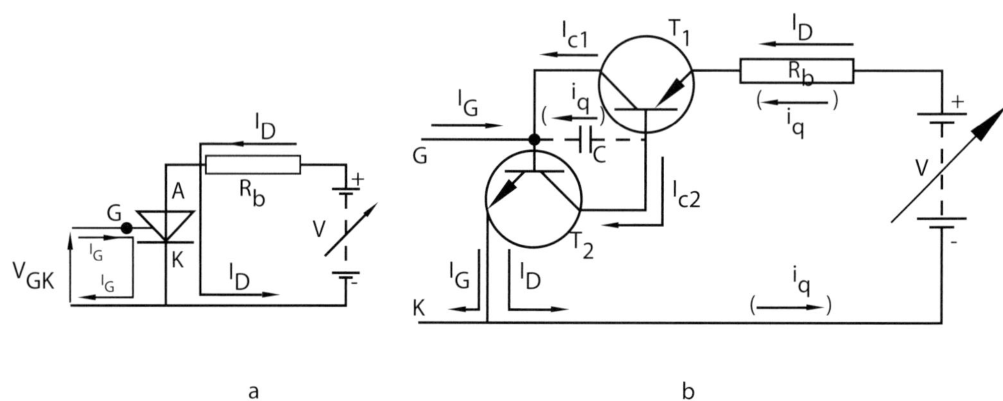

a   b

Fig. 4-7: Configuration and equivalent circuit of an SCR.

Fig. 4-8 shows the I_a - V_a characteristic of an SCR without gate current I_G.
In the I_a - V_a axis only the quadrant 1 and 3 are of interest.
Quadrant 1: has two states: a blocking or conducting SCR, see fig. 4-9.
Quadrant 3: reverse bias state of the SCR, as shown in fig. 4-10

A. Blocking state (Quadrant 1)

No current in the gate ($I_G = 0$) leads to the same situation as the Shockley diode.
As long as $V < V_{B0}$ the SCR is in blocking mode and only a small leakage current flows.
With $V > V_{B0}$ the thyristor conducts (fig. 4-8).
This manner of firing is impractical and undesirable (for explanation see no. 2.3)

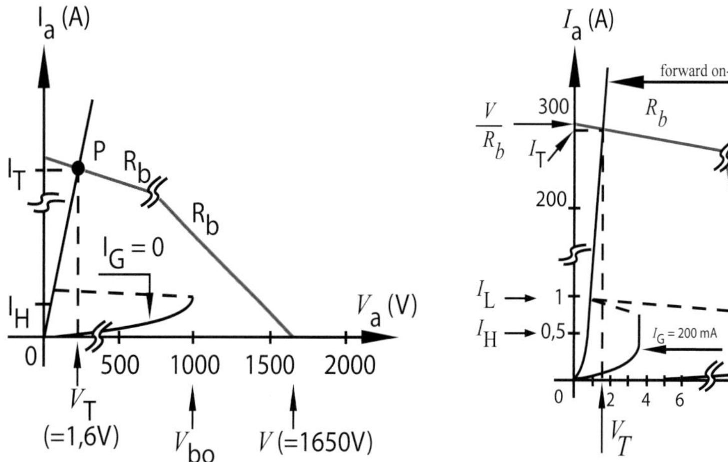

Fig. 4-8 : Blocking mode and undesirable firing of an SCR   Fig. 4-9: Firing an SCR with a gate current (I_G)

B. Conducting mode (Quadrant 1)

The anode current I_a that flows in the blocked thyristor is known to as the leakage current I_D. From fig. 4-7 b we can write a number of expressions (for now the charging current i_q of the barrier layer is equal to zero).

$$I_{C1} = \alpha_1 \cdot I_D + I_{CB01} \tag{4-4}$$

$$I_{C2} = \alpha_2 \cdot (I_D + I_G) + I_{CB02} \tag{4-5}$$

also :
$$I_D = I_{C1} + I_{C2} \tag{4-6}$$

$$I_{CB01} + I_{CB02} = I_{CB0} \tag{4-7}$$

After some rearranging it follows from (4-4), (4-5), (4-6) and (4-7) :

$$I_D = \frac{I_{CB0} + \alpha_2 \cdot I_G}{1 - (\alpha_1 + \alpha_2)} \tag{4-8}$$

Expression (4-8) provides a practical and useful method to make an SCR conduct. Even with a small V_a of a few volts we can increase the leakage current by injecting a gate current I_G. As a result of a cumulative effect the term ($\alpha_1 + \alpha_2$) quickly becomes unity so that the SCR conducts.

As seen in fig. 4-9 the anode current of the SCR is limited to $I_T < \dfrac{V}{R_b}$

With a thyristor current I_T there is a corresponding V_T (see fig. 4-9 and 4-11) .

C. Reverse bias mode (Quadrant 3)

If the polarity of the anode voltage is reversed in fig. 4-7b then the junctions J_1 and J_2 are blocking. This is similar to two reverse biased diodes in series. A small reverse current I_R flows as shown in fig. 4-10.

Remarks

1. *Notation I_D and I_T ?*

   D –index = off- state: I_D , V_{DRM}, ...

   T-index = on- state: I_T , V_T , ...

2. *Latching current (I_L) − Holding current (I_H)*

   As soon as the anode current exceeds the so called latching current the gate current may be removed and the thyristor remains conducting. Above that current the SCR (similar to the Shockley diode) will only switch-off when the current falls below the holding current.

   $I_H =$ holding current (e.g. 20 to 500mA depending on SCR type). I_H has almost always the same value as I_{GT}. For an explanation of I_{GT} see nr. 2.7.1.

   $I_L =$ latching current (in practice: $I_H \leq I_L \leq 2 \cdot I_H$)

2.3 Undesired firing methods

In addition to the practical firing method with **positive gate current** there are three firing methods to avoid.

We deal with them in brief:

- *Unplanned firing due to anode voltage V being too high:*
  With $V > V_{B0}$ the SCR will fire because I_{CB0} increases rapidly just as with the Shockley diode. This method is to be avoided because in practice V_{BO} is much larger than V_{DRM} as is shown in fig. 4-10. In this manner the thyristor may be permanently damaged.
  V_{DRM} = max. repetitive peak off-state voltage.

- *Unplanned firing due to an excessive T_j :*
  With an increasing junction temperature the leakage current increases (dubbles every $10°$ C temperature increase). When the junction temperature is too high (approximately from $T_j = 100°$ C) I_{CB0} rises quickly and the SCR can fire because of the avalanche effect. This can result in damage to the thyristor. The junction temperature therefore needs to be maintained (via cooling) below a certain permitted maximum value.

- *Unplanned firing due by a steep voltage gradient.*
  If an increasing anode voltage (dv/dt) is applied to a thyristor in blocking mode , then the capacitance of the barrier region will draw a current: $i_q = C \cdot dv/dt$. This current i_q flows in fact in the base of T_2 . If dv/dt is sufficiently large then this current i_q can cause the thyristor to fire without the application of an external gate signal. This is obviously not the intention. The thyristor specifications mention the critical dv/dt which may not be exceeded (critical rate-of-rise of off-state voltage) for a specific temperature.

  Example: $(dv/dt)_{crit.} = 200$V/µs at $T_j = 125°$ C

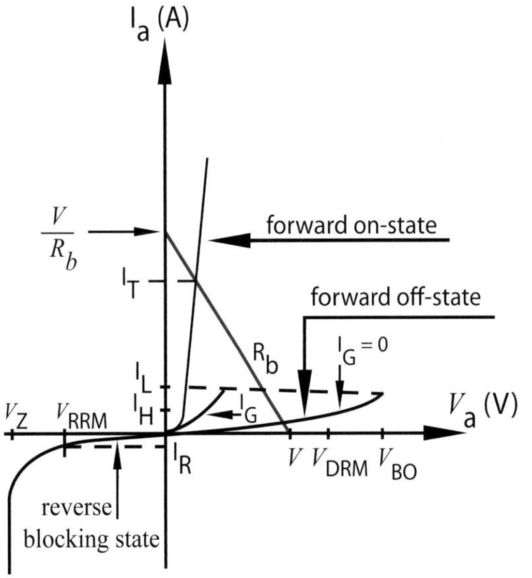

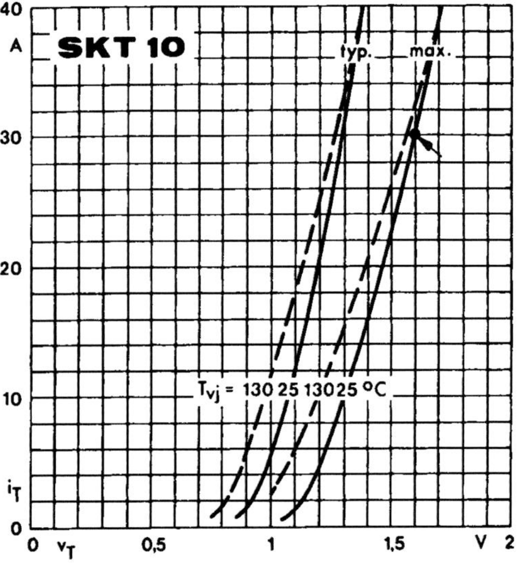

Fig. 4-10: $I_a - V_a$ curve of a unidirectional thyristor   Fig. 4-11: $I_T - V_T$ curve of a conducting SKT10 (Semikron)

Numeric example 4-1:

1. How large is R_b in fig. 4-9? How large is the anode current through the thyristor (fig. 4-9) with $V = 600$V and $R_b = 3\ \Omega$ if $V_T = 1.6$V?
2. What are the values of holding current and latching current of the SKT10? For the SKT600? $T_j = 25°$ C.
3. For a certain thyristor we know it fires with a latching current of 10mA (at $T_j = 50°$ C). The barrier capacitance is 50pF. What is the value of critical voltage gradient?
4. Find for the critical voltage gradient of an SKT10 and SKT600.

Solution:

1) $R_b = \dfrac{V}{I} = \dfrac{300}{300} = 1\ \Omega$; $I_T = \dfrac{300 - 1.6}{3} = 199.5$ A

2) SKT10: $I_H = 80$ mA ; $I_L = 150$ mA

    SKT600: $I_H = 150$ mA ; $I_L = 0.5$ A

3) $i = C \cdot dv/dt$; $dv/dt = 10 \cdot 10^{-3}/\ 50 \cdot 10^{-12} = 200 \cdot 10^6$ V/s $= 200$ V/µs

4) SKT10: $(\,dv/dt\,)_{crit.} = 500$ or 1000 V/µs according to type.

    SKT600: $(\,dv/dt\,)_{crit.} = 500$ or 1000 V/µs according to type

Numeric example 4-2:

Fig. 4-11 shows the on-characteristic of an SKT10.
Determine:

1. Forward voltage drop (V_T) with a nominal anode current (I_T) if $T_j = 25°$ C. Can you explain this T_j - value?
2. At which current strength does the junction temperature exercise little or no effect on the forward voltage drop?

Solution:

1) $I_{T(RMS)} = 1.57 \cdot I_{T(AV)} = 15.7$ A

    $V_T = 1.15$ V (typical) tot 1.4 V (max.).

    This is the forward voltage drop across the conducting junctions J_1 and J_3 (fig. 4-6a) in series.

2) Fig. 4-11 : $I_T = 38$ A !

2.4 Switching off a thyristor

When the anode-current falls below the I_H - value, then as in the case of the Shockley diode the thyristor will switch off. We distinguish two different cases:

1. AC-supply:

When the supply is AC, the SCR will switch off just before the zero cross-over of the anode AC-current (at I_H !). As the commutation occurs naturally via the zero cross-over points of the supply voltage, we speak of natural commutation.

An SCR is a controlled ON-switch:

- switch-on is controlled by the injection of a positive gate current I_G
- switch-off can not be controlled and occurs automatically when I_T falls below I_H

2. DC-supply:

In this case it is necessary to artificially reduce the anode-current to below I_H via a switch-off circuit. This is referred to in the literature as forced commutation. Essentially every commutation is forced. Therefore we refer to artificial commutation.

At the same time we have made an ON-OFF controllable switch from the SCR if we maintain control of the time of switch-off.

Fig. 4-12 shows a configuration of an artificial commutation circuit. During conduction by the SCR C is charged with the polarity indicated. By closing S a reverse voltage is applied across the conducting SCR and the capacitor will send a reverse current through the SCR so that it will switch-off. In practice the switch S is a transistor or a second SCR which is called the auxiliary switch-off thyristor.

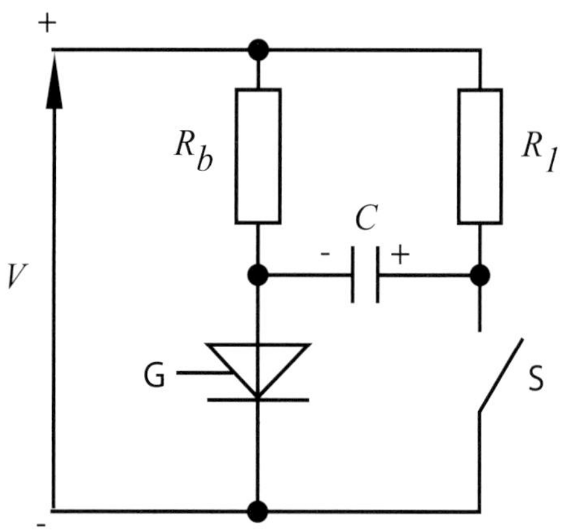

Fig. 4-12: Switch-off circuit for an SCR

2.5 Static characteristics of an SCR

2.5.1 I_a - V_a characteristic

The I_a - V_a characteristic is shown in fig. 4-10. At $I_G = 0$ the curve is similar to a Shockley diode. Once I_G is sufficiently positive the SCR will conduct with an anode-voltage of a few volts. Fig. 4-11 shows the detail of the CONDUCTING state (ON-state) at $T_j = 25°$ C and $130°$ C. The manufacturer indicates the temperature range where in the characteristic lies.

2.5.2 I_G - V_G characteristic

Fig. 4-13 shows the same configuration twice, except upper left we have used the SCR symbol. The control current I_G flows through a PN-junction. The gate characteristic (fig. 4-14) corresponds with an I_a - V_a curve of an Si-diode. Just as in fig. 4-11(I_T - V_T) the curve is influenced by manufacture determined ranges and temperature, therefore the extreme limits ($R_{Gmax.}$ and $R_{Gmin.}$). This should be so viewed that the R_G - line of a random sample of the SCR under consideration definitely falls within these limits. For some unclear reason the manufacturers have placed the voltage V_G along the ordinate axis instead of the current which is normally the case.

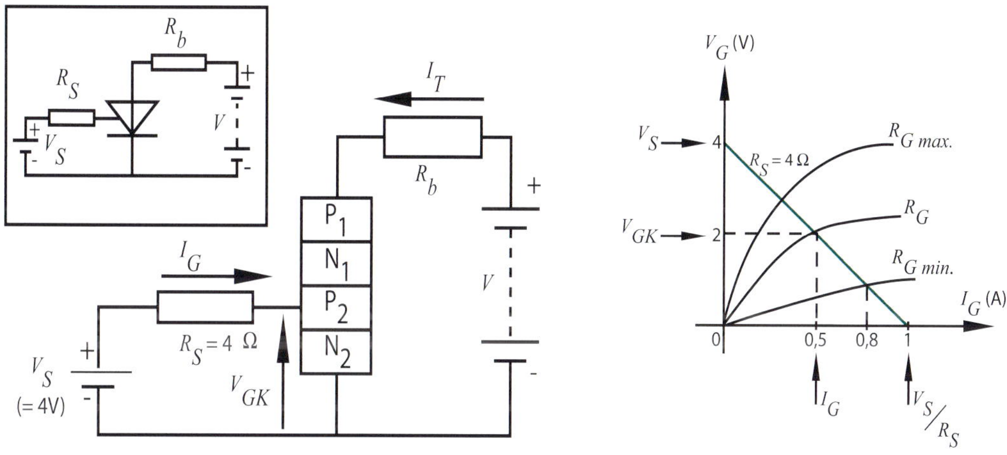

Fig. 4-13: DC control of an SCR

Fig. 4-14: Operating point in the gate characteristic

Consider a case with an R_G -line as shown in fig. 4-14. Let $V_s = 4V$ for example with an $R_s = 4\ \Omega$. The operating point of the gate circuit is determined by the intersection of the R_G -line with the load line ($R_s = 4\ \Omega$). The operating point is then $I_G = 0.5A$ and $V_{GK} = 2V$. If the gate control voltage remains is applied constantly then the gate needs to dissipate $V_{GK} \cdot I_G = 1W$.

For more details see p. 4.16.

Remark

In order for an SCR (I_{GT} , V_{GT}) to be correctly triggered the control circuit needs to comply with the following specifications (I_G , V_S): $I_G \geq 5\ x\ I_{GTmin.}$ and $V_S \geq 5\ x\ V_{GTmin.}$

Numeric example 4-3:

What power does the supply in the gate circuit need to provide in fig. 4-13?

Determine the power dissipated in the resistor R_S.

The V_G - I_G characteristic of the SCR is shown in fig. 4-14.

Solution:

In fig. 4-14 we see at $R_{Gmin.}$ an $I_{Gmax.} = 0.8\ A$

$V_S\ x\ I_{Gmax.} = 4\ x\ 0.8 = 3.2W$; $I^2_{Gmax.} \cdot R_S = 0.8^2\ x\ 4 = 2.56W$

2.6 Dynamic properties of an SCR

At high switching speeds the dynamic properties of an SCR are important. When a gate pulse is applied the junction J_2 does not immediately conduct across the entire junction. On the other hand when a thyristor is switched off quickly not all charge carriers can disappear immediately. An SCR has a turn-on time t_{on} and a turn-off time t_{off}.

Fig. 4-15 for example shows some details with a time axis. Voltage-, current- and time-values between brackets are valid for a 300A-2200V thyristor.

2.6.1 Turn-on and turn-off time

1. Turn-on time

The turn-on time t_{on} is the sum of the delay time and the rise time:

$$t_{on} = t_{gd} + t_{gr} \, (= t_{gt})$$

$t_{gd} = (t_{d}) = $ delay time = time that elapses from the instant the the gate pulse reaches 10% of its final value and the instant at which the thyristor voltage is still at 90 % of its initial value.

$t_{gr} = $ rise time = the time difference between the instant that $v_a = 0.9 \cdot V_D$ and $v_a = 0.1 \cdot V_D$

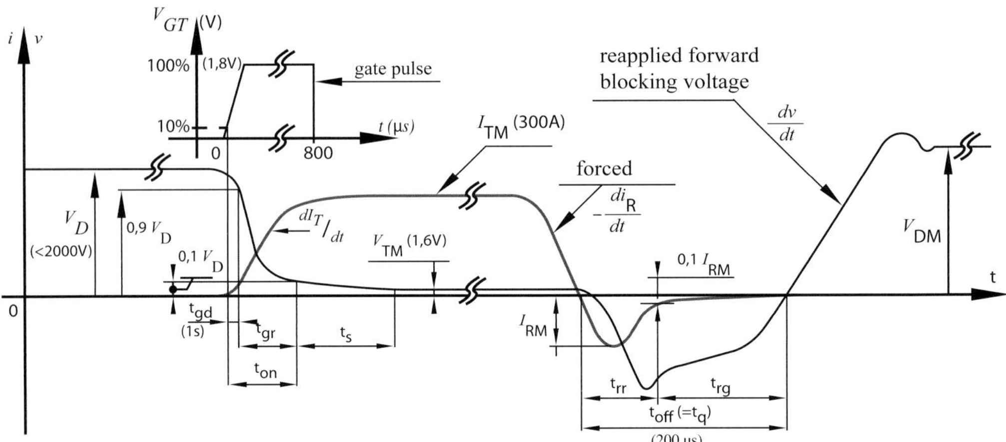

Fig. 4-15: Dynamic characteristics of an SCR

After the turn-on time t_{on} the thyristor current flows through a narrow channel in the vicinity of the gate contact, in other words the current is just connected. The dissipation speed of conduction in junction J_2 is approximately 0.1mm/µs so that, depending on the construction of the SCR, it can take t_s (delay dissipation time) before the junction J_2 is completely conducting and the voltage V_T is across the thyristor.

The time t_{gd} depends upon V_D, T_j and I_G. Increasing these factors reduces t_d. The time t_{gr} is dependent upon the anode-current during conduction and the rate of rise dI_T / dt. With suitable gate pulses t_{on} can have a value of 1 to 5 µs. The turn-on delay time can cause a problem switching thyristors in series and parallel. When switched in series the slowest thyristor will have to bear the full blocking voltage for a short time and with parallel switching the fastest thyristor will have to conduct the full current for a short time.

2. Turn-off time

When the anode current gradually decreases then the charge carriers that are no longer required for current transport have the time to recombine. The SCR turns off when the current falls below the holding current. This is the so called natural turn-off as occurs at normal supply frequencies (50, 60, 400 Hz). With choppers and inverters a DC-supply is used for the anode an so artificial commutation is needed. It is this turn-off behaviour that is shown in fig. 4-15. Via a test setup the voltage and current curves shown here are forced on the SCR. If via the test circuit an inverse voltage is applied then the current decreases. At the instant that I_T goes through zero there are still a large number of charge carriers present in the SCR. These temporarily result in a large inverse current under the influence of the inverse voltage. The time t_{rr} (reverse recovery time which is due to the anode current) corresponds to the restoration of the junctions J_1 and J_3 of the SCR. After this time the thyristor can withstand the inverse anode voltage. If after a time t_{rg} J_2 is restored, the thyristor can withstand the blocking voltage without turning on automatically. The time t_{rg} is called the gate recovery time.

The turn-off time = t_{off} = t_{gq} = t_q = t_{rr} + t_{rg} = time difference between the anode current I_T becoming zero and being able to re-apply the blocking voltage V_D. Depending on the type of SCR turn-off times can be between 5 and several hundred µs.

2.6.2 Critical current rate of rise (*di/dt*)

1. Hot spots

We have just noted that the dissipation speed of conduction of J_2 is approximately 0.1mm per µs. Since the voltage across the SCR at the start of turn-on remains high while the current only flows through a limited part of J_2 , the Si-tablet has to dissipate a large power and hot spots can result which permanently damage the tablet. To prevent this the manufacturer specifies a maximum permissible rate of rise current. For example: (*di/dt*) = 40A/µs.

If the anode load is resistive it may be necessary to place an inductor in series with the resistance to limit the rate of rise of current at turn-on. Instead of an inductor ferrite cores can be placed over current conductors.

2. Induced voltages

Quickly reducing the inverse current (I_{RM} to $0.1.I_{RM}$: see fig. 4-15) can cause high voltages if self inductance is present in the circuit ($e = L$. *di/dt*). In this case the voltage spike is in addition to the inverse applied voltage which can result in V_{RM} being exceeded. To prevent this an *RC*-network is placed in parallel with the SCR. This is called a snubber network.

2.6.3 Critical voltage rate of rise (*dv/dt*)

When the *dv/dt* across the terminals of an SCR is to high it can result in spurious firing. This is caused by the charge current of the barrier capacitance of J_2 . If a "shorted-emitter" construction is applied the *dv/dt* properties of the thyristor are dramatically improved. Where a "normal" SCR has a maximum rate of rise of for example 50V/µs a "shorted-emitter" version can take 500V/µs.

2.6.4 Power dissipation

Fig. 4-16 shows the temperature dependence of the breakdown voltage V_{BO} . With rising junction temperature the breakdown voltage decreases. This results in the SCR loosing its blocking properties.

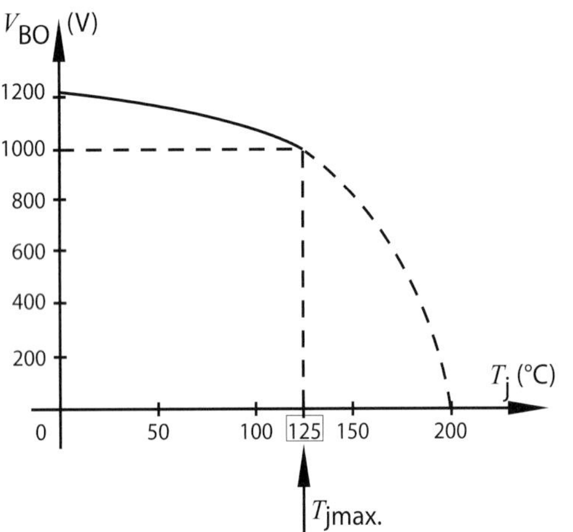

We need to ensure that the maximum junction temperature of the SCR is not exceeded. This T_{Jmax} is normally 125° C. In practice it will be necessary to cool the thyristor. The temperature rise of the thyristor is caused by:

- control losses
- cut-off losses
- switching losses
- leakage losses

Fig. 4-16: Breakdown voltage related to junction temperature of a thyristor

1. Control losses
The power that is dissipated in the gate is negligible at 50Hz. When higher frequencies occur (kHz) such as in choppers and inverters it may be necessary to consider these losses.

2. Cut-off losses
In non-conducting mode the cut-off losses are negligible since the currents I_D and I_R are very small.

3. Switching losses
At every junction between conducting and non-conducting or vice-versa there are switching losses. At 50Hz these losses are negligible but at higher frequencies they can play a role.

4. Forward Conduction losses

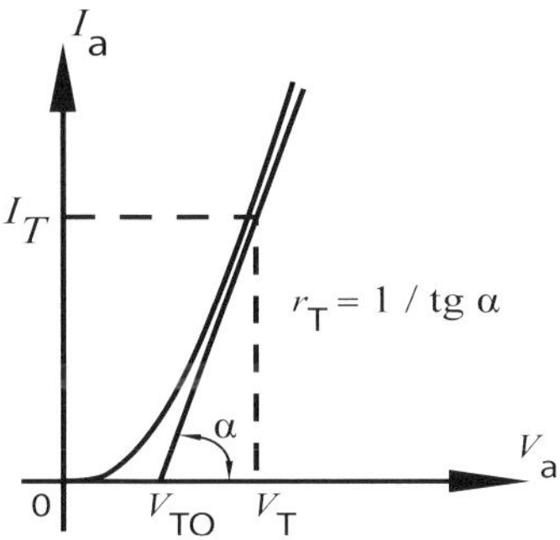

Fig. 4-17: Conduction characteristic of an SCR

Similar to the power diode (nr. 2-4 on p. 2.7) we find the dissipated power in the thyristor as follows:

$$P = V_{T0} \cdot I_{T(AV)} + r_T \cdot a^2 \cdot I^2_{T(AV)}$$

If V_{T0} and r_T are not provided we can determine their values using graphical means. (fig. 4-17). The manufacturer often provides curves that allow the dissipated power to be calculated, see fig. 4-19.

Numeric example 4-5:

Consider a rectangular shaped current flow through an SKT600.

The thyristor is part of a three phase bridge. Conduction angle per SCR is 120°.

Total DC bridge current = 750A. Determine the power dissipation per SCR.

Solution:

In a three phase bridge: $I_{T(AV)} = I_g / 3 = 750/3 = 250A$. Fig. 4-19: $P_{T(AV)} = 320W$

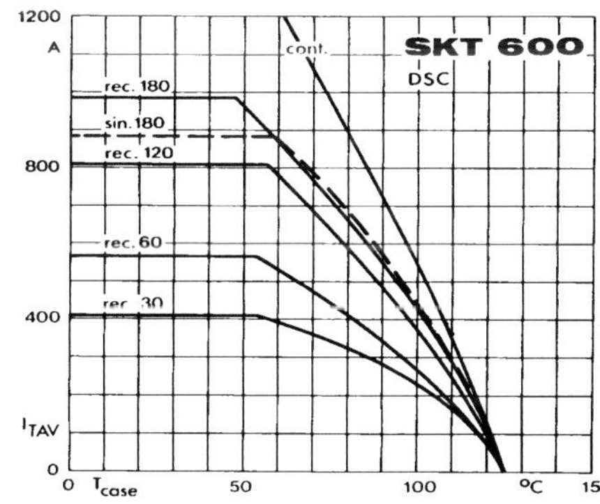

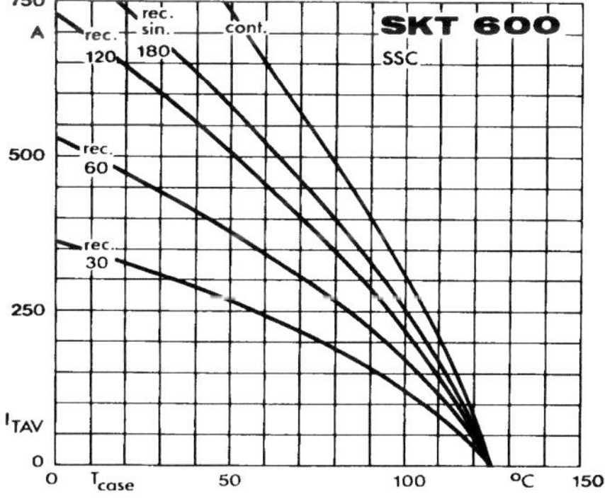

Fig. 4-18: Forward current as a function of case temperature for an SKT600 (Semikron)

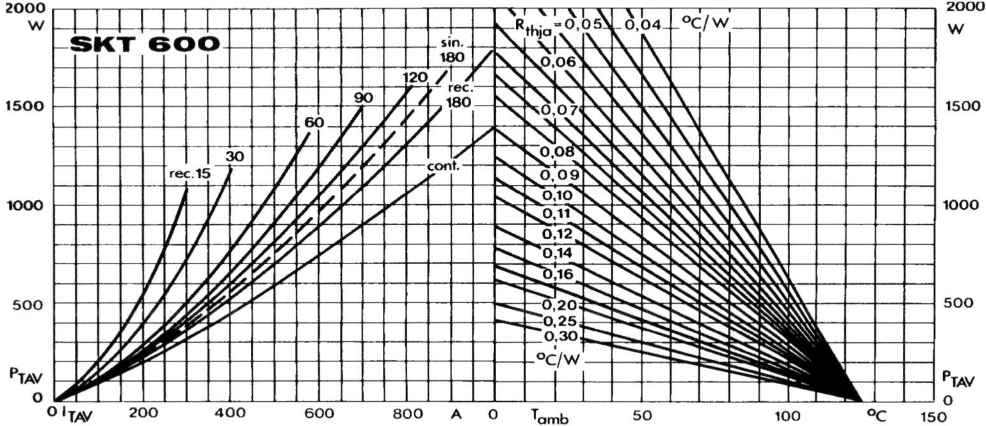

Fig.4-19: Power dissipation as function of forward current and ambient temperature for an SKT600 (Semikron)

Numerical example 4-4:

Find the turn-on time, turn-off time, critical rate of rise of current, critical rate of rise of voltage for an SKT10 and for an SKT600 from Semikron.

Solution:

SKT10: $t_{on} = t_{gd} + t_{gr} = 3\mu s$; $t_{off} = t_q = 80\mu s$; $(di/dt)_{crit.} = 50A/\mu s$;
$(dv/dt)_{crit.} = 500$ to $1000V/\mu s$ according to type

SKT600: $t_{on} = 3\mu s$; $t_{off} = 100$ to $200\mu s$; $(di/dt)_{crit.} = 125A/\mu s$;
$(dv/dt)_{crit.} = 500$ to $1000V/\mu s$ according to type

2.6.5 Determining the variables V_S and R_S of the control circuit

In view of the temperature and the distribution of the gate characteristics of a single SCR-type the manufacturer provides the surface area with all possible triggering points (this is the surface area of uncertain triggering). At $T_J = 25°$ C (fig. 4-20) with a set point of $V_G > 3V$ or $I_G > 0.15A$ will cause the SCR to trigger. In fig. 4-20 we note once again the limits $R_{Gmax.}$ and $R_{Gmin.}$. The characteristic of fig. 4-20 is visible as the green rectangle in the gate characteristic of fig. 4-22 and 4-23. Fig. 4-21 shows that with a rising gate current the turn-on time of a thyristor decreases.
If the anode load of the SCR causes a large di/dt it will be necessary to use large gate power in order to keep t_{on} small (avoids hot spots!)

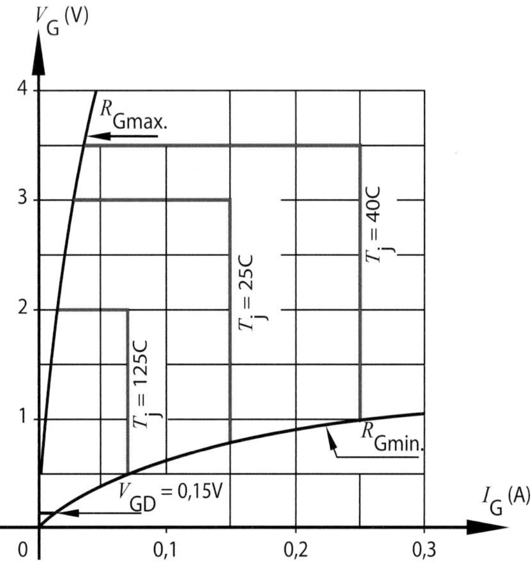

Fig. 4-20: Collection of all possible triggering points of an SCR

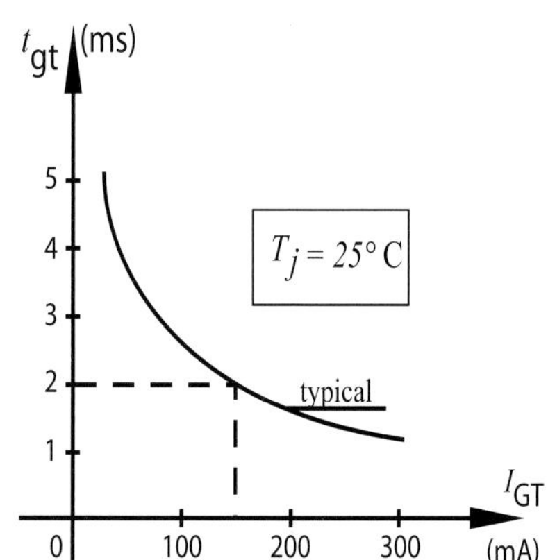

Fig. 4-21: Influence of the gate current on the turn-on time of an SCR

In addition to fig. 4-20 the manufacturer provides the curves shown in fig. 4-22.

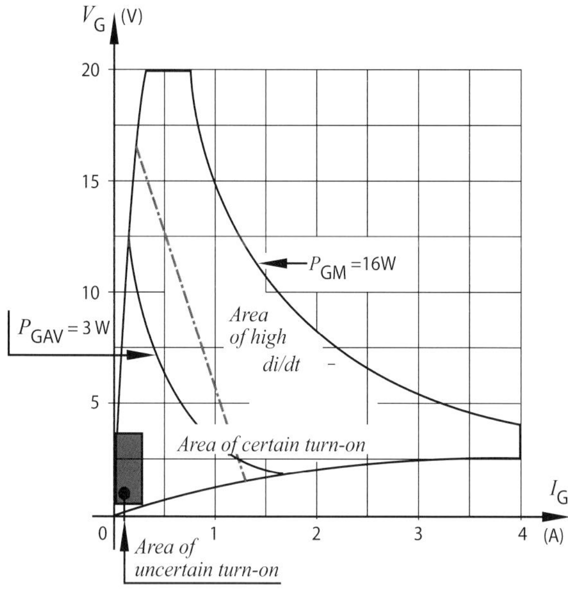

Fig. 4-22: Gate characteristic of an SCR

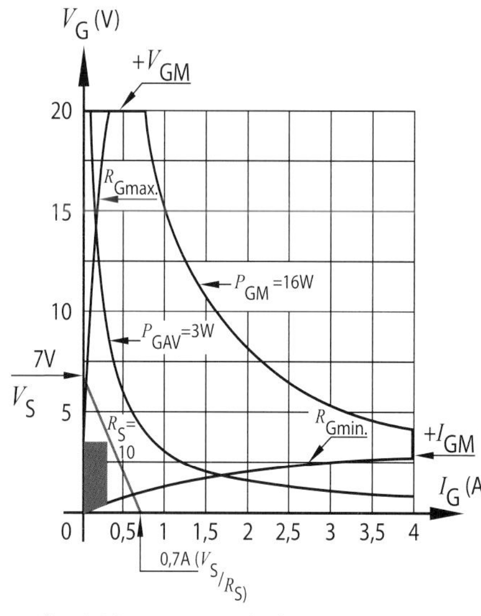

Fig. 4-23: DC control of an SCR

1. DC voltage control

The settings line in the gate characteristic (fig. 4-23) should be above the green area to achieve definite triggering. On the other hand the settings line should remain below the maximum allowable average gate power $P_{G(AV)}$. With V_S = 7V and R_S = 10Ω the SCR will trigger for all units of this model.

Numeric example 4-6:

Suppose the SCR of fig. 4-23 has an R_G - curve that coincides with the $R_{Gmax.}$ - curve.
If V_S = 15V and R_S = 7.5Ω , determine:

- the internal power dissipation (P_G) in the gate ,
- the dissipated power in the R_S - resistor (P_R),
- the voltage between cathode and gate (V_{GK}),
- the current in the gate (I_G).
- What happens if R_G of the SCR lies between
  $R_{Gmax.}$ and $R_{Gmin.}$?

Solution:

P_G = 13.8 x 0.17 = 2.35W

P_R = (15 − 13.8) . 0.17 = 0.2W

with $R_G = R_{G1}$ then $P_G > P_{G(AV)}$ then
(with V_S = 15V and R_S = 5.7Ω) we can only
apply impulse control (see next section)

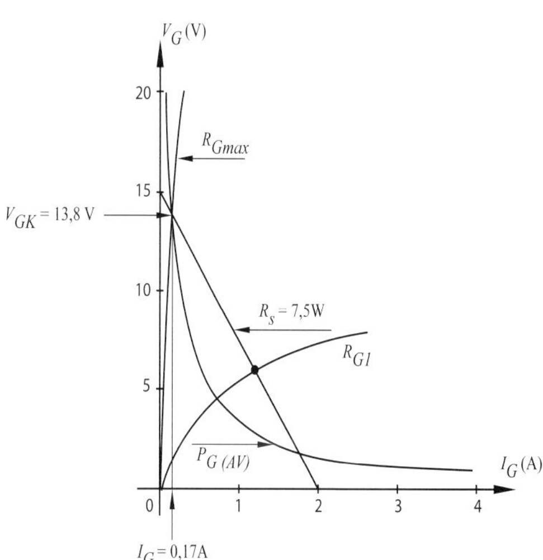

2. Pulse control

If an accurate trigger point is needed then pulse control may be used. The settings just discussed can be used if the pulse duration is sufficient to maintain I_T above the latching current I_L. The turn-on time of the SCR and the anode load play a role. We can do little about the anode load, we can attempt to make t_{on} as small as possible by using an as large as possible gate current. (fig. 4-21).

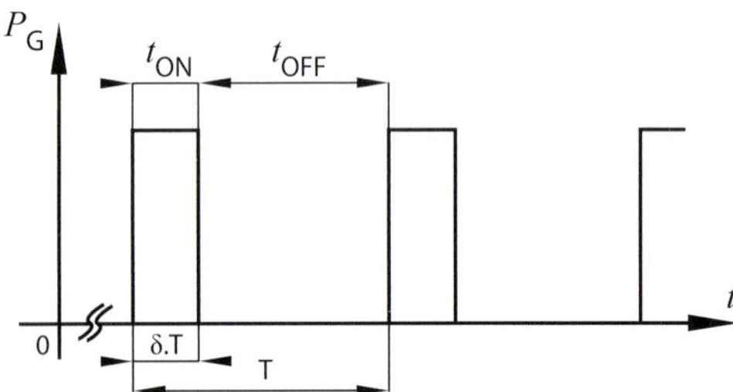

Fig. 4-24 : Pulse control of an SCR

With DC voltage control there is a limit imposed by the $P_{G(AV)}$ - hyperbola. With pulse control it is possible to reach P_{GM}, under the pre-condition that the average gate power is not exceeded. If the period time of the pulses is T and the duration of the pulse is t_{ON} then the duty cycle is (fig. 4-24):

$$\delta = \frac{t_{ON}}{t_{ON} + t_{OFF}} = \frac{t_{ON}}{T}$$

so that :
$$t_{ON} = \delta . T$$

With DC voltage control the allowable energy in the gate is: $P_{G(AV)} . T$. In this we call T the integration time (usually 10 ms). A certain power P_G ($\leq P_{GM}$!) can be permitted for a time $t_{ON} = \delta . T$ lead+ing to: $P_G . \delta . T = P_{G(AV)} . T$ or:

$$\delta = \frac{P_{G(AV)}}{P_G}$$

With $P_G = P_{GM}$ then δ_{min} is the same as: $\delta_{min} = \dfrac{P_{G(AV)}}{P_{GM}} = \dfrac{3}{16} = 0.1875$
With $T = 10$ ms, the power level can be 16W with a gate pulse of 1.875 ms.

Numeric example 4-7:

Suppose an SCR has $R_G = 10\ \Omega$ and $P_{G(AV)} = 3W$. Control is with pulses with an amplitude of 18V (V_S -source). The series resistance R_S is 9 Ω. Determine the gate power, the pulse duration (given $T = 10$ms), the amplitude of the pulse current which the controller has to supply.

Solution:

$$P_G = 9.5 \times 0.95 = 9.025 \text{ W}$$
$$\delta = \frac{P_{G(AV)}}{P_G} = \frac{3}{9.025} = 0.332 = 33.2\%$$
$$t_{ON} = \delta . T = 0.332 \times 10 = 3.32 \text{ ms} \; ; \; I = 0.95 \text{ A}$$

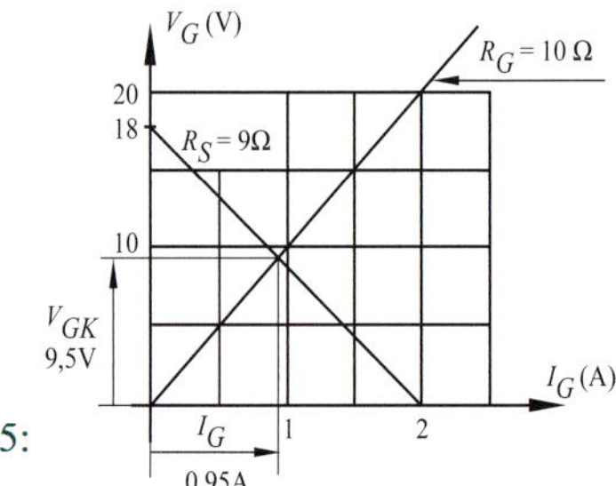

Fig. 4-25:

Remark

In fig. 4-22 we see trigger surface area for high di/dt operation. Indeed to cause an SCR to quickly fully conduct (avoiding hot spots!) we use a large gate power to cause t_{on} to be as small as possible (fig. 4-21). If you carefully observe fig. 4-22 then you will notice that the recommended surface area $P_{G(AV)} = 3W$ lies above the hyperbola. We are therefore compelled to use pulse control in cases with large di/dt such as for example with an anode load with little inductance.

2. 7 Thyristor current and voltage

2.7.1 Gate specifications notation

The manufacturer provides a number of gate specifications:

P_{GM} = maximum gate power

$P_{G(AV)}$ = maximum average gate power

$+I_{GM}$ = max. allowable positive gate current

$+V_{GM}$ = max. allowable positive gate voltage

$-V_{GM}$ = max. allowable negative gate voltage

V_{GD} = min. required gate voltage ($T_J = 125°$ C) to trigger the SCR

I_{GT} = required trigger current to turn on all examples of a particular type

V_{GT} = required trigger voltage to turn on all examples of a particular type.

2.7.2 Anode current notation

In the datasheet of an SCR multiple anode currents may be mentioned:

$I_{T(AV)}$ = maximum average current. This corresponds to the surface area of a half sinus (= conducting for $180°$!)

$I_{T(RMS)}$ = maximum effective current. This corresponds with the DC current value $I_T (= I_{DC})$ found on the $I_T - V_T$ curve.

I_{TRM} = maximum peak repetitive current. These peaks may be caused by transients, commutation etc.

I_{TSM} = maximum surge on-state current (due to short-circuit or inrush currents etc...)

I_H = holding current

I_L = latching current

I_R = reverse current

I_D = leakage current (in blocking mode).

In addition the $I^2.t$ value is stated as was the case with power diodes. This $I^2.t$ value is necessary for dimensioning the protection of the semiconductor.

Remark

We refer to an SCR with its I_{AV} -value or with its I_{RMS} -value , or with both.

We have already used the SKT600 (Semikron) as an example with a case temperature of $T_C = 57°$ C : $I_{T(AV)M} = 890A$ and $I_{T(RMS)M} = 1400A$ ($= 1.57$ x $890!$).

2.7.3 Anode voltage notation

V_{RRM} = reverse repetitive maximum voltage

V_{DRM} = off-state repetitive maximum voltage

V_{RSM} = reverse surge maximum voltage.

V_{TM} = max. thyristor (forward) voltage drop.

2.7.4 Subscripts with variables

The subscripts used in datasheets were highlighted on p. 2.9.

2.8 Voltage safety factor *g*

The maximum reverse voltage of a thyristor must be larger than every possible voltage peak (as a result of faults, transients etc...) in the circuit. As a result of industrial measurements a voltage safety factor of $g = 2$ to 2.3 is used.

$$g = \frac{\text{maximum reverse repetitive voltage SCR } (V_{RRM})}{\sqrt{2} \text{ x RMS supply voltage}} \qquad (4\text{-}9)$$

Example: In a 230 V circuit the reverse voltage of an SCR should be:

$V_{RRM} \geq 2.3 \times \sqrt{2} \times 230 = 748\text{V}$. With $g = 2$ we find $V_{RRM} \geq 650\text{V}$.

2.9 Temperature influence on a thyristor

The junction temperature effects the majority of thyristor specifications. With rising temperature:

1. V_{BO} decreases. The SCR needs to be cooled so that $T_J < T_{Jmax}$!
2. I_H and I_L decrease.
3. I_{GT} and V_{GT} will decrease.
4. The forward voltage drop V_T decreases across the conducting thyristor(fig. 4-11).
5. The allowable anode current decreases. This is also a reason to cool the SCR. We want the case temperature T_C (and consequently T_J) as low as possible.

2.10 Control methods of an SCR

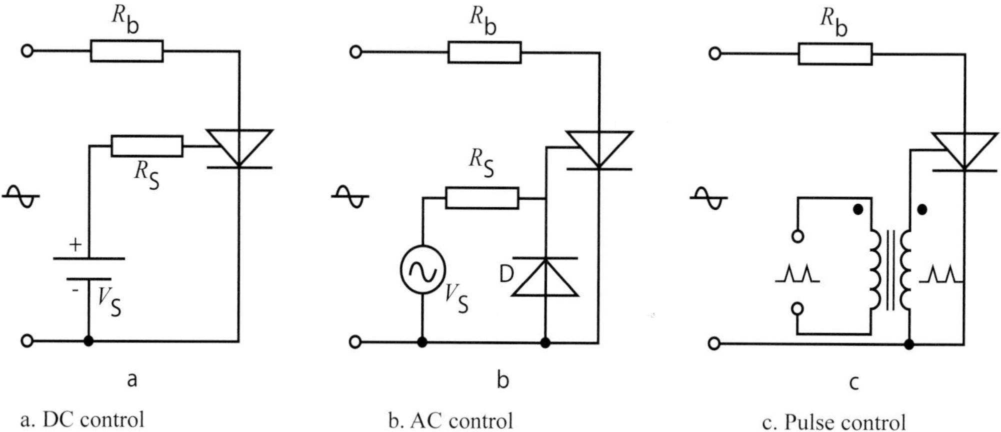

a. DC control   b. AC control   c. Pulse control

Fig. 4-26 : Three methods of SCR control

With control of a thyristor we mean that this semiconductor switch conducts at specific instants of time.

Fig. 4-26 shows the three principle control methods:

1. DC control.

We encountered this method earlier (fig. 4-26a) when discussing the SCR. A positive V_S causes gate current to flow. Once the supply voltage positively polarises the anode, the SCR conducts.

2. AC control.

During the positive half cycle of the control source the SCR can be made to conduct. Diode D is used in case the amplitude of V_S is greater than the allowable gate voltage of the SCR.

AC control may especially occur with triac control.

3. Pulse control.

If we ignore the (DC) control once the anode current exceeds the latching current then we have in fact a pulse control (fig. 4-26c). This control method is the most popular because amongst other reasons an extremely accurate firing time can be achieved. The pulse transformer in fig. 4-26c provides a galvanic separation between control circuit and power circuit.

In addition to the turn-on time of the SCR the pulse length is also determined by the anode load. The anode current needs to rise above the latching current I_L so that an inductive load will require a broader pulse than a resistive anode load (fig. 4-27b and - c).

In most cases a pulse train is used (fig. 4-27d) which means that a smaller pulse transformer can be used.

In a fully controlled three phase thyristor bridge a so called double pulse is required to cause the bridge to conduct. As shown in fig. 4-27e the second pulse is displaced 60° in relation to the first pulse.

To raise the *di/dt* of the thyristor a so-called "church tower" pulse is used. As a result of this large pulse the SCR is turned on quickly and after a few microseconds the gate pulse can return to its normal level. This is shown in fig. 4-27f.

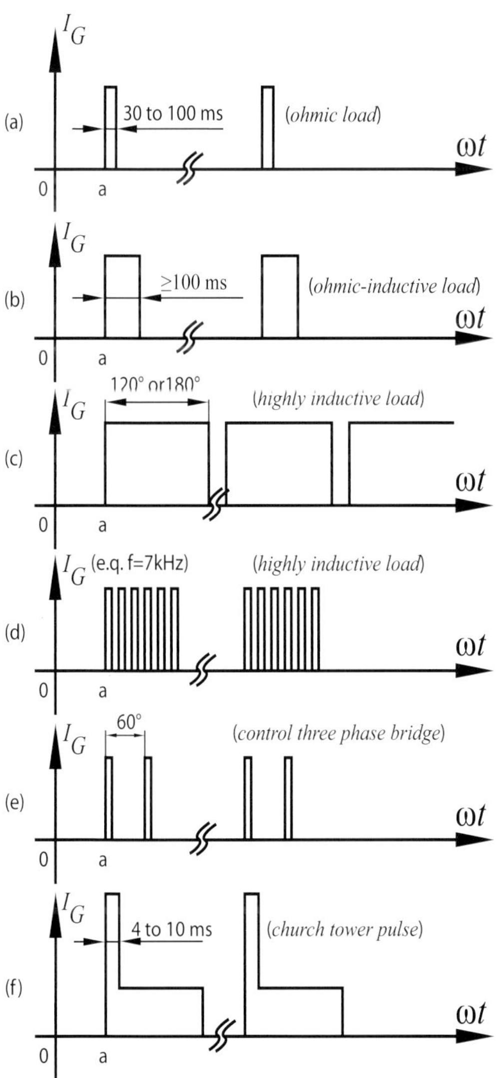

Fig. 4-27: Different types of thyristor trigger pulses

Numeric example 4-8:

a. Find the leakage current of an SKT600 (Semikron) if we:

   1) in off mode apply a voltage V_{DRM}

   2) with a reverse biased SCR we apply a voltage V_{RRM}

b) Is there any similarity between I_{GT} and I_H for a single thyristor?

   Answer this for the SKT10 and for the SKT600.

Solution:

a)   1) I_{DD} = 80mA

   2) I_{RD} = 80mA

b) SKT10: I_{GT} = 100mA ; I_H = 80mA to 150mA

   SKT600: I_{GT} = 200mA; I_H = 150mA to 500mA

Conclusion: the gate trigger current and the anode holding current are of the same order of magnitude.

Photo ABB (KWx): Waffle diffusion during the manufacture of semiconductors.

2.11 LTT (Light Triggered Thyristor)

The LTT is triggered by applying light to a specially constructed gate. Via a light channel an infrared pulse (e.g. 40mW during 10µs) to the gate triggers the LTT.

LTT's are used in high voltage applications where thyristors have to be connected in series.

An example is HVDC (High Voltage Direct Current) whereby the transport of electrical energy by means of high voltage DC. In contrast with an LTT a "normal" thyristor is sometimes referred to as an ETT (Electronic Triggered Thyristor).

2.12 Details of Thyristors

2.12.1 A 10A-400V(up to 1200V) SCR from Semikron: SKT10

Thyristors

SKT 10
SKT 16
SKT 24

V_{RSM}	V_{RRM} V_{DRM}	$(dv/dt)_{cr}$	I_{TRMS} (maximum values for continuous operation)		
			30 A	40 A	50 A
			I_{TAV} (sin. 180; T_{case} = . . . °C)		
V	V	V/µs	19 A (95 °C)	25 A (74 °C)	32 A (72 °C)
500	400	500	–	SKT 16/04 D	SKT 24/04 D
700	600	500	SKT 10/06 D	SKT 16/06 D*	–
900	800	500	SKT 10/08 D	SKT 16/08 D	SKT 24/08 D
1300	1200	1000	SKT 10/12 E	SKT 10/12 E*	SKT 24/12 E*
1500	1400	1000	–	SKT 16/14 E	SKT 24/14 E
1700	1600	1000	–	SKT 16/16 E	SKT 24/16 E*
1900	1800	1000	–	SKT 16/18 E♦	SKT 24/18 E♦

Symbol	Conditions	SKT 10	SKT 16	SKT 24	Units
I_{TAV}	sin. 180; (T_{case} = . . .)	10 (106)	16 (103)	24 (94)	A °C
I_{TSM}	T_{vj} = 25 °C; 10 ms	250	370	450	A
	T_{vj} = 130 °C; 10 ms	210	330	380	A
i^2t	T_{vj} = 25 °C; 8,35 ... 10 ms	310	680	1000	A^2s
	T_{vj} = 130 °C; 8,35 ... 10 ms	220	550	720	A^2s
t_{gd}	T_{vj} = 25 °C I_G = 1 A di_G/dt = 1 A/µs		typ. 1		µs
t_{gr}	V_D = 0,67 · V_{DRM}		typ. 2		µs
$(di/dt)_{cr}$	f = 50 ... 60 Hz		50		A/µs
I_H	T_{vj} = 25 °C; typ./max.		80 / 150		mA
I_L	T_{vj} = 25 °C; typ./max.		150 / 300		mA
t_q	T_{vj} = 130 °C; typ.		80		µs
V_T	T_{vj} = 25 °C; (I_T = . . .); max.	1,6 (30)	2,4 (75)	1,9 (75)	V A
$V_{T(TO)}$	T_{vj} = 130 °C	1,0	1,0	1,0	V
r_T	T_{vj} = 130 °C	18	20	10	mΩ
I_{DD}; I_{RD}	T_{vj} = 130 °C; V_{RD} = V_{RRM} V_{DD} = V_{DRM}	4	8	8	mA
V_{GT}	T_{vj} = 25 °C		3		V
I_{GT}	T_{vj} = 25 °C		100		mA
V_{GD}	T_{vj} = 130 °C		0,25		V
I_{GD}	T_{vj} = 130 °C		3		mA
R_{thjc}	cont.	1,2	0,8		°C/W
	sin. 180 / rec. 120	1,3 / 1,35	0,9 / 0,95		°C/W
R_{thch}		1,0	0,5		°C/W
T_{vj}			− 40 ... + 130		°C
T_{stg}			− 40 ... + 150		°C
M	SI units	2,0	2,5		Nm
	US units	18	22		lb. in.
a		5 · 9,81	5 · 9,81		m/s²
w		7	12		g
Case		B 1	B 2		

Features
- Hermetic metal cases with glass insulators
- Threaded studs ISO M5 and M6 or UNF 1/4-28
- International standard cases

Typical Applications
- DC motor control (e. g. for machine tools)
- Controlled rectifiers (e. g. for battery charging)
- AC controllers (e. g. for temperature control)

* Available with UNF thread 1/4-28 UNF2A, e.g. SKT 16/06 D UNF

♦ Available in limited quantities

2.12.2 A 600A-400V(up to 1200V) SCR from Semikron: SKT600

V_{RSM}	V_{RRM} V_{DRM}	$(dv/dt)_{cr}$	I_{TRMS} (maximum values for continuous operation)	
			1400 A	1600 A
			I_{TAV} (sin. 180; T_{case} = . . .; DSC)	
V	V	V/µs	890 A (57 °C)	1020 A (56 °C)
500	400	500	**SKT 600/04 D**	**SKT 760/04 D**
900	800	500	**SKT 600/08 D**	**SKT 760/08 D**
1300	1200	1000	**SKT 600/12 E**	**SKT 760/12 E**
1500	1400	1000	**SKT 600/14 E**	**SKT 760/14 E**
1700	1600	1000	**SKT 600/16 E**	**SKT 760/16 E**
1900	1800	1000	**SKT 600/18 E**	**SKT 760/18 E**

Thyristors

SKT 600
SKT 760

Symbol	Conditions	SKT 600	SKT 760	Units
I_{TAV}	sin. 180; (T_{case} = . . .); DSC	600 (85)	760 (80)	A °C
I_{TSM}	T_{vj} = 25 °C; 10 ms	11 500	15 000	A
	T_{vj} = 125 °C; 10 ms	10 000	13 000	A
i^2t	T_{vj} = 25 °C; 8,3 ... 10 ms	660	1 125	kA²s
	T_{vj} = 125 °C; 8,3 ... 10 ms	500	845	kA²s
t_{gd}	T_{vj} = 25 °C I_G = 1 A di_G/dt = 1 A/µs	typ. 1		µs
t_{gr}	$V_D = 0,67 \cdot V_{DRM}$	typ. 2		µs
$(di/dt)_{cr}$	f = 50 ... 60 Hz	125		A/µs
I_H	T_{vj} = 25 °C; typ./max.	150 / 500		mA
I_L	T_{vj} = 25 °C; typ./max.	0,5 / 2		A
t_q	T_{vj} = 125 °C; typ.	100 ... 200		µs
V_T	T_{vj} = 25 °C; I_T = 2400 A; max.	2,0	1,65	V
$V_{T(TO)}$	T_{vj} = 125 °C	1,0	0,92	V
r_T	T_{vj} = 125 °C	0,4	0,3	mΩ
I_{DD}; I_{RD}	T_{vj} = 125 °C; $V_{RD} = V_{RRM}$ $V_{DD} = V_{DRM}$	80		mA
V_{GT}	T_{vj} = 25 °C	3		V
I_{GT}	T_{vj} = 25 °C	200		mA
V_{GD}	T_{vj} = 125 °C	0,25		V
I_{GD}	T_{vj} = 125 °C	10		mA
R_{thjc}	cont. DSC	0,038		°C/W
	sin. 180; DSC/SSC	0,040 / 0,082		°C/W
	rec. 120; DSC/SSC	0,045 / 0,093		°C/W
R_{thch}	DSC/SSC	0,007 / 0,014		°C/W
T_{vj}		− 40 ... + 125		°C
T_{stg}		− 40 ... + 130		°C
F	SI units	10 ... 13		kN
	US units	2200 ... 2850		lbs.
w		240		g
Case		B 10		

Features

- Hermetic metal cases with ceramic insulators
- Capsule packages for double sided cooling
- Shallow design with single sided cooling
- International standard cases
- Off-state and reverse voltages up to 1800 V
- Amplifying gate

Typical Applications

- DC motor control (e. g. for machine tools)
- Controlled rectifiers (e. g. for battery charging)
- AC controllers (e. g. for temperature control)

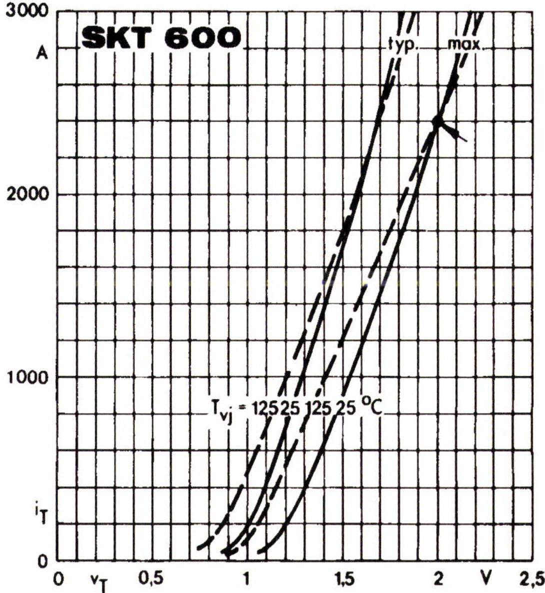

Fig. 4-28: Conduction characteristic of an SKT600 (Semikron)

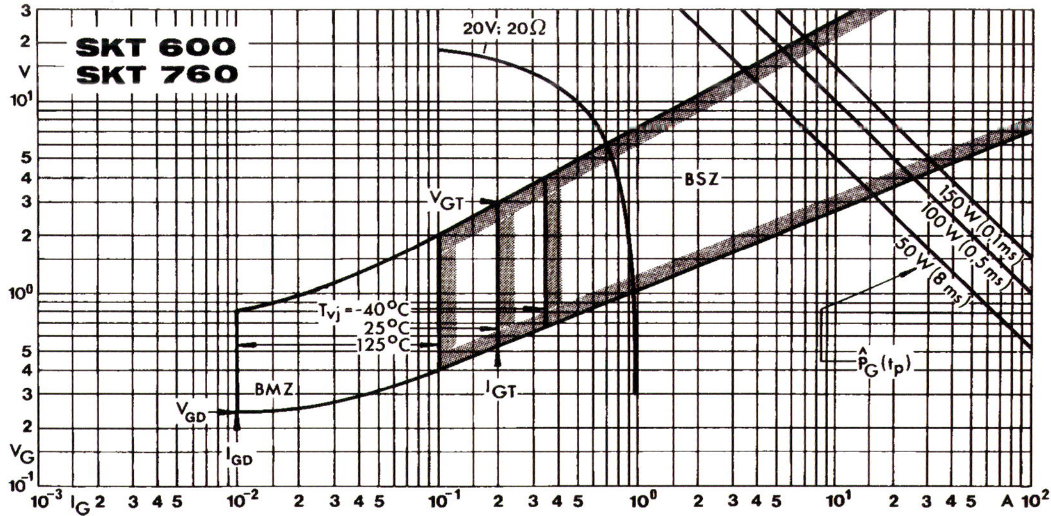

Fig. 4-29: Gate characteristic of an SKT600 (Semikron)

Remark

On p. 4-15 a number of curves from the SKT600 were provided.

3. DIAC

The diac is not a thyristor but we study the component here as it is almost exclusively used as a trigger diode for two directional thyristors (triacs).

3.1 Construction

The diac (diode for alternating current) is a trigger diode with a three layer structure.

The construction and symbol are shown in fig. 4-30. In fact a diac is a symmetrically doped PNP-transistor without a base connection.

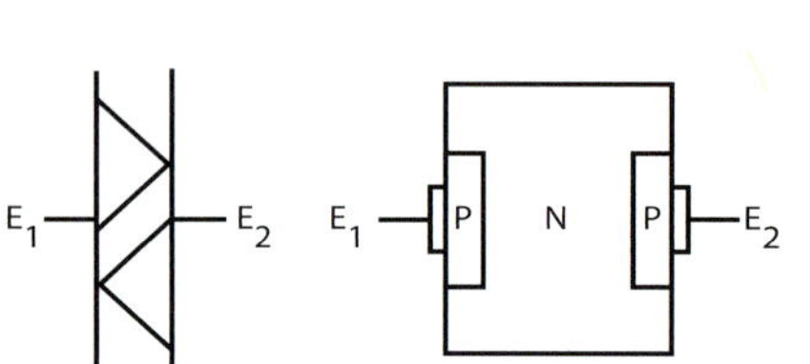

Fig. 4-30: Diac construction and symbol

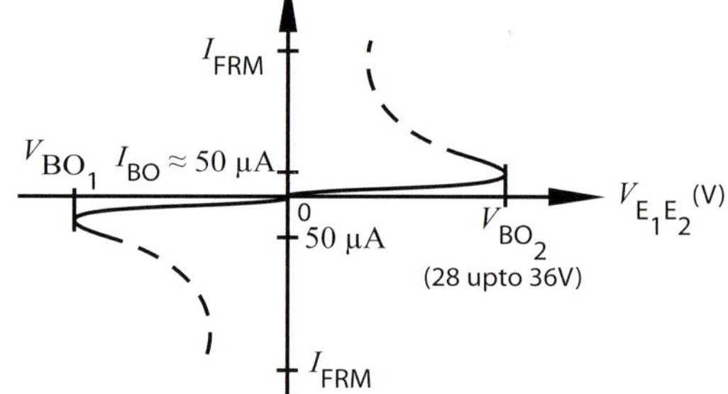

Fig. 4-31: *I-V* curve of a BR100-03 from Philips

3.2 Operation

Depending on the polarity of the voltage V_{E1E2} one of the junctions will be forward polarised while the other junction is reverse biased.

By raising V_{E1E2} to the break over voltage V_{B0} results in a rising field strength and avalanche effect so that the reverse biased junction conducts, comparable to the zener diode. If the dissipated power is maintained less than the maximum allowable then the diac will not be damaged and this break-over effect is reproducible. Fig. 4-31 shows the symmetric behaviour of such a diac (BR100-03 from Philips). The I_{FRM} is 2A for a maximum of 20 µs.

Since it is difficult to refer to anode and cathode we use the names electrodes (E_1 and E_2).

4. TWO DIRECTIONAL THYRISTOR (TRIAC)

Another important thyristor type is the triac (triode for alternating current) The triac with the help of a single gate electrode can be made to conduct in both directions. The triac is a bidirectional switch in contrast to the unidirectional SCR. The applications for the triac are exclusively in the field of AC energy converters up to a maximum power in the order of tens of kW.

4.1 Triac construction

Fig. 3-32 shows the construction and symbol for a triac. In addition to the gate electrode we also note the main electrodes E_1 and E_2 . The two oppositely directed "valves" in the symbol indicate that we are dealing with a bi-directional switch.

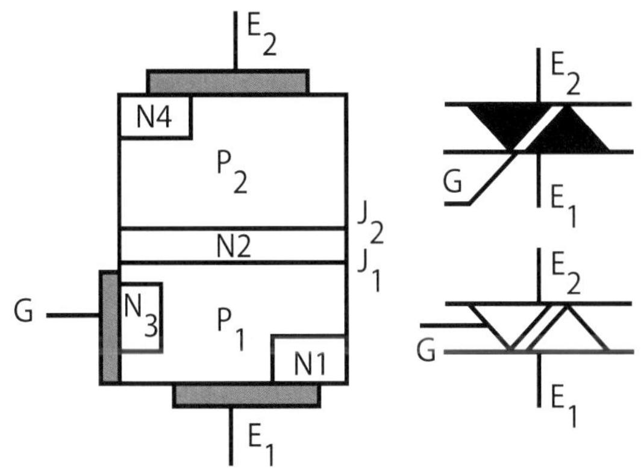

Fig. 4-32: Construction and symbols of a triac

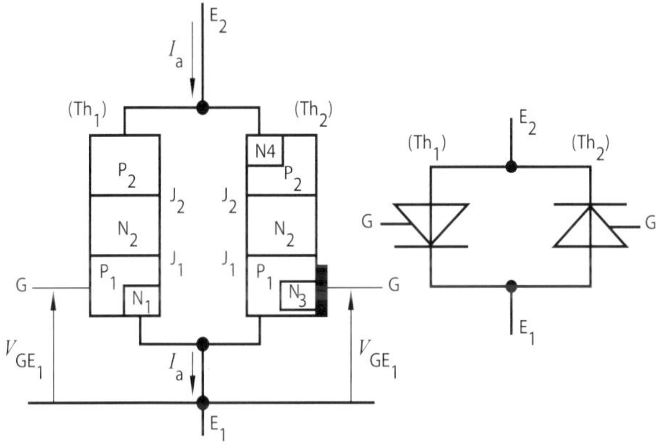

Fig. 4-33: Anti-parallel configuration of SCR's

4.2 Triac operation

Fig. 4-32 is somewhat redrawn in fig. 4-33. Between E_1 and E_2 we find two uni-directional thyristors (SCR's) Th_1 and Th_2 connected back to back. It is clear that the current can flow from E_2 to E_1 or visa versa depending on the polarity between the electrodes. Note that just as with the diac we no longer talk about anode and cathode but rather about electrodes.

With $V_{E2E1} > 0$ "Th_1" can conduct and with $V_{E2E1} < 0$ it is possible to trigger "Th_2".
In the $I_a - V_{E2E1}$ graphic of fig. 4-34 it is only possible to operate in quadrant I or III.
I^+ and III^+ possess a positive gate voltage and I^- and III^- a negative gate voltage.
Every quadrant corresponds to an SCR operating area.
The table provides an overview of the Quadrants related to the appropriate SCR, the junction in reverse bias and the gate diode that will initiate triac conduction.

Fig. 4-34: Operating quadrant of a triac

quadrant	SCR	barrier layer	gate diode
I^+	Th_1	J_1	P_1N_1
I^-			N_3P_1
III^+	Th_2	J_2	P_1N_1
III^-			N_3P_1

The free charge carriers in the gate diode partly recombine in the P_1 layer. The rest will, depending on the operating quadrant, cause the leakage current of the barrier layer J_1 or J_2 to increase. This increasing leakage current causes the mutually connected complementary transistors (see operation of an SCR: fig. 4-7b) to send each other into saturation. The result is that the triac conducts after a few µs.

Remark

Fig. 4-33 shows that a triac can be replace by two SCR's connected back to back. The advantage with the triac is that we only need one control electrode whereas the back to back configuration has two gates.

Since the triac can only operate with limited current (up to around 100A, see further) large AC power switches are constructed using SCR's connected back to back (see fig. 9-3 on p. 9.4).

4.3 Principal diagram of a triac circuit

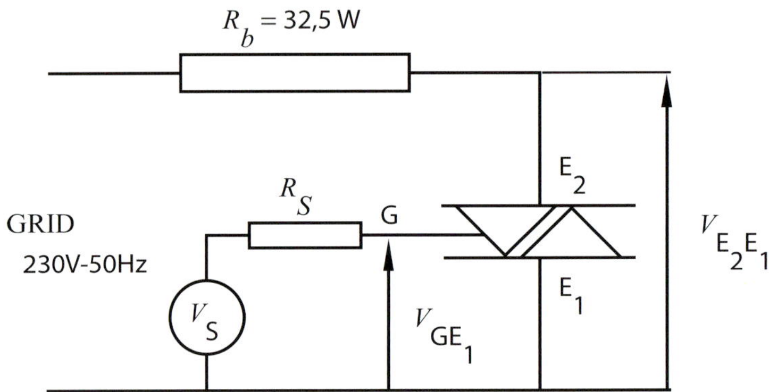

Fig. 4-35 : Principle configuration of a triac

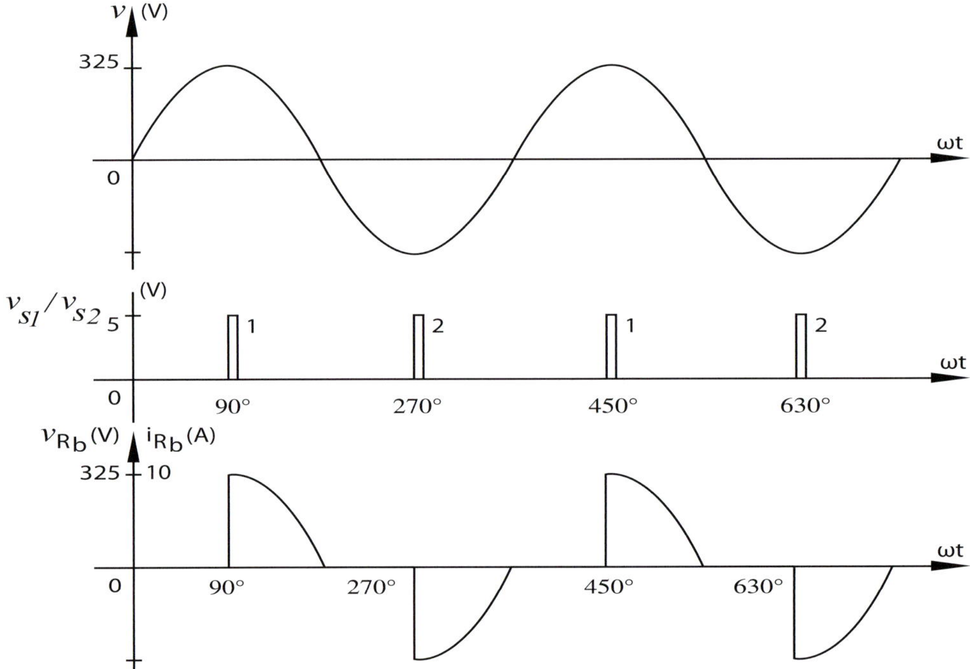

Fig. 4-36: Voltage and current waveforms of fig. 4-35

The circuit in fig. 4-35 shows the power circuit with supply, load resistor and triac. The control circuit contains a pulse generator V_S which generates 100 Hz control pulses. When a control pulse appears between the gate G and E_1 and a voltage is present V_{E2E1} (regardless of polarity) the triac will conduct. Since the bidirectional thyristor operates with an AC voltage no additional turn-off circuit is required for triacs. The triac turns-off automatically just before the zero cross over of its main current, more precisely at the instant the main current falls below the holding current I_H. As we see in fig. 4-36 the instant that a control pulse is applied can also be expressed as an angle. This angle is referred to as the firing angle α. In fig. 4-36 $\alpha = 90°$.

4.4 Static characteristics

4.4.1 I_a - V_a curve

Since the triac may be considered as two back to back SCR's the I_a - V_a curve in fig. 4-37 resembles in both directions the quadrant 1 curve of an SCR .

4.4.2 I_G - V_G curve

As discussed under operation the triac can trigger with any polarity of supply or gate voltage. As a result of the physical construction the largest gate power is required in quadrant III$^+$ to ensure triggering. For reasons of simplicity the manufacturer states the most unfavourable values of I_{GT} and V_{GT} so that these values are also used for all quadrants.

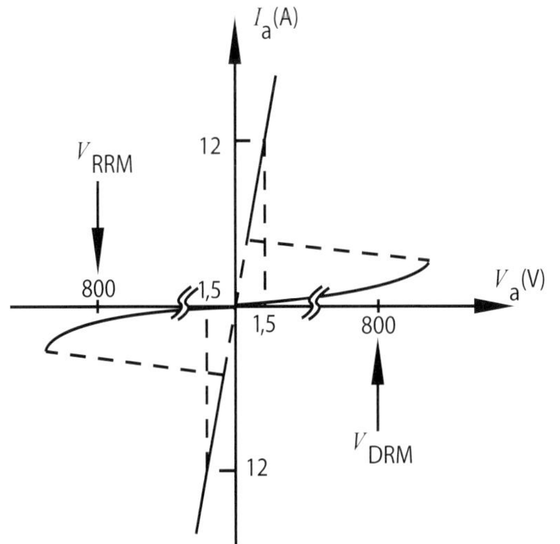

Fig. 4-37: I_a - V_a curves of a BT138-800(Philips)

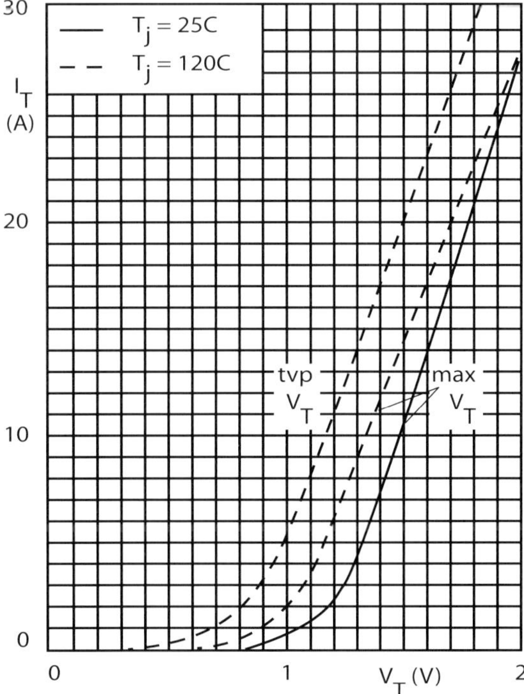

Fig. 4-38: On-characteristic of a BT138

4.5 Dynamic properties

4.5.1 Turn-on time

This varies between 1 and 20 µs and depends (also) on the applied gate current. The points discussed concerning the SCR are also valid for the triac.

4.5.1 Commutation of a triac

Turn-off time, voltage and current rate of rise of a triac are all closely related with the dominant issue of commutation. In addition to the critical rate of rise of voltage dv/dt of every blocking thyristor with the triac there is a second critical rate of rise of voltage called $(dv/dt)_{com.}$. What does this involve?

This $(dv/dt)_{com.}$ occurs close to the zero crossover at the end of current conduction. Consider a resistive-inductive loaded triac as shown in fig. 4-39. The firing angle is α. The current waveform is shown in fig. 4-40. At the instant t_2 the triac turns off and a steep commutation voltage dv/dt appears across its terminals. If this (dv/dt) is larger than the critical $(dv/dt)_{com.}$ then the triac remains conducting and control is no longer possible.

By including an RC-network in parallel across the triac the active dv/dt can be reduced. Such networks are referred to as snubber networks. If for example the current in Th$_1$ (fig. 4-33) falls too quickly ($- dI_T/dt$), then the large amount of charge carriers present during the recovery time cause the other SCR Th$_2$ to conduct as soon as the reverse voltage appears across Th$_2$, this is with $V_{E2E1} < 0$.

A triac is limited in terms of voltage rate of rise $(dv/dt)_{com.}$ and in current rate of rise ($- dI_T / dt$). This is the reason that triacs are only constructed for maximum currents of about 100A for applications at the normal net frequency.

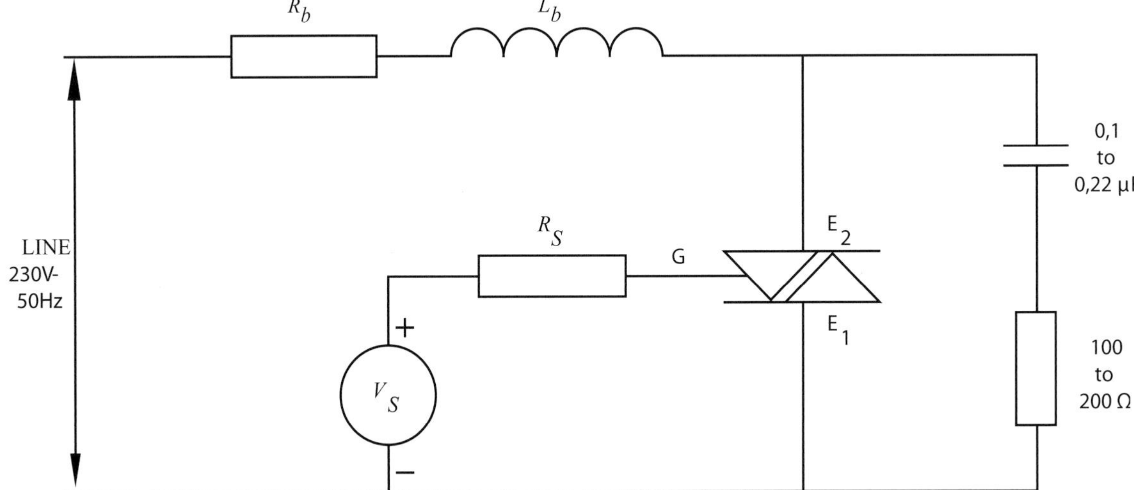

Fig. 4-39: Resistive-inductive loaded triac

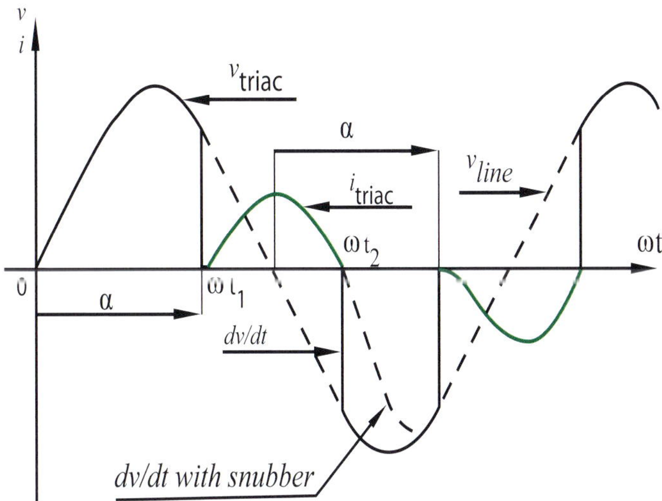

Fig. 4-40: Voltage rate of rise during commutation of a triac

Numeric Example 4-9:

a.   Find the ratio between I_{GT} for quadrants III$^+$ and I$^+$ for a BT138.

b.   Find for a BT138 (12A triac) the values of V_{DRM} / $(di/dt)_{max.}$ / $(dv/dt)_{normal}$ / $(dv/dt)_{com.}$.

Solution:

a)   III$^+$ = T$^-_2$ and G$^+$: $I_{GT} \geq 70mA$

      I$^+$ = T$^+_2$ and G$^+$: $I_{GT} \geq 35mA$

      ratio: 2

b)   BT138: V_{DRM} = 500 to 800V; $(di/dt)_{max.}$ = 30A/µs ; $(dv/dt)_{normal}$ = 100V/µs;

      $(dv/dt)_{com.}$ = 10V/µs

Remark

To control motors and other strongly inductive loads, triacs have been built with very large commutation rate of rise. Philips calls it a "hi-com"- triac and S.T. (SGS-Thompson) talks about an "alternistor".

4.6 Triac control methods

The control methods shown in fig. 4-26 for an SCR are valid for a triac. Pulse control is especially common with a triac. In some cases AC control may be used. Note that with a triac we need control pulses with twice the supply frequency (one pulse per half period of the supply!).

4.7 Temperature influence on a triac

Also here we can refer to the SCR. In addition a rising temperature will also cause the critical rate of rise of voltage $(dv/dt)_{com.}$ to decrease.

4.8 Extract from Philips data book

Philips BT138 triac specifications

RATINGS

Limiting values in accordance with the Absolute Maximum System (IEC134).

Voltages (in either direction)			BT138-500	600	800	
Non-repetitive peak off-state voltage						
$\quad$ ($t \leq 10$ ms)	V_{DSM}	max.	500*	600*	800	V
Repetitive peak off-state voltage						
$\quad$ ($\delta \leq 0.01$)	V_{DRM}	max.	500	600	800	V
Crest working off-state voltages	V_{DWM}	max.	400	400	400	V

Currents (in either direction)				
RMS on-state current (conduction angle 360°)				
$\quad$ up to $T_{mb} = 95°C$	$I_{T(RMS)}$	max.	12	A
Repetitive peak on-state current	I_{TRM}	max.	90	A
Non-repetitive peak on-state current;				
$\quad$ $T_j = 120°C$ prior to surge; full sinewave				
$\quad$ t = 20 ms	I_{TSM}	max.	90	A
$\quad$ t = 16.7 ms	I_{TSM}	max.	100	A
I^2t for fusing (t = 10 ms)	I^2t	max.	40	A^{2}s
$\quad$ Rate of rise of on-state current after triggering with $I_G = 200$ mA to $I_T = 20A$; $dI_G / dt = 0.2$ A/μs	dI_T/dt	max.	30	A/μs

Gate to terminal 1

Power dissipation

Average power dissipation				
$\quad$ (averaged over any 20 ms period)	$P_{G(AV)}$	max.	0.5	W
Peak power dissipation	P_{GM}	max.	5.0	W

Temperatures

Storage temperature	T_{stg}		- 40 to +125	°C
Operating junction temperature				
$\quad$ full-cycle operation	T_j	max.	120	°C
$\quad$ half-cycle operation	T_j	max.	110	°C

Photo ABB(KWx): Pulse power stack. Discharge voltage 20kV. Current up to 30kA. Duration 12 µs. Repetitive frequency 300Hz. Cooled with transformer oil.

5. GATE TURN-OFF THYRISTOR (GTO)

A gate turn-off thyristor (GTO) can be turned on and off via its gate. This is in contrast with an SCR and a triac which can only be turned on via their gate.

5.1 Operation

5.1.1 Triggering a GTO

Triggering occurs in the same manner as with the SCR. Due to the nature of its construction a GTO will require a larger level of gate power than an SCR of the same (switching) power. The GTO is from this viewpoint a robust thyristor (see numeric example 4-10).

5.1.2 GTO turn-off

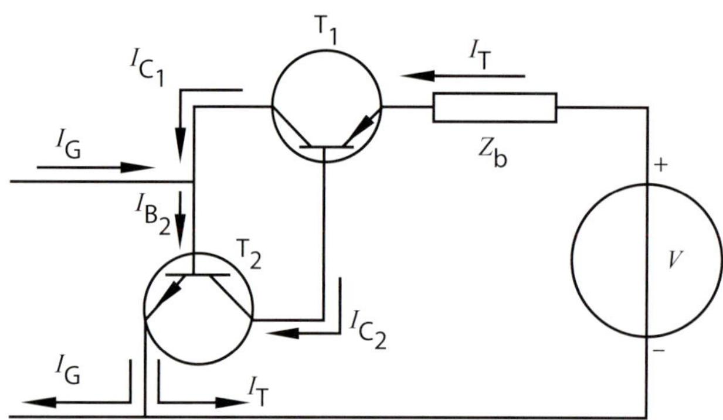

To turn-off a thyristor with a negative gate current the PNP current gain factor α_1 from the double transistor configuration shown in fig. 4-42 needs to to have a low value. We can demonstrate this with the following simple formulas.

Fig. 4-42: Equivalent circuit of a GTO

For a transistor the following apply:

$$\beta = I_C / I_B \; ; \quad \alpha = I_C / I_E \; ; \quad \beta = \frac{\alpha}{1-\alpha} \qquad \text{or:} \qquad \alpha / \beta = 1 - \alpha \qquad (4\text{-}10)$$

If the GTO is conducting, then T_1 and T_2 are saturated. To turn-off the GTO via the gate we need to inject a negative current I_G in the gate. As a result T_2 blocks with as a result T_1 blocks.

The pre-condition for T_2 to block is that I_{B2} is too small to maintain I_{C2} : $\quad I_{B2} < I_{C2}/\beta_2$ (4-11)

In fig.4-42 we see: $\qquad I_{B2} = I_{C1} + I_G = \alpha_1 . I_T + I_G \quad$ and $\quad I_{C2} = \alpha_2 . (I_G + I_T)$

so that (4-11) can be rewritten as :

$$\alpha_1 . I_T + I_G < \alpha_2 . (I_G + I_T) / \beta_2 \; ; \qquad I_G - \frac{\alpha_2}{\beta_2} . I_G < \frac{\alpha_2}{\beta_2} . I_T - \alpha_1 . I_T .$$

With (4-10): $\alpha_2 . I_G < (1 - \alpha_2) . I_T - \alpha_1 . I_T \; ; \qquad - I_G > \frac{\alpha_1 + \alpha_2 - 1}{\alpha_2} . I_T$

$$\rightarrow \rightarrow \rightarrow \qquad \boxed{- I_G > I_T / \beta_{\text{off}}} \qquad (4\text{-}12)$$

A current I_T can be switched off with a smaller current $-I_G$ if $\beta_{off} > 1$

$$\beta_{off} = \frac{\alpha_2}{\alpha_1 + \alpha_2 - 1} > 1 \quad \text{if } \alpha_2 \approx 1 \text{ and } \alpha_1 \text{ is small.}$$

To ensure α_1 is small we can:

- include gold diffusion in a thick internal N-layer, as a result V_{RRM} and V_T rise.
- implement a shorted emitter construction (in external P-layer) with as a result V_{RRM} and V_T of the GTO will decrease.

To ensure α_2 will be large:

- include a thin P-layer
- include an external N-layer that contains a lot of impurities (N^+ !).

Fig. 4-43 shows the construction of a GTO, taking into account our former decisions.
The geometry of the GTO gate is different from that of an SCR. The control electrode of an SCR is concentrated in one point of the thyristor disc, while with the GTO the gate is divided over the entire surface.

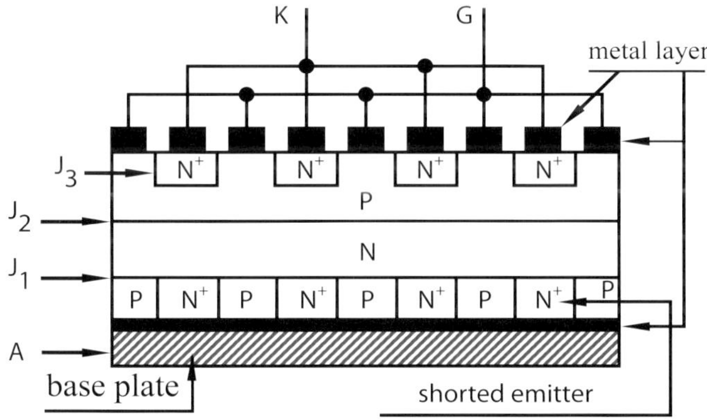

Fig. 4-43: Construction of a GTO

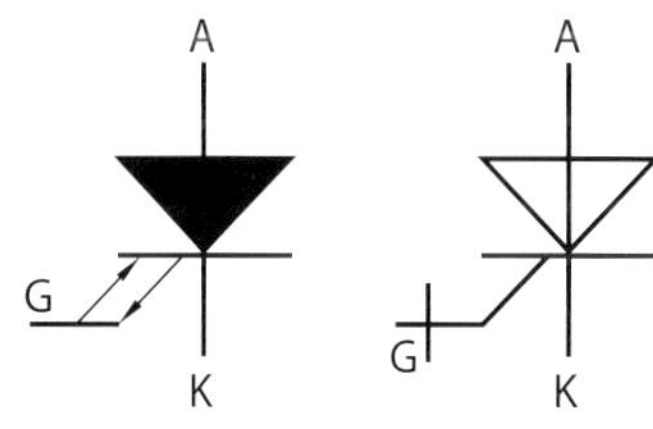

Fig. 4-44: GTO symbols

At turn-off a part of the I_{C1} current (fig. 4-42) flows via the gate so that I_{B2} drops. As a result I_{C2} decreases with as a result I_{C1} decreases etc. In this way the anode-cathode current is brought below the holding current so that the GTO turns-off. We may therefore view the anode current of the GTO as flowing for a short time (in the order of µs) through the gate. Note further that the GTO requires a high inverse break over voltage of the gate-cathode space. This is a property we do not find with SCR's.

In the elementary control circuit of fig. 4-45 we see an inductor L in the turn-off circuit. Now we discuss the eventual necessity of this inductor. If we take a negative gate current I_{GQ} which complies with the requirement to turn off the anode current, then with a GTO: $\beta_{off} = I_T / I_Q$ = turn-off gain. This β_{off} is typically 20 to 25.

The higher the - di/dt in the gate circuit, the quicker the GTO turns off but the lower β_{off} will be (drops for example from 20 to 2).

A small coil L has a large $- di/dt$ in the gate so that t_{off} is small and at he same time β_{off} is low so that $I_G \approx I_T$.

With a larger self inductance L the - di/dt will be less and t_{off} larger and at the same time β_{off} is for example 20 with as result $I_{GT} \approx I_T / 20$. Since the turn-off time t_{off} is larger the switching losses rise.

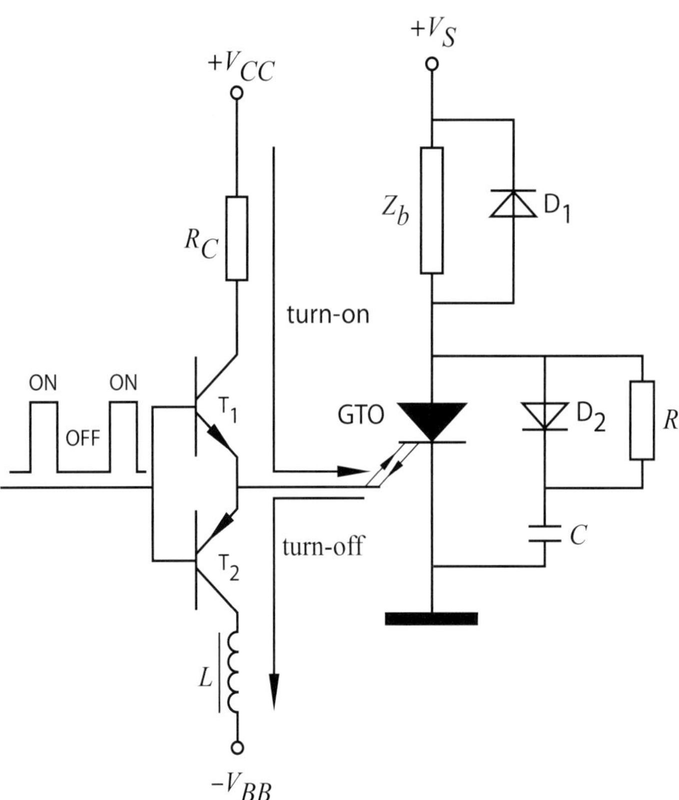

Fig. 4-45: Principal configuration controlling a GTO

5.2 Characteristics and specifications

Static and dynamic properties and specifications remain valid as for other thyristors. Fig. 4-46 shows an I_a - V_a curve of a 1570A-GTO (FG3300AH-50DA from Mitsubishi). There are also asymmetric GTO's (e.g. V_{RRM} = 100V and V_{DRM} = 1800V) and symmetric GTO's (V_{DRM} = V_{RRM}) .

The manufacture of asymmetric GTO's seems to be more efficient than that of symmetric GTO's when the blocking voltage is required to be larger than 2000V.

Numeric example 4-10:

We consider an SCR (SKT600) and a GTO (FG3300AH-50DA).

Compare each of the following V_{RRM}, V_{DRM}, I_{GT}, $P_{G(AV)}$, t_{on}, t_{off}, dv/dt, di/dt.

Solution:

	SCR (Semikron) SKT600	GTO (Mitsubishi FG3300AH-50DA	units
V_{RRM}	400 to 1800	17	V
V_{DRM}	400 to 1800	2500	V
I_{GT}	0.2	4 (for I_T = 200 A !)	A
$P_{G(AV)}$	?	100	W
t_{on}	3	10	μs
t_{off}	100 to 200	30	μs
dv/dt	500 to 1000	1000	V/μs
di/dt	125	500	V/μs

From the solution to numeric example 4-10 we see the robustness of the GTO in comparison to an SCR as far as turn-on is concerned. The GTO switches faster than an SCR, see t_{off}! On the other hand we can conclude that this GTO has practically no inverse voltage capability.

With a DC voltage source an asymmetric GTO has no problem, but if we work with an AC voltage source a diode needs to place in series (fig. 4-47b). If the power switch has to conduct inversely then a free wheel diode shown in fig. 4-47c can be a solution.

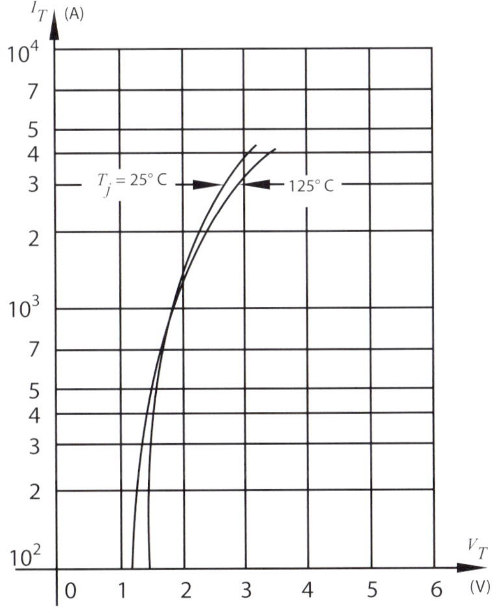

Fig. 4-46 : "ON-STATE" of a FG3300AH-50DA (Mitsubishi)

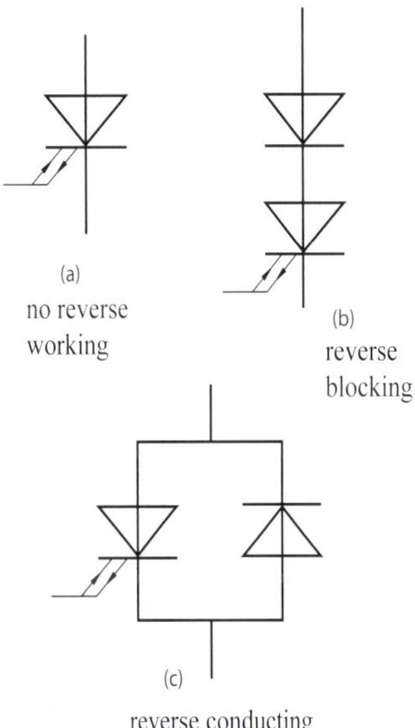

Fig. 4-47: Protective diodes

Fig. 4-48 shows the switching times t_d...t_t. The tail time t_t is a result of the current drawn as a result of charges in other parts of the GTO than the gate-anode space. These charges are not influenced by the negative gate voltage, they disappear as a result of natural recombination.
Just as with the SCR we write $t_{on} = t_{gt}$ and $t_{off} = t_{gq}$.

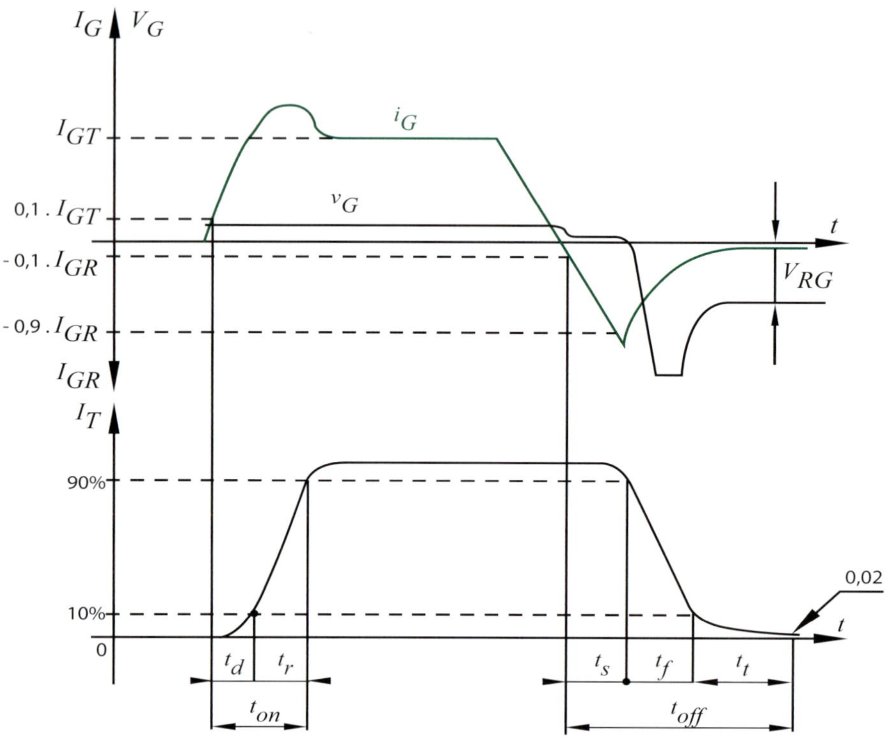

Fig. 4-48: GTO switching times

Remarks

1. *Switch off losses:*
    These are high for a GTO. The turn-off phase is the most delicate phase for the GTO.
    To limit the *dv/dt* and switch off losses snubbers are used.

2. *Snubber:*
    Is normally of the *RCD*-type (see p. 3-10).
    The resistor limits the discharge current of the capacitor during a retriggering of the GTO.
    The snubber capacitor has to meet high requirements:
    - the self inductance of the capacitor has to be very small.
    - it has to be resistant to high current peaks and high repeat frequencies.

    The wiring of a snubber circuit should be very short, for example a wiring self inductance of 1 to 2 µH will strongly reduce the effectiveness of the snubber especially with rapid switching. Therefore the snubber should be placed as close as possible to the GTO.

3. *Large power levels:*
    In the main traction motors of metros (1500V DC) we can encounter GTO's rated 2500A.
    Industrial applications such as AC-drives up to 3MW can use 2500V - 1000A GTO's.

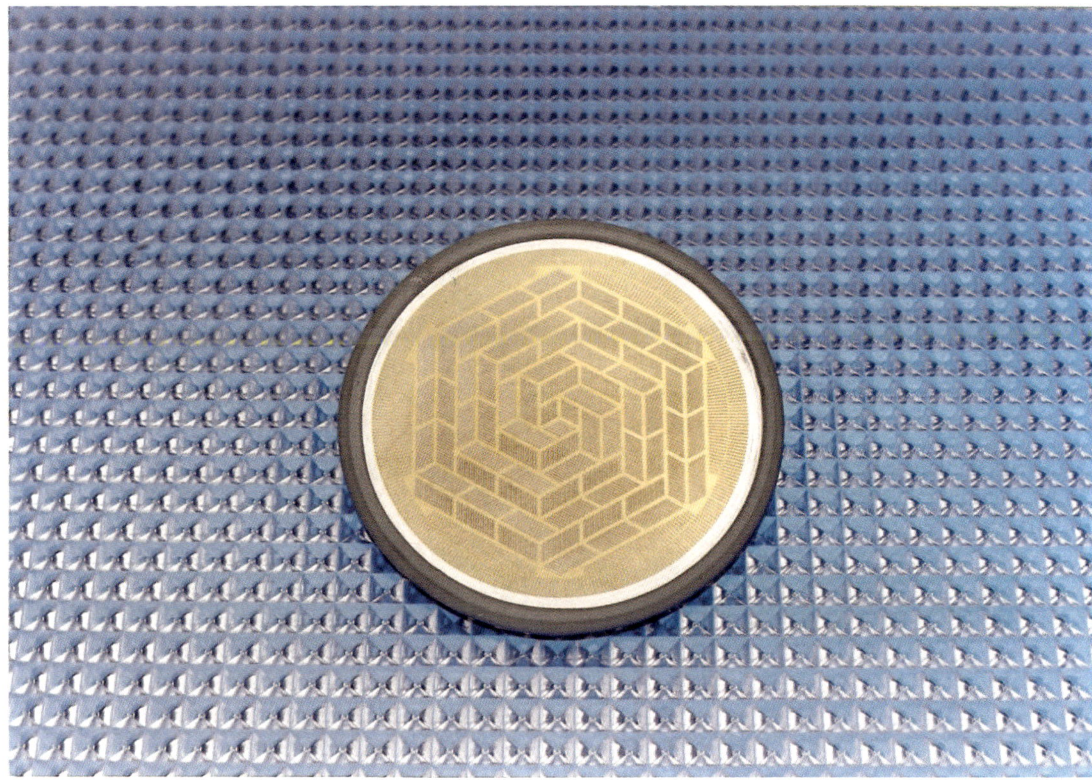

Photo Mitsubishi: Wafer of an asymmetric GTO. This is a 4000A-4500V GTO

5.3 Details of a high Power GTO: FG3300AH-50DA (Mitsubishi)

MAXIMUM RATINGS

SYMBOL	VOLTAGE CLASS	UNITS
	50DA	
V_{RRM}	17	V
V_{RSM}	17	V
$V_{R\,(DC)}$	17	V
V_{DRM}	2500	V
V_{DSM}	2500	V
$V_{D\,(DC)}$	–	V

FG3300AH-50DA (Mitsubishi)   MAXIMUM RATINGS (continued)

SYMBOL	CONDITONS	LIMITS	UNITS
I_{TQRM}	U_D=1250V, U_{DM}=1875V, Cs = 6μF, Ls = 0.2μH, di_G/dt = 40A/μs, T_j = 125°C	3300	A
$I_{T(RMS)}$		1570	A
$I_{T(AV)}$	Single phase half sine wave, 180° conduction, T_f = 80°C	1000	A
I_{TSM}	60Hz half sine wave, non repetitive peak value	24.000	A
I^2t	One cycle at 60Hz	$2,4 \cdot 10^6$	A²s
di_T/dt	U_D = 1250V, I_T = 3300A, I_{GM} = 40A, di_G/dt = 10A/μs, T_j = 125°C	500	A/μs
P_{FGM}		400	W
P_{RGM}		27.000	W
$P_{GF(AV)}$		100	W
$P_{RG(AV)}$		230	W
U_{FGM}		10	V
U_{RGM}		17	V
I_{FGM}		100	A
I_{RGM}		900	A
T_j		- 40...125	°C
T_{stg}		- 40...150	°C

ELECTRICAL CHARACTERISTICS: FG3300AH-50DA (Mitsubishi)

SYMBOL	TEST CONDITIONS	MIN	TYP	MAX	UNITS
I_{RRM}	U_R=17V, T_j=125°C	–	–	300	mA
I_{DRM}	U_D = 2500V, T_j=125°C, U_{GK} = 2V	–	–	150	mA
I_{RG}	U_{RG} = 17V, T_j = 125°C	–	–	300	mA
U_{TM}	I_{TM} = 3300A, T_j = 125°C	–	–	3	V
du/dt	U_D = 1250V, U_{GK} = –2V, T_j = 125°C	1000	–	–	V/μs
U_{GT}	I_T = 25 tot 200A, U_D = 5 tot 20V, T_j = 25°C	–	–	1,5	V
I_{GT}	I_T = 25 tot 200A, U_D = 5 tot 20V, T_j = 25°C	–	–	4	A
t_{gt}	I_T = 3300A, U_D = 1250V, di/dt = 500A/μs, I_{GM} = 40A, T_j = 125°C	–	–	10	μs
t_{gq}	I_T = 3300A, U_{DM} = 1875V, U_D = 1250V, Cs = 6μF, Ls = 0,2μH, T_j = 125°C, di_{GQ}/dt = 40A/μs, U_{GK} = 17V	–	–	30	μs
		–	–	30	μs
$R_{th(j-f)}$	Junction to fin	–	–	0,015	°C/W

6. THE MOS CONTROLLED THYRISTOR OR MCT

The Mos Controlled Thyristor is a combination of a PNPN-thyristor and two MOSFETS. These MOSFETS serve to turn-ON and turn-OFF the MCT. The MCT has thyristor properties at the output (V and I) and MOSFET properties at the input (control power and ON-OFF switching via the gate).

Fig. 4-49 shows the four-transistor model and fig. 4-50 indicates the common symbols for an MCT.

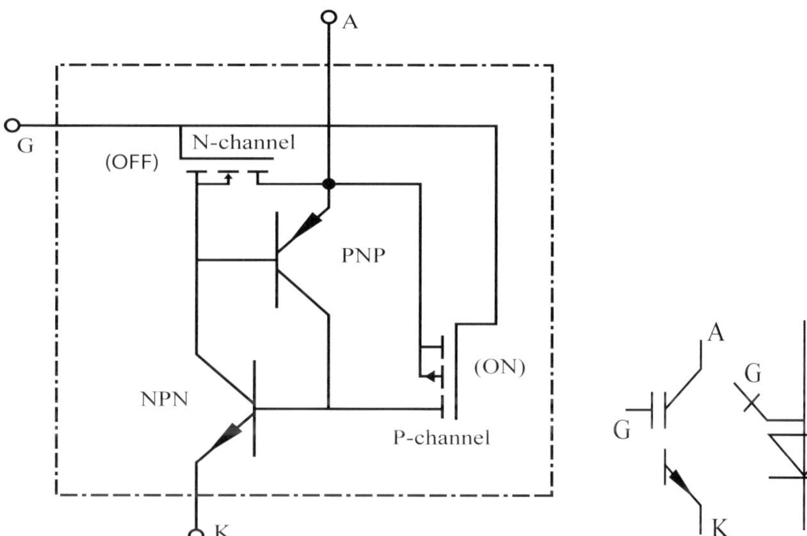

Fig. 4-49: Equivalent circuit of an MCT

Fig. 4-50: MCT symbols

Consider V_{AK} as positive and the MCT is not conducting. A negative gate voltage V_{GA} of for example -15V causes the P-channel MOSFET to conduct so that the NPN-transistor is controlled. The known regenerative effect between PNP- and NPN-transistor causes the thyristor (PNPN) to conduct between A and K.

A conducting MCT can be turned off by a positive V_{GA} (e.g. +18V). This positive anode voltage causes the conducting ON-mosfet to turn-off and the OFF-mosfet to conduct. As a result the base-emitter diode of the PNP-transistor of the thyristor equivalent is short circuited, resulting in an interruption of the regenerative effect between NPN- and PNP transistor and the thyristor turns off. To create an effective short circuit of the PNP base-emitter diode $R_{DS(ON)}$ from the OUT-mosfet need to be sufficiently small. With rising temperature this $R_{DS(ON)}$ increases so that the controllable MCT current drops with rising temperature.

The first MCTs in Europe, available from 1994, were from Harris semiconductors.

Their first models belonged to the 600V-generation (75P60) suitable for 75A. This was an asymmetric P-type MCT.

Now there are MCTs available for 1000V blocking voltage and currents of 35 to 75A. Special control IC's have been developed by Harris and also Unitrode.

7. INTEGRATED GATE COMMUTATED THYRISTOR (IGCT) - FIRM ABB

7.1 From GTO to GCT (Gate Commutated Thyristor)

In the 1990's from the viewpoint of performance, reliability and cost price GTO's and IGBT's were the most important ON-OFF controllable semiconductors for large power levels.

A number of interesting comparison points between the thyristor type (GTO) and the transistor type (IGBT) are shown in the following table.

Switch	Advantage	Disadvantage
GTO	• low voltage drop	• difficult snubber • wide spread during switch-off time produces complications with the series switching of GTO's
IGBT	• no snubber required	• larger losses than a GTO • less power than a GTO

A break through in GTO-technology came with the development of the snubber free " Hard Driven GTO's". This new switching technology is characterised by extremely fast commutation of the cathode current to the gate, hence the name gate commutated thyristor (GCT). This commutation takes place within 1 µs.

7.2 Operation of GCT

Before considering the construction we first will consider the working of the GCT, from the start point of a thyristor which is formed by a classic two transistor model. In fig. 4-51 a GCT in conducting mode is shown. The GCT behaves exactly as an SCR and GTO. We mean amongst other things the possibility to conduct a large current with a low forward voltage drop.

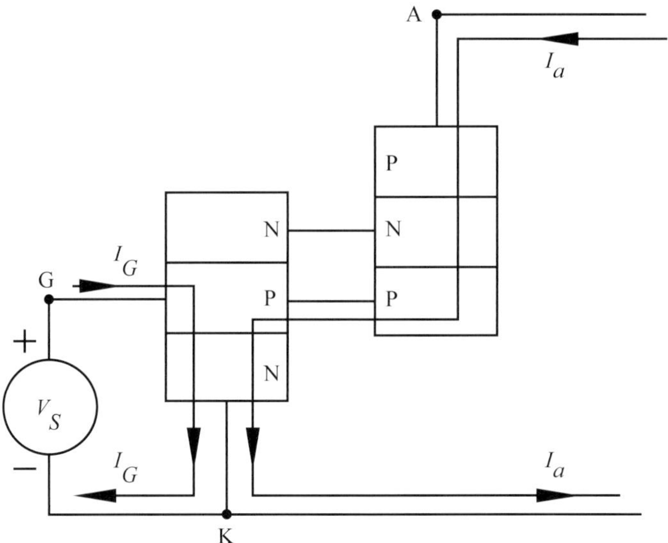

Fig. 4-51: Conducting GCT

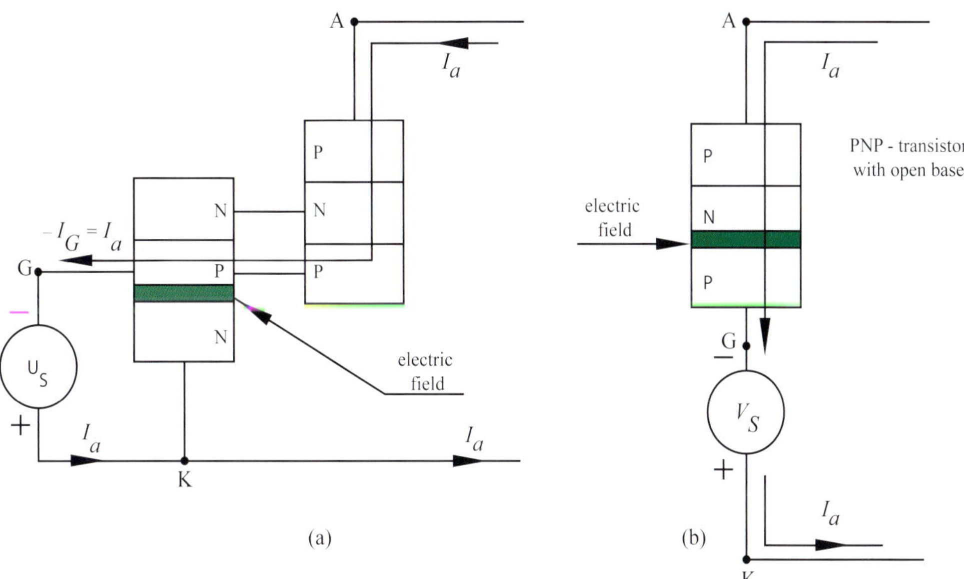

Fig. 4-52: Turn-off of an GCT

Fig. 4-52a shows the blocking mode of the GCT. At first glance it looks like the turn-off of a GTO. The main difference lies in the " instantaneous" blocking of the cathode PN-junction, this is the base-emitter-junction of the cathode-NPN transistor. With the GTO turn-off occurs via an intermediate step, namely by bringing the two interconnected transistors out of conduction. These two transistors bring each other out of saturation, in exactly the opposite manner that the GTO was made to conduct. It is this GTO turn-off behaviour that requires a large snubber to minimise the effect of the large dv/dt.

With the GCT this GTO-phase is avoided so that at turn-off the situation of fig. 4-52b exists almost instantaneously. Here only the "anode-" PNP-transistor remains conducting. The NPN-transistor is already turned off without the PNP-transistor "being aware of anything". The GCT is a transistor during turn-off while the GTO switches off as a "thyristor".

7.3 From GCT to IGCT

The automatic turn-off of the PNP-transistor with its open base as shown in fig. 4-52b requires only a very small snubber and in most cases no snubber at all. Fig. 4-52b shows also that with the GCT, the complete anode current flows through the gate circuit during the turn-off process. To allow this to happen quickly, in other words to have a large di_{gate}/dt the self induction of the gate circuit needs to be very small. To this end ABB developed a new casing with extremely low self inductance for the GCT. This casing is integrated together with the gate-control circuit.

The GCT now becomes an IGCT (integrated gate commutated thyristor). This new casing together with the integration of the control circuit results in a total gate-self-induction of 4 nH according to ABB. In this way gate turn-off peaks of more than 4kA/µs are realised with a control voltage of no higher than 20V. The low gate-self-inductance also has the advantage that the IGCT now also turns on quickly. A standard GTO-gate circuit has a self-inductance of about 300nH.

Fig. 4-53 shows a block diagram of the IGCT while fig. 4-54 shows the snubber free turn-off of an IGCT. On the right hand side of fig. 4-53 the symbol of an IGCT is drawn with internal free-wheel diode.

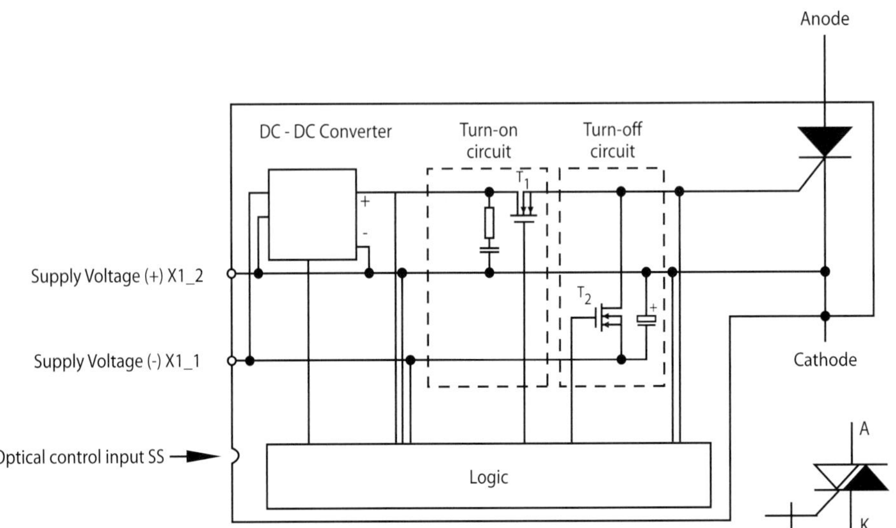

Fig. 4-53: Block diagram of a 5SHX 14H4502 (ABB)

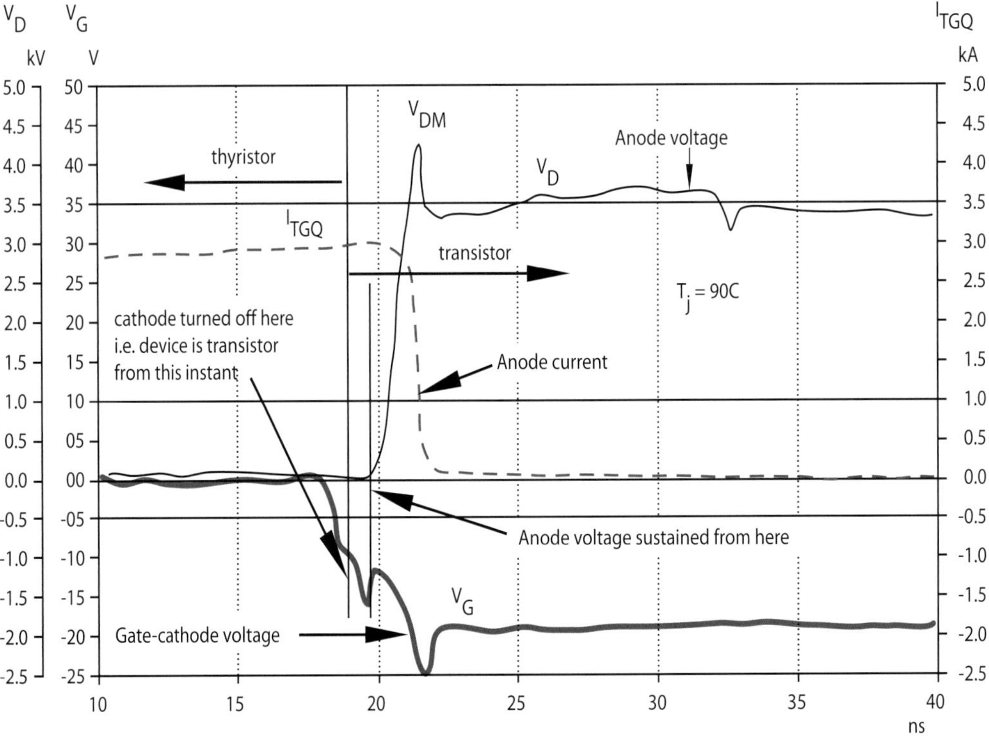

Fig. 4-54 : Turn-off 3kA/4.5kV IGCT (press conference ABB, Milwaukee 20 May 1997)

7.4 Static characteristics

Fig. 4-55 shows the ON-characteristics of a 4kA - 4.5kV IGCT. This is an asymmetric IGCT with a calibre of 91mm. The voltage drop across the conducting transistor is typically 2.8V at 5kA and $T_J = 125°$ C.

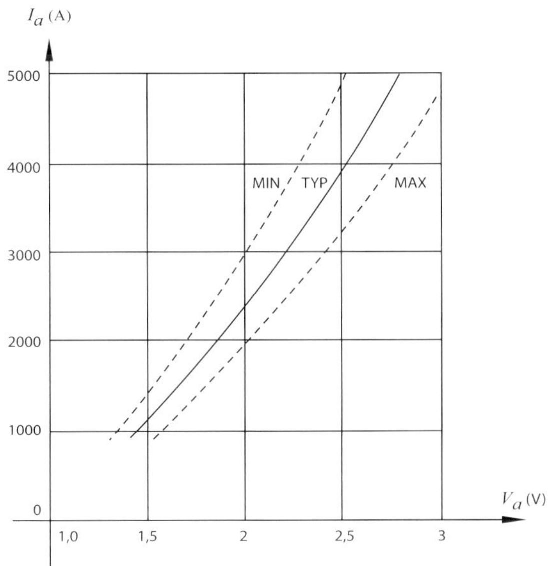

Fig. 4-55: ON-characteristic of a 5SHY 35L4502 (ABB)

7.5 Dynamic properties

The manner in which the firm ABB presents the rise-time and fall-time of an IGCT is shown in fig. 4-56.

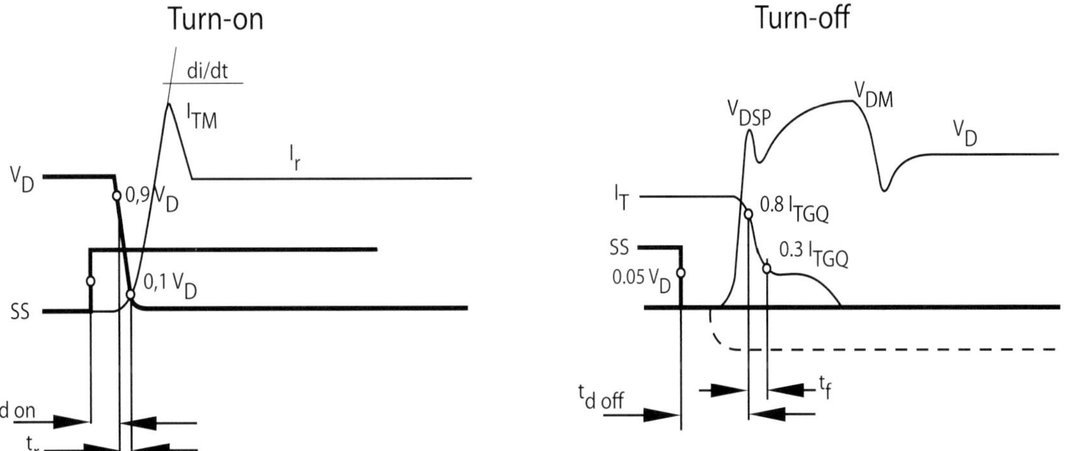

Fig. 4-56: Voltage and current waveforms of an IGCT

7.6 Extract of specifications of 5SHX 14H4502 (ABB semiconductors AG)

V_{DRM}	=	4500 V	**Reverse Conducting Integrated**
I_{TGQM}	=	1100 A	**Gate-Commutated Thyristor**
I_{TSM}	=	8.8 kA	**5SHX 14H4502**
V_{T0}	=	1.65 V	
r_T	=	1.2 mΩ	**ABB**
V_{DClink}	=	2800 V	

Blocking

V_{DRM}	Repetitive peak off-state voltage	4500 V	$V_{GR} \geq 2V$
I_{DRM}	Repetitive peak off-state current	$\leq$ 20 mA	$V_D = V_{DRM}$ $V_{GR} \geq 2V$
V_{DClink}	Permanent DC voltage for 100 FIT failure rate	2800 V	$0 \leq T_j \leq 115\,C$. Ambient cosmic radiation at sea level in open air.

GCT Data

On-state

I_{TAVM}	Max. average on-state current	415 A	Half sine wave, $T_C = 85\ °C$			
I_{TRMS}	Max. RMS on-state current	655 A				
I_{TSM}	Max. peak non-repetitive surge current	8.8 kA	t_p =	10 ms	$T_j = 115\ °C$	
		15.7 kA	t_p =	1 ms	After surge: $V_D = V_R = 0V$	
I^2t	Limiting load integral	391×10^3 A^{2}s	t_p =	10 ms		
		124×10^3 A^{2}s	t_p =	1 ms		
V_T	On-state voltage	$\leq$ 3 V	I_T =	1100 A		
V_{T0}	Threshold voltage	1.65 V	I_T =	200 - 2000 A	$T_j = 115\ °C$	
r_T	Slope resistance	1.2 mΩ				

5SHX 14H4502 (ABB semiconductors AG) continued

Turn-on switching

di/dt_{crit}	Max. rate of rise of on-state current		425 A/µs	f = 500 Hz	T_j = 115 C		
				I_T = 1100 A	V_D = 3200 V		
t_{don}	Turn-on delay time	≤	3 µs	V_D = 2700 V	T_j = 115 C		
t_r	Rise time	≤	1 µs	I_T = 1100 A	di/dt = 360 A/µs		
$t_{on\ (min)}$	Min, on-time		10 µs	R_S = 1.25 Ω	L_i = 7.5 µH		
E_{on}	Turn-on energy per pulse	≤	0.45 J	C_{CL} = 1 µF	L_{CL} = 0.6 µH		

Turn-off switching

I_{TGQM}	Max. controllable turn-off current		1100 A	V_{DM} ≤ V_{DRM} $\quad$ T_j = 115 C	
				V_D = 2700 V $\quad$ L_{CL} ≤ 0.6 µH	
I_{TGQM2}	Max. controllable turn-off current		480 A	V_{DM} ≤ V_{DRM} $\quad$ T_j = 115 C	
				V_D = 3200 V $\quad$ L_{CL} ≤ 0.6 µH	
t_{doff}	Turn-off delay time	≤	6 µs	V_D = 2700 V	V_{DM} ≤ V_{DRM}
t_f	Fall time	≤	1 µs	T_j = 115 C	R_s = 1.25 Ω
$t_{off\ (min)}$	Min. off-time		10 µs	I_{TGQ} = I_{TGQM}	L_i = 7.5 µH
E_{off}	Turn-off energy per pulse	≤	5 J	C_{CL} = 1 µF	L_{CL} ≤ 0.6 µH

8. EVALUATION

4.1 What is the most important difference between a GTO and an SCR?

4.2 Discuss the commutation of a triac.

4.3 What are the undesirable triggering methods of an SCR.

4.4 Name a triac specification that does not occur with an SCR and upon which T_J exercises an influence?

4.5 Which subscript do we use for anode voltage and current to indicate that an SCR is:
1) blocked $\qquad$ 2) conducting $\qquad$ 3) inversely polarised ?

4.6 What can be the consequences if the critical rate of rise of current is exceeded by a unidirectional thyristor? Explain.

4.7 Explain the following: V_{GT}, I_{GT}, V_{GD} ?

4.8 For what reason will T_J rise with an SCR?

4.9 Can you name a bidirectional thyristor? Which?

4.10 What is a church tower pulse?

4.11 Which notations for the anode current of a thyristor do you know? Indicate the notation and the explanation thereof?

4.12 Why does the controllable MCT - current fall with rising temperature?

4.13 The current through a BT138 triac falls according to $- dI_T/dt = 12A/ms$ and the junction temperature is 40° C. What is the critical dv/dt commutation?

4.14 What is the difference between a GTC and an IGCT?

4.15 Do you know a semiconductor switch with thyristor properties at the output and MOSFET-properties at the input?

4.16   Listed with the specifications of the SKT600 on p.4-24 is amongst others DSC. What does this mean?

4.17   Explain the function of the diodes D_1 and D_2 in fig. 4-45.

4.18   What is the advantage of an IGBT compared to a GTO?

4.19   For which applications do we use a triac?

4.20   How is t_{on} determined for a GTO? And also t_{off}?

4.21   What is the difference between a symmetrical and asymmetrical GTO?

4.22   A certain SKT10 has a junction temperature of $130°$ C and the RMS value of the current is 10A. Determine the forward voltage drop.

4.23   What do the initials SCR, DIAC, TRIAC, GTO, GTC, MTC mean?

4.24   What is the effect of a thin internal P-layer in a GTO?

4.25   What is the purpose of shorted-emitter model of a GTO?

4.26   Find the maximum values of $t_{d\,on}$, $t_{d\,off}$, rise time, fall time, di/dt for a 5SHX14H4502 ? How many amperes can this IGCT control as a maximum? A current peak lasts 1ms and the junction temperature is $115°$ C. What is the maximum acceptable value of this peak?

4.27   a.   On p. 4-5 we mentioned an I.E.C. definition. What does I.E.C. mean?

       b.   What are complementary transistors?

4.28   Explain the meaning of all symbols used in fig. 4-10 and 4-11? Note that we became familiar with the meaning of V_{RRM} during the study of power diode in chapter 2.

4.29   a.   What do the following letters used as subscripts in thyristor specifications mean:
         A / AV / C / G / H / J / L / M / R / RMS / S / T.

       b.   Look at fig. 2-7 (p. 2.8). Draw a similar figure for the voltages and currents of an SCR.

4.30   At which anode-cathode voltage and junction temperature is the gate trigger voltage and trigger current specified?

4.31   What does the technical term "to trigger" mean?

4.32   What does the term "snubber" mean and explain what a snubber network is?

4.33   With a triac we have no $I_{T(AV)}$ - and I_R - values like we had with the SCR. Can you explain why?

4.34   With a triac we have no V_{RSM} - and normally no V_{RRM} - value. Do you have an explanation for this?

4.35   In fig. 4-45 why does the current flow through T_2 instead of straight through the cathode to ground during turn-off of the GTO?

4.36   What is a "stud-mounted" thyristor and a "capsule-" thyristor?

4.37   Which elements of the block diagram in fig. 4-53 start working at turn-on? Which elements start working at turn-off ? Provide a short explanation.

4.38   How long does it take to turn off 3kA with the IGCT shown in fig. 4-54?

4.39   A 5SHY 35L4502 is conducting with an anode current of 2000A. What is the maximum voltage drop across this IGCT?

5   NOTES

CONTENTS

1. ELECTRICAL AND MATHEMATICAL NOTATION

1.1 Effective and average values

A. Effective voltage-current

If a current i flows for a time Δt through a resistor R, then the electrical energy that is converted to heat is given by: $\quad W = i^2 . R . \Delta t = \dfrac{v^2}{R} . \Delta t$.

This energy transformation is the so called Joule-effect because the expression was developed by James Prescott Joule. We will continue by focusing on the expression for the value of the current. The expression is equally valid for variables that vary with time under the pre-condition that we take a sufficiently small time interval (dt) such that i remains practically constant:

$dW = i^2 . R . dt$.

In a time span between 0 and T (period) we find the total energy: $W = R \int_0^T i^2 \, . \, dt$

1. Constant direct current: $W_1 = I^2 \, . \, R \, . \, T$

2. Changing current : $W_2 = R \, . \int_0^T i^2 \, . \, dt$

Effective Value:

We define the effective value I_{eff} of an alternating current as the value of a constant direct current I that will produce the same amount of heat in the same resistor R in the same time ($W_1 = W_2$) :

$$I^2 \, . \, R \, . \, T = R \int_0^T i^2 \, . \, dt \qquad \text{so that:} \qquad I_{eff.} = \sqrt{\frac{1}{T} \int_0^T i^2 \, . \, dt} \qquad (5\text{-}1)$$

Since we take the square root from the quadratic average this is most often referred to as RMS (root mean square): I_{RMS} ($= I_{eff.}$).

In a similar manner we find :

$$V_{eff} = \sqrt{\frac{1}{T} \int_0^T v^2 \, . \, dt} \qquad (5\text{-}2)$$

3. Calculation examples
    a. Sine wave (fig. 5-1)

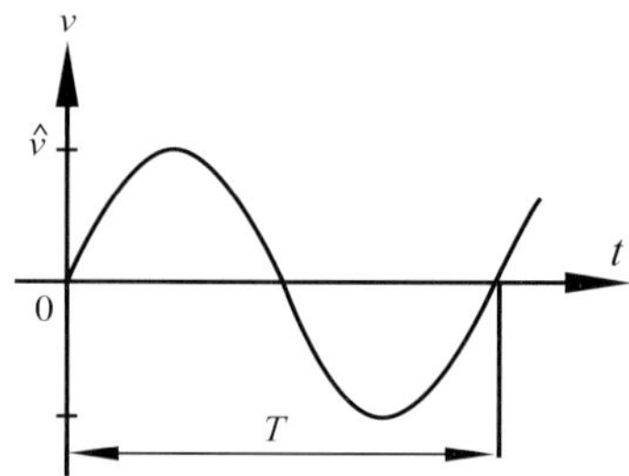

$$V_{eff.} = V_{RMS} = \sqrt{\frac{1}{T} \int_0^T v^2 \, . \, dt} = \sqrt{\frac{1}{T} \int_0^T (\hat{v} \, . \, sin \, \omega \, t)^2 \, . \, dt}$$

$$V_{RMS} = \sqrt{\frac{1}{2\pi} \int_0^{2.\pi} \hat{v}^2 \, . \, (\frac{1 - cos \, 2\omega t}{2}) \, . \, d\omega t} = \frac{\hat{v}}{\sqrt{2}} \qquad (5\text{-}3)$$

Fig. 5-1: Sine wave

    b. Half wave rectified sine wave (fig. 5-2)

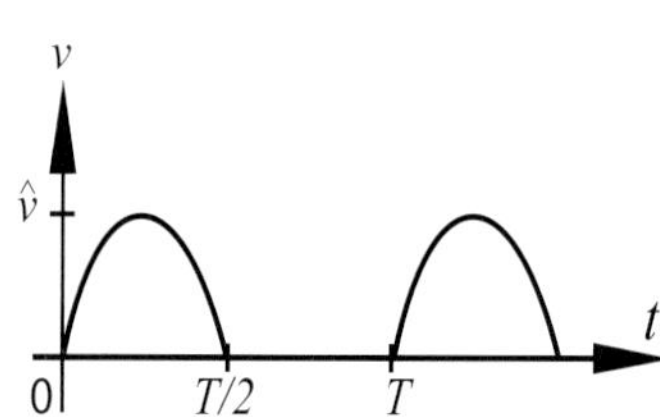

$$V_{RMS} = \sqrt{\frac{1}{T} \int_0^{T/2} v^2 \, . \, dt} = \sqrt{\frac{1}{2.\pi} \int_0^{\pi} (\hat{v} \, . \, sin \, \omega \, t)^2 \, . \, d\omega t}$$

$$V_{RMS} = \sqrt{\frac{\hat{v}^2}{2.\pi} \int_0^{\pi} (\frac{1 - cos \, 2 \, \omega t}{2}) \, . \, d\omega t} = \frac{\hat{v}}{2} \qquad (5\text{-}4)$$

Fig. 5-2: Half wave rectified

c. Full wave rectified sine wave (fig. 5-3)

We can consider this as the sum of two waveforms (v_1 and v_2) whereby the effective value is $\hat{v}\,/\,2$. The total power is: $V_1^2\,/R + \;V_2^2\,/R = V_{RMS}^2\,/\,R$

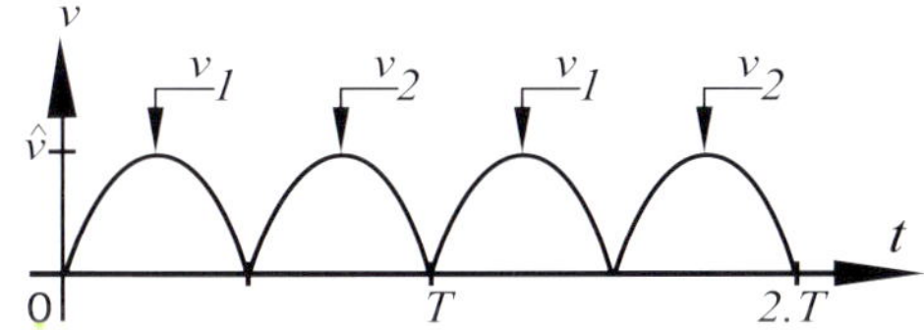

$$V_{RMS} = \sqrt{V_1^2 + V_2^2} = \sqrt{\left(\frac{\hat{v}}{2}\right)^2 + \left(\frac{\hat{v}}{2}\right)^2} = \frac{\hat{v}}{\sqrt{2}} \qquad (5\text{-}5)$$

Fig. 5-3: Full wave rectification

Remark

If we folded v_2 (fig. 5-3) about the t-axis, then v_1 and v_2 form once again a perfect sine wave. The effective value of the full wave rectified sine wave ($\hat{v}/\sqrt{2}$) is the same as that of a perfect sine wave since in the formula we use v_2^2 and as a result the polarity of v_2 plays no role.

d. Pulse voltage (fig. 5-4)

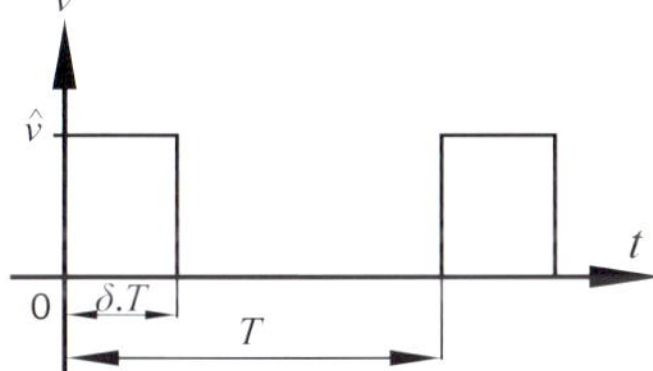

$$V_{RMS} = \sqrt{\frac{1}{T}\int_0^{\delta.T} V^2\,.\,dt} = \sqrt{\frac{1}{T}\,.\,V^2\,.\,\delta\,.\,T} = V\,.\,\sqrt{\delta} \qquad (5\text{-}6)$$

Fig.5-4: Pulse voltage

e. Saw tooth voltage (fig. 5-5)

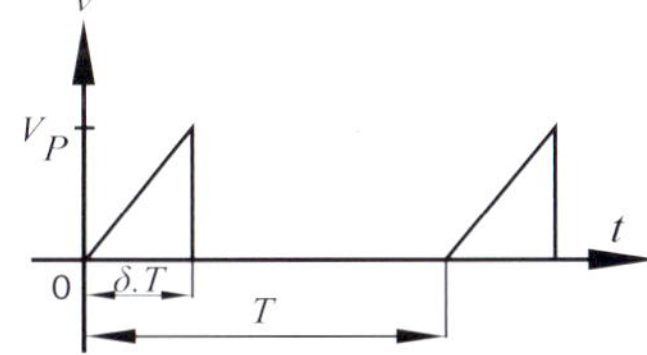

$$V_{RMS} = \sqrt{\frac{1}{T}\int_0^{\delta T} v^2\,.\,dt} = \sqrt{\frac{1}{T}\int_0^{\delta T}\left(\frac{V_p}{\delta T}\,.\,t\right)^2\,.\,dt} = \frac{V_p\,.\,\sqrt{\delta}}{\sqrt{3}} \qquad (5\text{-}7)$$

Fig.5-5 : Saw tooth voltage

Remarks (saw tooth voltage)
1. $\delta = 1$ (fig. 5-6)

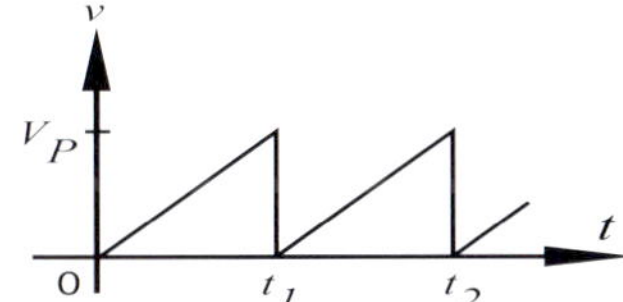

$$V_{RMS} = \frac{V_p}{\sqrt{3}} \qquad (5\text{-}8)$$

Fig. 5-6: Saw tooth voltage

2. $\delta = 1$, and combining saw tooth forms in different combinations

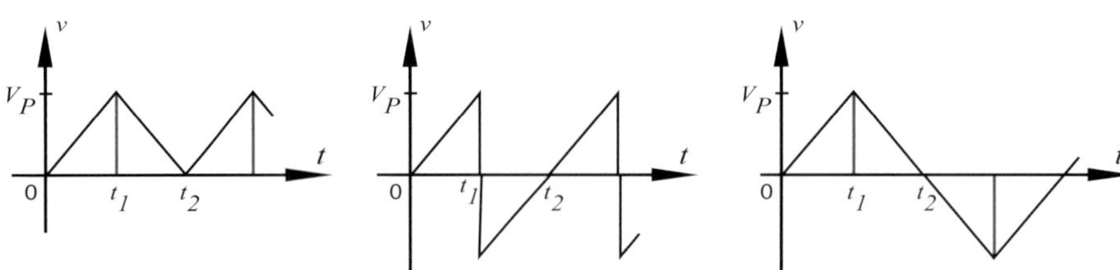

Fig. 5-7: Saw tooth and triangular waves

Since in the formula for V_{RMS} the square of the instantaneous values is taken all these waveforms have the same effective value: $V_{RMS} = \dfrac{\hat{v}}{\sqrt{3}}$ (5-9)

   e. Effective value of current.

In exactly the same manner as shown above we can find the effective value of the current: $I_{RMS} = \dfrac{\hat{i}}{\sqrt{2}}; \quad I_{RMS} = I\sqrt{\delta} \quad$ etc ...

B. Average value

1. The average value I_{AV} of a random current waveform i is the value of the constant direct current $I\,(=I_{AV})$ that moves the same amount of electricity $Q = I \cdot t_1$ in the same time period t_1.

In fig. 5-8 we see $Q = I \cdot t_1$ = surface area between the wave form and the t - axis between 0 and T. The height of the rectangle is therefore I_{AV}.(average).

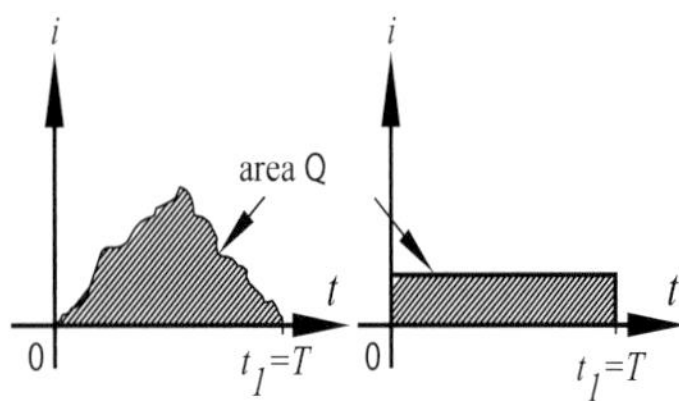

Fig. 5-8: Average value

$$I_{AV} = \frac{1}{T}\int_0^T i.dt \qquad (5\text{-}10)$$

2. Similarly: $V_{AV} = \dfrac{1}{T}\displaystyle\int_0^T v \, . \, dt$

3. Calculation examples:

a. Sine wave (fig. 5-9)

$$V_{AV} = \frac{1}{T}\int_0^T v \, . \, dt = \frac{1}{T}\int_0^T \hat{v} \, . \, sin\,\omega.t \, . \, dt = 0 \qquad (5\text{-}11)$$

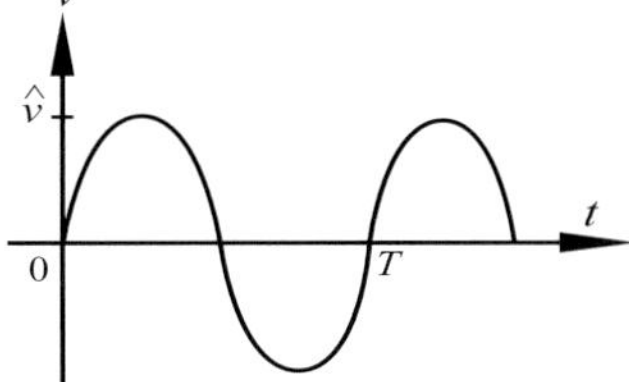

Fig. 5-9: Sine wave

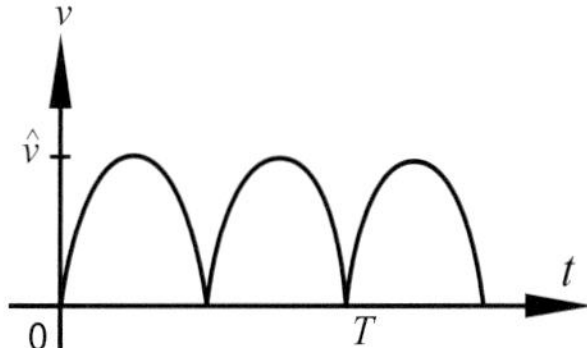

Fig. 5-10: Half wave rectified

Fig. 5-11: Full wave rectified

b. Half sine wave (fig.5-10)

$$V_{AV} = \frac{1}{T} \int_0^{T/2} v \cdot dt = \frac{1}{2\pi} \int_0^{\pi} \hat{v} \cdot \sin \omega t \cdot d\omega t = \frac{\hat{v}}{\pi} \qquad (5\text{-}12)$$

c. Full wave rectified sine wave (fig. 5-11)

$$V_{AV} = \frac{1}{T/2} \int_0^{T/2} v \cdot dt = \frac{1}{\pi} \int_0^{\pi} \hat{v} \cdot \sin \omega t \cdot d\omega t = 2\,\frac{\hat{v}}{\pi} \qquad (5\text{-}13)$$

Remark

To indicate that the dc current is the zeroth harmonic of a waveform often in the literature the notations I_0 and V_0 are used for the average value (= zeroth harmonic). In this book we use I_d and V_d for the DC component in circuits (e.g. output current of a rectifier). The notation I_{AV} and V_{AV} are reserved for the average values of semiconductors (e.g. diode current I_{AV}).

C. True RMS

1. If we take a waveform with an RMS value V_1 and add a DC component V_g then we can add

the power so that: $P = \dfrac{V_1^2}{R} + \dfrac{V_g^2}{R} = \dfrac{V_{RMS}^2}{R}$ or: $V_{RMS} = \sqrt{V_1^2 + V_g^2}$ $\qquad (5\text{-}14)$

To distinguish between the RMS value of the whole waveform and that of the AC component only we use true RMS.

True RMS = effective value of the entire waveform (AC + DC component).

AC (RMS) = effective value of the AC part only. The results in the calculation examples up to now were the true RMS values.

2. Calculation examples:

a. Fig. 5-12

Determine the RMS value of the voltage waveform which is composed of a sine wave $v_1\,(=\hat{v}\cdot\sin \omega t)$, superimposed on a DC voltage $V_g\,(=\hat{v})$.

$$V_{RMS} = \sqrt{V_1^2 + V_g^2} = \sqrt{\left(\frac{\hat{v}}{\sqrt{2}}\right)^2 + \hat{v}^2} = \hat{v} \cdot \sqrt{\frac{3}{2}} \qquad (5\text{-}15)$$

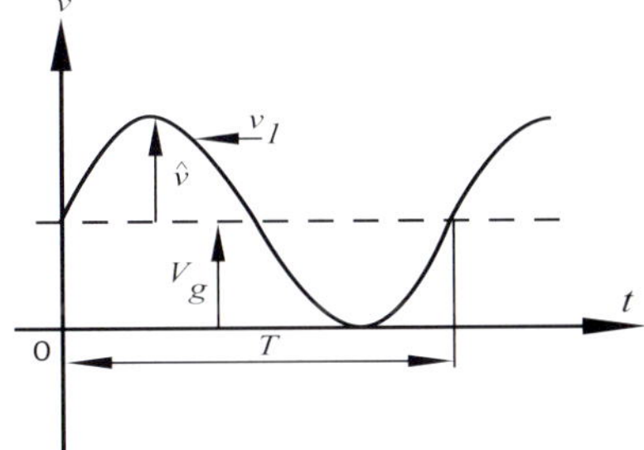

Fig. 5-12: Sine wave superimposed on a DC-voltage

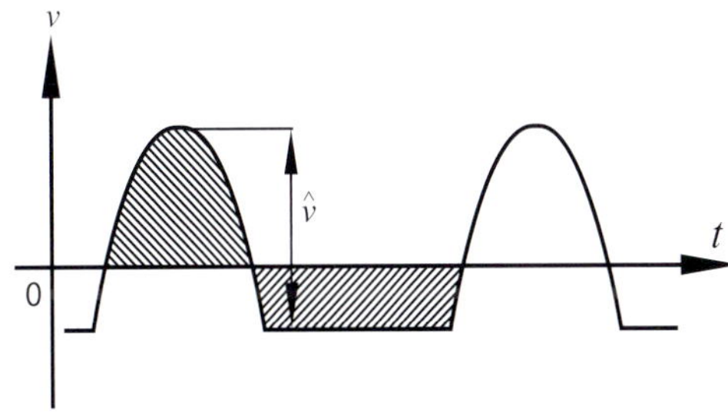

Fig. 5-13: Half sine wave without DC component

b. Half sine wave without DC component (fig. 5-13)

A half sine wave (fig. 5-10) has $V_g = \hat{v}/\pi$ and $V_{RMS} = \hat{v}/2$. We slide the half sine wave down the t-axis (fig. 5-13) so that the average value is zero (the same surface area above and below the t-axis). The new waveform (by removing V_g) has now only an AC component (RMS value V_r)

Originally we had: $V_{RMS} = \sqrt{V_r^2 + V_g^2}$.   so that: $V_r^2 = V_{RMS}^2 - V_g^2$

or: $V_r = \sqrt{V_{RMS}^2 - V_g^2} = \sqrt{\left(\dfrac{\hat{v}}{2}\right)^2 - \left(\dfrac{\hat{v}}{\pi}\right)^2} = \hat{v} \cdot \sqrt{\dfrac{1}{4} - \dfrac{1}{\pi^2}}$ (5-16)

With a resistive loaded rectifier we define the ripple factor

$$r = \frac{\textit{effective value AC components (on } R_b)}{\textit{DC component on } R_b}$$ (5-17)

Applied to a half wave rectifier we find:

$$r = \frac{V_r}{V_g} = \hat{v} \cdot \frac{\sqrt{\dfrac{1}{4} - \dfrac{1}{\pi^2}}}{\hat{v}/\pi} = 1.21 = 121\%$$ (5-18)

D. Numeric example power:

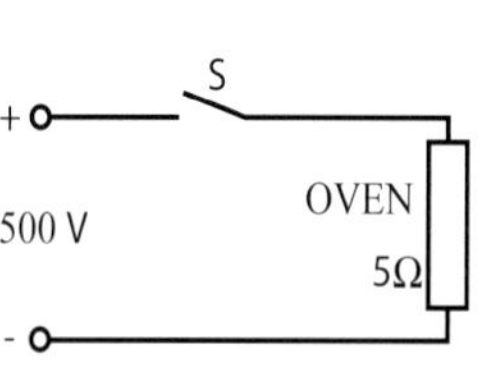
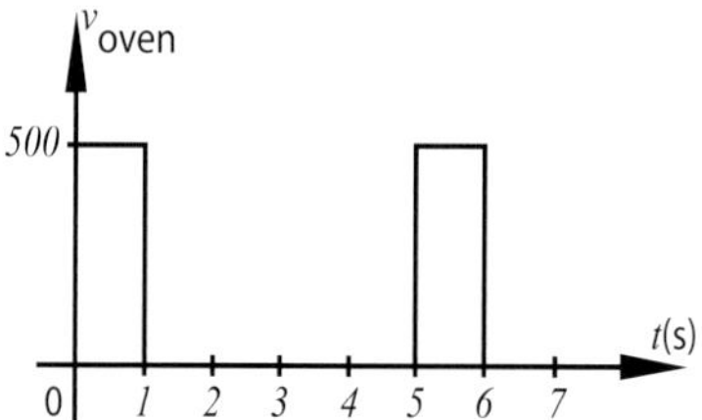

Fig. 5-14 : Pulse controlled oven

Given: switch closes according to the indicated time diagram

Question: determine the dissipated power in the oven

Solution:

1.   With average values: $P = \dfrac{V^2}{R} = \dfrac{500^2}{5} = 50\text{kW}$

This power is on for 1/5 of the time: $P_{average} = 10\text{kW}$

2.   Calculating using the RMS value:

$$P = \frac{V_{RMS}^2}{R} = \frac{(V\sqrt{\delta})^2}{R} = \frac{\left(500 \cdot \sqrt{\dfrac{1}{5}}\right)^2}{5} = 10\text{kW}$$

3.   Calculating using the average value:

$$P = \frac{100^2}{5} = 2\text{kW} \text{ (WRONG !!!)}$$

Table 5-1: Average and RMS values

Waveform	RMS value	Average value	Form Factor
	$X_{RMS} = \sqrt{\dfrac{1}{T} \displaystyle\int_0^T x^2 \cdot dt}$	$X_{AV} = \dfrac{1}{T} \displaystyle\int_0^T x \cdot dt$	$a = \dfrac{X_{RMS}}{X_{AV}}$
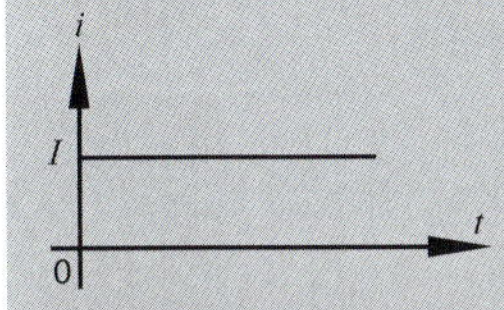 Fig. 5-15	$I_{RMS} = I$	$I_{AV} = I$	$a = 1$
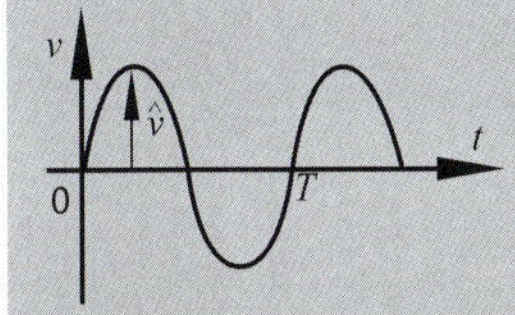 Fig. 5-16	$V_{RMS} = \dfrac{\hat{v}}{\sqrt{2}}$	$V_{AV} = 0$	————
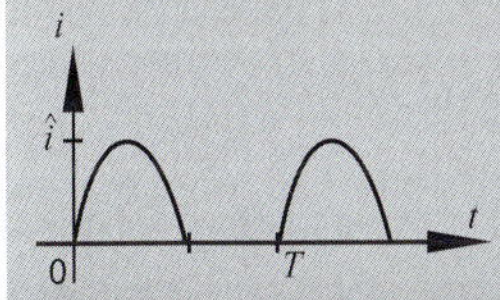 Fig. 5-17	$I_{RMS} = \dfrac{\hat{i}}{2}$	$I_{AV} = \dfrac{\hat{i}}{\pi}$	$a = 1.57$
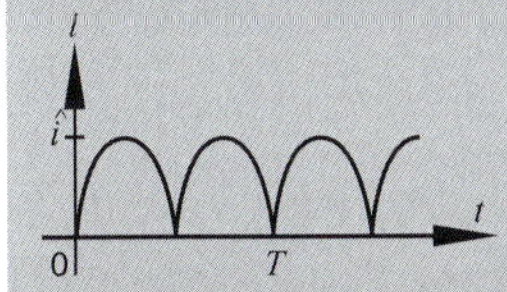 Fig. 5-18	$I_{RMS} = \dfrac{\hat{i}}{\sqrt{2}}$	$I_{AV} = 2 \cdot \dfrac{\hat{i}}{\pi}$	$a = 1.11$
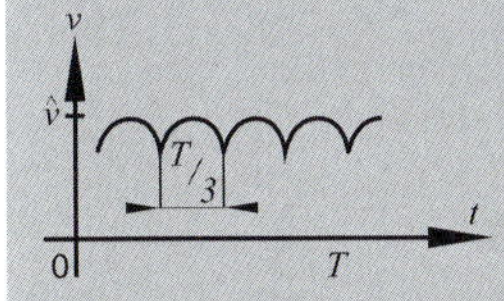 Fig. 5-19	$V_{RMS} = 0.84 \cdot \hat{v}$	$V_{AV} = 0.83 \cdot \hat{v}$	$a = 1.02$
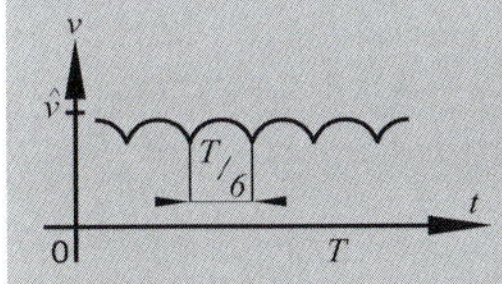 Fig. 5-20	$V_{RMS} = 0.9557 \cdot \hat{v}$	$V_{AV} = 0.9549 \cdot \hat{v}$	$a = 1$

Waveform	RMS value	Average value	Form Factor
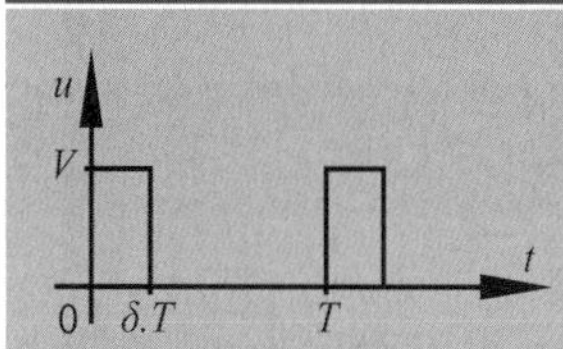 Fig. 5-21	$V_{RMS} = V . \sqrt{\delta}$	$V_{AV} = V . \delta$	$a = \dfrac{1}{\sqrt{\delta}}$
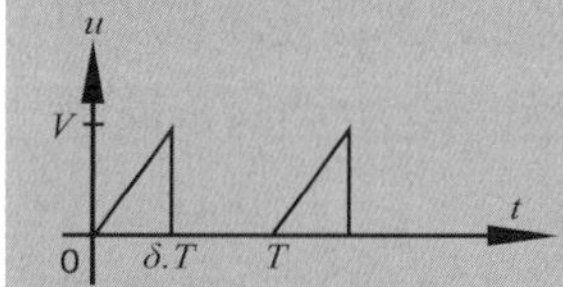 Fig. 5-22	$V_{RMS} = \dfrac{V . \sqrt{\delta}}{\sqrt{3}}$	$V_{AV} = \dfrac{V . \delta}{2}$	$a = \dfrac{2}{\sqrt{3.\delta}}$
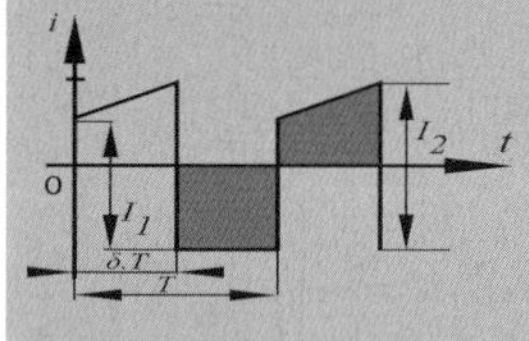 Fig. 5-23	$I_{RMS} = \sqrt{\delta \left[\dfrac{I_1^2 + I_1 . I_2 + I_2^2}{3} - \dfrac{\delta}{4} (I_1 + I_2)^2 \right]}$	$I_{AV} = 0$	
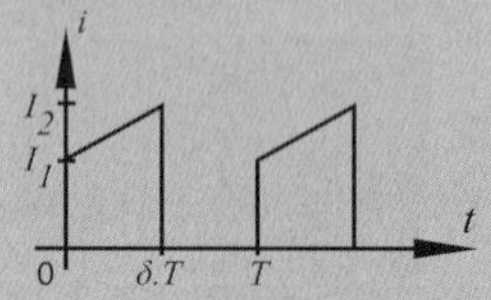 Fig. 5-24	$I_{RMS} = \delta . \sqrt{\dfrac{I_1^2 + I_1 I_2 + I_2^2}{3}}$	$I_{AV} = \delta \dfrac{(I_1 + I_2)}{2}$	$a = \dfrac{I_{RMS}}{I_{AV}}$
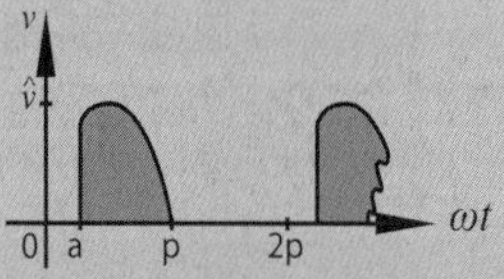 Fig. 5-25	$V_{RMS} = \dfrac{\hat{v}}{2} \sqrt{\dfrac{180° - \alpha}{180°} + \dfrac{\sin 2\alpha}{2.\pi}}$	$V_{AV} = \dfrac{\hat{v}}{2.\pi} (1 + \cos \alpha)$	$a = \dfrac{V_{RMS}}{V_{AV}}$
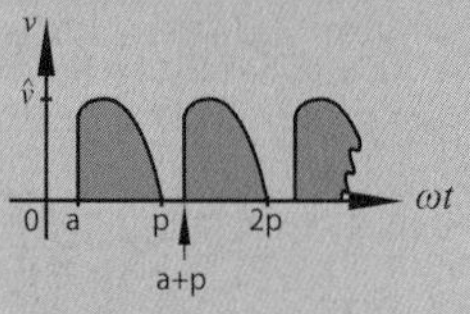 Fig. 5-26	$V_{RMS} = \dfrac{\hat{v}}{\sqrt{2}} \sqrt{\dfrac{180° - \alpha}{\alpha} + \dfrac{\sin 2.\alpha}{2.\pi}}$	$V_{AV} = \dfrac{\hat{v}}{\pi} (1 + \cos \alpha)$	$a = \dfrac{V_{RMS}}{V_{AV}}$
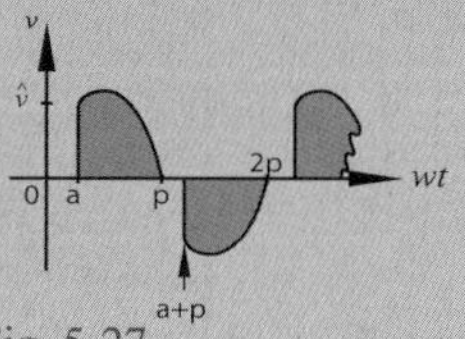 Fig. 5-27		$V_{AV} = 0$	

Three phase converters : Calculations and formulas chapter 8

1.2 Impedance transformer

Consider a loss free transformer (fig. 5-28) with transformation ratio $n = \dfrac{N_p}{N_s}$

Since we assume no losses the emf and terminal voltage are the same: $V_p = E_p$ and $V_s = E_s$ so that:

$$n = \frac{N_p}{N_s} = \frac{E_p}{E_s} = \frac{V_p}{V_s} \; .$$

The power absorbed by the secondary load (R_s) is:

$$P_S = \frac{V_S^2}{R_S}$$

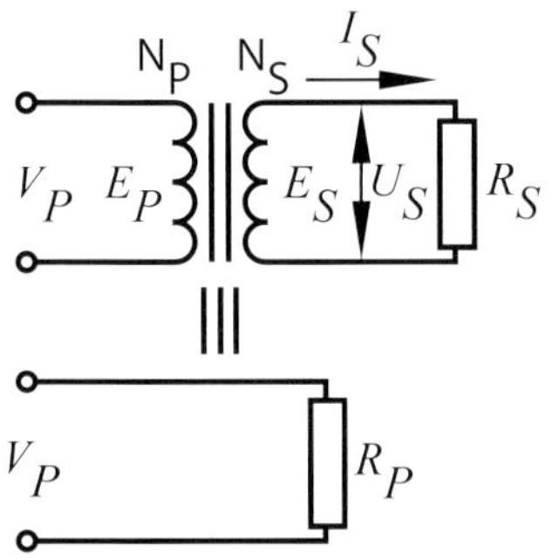

This power has to flow in the primary side drawn from the supply. The entire circuit can be replaced with an equivalent resistor (R_p), so that: $\quad P = V_p^2/R_p = V_S^2/R_S$

or:

$$R_p = R_S \cdot \left(\frac{V_p^2}{R_S^2} \right) = R_S \cdot n^2 \qquad (5\text{-}19)$$

Fig. 5-28: Impedance transformer

We can imagine a resistor transformed from secondary to primary ($R_p = n^2 \cdot R_S$) or from primary to secondary ($R_S = R_p / n^2$) .

1.3 Harmonics

A. Fourier series

A periodic function with frequency f (pulsating ω) can according to Fourier be written as a series:

$$a = A_0 + \sum_{k=1}^{\infty} X_K \cdot sin\,(\,k\omega t + \phi_K) \qquad (5\text{-}20)$$

Here in:

1. k = every positive whole number (1,2,3...)
2. A_0 = the average value of the function (in electrical theory called the DC-component)
3. Harmonics (= sinusoidal components):

    $X_1 \cdot sin\,(\,\omega t + \phi_1\,)$ = first harmonic (= the fundamental)

    $X_2 \cdot sin\,(\,2\omega t + \phi_2\,)$ = second harmonic

    $X_3 \cdot sin\,(\,3\omega t + \phi_3\,)$ = third harmonic

    etc....

Solving for these harmonics gives:

$$X_k \cdot sin\,(\,k.\omega t + \phi_k\,) = (\,X_k \cdot cos\,\phi_k\,) \cdot sin\,k\omega t + (X_k \cdot sin\,\phi_k\,) \cdot cos\,k\omega t$$

We can now rewrite the Fourier series:

$$a = A_0 + \sum_{k=1}^{\infty} (\,A_K \cdot sin\,k\omega t + B_K \cdot cos\,k\omega t\,) \qquad (5\text{-}21)$$

According to Fourier a periodic function can be considered the sum of:

1. DC component $A_0 = \dfrac{1}{2\pi} \displaystyle\int_{-\pi}^{+\pi} a \cdot d\,(\omega t)$ (5-22)

2. Serie of sine waves with amplitude: $A_k = \dfrac{1}{\pi} \displaystyle\int_{-\pi}^{+\pi} a \cdot \sin k\omega t \cdot d(\omega t)$ (5-23)

3. Serie of cosine waves with amplitude: $B_k = \dfrac{1}{\pi} \displaystyle\int_{-\pi}^{+\pi} a \cdot \cos k\omega t \cdot d(\omega t)$ (5-24)

B. Simplifications

1. Symmetry with respect to the origin.

If the time function is symmetrical with respect to the origin (fig. 5-29) then the series will only

contain sine wave terms. Indeed, since $a(t) = - a(-t)$ then $A_0 = 0$ and $B_k = 0$

so that: $a = \displaystyle\sum_{k=1}^{k=\infty} A_k \cdot \sin k\omega t$

2. Symmetry with respect to the ordinate-axis.

Since now $a(t) = a(-t)$ then $A_k = 0$ and the series contains no sine wave terms.

$$a = A_0 + \sum_{k=1}^{k=\infty} B_k \cdot \cos k\omega t \,.$$

Fig. 5-30 shows an even symmetrical function with respect to the ordinate-axis.

3. Half wave symmetry.

Calculation shows that this series for a function as shown in fig. 5-31 only has uneven harmonics.

$$a = \sum_{k=1}^{k=\infty} (A_k \cdot \sin k\omega t + B_k \cdot \cos k\omega t)$$

Here in $k =$ uneven number

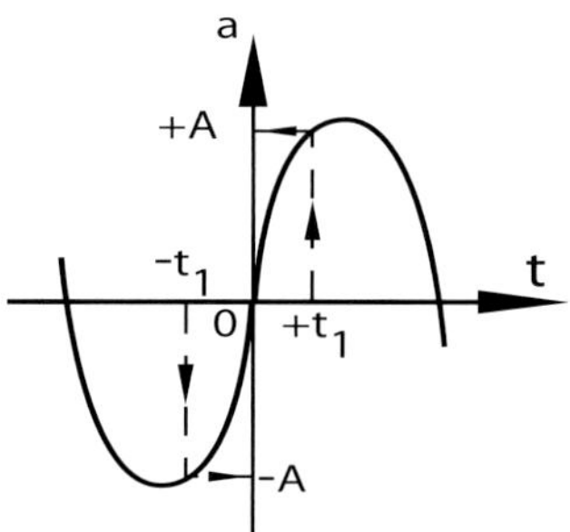

Fig. 5-29: Symmetrical function with
respect to the origin

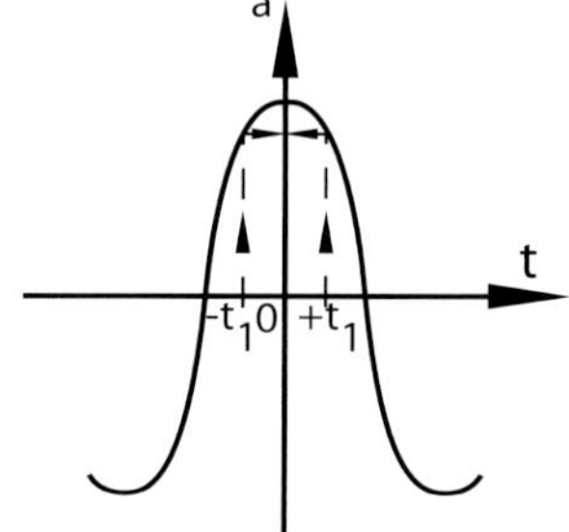

Fig. 5-30: Function, even symmetrical
with respect to ordinate-axis

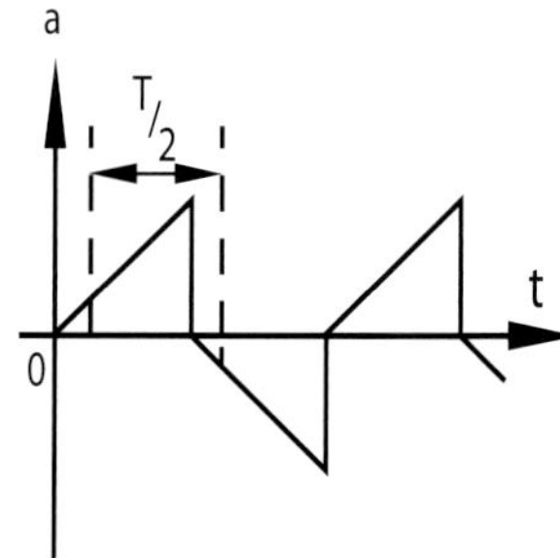

Fig. 5-31: Half wave symmetry

C. Example of a series expansion

Fig. 5-32 shows a square wave which contains "even" symmetry (no sine terms!) and also posses half-wave symmetry (contains no DC component and only uneven harmonics!).
Calculation gives:

$$A_0 = A_k = 0 \ \text{ and: } \ B_k = \frac{1}{\pi}\int_{-\pi}^{+\pi} V \cdot \cos k\omega t \cdot d\omega t = 4 \cdot \frac{1}{\pi}\int_{0}^{\pi/2} V \cdot \cos k\omega t \cdot d\omega t$$

$$= \frac{4 \cdot V}{k \cdot \pi}\int_{o}^{\pi/2} \cos k\omega t \, d\omega t = \frac{4 \cdot V}{k \cdot \pi}\left(\sin k\omega t\right)\Big|_{0}^{\pi/2} = \pm\frac{4 \cdot V}{\pi \cdot k}$$

$+$ for k = 1, 5, 9,...

$-$ for k = 3, 7, 11, ...

We can now write the series as:

$$v = \sum_{k=1}^{k=\infty} B_k \cdot \cos k\omega t = \frac{4}{\pi} \cdot V \cdot \left(\cos \omega t - \frac{\cos 3\omega t}{3} + \frac{\cos 5\omega t}{5} - \ldots \right)$$

The graphical representation whereby the lines perpendicular to frequency axis represent the amplitude of components from the series expansion is referred to as a **frequency spectrum**.
Fig. 5-33 shows the frequency spectrum of the square wave from fig. 5-32

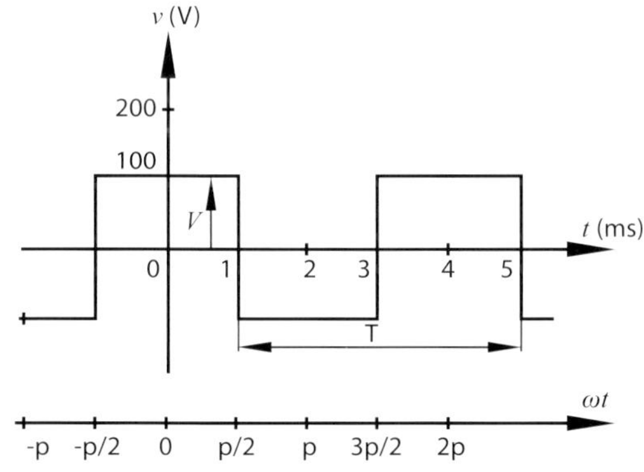

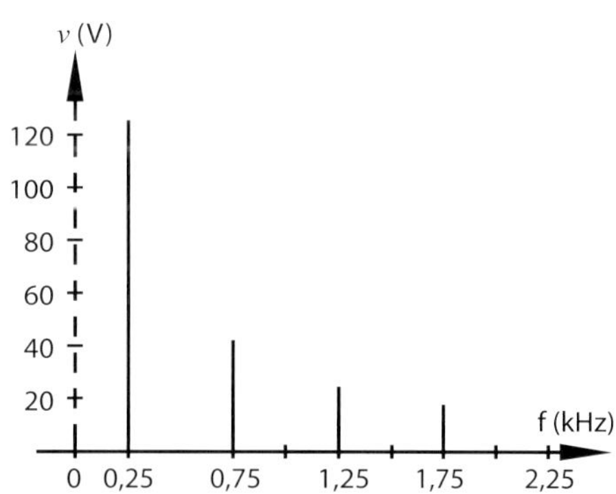

Fig. 5-32: Square wave

Fig. 5-33: Frequency spectrum of fig. 5-32

1.4 The s-domain

A. Laplace-transformation

- With the Laplace-transformation an independent variable (time t) is converted to another independent variable (s).

Definition:
$$\mathcal{L}[f(t)] = F(s) = \int_{o}^{\infty} e^{-st} \cdot f(t) \cdot dt \qquad (5\text{-}25)$$

- From the Laplace- transformed $F(s)$ we can find the original function again with an inverse Laplace-transformation : $\mathcal{L}^{-1}[F(s)] = f(t)$.
- A linear differential equation is converted by a Laplace-transformation into an algebraic expression.
  This expression is easy to solve after which the inverse Laplace-transformation is applied.
- To transform a function with Laplace we can use the definition integral or make use of table 5-2. This table also enables rapid determination of the inverse Laplace-transformation of a number of functions.
- Differentiation in the time domain is equivalent to multiplication with s in the s-domain.

$$\mathcal{L}[\frac{df(t)}{dt}] = s \cdot F(s) - f(0) \qquad (5\text{-}26)$$

Integration in the time domain corresponds to dividing by s in the s-domain.

- Linear electrical elements

time domain	s-domain
$v_R = i_R \cdot R$	$V_R(s) = I_R(s) \cdot R$
$v_L = L \cdot di_L/dt$	$V_L(s) = L \cdot s \cdot I_L(s)$
$v_C = \frac{1}{C}\int_0^t i_C \cdot dt$	$V_C(s) = \dfrac{I_C(s)}{s \cdot C}$

- If we divide (in the s-domain) voltage by current then we find the impedances R , s.L and $\dfrac{1}{s.C.}$

element	s-domain	sinusoidal service
R	R	R
L	$s\,L$	$j\,\omega\,L$
C	$\dfrac{1}{s\,C}$	$\dfrac{1}{j\,\omega\,C}$

B. Laplace-transformations of some functions

Table 5-2

$f(t)$	$F(s)$	$f(t)$	$F(s)$
1 (Heaviside's unit step)	$1/s$	$\sin \omega t$	$\dfrac{\omega}{s^2 + \omega^2}$
A (step A)	A/s	$\cos \omega t$	$\dfrac{s}{s^2 + \omega^2}$
δ (unit pulse) = Dirac pulse	1	$\sin(\omega t + \phi)$	$\dfrac{s\,.\sin\phi + \omega\,.\cos\phi}{s^2 + \omega^2}$
$e^{+\alpha t}$; $e^{-\alpha t}$	$\dfrac{1}{s-\alpha}$; $\dfrac{1}{s+\alpha}$	$e^{-\alpha t}\,.\sin\omega t$	$\dfrac{\omega}{(s+\alpha)^2 + \omega^2}$
$1 - e^{-\alpha t}$	$\dfrac{\alpha}{s\,(s+\alpha)}$	$e^{-\alpha t}\,.\cos\omega t$	$\dfrac{(s+\alpha)}{(s+\alpha)^2 + \omega^2}$
t (ramp)	$1/s^2$	$\dfrac{t}{2.\alpha}\,.\sin\alpha t$	$\dfrac{s}{(s^2 + \alpha^2)^2}$
$A.t$ (ramp)	A/s^2	$\dfrac{1}{2.\alpha^3}(\sin\alpha t - \alpha t.\cos\alpha t)$	$\dfrac{1}{(s^2 + \alpha^2)^2}$
t^n (n = positive integer)	$\dfrac{n!}{s^{n+1}}$	$\dfrac{1}{\alpha - \beta}\,.\,(e^{\alpha t} - e^{\beta t})$	$\dfrac{1}{(s-\alpha).(s-\beta)}$
$t\,.\,e^{-\alpha t}$	$\dfrac{1}{(s+\alpha)^2}$	$\dfrac{1}{\alpha^2}\,.\,(1 - \cos\alpha t)$	$\dfrac{1}{s\,.\,(s^2 + \alpha^2)}$
$t^n\,.\,e^{-\alpha t}$	$\dfrac{n!}{(s+\alpha)^{n+1}}$	$\sinh\alpha t$	$\dfrac{\alpha}{s^2 - \alpha^2}$
$-\dfrac{1}{\alpha^2}\,.\,e^{-t/\alpha}$	$\dfrac{s}{1 + \alpha.s}$	$\cosh\alpha t$	$\dfrac{s}{s^2 - \alpha^2}$
$e^{-\alpha t} - e^{-\beta t}$	$\dfrac{\beta - \alpha}{(s+\alpha)(s+\beta)}$	$e^{-\beta t}\,.\,\sinh\alpha t$	$\dfrac{\alpha}{(s+\beta)^2 - \alpha^2}$
$\dfrac{1}{(\beta^2 - \alpha^2)}\,.\,(\cos\alpha t - \cos\beta t)$	$\dfrac{s}{(s^2 + \alpha^2)\,(s^2 + \beta^2)}$	$e^{-\beta t}\,.\,\cosh\alpha t$	$\dfrac{s+\beta}{(s+\beta)^2 - \alpha^2}$

C. Theorems

In addition to table 5-2 we can also make good use of a number of theorems.

Table 5-3

theorem	t - domain	s - domain
1. Damping property	$e^{-\alpha t} . f(t)$	$F(s + \alpha)$
2. Linearity	$a . f_1(t) \pm b . f_2(t)$	$a . F_1(s) \pm b . F_2(s)$
3. Scaling property	$if\ F(s) = \mathscr{L}\,[f_1(t)\,]\ then: \mathscr{L}\,[f_1(t/\lambda)\,] = \lambda . F(\lambda.s)$	
4. Time delay	$f(t - \lambda)$	$e^{-\lambda s} . F(s)$
5. Initial value theorem	$f(0) = \lim_{t \to 0} f(t)$	$f(0) = \lim_{s \to \infty} s. F(s)$
6. Final value theorem	$f(\infty) = \lim_{t \to \infty} f(t)$	$f(\infty) = \lim_{s \to 0} s. F(s)$

1.5 Power in sinusoidal service

We calculate the useful power delivered by a sinusoidal voltage and sinusoidal current that occur simultaneously. Both signals do not necessarily to have the same frequency. We assume in addition a phase difference ϕ at $t = 0$.

$$v = \hat{v} . \cos n.\omega.t$$
$$i = \hat{\imath} . \cos (m.\omega.t - \phi)$$

The instantaneous power is: $p = u . i$.

Per period of the pulsation we find the the average or RMS power:

$$P = P_{AV} = P_{RMS} = \frac{1}{2\pi} \int_0^{2\pi} p . d\omega t = \frac{1}{2\pi} \int_0^{2\pi} v . i . d\omega t$$

$$P = \frac{1}{2\pi} \int_0^{2\pi} \hat{v} . \cos n\omega t . \hat{\imath} . \cos (m\omega t - \phi) . d\omega t$$

$$P = \frac{\hat{v} . \hat{\imath}}{2\pi} \int_o^{2\pi} \cos n\omega t . \cos (m\omega t - \phi) . d\omega t$$

With $\cos a \cdot \cos b = \dfrac{\cos(a+b) + \cos(a-b)}{2}$ becomes:

$$P = \frac{\hat{v}\cdot\hat{i}}{4\pi} \int_{0}^{2\pi} \cos\left[(n+m)\cdot\omega t - \phi\right].d\omega t + \frac{\hat{v}\cdot\hat{i}}{4\pi} \int_{0}^{2\pi} \cos\left[(n-m)\cdot\omega t + \phi\right].d\omega t$$

1. $\boxed{n = m}$

$$P = \frac{\hat{v}\cdot\hat{i}}{4\pi} \int_{0}^{2\pi} \cos\left[2n\omega t - \phi\right].d\omega t + \frac{\hat{v}\cdot\hat{i}}{4\pi} \int_{0}^{2\pi} \cos\phi \; d\omega t$$

$$P = 0 + \frac{\hat{v}\cdot\hat{i}}{4\pi} \cdot (\cos\phi)\cdot\omega t \,\Big|_{0}^{2\pi} \;\; = \frac{\hat{v}\cdot\hat{i}}{2}\cdot\cos\phi$$

$$\boxed{P = V \cdot I \cdot \cos\phi}$$

2. $\boxed{n \neq m}$

$$P = \frac{\hat{v}\cdot\hat{i}}{4\pi} \int_{0}^{2\pi} \frac{\cos\left[(n+m)\omega t - \phi\right]}{(n+m)} \cdot d(n+m)\omega t \;\; + \;\; \frac{\hat{v}\cdot\hat{i}}{4\pi} \int_{0}^{2\pi} \frac{\cos\left[(n-m)\omega t + \phi\right]}{(n-m)} \cdot d(n-m)\omega t$$

$$P = \frac{\hat{v}\cdot\hat{i}}{4\pi} \cdot \frac{\sin\left[(n+m)\omega t - \phi\right]}{(n+m)} \Big|_{0}^{2\pi} \;\; + \;\; \frac{\hat{v}\cdot\hat{i}}{4\pi} \cdot \frac{\sin\left[(n-m)\omega t + \phi\right]}{(n-m)} \Big|_{0}^{2\pi}$$

$$P = \frac{\hat{v}\cdot\hat{i}}{4\pi} \left[\frac{\sin(2\pi - \phi) - \sin(-\phi)}{(n+m)} \right] \;+\; \frac{\hat{v}\cdot\hat{i}}{4\pi} \left[\frac{\sin(2\pi + \phi) - \sin\phi}{(n-m)} \right]$$

$$\boxed{P = 0}$$

Conclusion: *Only a current and voltage with the same frequency acting simultaneously can form active power.*

2. OPTO-ELECTRONICS

2.1 LED (Light Emitting Diode)

The light emitting diode (LED) has been used for years as a signal lamp, as number indicator and as a light emitting transmitter in an opto-coupler. In recent years the LED is becoming ever more popular as a light source. This aspect is dealt with in chapter 15 in the applications of power electronics. The operating principle of the LED rests on the release of energy by direct recombination of electrons and holes in a PN-semiconductor. This energy is released in the form of electromagnetic radiation. In general this is:

$$W = h \cdot f \qquad\qquad (5\text{-}29)$$

with: W = energy (eV)

$\quad\;\; h$ = Planck constant = $4.133 \times .10^{-15}$ eVs

$\quad\;\; f$ = frequency of electromagnetic waves in Hz.

In the specific case of a semiconductor we find: $W = W_g$ = energy gap or band gap between the valence and conduction band. This energy gap is also temperature dependent. Si has a $W = 1.12\text{eV}$ at 0 K($= -273°$ C) and this becomes 1.1 eV at 300K (room temperature). The wavelength of the released radiation is given by:

$$\lambda = \frac{c}{f} = \frac{300 \cdot 10^6}{f} = \frac{300 \cdot 10^6}{W_g} \cdot h = \frac{300.10^6 \times 4.133 \times 10^{-15}}{W_g} = \frac{1239}{W_g} \;\text{(nm)}$$

Herein $c = 300{,}000\text{km/s}$ = speed of light.

To have continuous radiation we require continuous recombination to occur between electrons from the conduction band and holes from the valence band. This continuous process can be obtained by a forward polarised PN-junction (e.g. with $I_F = 20\text{mA}$).

To obtain visible light ($\lambda = 400$ to 700nm, see fig. 5-36) W_g must lie between 1.77 and 3.1eV. The semiconductor material most often used for the construction of LED's is 3-5 crystals with the following configuration: ● group 3: Ga, Al   ● group 5: As, P, N.

The different combinations and impurities lead to different wavelengths of the radiated light. The spectrum of a galliumphosphide - LED (GaP) can be yellow ($\lambda = 575\text{nm}$) to green ($\lambda = 560\text{nm}$). Gallium-arsenidephoshide (GaAsP) emits yellow ($\lambda = 590\text{nm}$) to red light ($\lambda = 650\text{nm}$). GaAs has an energy gap of about 1.43eV so that $\lambda = 900\text{nm}$ and GaAs can operate as an infrared transmitter (IRED = infrared emitter). Fig 5-35 shows the curves that accompany a GaAsP - LED.

Remarks

1. With nominal current the forward voltage drop of a LED is between 1.3 to 3V depending on the the type of semiconductor and the impurities.

2. With the passing of time LED's start to age: the radiated power decreases. High currents and high ambient temperatures have a disastrous effect on the aging process. A load operates classically with $I_F = 20\text{mA}$, but if a lower light intensity is required an $I_F = 5$ or 10mA is used.

3. Fig. 5-35d shows a LED as indicator. With $V = 12\text{V}$; $I_F = 10\text{mA}$ and $V_F = 1.6\text{V}$ the value of the current limiting resistor becomes: $R_V = \dfrac{V - V_F}{I_F} = \dfrac{12 - 1.6}{10} \approx 1\ \text{k}\Omega$

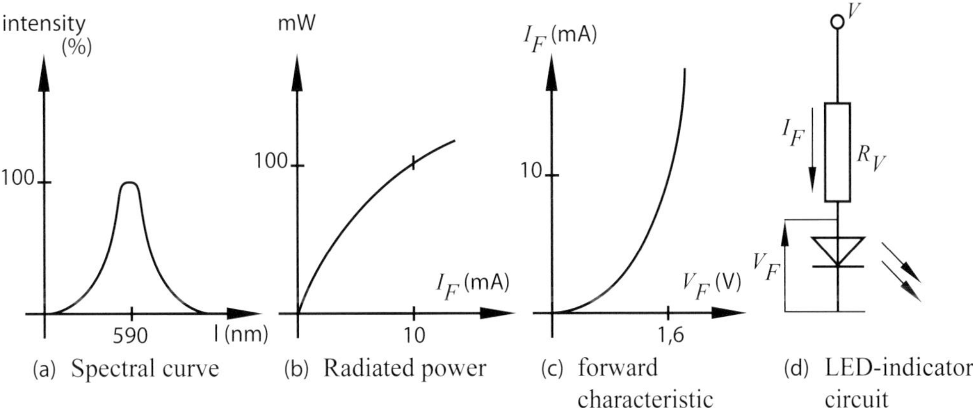

Fig. 5-35: Curves of a GaAsP-LED

2.2 Spectral sensitivity

In fig. 5-36 we draw:

1. The standard human eye curve prepared by the Commission Internationale de l'Eclair (the CIE-curve). This curve not only provides the sensitivity of the eye at different wavelengths but also the lumens/watt conversion of the light source at that particular frequency.
2. The spectral curve of a Ga-As transmitter (LED).
3. The wavelength of maximum sensitivity of other LED-materials.
4. The sensitivity of a silicon detector (diode, transistor). It is clear why Ga-AS LED's are used with an (Si-) transistor or diode as opto-couple elements.

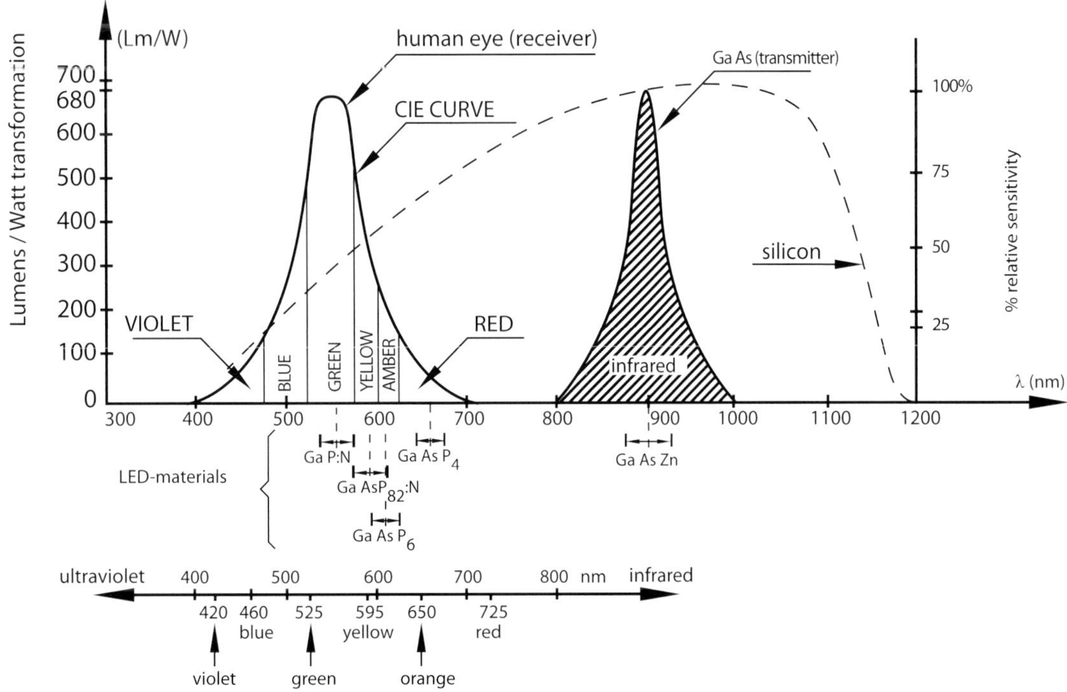

Fig. 5-36: Human eye curves and spectral sensitivity of LEDs and Si-detectors

2.3 Optical couple elements

The inverse current through a PN-junction also changes through exposure. This property is used by photo-diodes and photo-transistors. In a photo-transistor the base collector diode is exposed to light. A combination of LED and photo transistor is a so called optocoupler (opto-isolator or couple element: see fig. 5-37a).

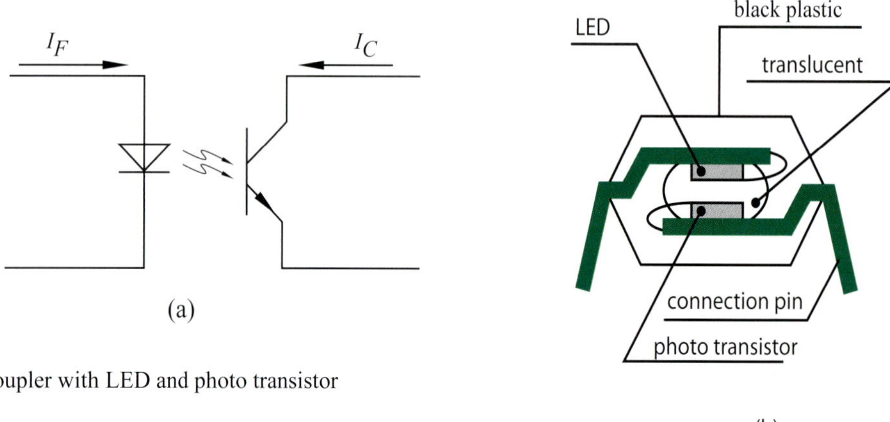

Fig. 5.37: Optocoupler with LED and photo transistor

Information is transmitted via an optical route, without galvanic contact between in and output. We send a current I_F (e.g. 10mA) through the LED (light emitting diode). This LED (transmitter) will convert I_F in radiated light that via a translucent material hits a detector (e.g. a photo transistor) (fig. 5-37b). The detector transforms the light into electric current I_C (with a connected transistor).

The current transfer ratio is important $\eta = \dfrac{\Delta I_C}{\Delta I_F} = \dfrac{I_C}{I_F}$ (= 20 to 600 %).   (5-28)

Sender and receiver can be constructed in a DIP-housing. Equally easily they can be connected via fibre optic cables from a few meters to a few kilometers in length. For very large distances a laser can be used as transmitter. The larger the distance between the LED and photo transistor, the larger the break over voltage (1 to 10kV) between input and output. In the case where an opto coupler is used as detection element to record fast current and voltage changes the special attention needs to be paid to the choice of type. An opto-coupler is a slow element. In addition there is capacitive feedback between output and input that can cause problems at high switching frequencies. In addition to much used types with LED - photo transistor and LED photo - Darlington's there are also a number of other combinations possible such as LED - photo diode, LED – photo thyristor, LED - photo triac, LED - photo FET, etc...

2.4 Photovoltaic cell

Photovoltaic cells convert visible and invisible light into an emf. The currently used cell is in fact a Si-diode. In fig. 5-38a we have drawn such a diode with the contact potential between metal and silicon and the diffusion voltage of the barrier layer in no-load conditions. If light hits the junction free pairs of electrons and holes appear. The minority charge carriers (electrons in the P-area) are attracted to the positive potential (N-area) so that V_d reduces. Since the contact potential between metal and silicon does not change, a photo - emf (approximately 0.5V) results across the diode depending on the luminous flux ϕ .

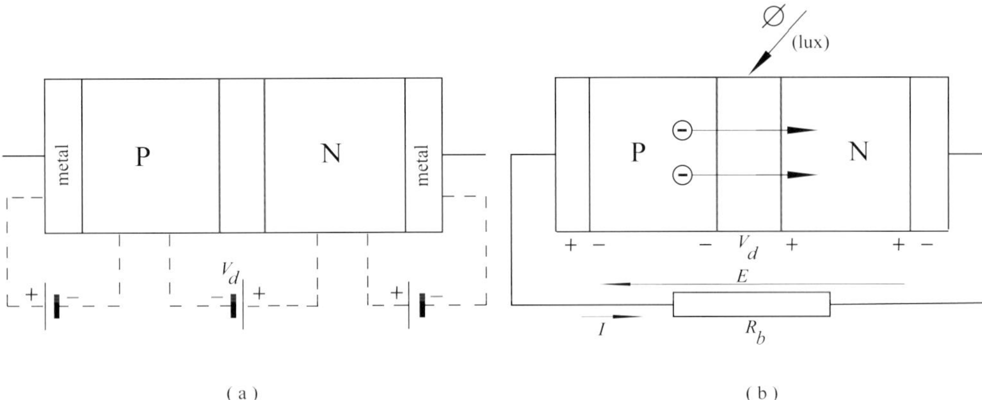

(a)   (b)

Fig. 5-38: Si-photo-voltaic cell

From fig. 5-36 it follows that Si-photo-voltaic cells are sensitive for light with a wavelength of between 400 and 1100 nm. This cells are especially sensitive for red and infrared light. A logarithmic scale links photo-emf (V) and illuminance (lux).

If we connect a resistor R_b across a photo-voltaic cell then we can determine the operating point of this circuit (fig. 5-39). From the product of the photo-emf and the current produced we find the power that the cell gives. In the example of fig. 5-39 we find a maximum power of 27.8mW with a load resistor of 2.52 kΩ . We always attempt to load a PV - cell with an optimal resistor so that maximum power is obtained at the existing illumination. This operating point is indicated as the MPP (maximum power point).

Note the PV-cell can be short circuited and then has a current of I_{SC} . In fig. 5-29 this short circuit current I_{SC} = 0.15 mA and the open circuit photovoltaic emf is about 0.45V. To increase the voltage and required current we form batteries of photovoltaic cells. These are called solar batteries or sun panels. These days they are being more and more promoted as a clean energy source. This will be dealt with in chapter 15. Photovoltaic cells are used in control circuits, light measurement of visible and almost infrared light, to power calculators and photo-electric relays, etc.

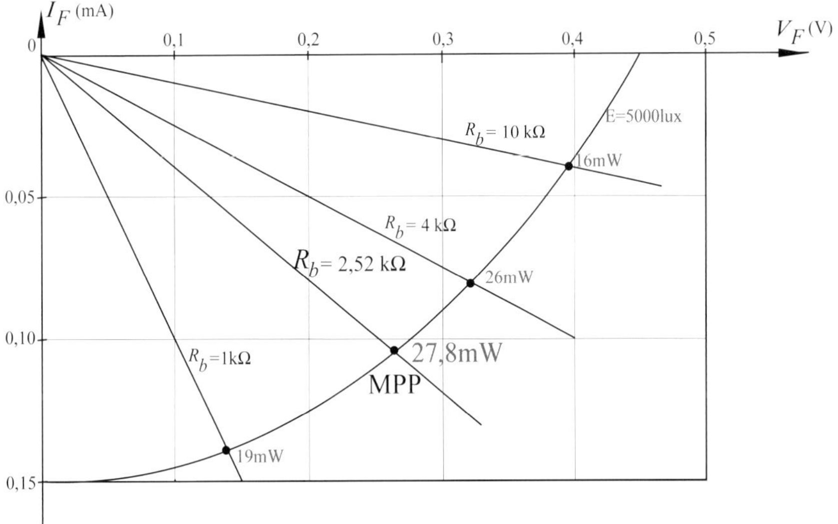

Fig. 5-39: Optimum operating point of a loaded cell

Photo Ixys (KWx): Solar BIT's are monocrytalline PV-cells with a high degree of efficiency (17%).
Spectral sensitivity 300 to 1100nm. The solarbit exists in different voltage-current configurations (see www.ixys.com). Examples: XOB17-12x1: 0.69V - 39mA with an MPP of 0.51V - 39mA. XOB17-04x3: MPP of 1.51V - 11.7mA.

3. HALL-EFFECT SENSORS

3.1 Hall-effect

If we place a current carrying conductor or semiconductor in a perpendicular magnetic field B (fig. 5-40a) then an electric field arises perpendicular to the I-B surface. This effect is known as the Hall-effect. This effect was discovered in 1879 by the American physicist Edwin Herbert Hall.

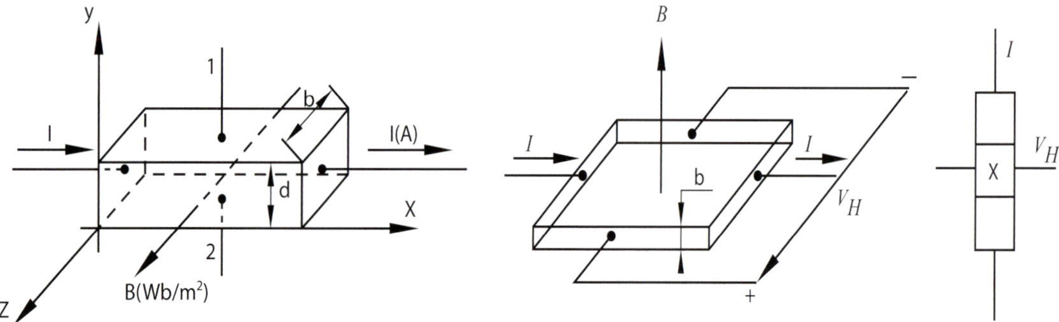

Fig 5-40: Hall-effect

Consider a current composed of holes (I) flowing in the positive x-direction. Due to the magnetic field the holes are forced towards the lower surface (2). Side 2 of the conductor is positive relative to surface 1, in other words an electric field E exists in the positive y-direction, and therefore perpendicular to the x-z surface (*I-B* surface).

A Hall voltage is generated $V_H = E.d$ between the electrodes 1 and 2.

In equilibrium the electric field will exercise a force (with opposite polarity) equal to the force of the magnetic field on the flowing charge carriers: $q . E = B . q . v$ with q = charge and v = drift velocity. From this it follows: $E = B . v$ and $V_H = E . d = B . v . d$.

If there are n holes per m^3 then the charge density: $r = n . q$ (C/m^3). If this charge is displaced with a drift velocity v (m/s) then the current density: $J = n . q . v = \rho . v$ (A/m^2).

The current density may also be written as: $J = \dfrac{I}{d . b}$, so that : $V_H = B . v . d = B . \dfrac{J}{\rho} d = \dfrac{B I}{\rho . b}$.

With the Hall constant of the material: $R_H = 1/\rho$, this becomes :

$$V_H = R_H . \frac{B . I}{b} = k . I . B \tag{5-29}$$

Table 5-4

Hall constant R_H (cm^3 / As)			
metals	10^{-4}	InSb	380
Ge	10^{+3}	GaAs	10^{+4}
Si	10^{+6}		

In contrast to metals some semiconductors (e.g. indium based) have an important R_H-value (see table 5-4).

From $V_H = k . I . B$ the possible applications follow:

3.2 Applications

1. With a constant current a magnetic field can be measured: $V_H = k_1 . B$
2. With constant magnetic field, e.g. from a permanent magnet, it is possible to measure currents:
   $V_H = k_2 . I$
3. If I is made proportional to the first input signal (V_1) and B is proportional to a second input signal (V_2), then $V_H = k_3 . V_1 . V_2$ and we have a Hall-effect multiplier. In this way we can measure power $\left[V_1 = f_1 (v) \text{ and } V_2 = f_2 (i) \right]$

3.3 Technical implementation

A thin layer of a few µm of semiconductor material placed on a ceramic substrate is sufficient to make a sensor for detecting magnetic fields.

The semiconductor can be InSb (= indiumantimonide). A constant current is caused to flow through the semiconductor. If a perpendicular magnetic field is present at the sensor, then there is an output voltage (V_H).

In practical sensors this voltage is 0.2 to 1V/T , hereby: 1 T (Tesla) = 1Wb/m^2 .

A differential amplifier is usually integrated into the sensor to produce a useful output voltage.

4. HEAT DISSIPATION OF SEMICONDUCTORS

4.1 General

If a certain voltage is present across the junction of a semiconductor (diode, thyristor, transistor) and a current is flowing through this junction then a power P is converted to heat. The heat produced is dissipated to the environment and in equilibrium the junction in question assumes a certain temperature.

The manufacturer provides the maximum allowable junction temperature T_{Jmax}. This is typically 125 to 150°C. With the most unfavourable temperature peak 200°C may be allowed and for designs with high reliability 120°C is used. On route from junction to the ambient air (fig. 5-41) the heat flow encounters several thermal resistances R_{th}, usually classified as follows:

- R_{JB} = internal thermal resistance between junction and casing of the semiconductor, in the literature and data sheets also referred to as $R_{thj-c} = R_{thj-mb}$ or Θ_{JC} (J = junction, C = case = mb = mounting base)
- R_{BK} = thermal resistance of the case to the heat sink = $\Theta_{C-HS} = R_{th\ mb-h}$ (HS = h = heat sink)
  A mica plate between heat sink and semiconductor has a typical R_{BK} = 1.25°C/W. If thermal paste is used between heat sink and semiconductor, R_{BK} can be as low as 0.1 to 0.5°C/W
- R_{KA} = thermal resistance of the heat sink to surrounding air. Also indicated as $R_{thK} = \Theta_{HS-A}$ (A = a = ambient).

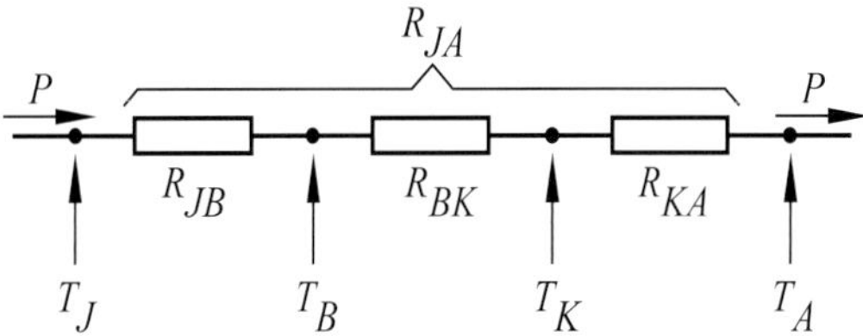

Fig. 5-41: Electrical equivalent of thermal resistance

The sum of $R_{JB} + R_{BK} + R_{KA} = R_{JA}$ = thermal resistance between junction and surrounding air. This thermal resistance is expressed in °C/W and is a measure of a temperature rise between start and end point of a thermal heat flow related to the radiated calorific power.
In fig. 5-41 we can write:

$$P = \frac{T_J - T_A}{R_{JA}} = \frac{T_J - T_B}{R_{JB}} = \frac{T_B - T_K}{R_{BK}} = \frac{T_K - T_A}{R_{KA}} \tag{5-30}$$

4.2 Low power semiconductors

These are not suitable for use with a heat sink or thermal radiator. The manufacturer provides the thermal resistance between junction and air (R_{JA}).

Numeric example:

From the data sheet of a certain diode we find: $T_{Jmax} = 200°C$; $R_{JA} = 0.38°C/mW$.
The ambient temperature is 50°C. The electrical power that can be dissipated in the diode is:

$$P_{max} = \frac{T_J - T_A}{R_{JA}} = \frac{200 - 50}{0.38} = 394 mW$$

4.3 Choice of heat sink

With large power levels a semiconductor will be mounted on a heat sink. This is required for

optimum use of the component. The previously noted general expression (5-30) can be used.

$$P = \frac{T_J - T_A}{R_{JA}} = \frac{T_J - T_B}{R_{JB}} = \frac{T_B - T_K}{R_{BK}} = \frac{T_K - T_A}{R_{KA}} \quad \text{with: } R_{JA} = R_{JB} + R_{BK} + R_{KA}$$

In the above mentioned general expression we see clearly that the thermal resistance is expressed in °C/W or K/W.

4.4 Numeric example:

1. A voltage regulator (µA7805) has to dissipate $P_D = 3.936W$ and in a TO-3 casing
   $R_{JB} = 5.5°C/W$ and $R_{JA} = 45°C/W$.
   Determine the heat sink (R_{KA}) if $T_{Jmax} = 125°C$ and $T_{Amax} = 60°C$.

Solution:

a. $R_{JA} = \dfrac{T_J - T_A}{P_D} = \dfrac{125 - 60}{3.96} = 16.41 \ °C/W$

b. $R_{KA} = R_{JA} - R_{JB} - R_{BK}$

   Consider the situation where heat conducting paste is used between casing and heat sink
   $R_{BK} = 0.5°C/W$. As a result: $R_{KA} = 16.41 - 5.5 - 0.5 = 10.41°C/W$.
   An FK205/SA/L from Fischer Electronics (see p. 5.26) with a thermal resistance
   $R_{KA} = 9°C/W$ is sufficient.

c. Let us also calculate the acceptable power that can be dissipated by a µA7805
   without heat sink. We know that $R_{JA} = 45°C/W$.
   With the same temperatures as previously: $P_D = \dfrac{125 - 60}{45} = 1.44W$.
   With a voltage difference $V_{in} - V_{out}$ of for example 3V, then a maximum current of
   $I = 1.44/3 = 480$ mA may flow through the regulator.

2. Consider a transistor in a TO-3 casing with an $R_{JB} = 7.5°C/W$ and a power dissipation of 10 W with a $T_{Jmax} = 150°C$.

Required: Determine the heat sink (R_{KA}) for a maximum ambient temperature of 45°C. Assume an $R_{BK} = 0.5°C/W$.

Solution: 1. $P = \dfrac{T_J - T_A}{R_{JA}}$ gives: $R_{JA} = \dfrac{150 - 45}{10} = 10.5°C/W$

2 $R_{JA} = R_{JB} + R_{BK} + R_{KA}$ gives us:

$$R_{KA} = R_{JA} - R_{JB} - R_{BK} = 10.5 - 7.5 - 0.5 = 2.5°C/W.$$

In the catalogue of Fischer Elektronik we find that an SK−04 heat sink of at least 45 mm long is required. With a safety margin of 20% (see remark 1 below) we require 5.4 cm.

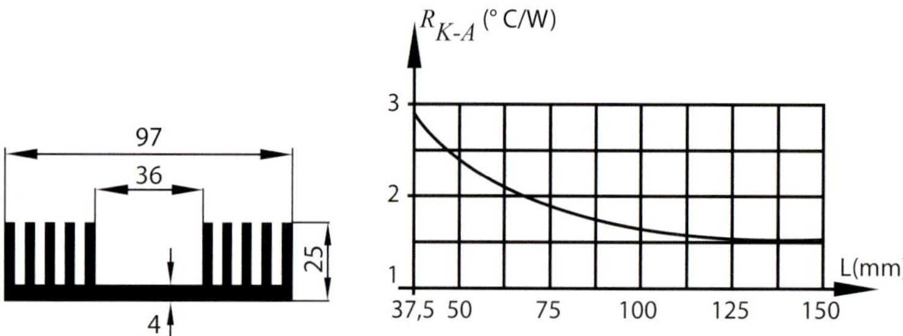

Fig. 5-42: SK-04 heat sink element from Fischer Elektronik.

Remarks

1. The data sheets of the manufacturers of heat sinks are valid for vertical mounting of black anodised radiators or sinks.

    For untreated heat sinks the heat sink would need to be dimensioned 10 to 15% larger and for horizontal mounting also a 10 to 15% increase.

    In addition the ambient temperature may only be considered constant if the ambient area is ventilated, which is often not the case. Therefore it is usual to increase the dimensions of the theoretically calculated heat sink (e.g. +20%).

2. By using expression (5-30) from the previous example it is also possible to determine the casing temperature and the heat sink temperature:

    $P = \dfrac{T_J - T_B}{R_{JB}}$ gives: $T_B = T_J - P \cdot R_{JB} = 150 - 10 \times 7.5 = 75°C.$

    and: $P = \dfrac{T_B - T_K}{R_{BK}}$ gives : $T_K = T_B - P \cdot R_{BK} = 75 - 10 \times 0.5 = 70°C.$

3. To determine the heat sink some manufacturers give $\Delta T = T_B - T_A$ and P together in one diagram. With the previous numeric example we find: $\Delta T = T_B - T_A = 75 - 45 = 30°C$ and $P = 10W$. In the catalogue of Austerlitz-elektronik we choose for example a KS95.1 (see fig. 5-43). We find a required length of about 50mm which corresponds with the choice from Fischer. Also here we add a 20% safety margin.

4. Subscripts with heat sinks
   A = a = ambient
   C = case (also used is B)
   H = HS = heat sink (also used is K)
   J = junction
   mb = mounting base
   th = thermic

5. Excerpts from Fischer Elektronik (see table 5-5).

Table 5-5

	calibre bore	dimensions	surface	thermal resistance
FK201SA	TO3	45x45x25.4	anodized	6° C/W
FK202SA	TO3	45x45x12.5	anodized	8° C/W
FK205/SA/L	TO3/9/66/32/220	41.3x33x31.8	anodized	9° C/W
FK206/SA/L	TO3/9/66/32/220	41.3x33x25.4	anodized	10.5° C/W
FK207/SA/L	TO3/9/66/32/220	41.3x33x19.1	anodized	12° C/W
FK208/SA/L	TO3/9/66/32/220	41.3x33x12.7	anodized	14° C/W

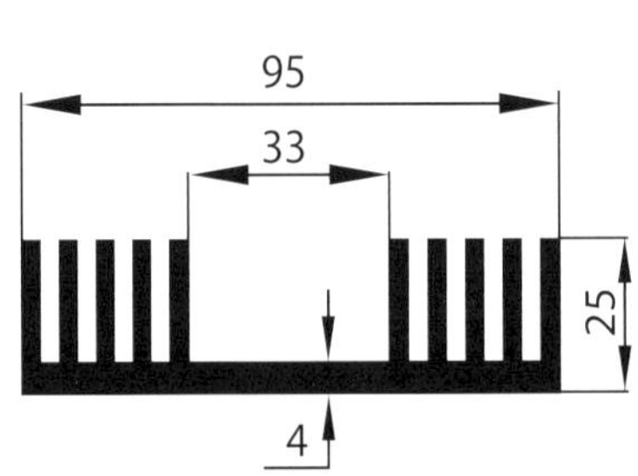

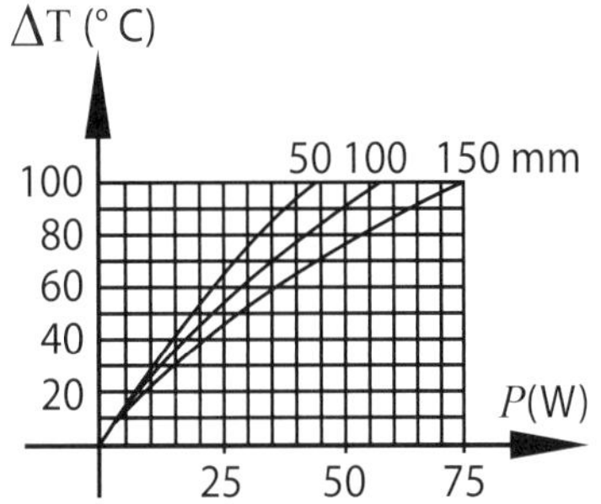

Fig. 5-43: Heat sink from Austerlitz electronic

4.5 Heat sinks for large semiconductors

The data sheets of high power diodes and thyristors usually contain nomograms that allow the heat sink to be determined.

We consider the example of the diode SKN400 (Semikron) and redraw fig. 2-10 in fig. 5-44.

As determined on p. 2.9 , with a half wave rectifier (sin 180°) and $I_{F(AV)}$ = 300A the dissipated power P = 385 W and the case temperature is 118°C.

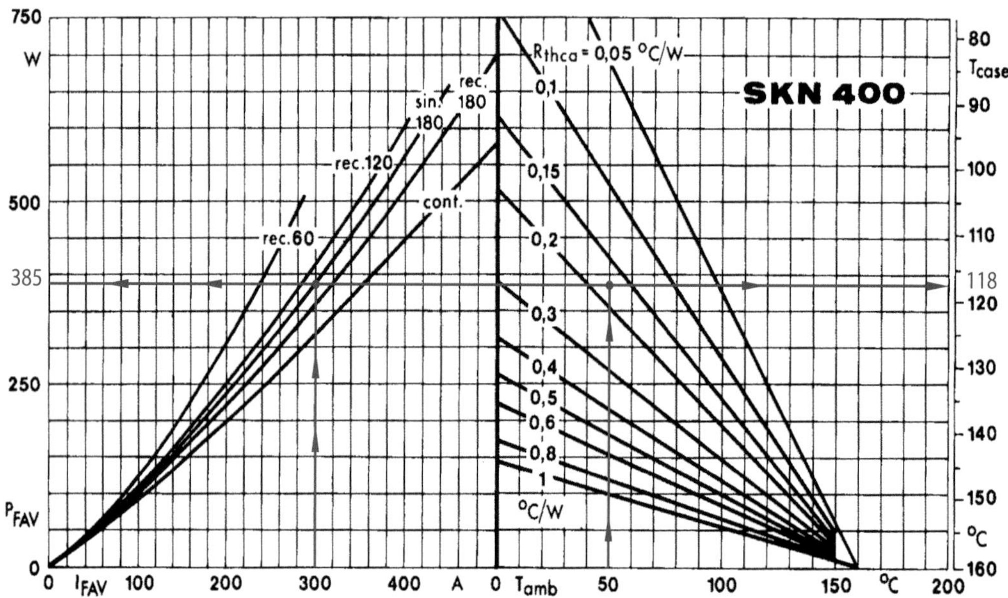

Fig. 5-44: Relationship between power dissipation / diode current /case temperature

With an ambient temperature of 50°C we need to use a heat sink with $R_{th\ c-a}$ = 0.18°C/W.

This can be confirmed using expression (5-30) with the details of the SKN400 diode that can be found on p. 2.12: $R_{th\ j-c}$ = 0.11°C/W; T_{vjmax} = 160°C.

With P = 385W and T_{amb} = 50°C we find:

$$P = \frac{T_J - T_a}{R_{ja}} = 385 = \frac{160° - 50°}{R_{ja}} \quad \text{so that:} \quad R_{ja} = 0.2857 \text{ °C/W}$$

$$R_{ja} = R_{jc} + R_{ca} \text{ gives:} \quad R_{ca} = R_{ja} - R_{jc} = 0.2857 - 0.11 = 0.1757 \text{ °C/W}$$

In fig. 5-44 we see for example that with a case temperature of 140°C, and an ambient temperature of 70°C the thermal resistance between semiconductor and the air should be: $R_{th\ c-a}$ = 0.39°C/W. Here the transition resistance between heat sink and semiconductor casing ($R_{th\ c-h} \approx 0.01$°C/W) is included so that the thermal resistance of the heat sink should be a maximum of 0.38°C/W.

5.   A. COLOUR CODE FOR RESISTORS

On the body of resistors we find four to six coloured rings. The explanation of the colours is provided in the following table. As memory aid ROYGBIV [pronounced (roy-gee-biv)] can help to remember the colour sequence (apart from black-brown and grey-white !).

Table 5-6

COLOUR	A FIGURE	B FIGURE	C FIGURE	C/D number of ZEROS	D/E TOLE-RANCE %	TEMP COEFFIC. $\times 10^{-6}$/C
black	0	0	0	0	-	200
brown	1	1	1	1	1	100
red	2	2	2	2	2	50
orange	3	3	3	3	-	15
yellow	4	4	4	4	-	25
green	5	5	5	5	0,5	-
blue	6	6	6	6	0,25	-
violet	7	7	7	7	0,1	-
grey	8	8	8	8	-	-
white	9	9	9	9	-	-
				gold $\times 10^{-1}$	gold 5	
				Ag $\times 10^{-2}$	Ag 10	

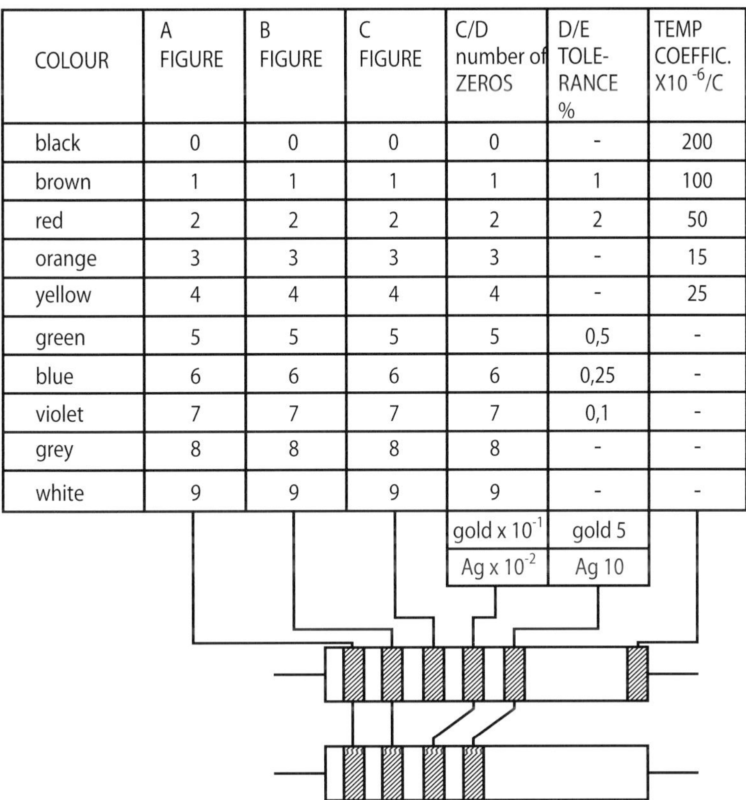

Example: A = brown / B = red / C = black / D = gold. $R = 12\ \Omega \pm 5\%$
(this resistor can have a value between 11.4 and 12.6 Ω !).

B. NON-INDUCTIVE RESISTORS

We need to be able to use non-inductive resistors as current sensors in power electronics due to current waveforms that we encounter. Caddock(.com) is for example a specialist in the field of power film resistors from 0.022 Ω (15W) to 40 MΩ (up to 200 V operating voltage !).

6. NORMALISATION OF RESISTOR AND CAPACITOR VALUES

The following normalised sequences of resistors (Ω) and capacitors (pF) occur frequently:
E24 (5%): 10/11/12/13/15/16/18/20/22/24/27/30/33/36/39/43/47/51/56/62/68/75/82/91.
E12 (10%): 10/12/15/18/22/27/33/39/47/56/68/82.
All decimal multiples of these values are manufactured with the lower and upper limits being 10 Ω and 22 MΩ. Capacitors range from 1pF to 220,000 μF (industrial capacitors).

7. COLOUR CODE FOR CAPACITORS

This code corresponds in large measure with that of resistors. The values associated with the different colour bands are shown in the following table

Table 5-7

COLOUR	A	B	C	D	E	
			multiplication factor (pF)	tolerance > 10 pF	tolerance < 10 pF	operating voltage (V)
black	––	0	1	± 20%	––––	–––
brown	1	1	10	––––	––––	100
red	2	2	100	––––	± 0.25 pF	250
orange	3	3	1000	––––	––––	–––
yellow	4	4	10000	––––	––––	400
green	5	5	100000	± 5%	± 0.5 pF	–––
blue	6	6	––––	––––	––––	630
violet	7	7	––––	––––	––––	–––
grey	8	8	0.01	––––	––––	–––
white	9	9	0.1	± 10%	––––	–––

8. TYPE-INDICATION OF SEMICONDUCTORS AND IC'S

8.1 Semiconductors

The type number consists of two letters and a serial number

First Letter
This is related to the material
A =   germanium (or material with a band gap of 0.6 to 1eV)
B =   silicon (or material with a band gap of 1 to 1.3eV)
C =   gallium-arsenide (or material with a band gap of 1.3eV or higher)
R =   impure materials (hall-generators, photo-conducting elements)

Second Letter
This is related to the function of the semiconductor.
A =   detection/switching/mixing: BA 114
B =   varicapdiode: BB105
C =   LF-transistor (low power): BC109
D =   LF-transistor (large power): BDX67
E =   tunnel diode: AEY23
F =   HF transistor (small power):BFW45

H =   magnetic sensitive diode

K =   hall generator in an open magnetic circuit

L =   HF-transistor (large power): BLY93

M =   hall generator in a closed electrically power magnetic circuit

N =   photo-coupling: CNY62

P =   light sensitive semiconductor: BPX70; solar cell

R =   controlled diode (small power) /diac / trigger diode: BR100 / uni-junction transistor, etc...

S =   switching transistor (small power): BSX21

T =   thyristor: BTW34

U =   switching transistor (large power): BU208

Y =   power diode: BYX97

Z =   zener diode: BZY79-C5V6

Serial number

Semiconductors for professional use have 1 letter and 2 numbers (BDX67). Other semiconductors have 3 numbers. (BC109)

Remark: COLOUR CODE FOR DIODES

With industrial diodes sometimes four coloured rings are used with two being wide and two narrow rings.

The first wide ring determines the semiconductor material and the cathode size.
Brown = germanium. Red = silicium.

Outline	wide ring	narrow ring	distance
DO -7	0.6 mm	0.4 mm	0.4 mm
DO - 35	0.5 mm	0.3 mm	0.3 mm

The second wide ring indicates the industry type and the colour corresponds with a certain letter

R = brown	T = orange	V = green	X = violet
S = red	U = yellow	W = blue	Y = grey

The narrow rings correspond to an international colour code to indicate a specific number:

0 = black	3 : orange	6 = blue	8 = grey
1 = brown	4 = yellow	7 = violet	9 = white
2 = red	5 = green		

Example: BAW75 = red (wide), blue (wide), violet (narrow), green (narrow)

8.2 Integrated Circuits

A. PRO-ELECTRON CODE (= European classification type)

The basic type number consists of three letters and a serial number of four letters or letters and numbers.

First and second letter:

1. *Digital family:*

FA...FZ, GA...GZ,HA...HZ,PC...PZ indicate the family

2. *Standalone types:*

First letter: S = digital circuits

T = analogue circuits

U = mixed circuits (analogue/digit.)

Second letter: no significance, unless H = hybrid.

Third letter:

Indicates the temperature range wherein the IC can operate normally.

A =	no temperature indicated	E = − 25°C to +85°C
B =	0° to 70°C	F = − 40°C to +85°C
C =	- 55°C to +125°C	G = − 55°C to +85°C
D =	- 25°C to +70°C	

Serial number:

Follows the first three letters and can be made up of four numbers or a combination of letters and numbers.

Addition of suffixes:

If two letters are added, they have the following significance

First letter: (appearance)		**Second letter:** (case material)	
C =	cylindrical	S =	single-in-line (TO220...)
D =	dual in line (DIL)	C =	ceramic (metal)
E =	power DIL	G =	ceramic (glas)
	(with external heat sink)	M =	metal
F/G =	flat casing	P =	plastic
K =	TO3 casing		

B. AMERICAN CLASSIFICATION

1. *DIGITAL IC's*

American manufacturers often use numbers for their digital IC's with the following significance:

54...:   suitable for −55°C to +125°C (military specifications)

74...:   suitable for 0°C to +70°C (commercial application)

54/74: saturated TTL logic for general applications. 10mW per gate;

fan-out = 10; 1 fan-in or 1 fan-out; 1.6mA ; 10ns

54H/74H:	High-speed saturated TTL for high speed high noise margin; 23mW/gate; fan-out = 10; 6ns.
54S/74S:	Schottky TTL for ultra-high speed; 19mW/gate ; 3ns.
54L/74L:	Low power TTL; 1mW/gate; 33ns.
54LS/74LS:	Low power Schottky TTL; 2mW/gate; 10ns; fan-out = 20.
54/74AS:	Advanced Schottky; 22mW/gate; 1.5ns.
5474ALS:	Advanced LP Schottky; 1mW/gate; 4ns.
7400J:	ceramic
N:	plastic
4000:	CMOS-type; 10mW/gate; 25ns; $-40°C$ to $+85°C$; fan-out = 50; 3 to 15V

2. *LETTERS FOR THE TYPE INDICATION*

Most manufacturers use a one or two letter code before the type indication code to indicate the manufacturer.

AD	=	circuits from Analog Devices
μA	=	micro amplifier (Fairchild); for linear IC's
CA	=	circuits from RCA
CD	=	digital circuits from RCA (MOS-technology)
HEF	=	LOCMOS family from Philips (= locally oxidised complimentary MOS)
IP	=	integrated power semiconductors Ltd.
LM	=	linear monolithic (National Semiconductor)
LTC	=	Linear Technology
MC	=	Motorola components
NE	=	Nippon Electronic Corporation
SG	=	Silicon General
SL	=	IC's from Plessey Semiconductors
TI and SN	=	Texas Instruments components
TIP	=	Texas Instruments Power
TISP	=	Texas Instruments transistors with surge protection
UC	=	Unitrode components
XR	=	EXAR

Example: μA741, LM741, CA741 are equivalent opamps.

3. *MOTOROLA:* this company adds a "1". An MC14001 is equivalent to a CD4001.

4. *SERIAL LETTERS*

The first letter (after the numbers of the type indication) refer to the casing (P = plastic...), a second letter indicates temperature range.

5. *FIRST DIGIT OF LINEAR IC's.*

The first digit of linear IC's indicate temperature range:
1 = military temperature range ($-55°C$ to $+125°C$)
2 = industrial temperature range ($-25°C$ to $+85°C$)
3 = commercial temperature range ($0°C$ to $70°C$)
An LM301 is electrically equivalent to an LM101

C. Casing of semiconductors: terminology

DO	=	diode outline = diode casing (outline = circumference)
DIL	=	dual in-line
DIP	=	dual in-line plastic (MINI-DIP = 8 pins)
F-pack	=	complete electrically isolated casing, but thermically conducting
hockey puck	=	capsule
MIL-SPEC	=	military specifications
QFJ (PLCC)	=	quad flat J-lead (plastic lead carrier)
SIP	=	single in-line
STUD-mounted	=	constructed with thread
SO	=	small outline (small casing from Philips). We also find SOD (diodes), SOT (transistors), SOP (package by ICs), SSO = stretched small outline
TO	=	transistor outline; TO-can
ZIP	=	zigzag in-line package.

9. THE SKIN EFFECT

9.1 General

If a DC current I flows through a conductor with cross-sectional area d, then the current density is $J = \dfrac{I \cdot 4}{\pi \cdot d^2}$ .and is constant throughout the cross-sectional area. In the case of an AC current this is not so. We can visualise this by imagining a conductor (fig. 5-45) is being composed of n concentric cylinders in which a part of the total current I flows. The current that flows in each cylinder produces a magnetic field as shown for example in fig. 5-46 which is drawn for cylinders 1 and 2.

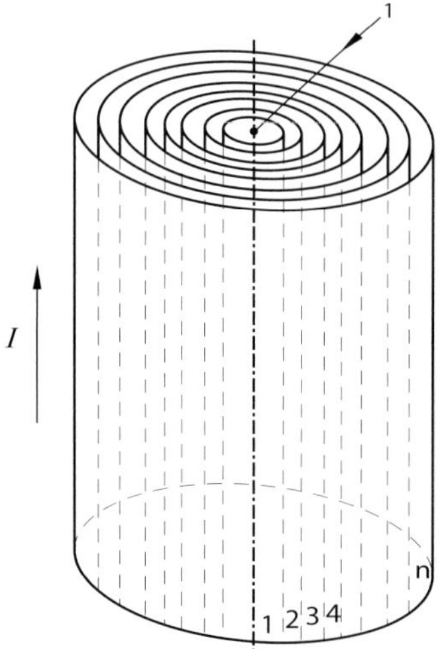
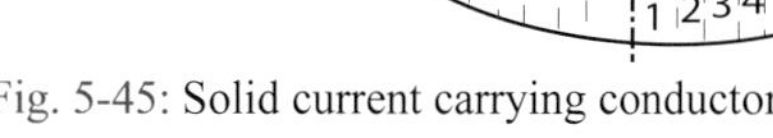

Fig. 5-45: Solid current carrying conductor

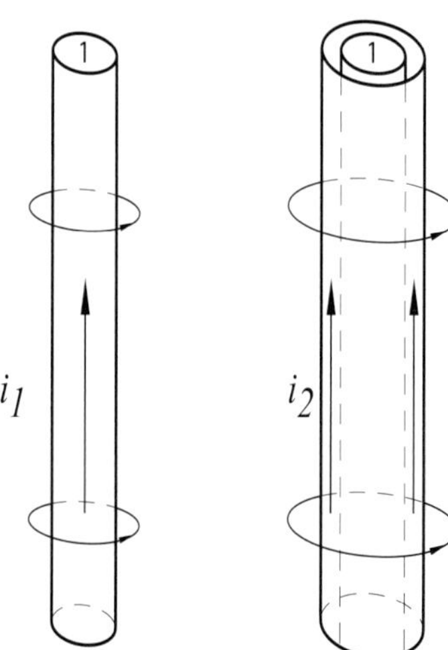

Fig. 5-46: Cylinders one and two from fig. 5-45

The smaller the follow-on number (1, 2, ... n) of the cylinders in fig. 5-45, the more magnetic field lines that are linked and therefore the larger the cell conductance of these sub cylinders. And in the case of an AC current the impedance of one cylinder would therefore be greater than that of a greater sized, in other words the impedance rises as the cylinder numbers decrease. The current density will be greatest near the circumference of the conductor: we call this the **skin effect**.

The higher the frequency, the more pronounced this effect. The useful cross-sectional area of current carrying conductors decreases with an increase of the frequency of the current. It may be necessary for example with switch mode power supplies that operate at higher frequencies (50 to 100kHz) to use (instead of solid conductors) a number of thinner conductors in parallel for the windings of coils and transformers. Above 100kHz a so-called Litze wire is used to wind coils. This Litze wire is in fact a cable, composed of a large number of thin wires switched in parallel. The currents carrying cross-sectional area of this Litze wire is obviously much greater than a solid conductor of the same outside diameter.

With extremely high frequencies the current flows only in a thin layer on the outside edge of the conductor. We may just as well use a hollow conductor under such circumstances (waveguides in radar applications: 3 to 10 GHz!).

9.2 Pronounced and non-pronounced skin effect

Fig. 5-47 shows the current density respectively for non-pronounced and pronounced skin effect. We can determine the factor u which indicates if the skin effect is pronounced are not:

$$u = d . \sqrt{\frac{\pi . \mu_0 . \mu_r . f}{2 . \rho}} .$$ If $u > 6$ there is pronounced skin effect (S.E.)

d = diameter of conductor (m)

μ_0 = $4 . \pi . 10^{-7}$ (H/m)

μ_r = relative permeability

ρ = specific resistance (Ωm)

f = frequency (Hz)

For copper wire ($\rho = 1.72 \times 10^{-8}$ Ωm) and with d in mm and f in kHz, gives :

$$u = 0.33868 . d . \sqrt{f}$$

The ohmic AC-resistance is to calculate with: $$R_{AC} = R_{DC} . K$$ (5-31)

$u > 6$ gives: $$K = \frac{u}{2. \sqrt{2}} + 0.25$$

$u < 6$: K is given in table 5-8

u	K	u	K
0.25	1	2.25	1.1205
0.50	1.0005	2.50	1.1754
0.75	1.0033	2.75	1.2416
1	1.0104	3	1.318
1.25	1.0125	3.50	1.4919
1.50	1.0254	4	1.6775
1.75	1.0470	5	2
2	1.0781		

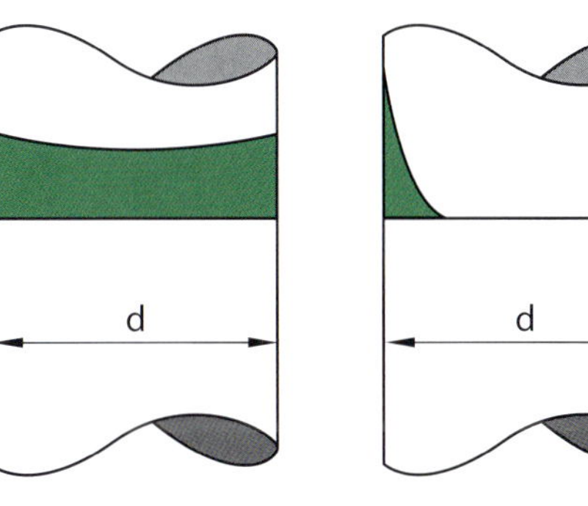

Table 5-8: K values for non-pronounced S. E.   Fig. 5-47: Current density in conductor

9.3 Skin depth figure

Fig. 5-48 shows the current distribution inside a conductor in relation to the current density on the circumference of the conductor.

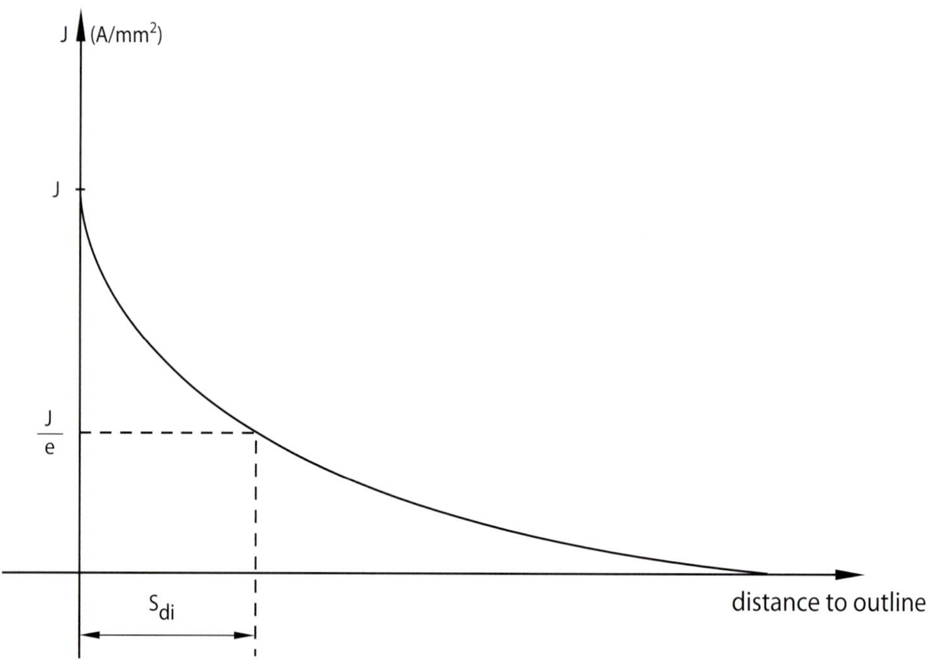

Fig. 5-48: Current density with pronounced skin effect

An interpretation for the skin depth is the distance to the circumference whereby the current density is reduced to *1/e* times the current density of the circumference.

Another approach is the replacement of the distributed current density by a fictive constant current density (amplitude and phase!) in a layer S_{di} at the circumference of the conductor. We consider then the useful current carrying part of the conductor, a cross-sectional area given by $S_{di} . \pi . d$.

The value for the skin depth is given by: $S_{di} = \sqrt{\dfrac{\rho}{\pi . \mu_0 . \mu_r . f}}$

$$(5-32)$$

In the case of copper conductors and with f in kHz then S_{di} in mm: $\quad S_{di} = \dfrac{2.08}{\sqrt{f}}$

$$(5-33)$$

9.4 Numeric example 5-2:

1. **Given:** $f = 100\text{kHz}$; respective cross-sectional area of 0.76mm (AWG21) and 0.56mm (AWG24)

    Question: pronounced or non-pronounced skin effect?

    - $R_{AC} = ?$
    - $S_{di} = ?$

Solution: **a) wire 0.76 mm:**

. $u = 0.338688. \, 0.76 . \sqrt{100} = 3.01; \quad u < 6 \rightarrow\rightarrow$ non-pronounced S. E.

. Table 5–8: K = 1.318; $\quad R_{AC} = 1.318. \, R_{DC}:$ the AC- resistance is 31.5% larger

. than the DC-resistance of the wire.

$$S_{di} = \frac{2.08}{\sqrt{100}} = 0.208 \text{ mm}$$

b) wire of 0.56 mm:

. $u = 0.33868. \, 0.56 . \sqrt{100} = 1.89 \rightarrow\rightarrow$ non-pronounced S. E.

. Table 5–8: K = 1.067: $\quad R_{AC} = 1.067. \, R_{DC}$

. $S_{di} = 0.208$ mm

2. Given: a copper wire with $d = 1.35$ mm (AWG16). We use frequencies of 100kHz and 1MHz

Question: determine for both frequencies if there is pronounced S. E.

Solution: a) $f_1 = 100$**kHz:**

. $u = 0.33868. \, 1.35. \sqrt{100} = 4.572 \rightarrow\rightarrow$ non-pronounced S. E.

. Table 5–8: K = 1.86 ; $\quad R_{AC} = 1.86. \, R_{DC}$

. $S_{di} = \dfrac{2.08}{\sqrt{100}} = 0.208$ mm

b) $f_2 = 1$**MHz:**

. $u = 0.33868. \, 1.35. \sqrt{10^4} = 45.72 \rightarrow\rightarrow$ pronounced S. E.

. $K = \dfrac{u}{2 . \sqrt{2}} + 0.25 = 16.41; \quad R_{AC} = 16.41. \, R_{DC}$

. $S_{di} = \dfrac{2.08}{\sqrt{10^4}} = 0.0208$ mm

If we use the model for the skin depth with constant current division, then we find
a cross-sectional area for current conduction:

$S_{di} . \pi . d = 0.0208.\pi.1.35 = 0.08821$ mm²

The cross-sectional area for DC current would be: $\dfrac{\pi . d^2}{4} = \dfrac{\pi . 1.35^2}{4} = 1.4314$ mm²

The ratio $\dfrac{1.4314}{0.0882} = 16.23$ approaches the value for K!

10. EMC – EMI

10.1 General

Originally electronics was synonymous with telecommunications (radio, television, telephony). During operation these electronic devices often produced frequencies and harmonics which interfered with the correct operation of other equipment. These undesirable effects were known as RFI (RFI = radio frequency interference = high-frequency disturbance). Since the last few decades the number of electrical and electronic devices has become so large that it is no longer only a problem of high-frequency suppression (radio-frequency suppression) but a much broader definition needs to be used. That is why we now talk about EMC (electromagnetic compatibility). With EMC we mean that electrical setups can function normally in the electromagnetic environment and that in addition they do not cause unacceptable effects on equipment in the vicinity. The electromagnetic transmission between the "source" of the disturbance and the "receiver" can occur via radiation (RE = radiated emission and RS = radiated susceptibility) or via conduction (CE = conducted emission and CS = conducted susceptibility). Undesired radiation is counteracted via screening of electrical equipment and cables. The undesired conduction via mains-, control- or signal lines is counteracted by the use of EMI–filters (EMI = electromagnetic interference).
These EMI–filters are most frequently LC–filters which are designed as low pass filters. Some manufacturers provide filter modules which only need to be connected between the mains and the equipment.

With EMC we also make a distinction with respect to the frequency spectrum of the interference:

A. **Source of disturbance**
 1. Discreet frequency spectrum (high-frequency generator, μ-processor systems, radio and TV receivers, SMPS,...)
 2. Continuous frequency spectrum (household appliances, power semiconductors, discharge lamps,...)

B. **Receiver (of disturbances)**
 1. small band sensitivity (radio and TV receivers, modems,...)
 2. wide band sensitivity (control systems, analogue and digital systems, process computers,...)

In fig. 5-9 the above information is graphically represented.

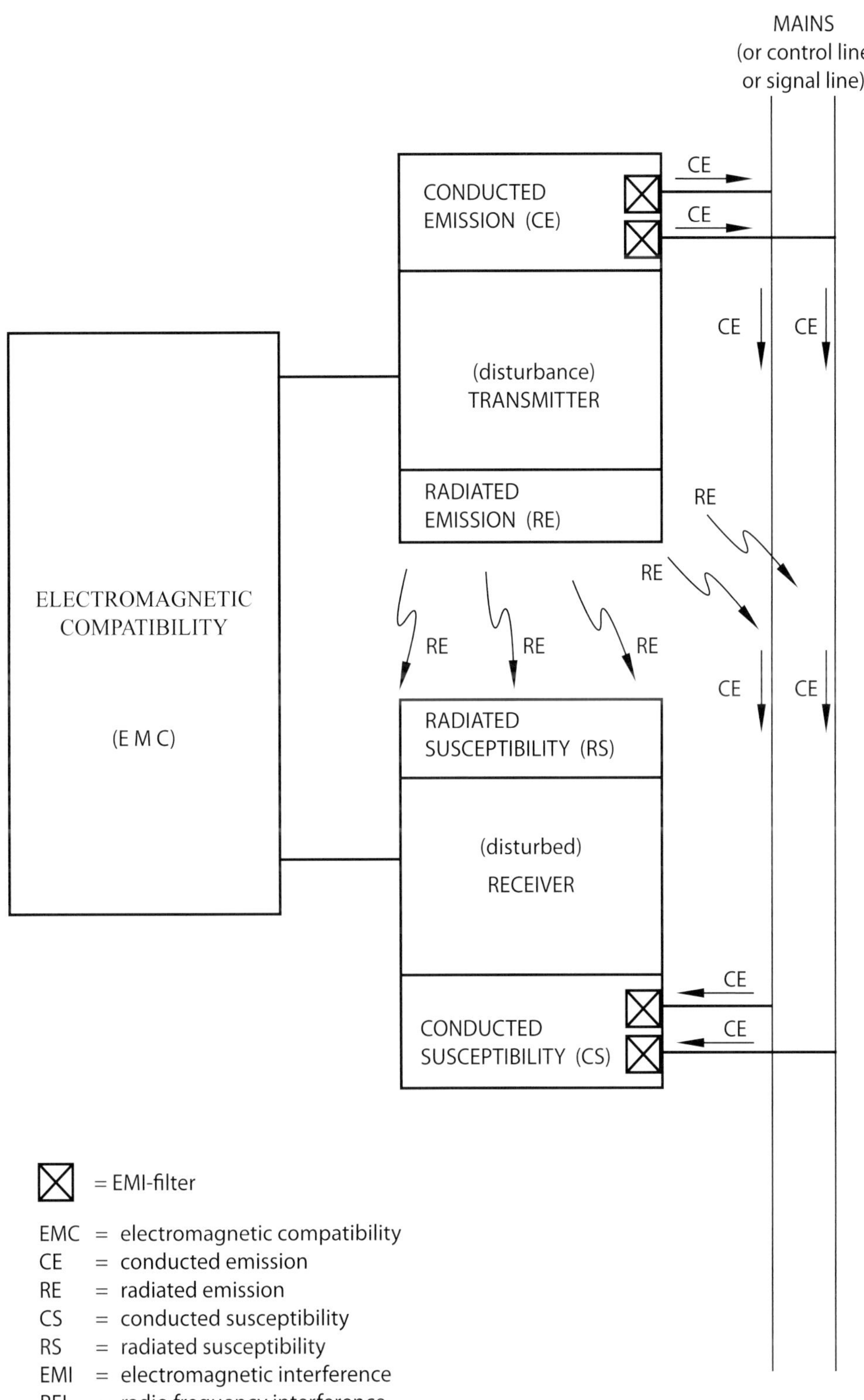

Fig. 5-49: Electromagnetic interaction and location of the EMI-filters

10.2 EMI - filter

We distinguish two types of disturbance according to the current direction of the disturbance. The first type is the differential mode noise that occurs in the signal line and in the ground line in opposite directions as shown in fig. 5-50. To suppress this disturbance we use capacitor C_x in parallel between the two lines (fig. 5–51). Additionally a small coil (= choke; e.g. 1µH) can be placed in the signal line but this does produce an undesired voltage drop. In addition this choke needs to be suitable for the nominal consumer current, a reason to avoid using this choke if possible.

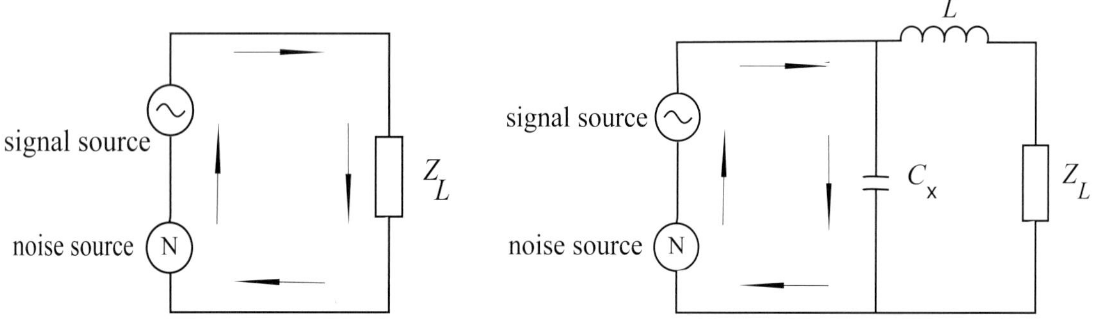

Fig. 5-50: Differential mode disturbance

Fig. 5-51: Suppression of differential mode disturbance

A second sort of disturbance is the **common mode noise** that operates in all lines in the same direction. In the case of an AC-supply line the disturbance operates in both wires in the same direction (fig. 5-52) and in the case of a signal cable the disturbance is in all cores of the cable in the same direction.

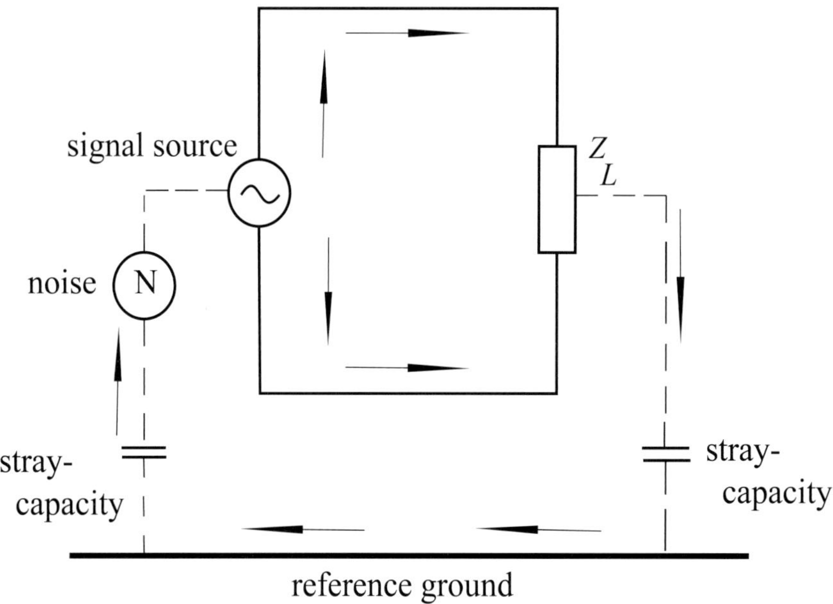

Fig. 5-52: Common mode disturbance

To counteract the **common mode noise** we combine two methods. In the first solution (fig. 5–53) we connect an inductor in both lines.

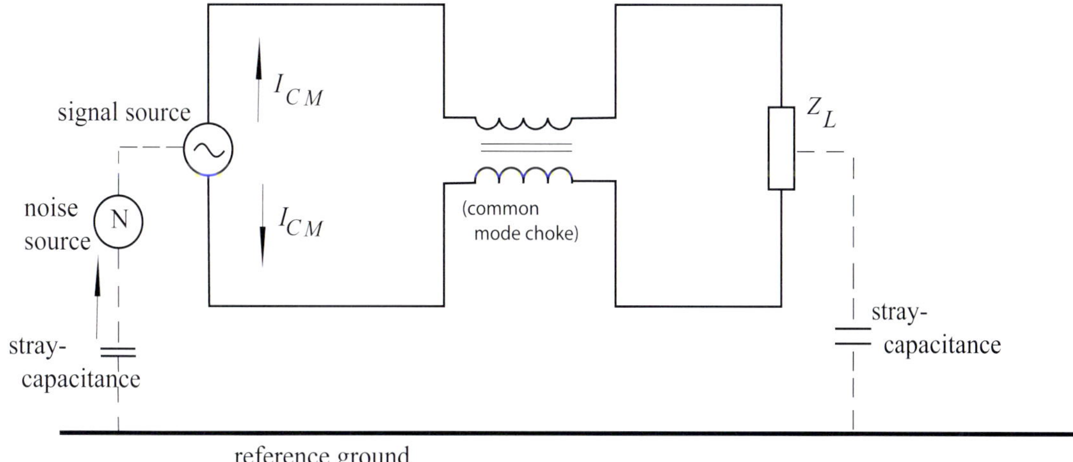

Fig. 5-53: Common mode choke

Both coils are wound on a ferrite core in such a way that they are in series for the influence of I_{CM}, but in opposite directions for the **differential mode current** I_D. The self induction has no effect on the current I_D.

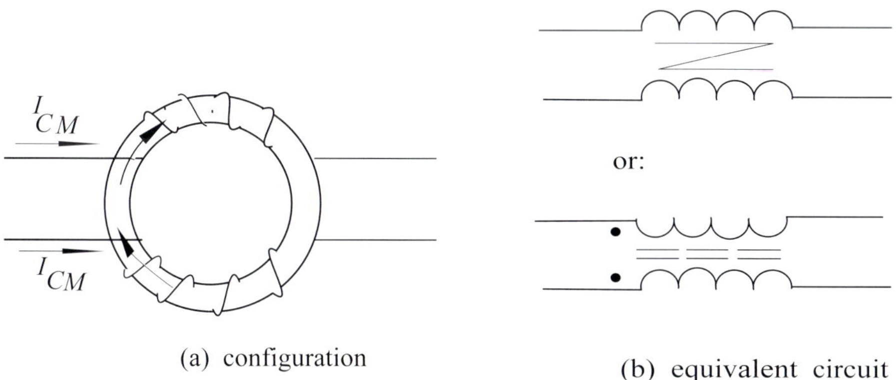

(a) configuration   (b) equivalent circuit

Fig. 5-54: Configuration and equivalent circuit of a common mode choke

Fig. 5-54b shows the equivalent circuit of such a common mode choke. This common mode choke produces no distortion of the signal, which is important if the signal lines are carrying video signals.

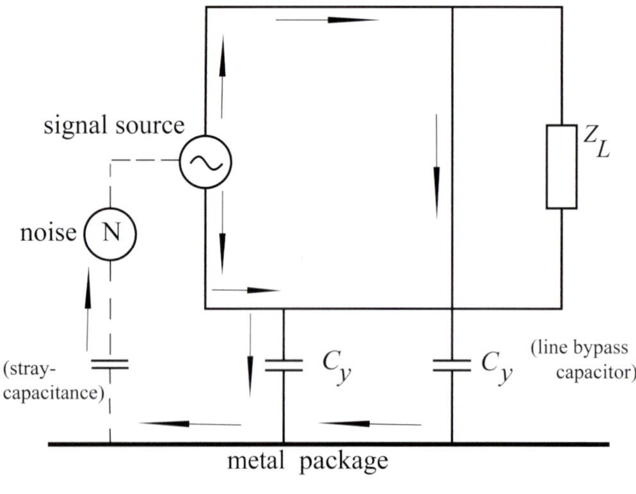

reference ground

Fig. 5-55: Application of a metal enclosure

A second method involves a metal enclosure which includes a bridging capacitor (line bypass capacitor) C_y as shown in fig. 5–55. The disturbance flows via the signal lines and via C_y back to the casing and via the stray capacitance to the source of the disturbance.

Fig. 5–56 shows an example of an EMI-filter commonly used in the supply cable on the AC-input side of equipment (for example a computer).

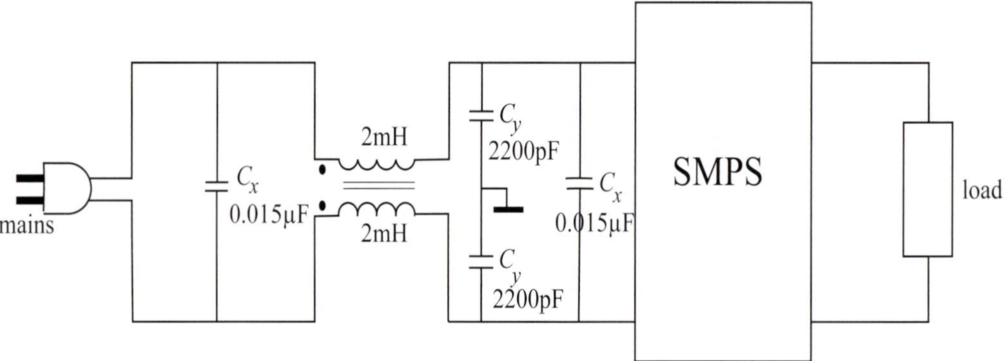

Fig. 5-56: EMI-filter at the supply-side of equipment

Remarks

1. In the case of a single phase mains the differential mode current consists of the 50Hz-current together with the disturbance. The common mode disturbances normally have higher frequencies (10kHz to 10MHz). They flow via parasitic capacitances back to the source of noise.

2. A C_y is placed between every line and the metal enclosure. This "line bypass capacitor" is usually only a few nF, and this is to limit the leakage current $V.\omega.C_y$.

3. Practical values for the common mode choke are for example 500µH up to 2mH.

4. C_x (across-the-line capacitor) suppresses the differential mode disturbance. A practical value is for example 0.015µF.

5. In the case of a frequency converter, depending on whether it is a single phase or three-phase supply an appropriate filter is included in the input.

11. SI-SURFACE AREA OF POWER SWITCHES

There is a difference between the surface area of Si-tablets of various power switches. Consider the Si-tablet in fig. 5-57. According to General Electric the surface area A is given by the expression:

$$A = K \cdot V_\beta^\alpha$$

Herein: K = a factor proportional to the ratio between the active surface area and the total chip surface area.

V_β = the assumed blocking voltage of the component

α = an exponent which is determined by the accepted current density in the semiconductor switch.

General Electric then provides the following expressions:

Thyristor (PNPN): $A = K \cdot V^{1.1}$

Bipolar transistor: $A = K \cdot V^{2.1}$

Bipolar Darlington: $A = K \cdot V^{1.9}$

MOSFET: $A = K \cdot V^{2.6}$

I G (B) T: $A = K \cdot V^{\sqrt{2}}$

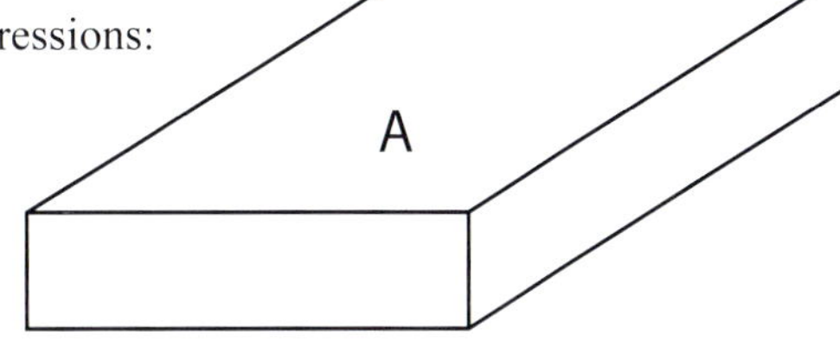

Fig. 5-57: Si tablet

From this it is clear that the thyristor is the most interesting switch since the surface area increases linearly with the permissible voltage.

Numeric example 5-3:

Given: bipolar (10 A; 1000V; V_{CEsat} = 1V);
power MOSFET (10 A; 1000 V; $R_{DS(ON)}$ = 0.1Ω).

Required: surface area ratios of transistor chips

Solution: $\dfrac{V^{2.6}}{V^{2.1}}$ = 31.62. The chip surface area of the power MOSFET is 31.62 times that of

the BJT.

12. SIZE WINDING WIRE (AWG = American Wire Gauge)

Table 5-9 Heavy film-insulated magnet wire specifications

AWG	Diameter over insulation (inches)		Nominal circular mil area	Resistance per 1000 ft	Current capacity in milliamperes based on 1000 c.m./A	AWG
	Min.	Max.				
8	0.130	0.133	16510	0.6281	16510	8
9	0.116	0.119	13090	0.7925	13090	9
10	0.104	0.106	10.380	0.9985	10380	10
11	0.0928	0.0948	8230	1.261	8226	11
12	0.0829	0.0847	6530	1.588	6529	12
13	0.0741	0.0757	5180	2.001	5184	13
14	0.0667	0.0682	4110	2.524	4109	14
15	0.0595	0.0609	3260	3.181	3260	15
16	0.0532	0.0545	2580	4.020	2581	16
17	0.0476	0.0488	2050	5.054	2052	17
18	0.0425	0.0437	1620	6.386	1624	18
19	0.0380	0.0391	1290	8.046	1289	19
20	0.0340	0.0351	1020	10.13	1024	20
21	0.0302	0.0314	812	12.77	812.3	21
22	0.0271	0.0281	640	16.20	640.1	22
23	0.0244	0.0253	511	20.30	510.8	23
24	0.0218	0.0227	404	25.67	404	24
25	0.0195	0.0203	320	32.37	320.4	25
26	0.0174	0.0182	253	41.02	252.8	26
27	0.0157	0.0164	202	51.44	201.6	27
28	0.0141	0.0147	159	65.31	158.8	28
29	0.0127	0.0133	128	81.21	127.7	29
30	0.0113	0.0119	100	103.7	100	30
31	0.0101	0.0108	79.2	130.9	79.21	31
32	0.0091	0.0098	64	162	64	32
33	0.0081	0.0088	50.4	205.7	50.41	33
34	0.0072	0.0078	39.7	261.3	39.69	34
35	0.0064	0.0070	31.4	330.7	31.36	35

6 COMPUTER SIMULATIONS

CONTENTS

1. Introduction

2. History: analogue simulations

3. Simulation with block diagrams: SIMULINK

4. Simulation of an electrical network: SPICE

5. Multilevel model and simulation: CASPOC

The majority of numeric examples in this book can also be simulated using a PC. The solutions can be found on the website http://www.caspoc.com/education under the book "Pollefliet". From this website you can also download a free version of CASPOC.

A part of this chapter is taken with permission from articles written by Dr. ir. P.J. van Duijsen of Simulation Research – Alphen aan den Rijn – The Netherlands. The simulation packet CASPOC was developed by Dr. Van Duijsen.

1. INTRODUCTION

The origins of computer simulations date back to the analog computers of the 1940s of the previous century. In the 1970s analog computers were replaced by digital computers. Simulations allow us to study amongst other things the waveforms, dynamic behaviour and the steady state of systems and components. Modelling and computer simulation play an important role in the analysis, the design and the study of systems of electronic power control. Almost all commercially available programs operate in the time domain.

To describe physical systems it is usually possible to create a system of differential equations (DE) and algebraic relationships.

The solution of differential equation can usually be determined by one of the following methods:

1. Solution by human operator (mathematician, engineer...). This is a possibility but would you choose to drive in a nail with your fist ? No ! People use tools that are available for the job. The tool available is a computer!

2. Solution with a digital computer. Numerical methods are used.

In education we can consider computer simulations as an additional teaching tool. Simulations in industry allows us to dramatically shorten the development phase. An additional benefit is that if we simulate we cannot "blow up" any components. It is safer to study the behaviour of an IGBT and the new circuit in a simulation first. In the case of power electronics computer simulation is, in contrast to analogue electronic circuits at signal level, complimentary with the creation of proto-types in the lab. The simulation does not replace the prototype.

In the following figures we see the advantage obtained by implementing a simulation in the development process.

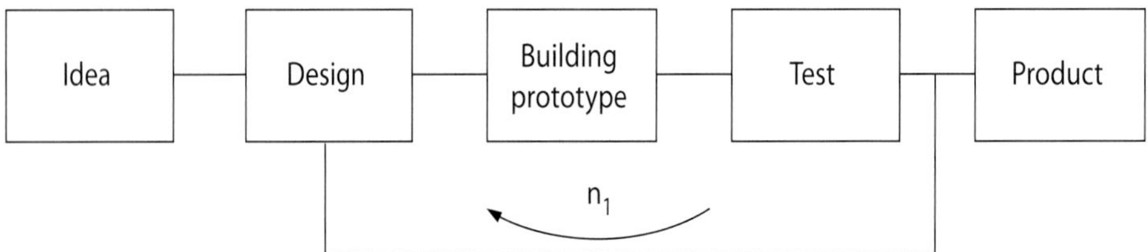

Fig. 6-1: Traditional development process

The loop indicated by n_1 is repeated several times during the development-build-test phase. The most important delay in the loop is caused by building and adjusting the prototype. The duration time of loop n_1 can vary from a few hours to a few weeks or months.

The introduction of model forming simulation in the development process has the goal of shorten-ing the circulation time t_1 of loop n_1 . In figure 6-2 the development process is shown in the case where model forming and simulation is applied.

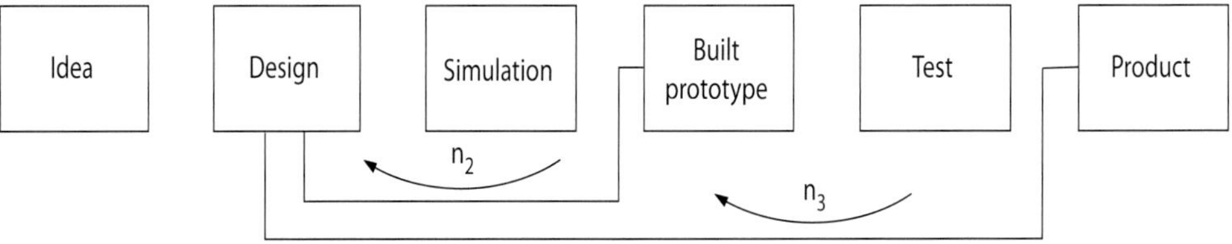

Fig. 6-2: Design process using simulation

There are now two loops in the development process. Loop n_2 occurs during simulation. The circulation time t_2 of loop n_2 is short in comparison to the circulation time t_1 of loop n_1 from fig. 6-1. In fig. 6-2 loop n_1 is replaced by loop n_3. The number of iterations in the development process can be the same: $n_1 = n_2 + n_3$. The difference occurs in the fact that the largest amount of iterations in the development process occur in the simulation, indicated by loop n_2 with the shorter circulation time t_2.

The total time $t_2 + t_3$ consisting of the circulation time of the simulation (t_2) and building and testing (t_3) is shorter than the circulation time t_1 which is required for building the circuit. This is indicated in fig.6-3. The most important reason for this is that with the assistance of the simulation the new design can be tested quicker. A test with eventual adjustment is much quicker in a simulation then will be in the case of building a prototype. Optimisation can also be carried out in the simulation, included in this is the component choice and the configuration parameters of the system. In the first instance a number of different components can be compared with each other. In the second instance for the selected system the configuration parameters can be chosen.
An optimisation criterion, for example the smallest error or the quickest response, can be added to the model. On the basis of the optimisation criteria calculated in the simulation the design can be adjusted.

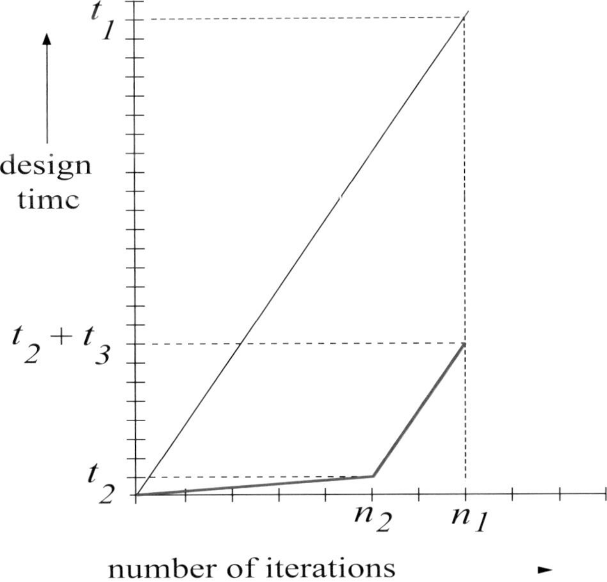

Fig. 6-3: Reduction of development time through the use of simulation

2. HISTORY: analogue simulations

2.1 Analogies

Although analog simulators are not popular any more it is still interesting to talk briefly about these analogs. This will certainly help the student to develop "simulation" insight.

A number of universal physical laws are applicable in the sub areas of physics (electricity, mechanics, hydraulics, pneumatics, acoustics).
The law of the moment of inertia in mechanics ($F = m.a = m \frac{dv}{dt}$) has exactly the same meaning as the electrical law $i = C . \frac{dv}{dt}$.
We can this universality recognise in the written formulas if we compare force with an electric current, mass with capacitance (inertia) and speed with electrical voltage. These transitions are called electro-mechanical analogues. Clearly there are electro-hydraulic and electro-pneumatic analogues also. These analogs are often used to study a system from the perspective of automatic control engineering. Usually we attempt to realise the electrical analogs since the elements (coil, resistor and capacitor) are easily adjusted and calibrated. We can then apply step functions, pulse functions,... to the electrical analog and see what the result would be on the physical system. This is a direct analog simulation. Fig. 6-4 shows the comparison between a spring system and an *RLC* parallel network. In fig. 6-5 an overview of analogue simulations is provided.

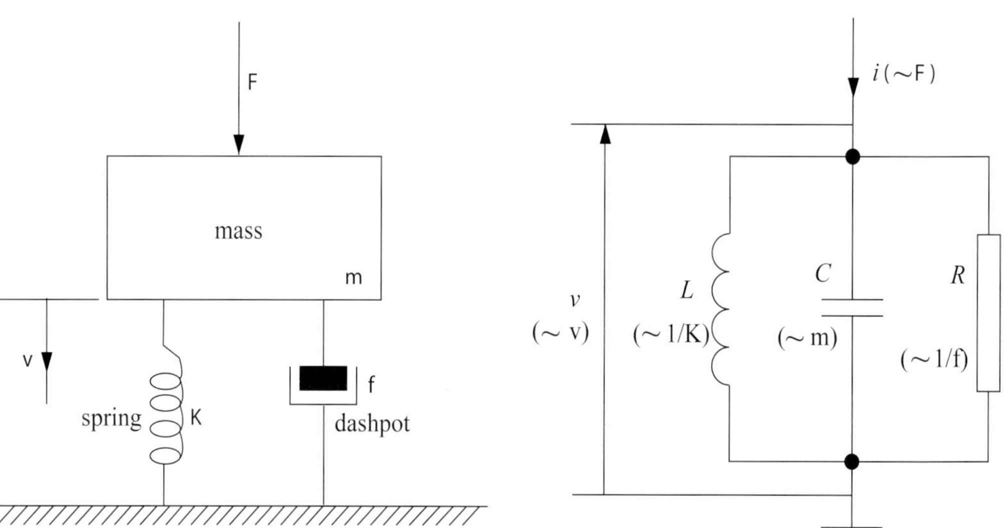

Fig. 6-4: Direct electromechanical analogs

For the mechanical system in fig. 6-4 we find:

$$F(t) = m . \frac{dv}{dt} + f . v + K . \int_0^t v . dt \qquad (6\text{-}1)$$

and for the electrical system:

$$i(t) = C . \frac{dv}{dt} + \frac{1}{R} . v + \frac{1}{L} . \int_0^t v.dt \qquad (6\text{-}2)$$

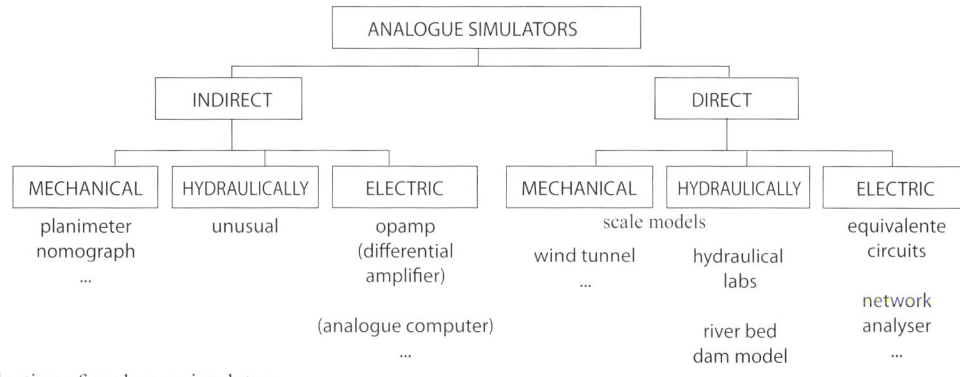

Fig. 6-5: Classification of analogue simulators

2.2 Analogies between physical systems

2.2.1 Table 6-1: Similarities between different physical quantities

Parameter	Mechanical		Electric	Hydraulically	Acoustic
	Translation	Rotation			
Applied force	F	M	i	p	P
Speed	v	$\omega = \dfrac{d\Theta}{dt}$	v	Q (mass flow)	$c = \dfrac{dx}{dt}$
Displacement	$s = \int_0^t v.\,dt$	Θ	$\int_0^t v.\,dt$	Q (mass)	x
Acceleration	$a = \dfrac{dv}{dt}$	$\dfrac{d\omega}{dt}$	$\dfrac{dv}{dt}$		$\dfrac{dc}{dt}$
Pulse	$\int_0^t F.\,dt$	$\int_0^t M.\,dt$	$\int_0^t i.\,dt$	$\int_0^t p.\,dt$	$\int_0^t P.\,dt$
Power	$F.v$	$M.\omega$	$v.i$	$q.p$	$P.c$
Inertia	$m.\dfrac{dv}{dt}$	$J.\dfrac{d\omega}{dt}$	$C.\dfrac{dv}{dt}$		$M_a.\dfrac{dc}{dt}$
Elasticity	K	K	L	C_H	C_a
Coefficient of friction	f	f	R	R_H	r_a
Momentum (of torque)		$J.\dfrac{d\omega}{dt} ; f.\dfrac{d\Theta}{dt} ; K.\Theta$			
Kinetic energy	$\tfrac{1}{2}.m.v^2$	$\tfrac{1}{2}.J.\omega^2$	$\tfrac{1}{2}.C.v^2$		$\tfrac{1}{2}.M_a.c^2$
Potential energy	$\tfrac{1}{2}.K.s^2$	$\tfrac{1}{2}.K.\Theta^2$	$\tfrac{1}{2}.L.i^2$	$\tfrac{1}{2}C_H.p^2$	$M_a.x$
Dissipated energy	$\tfrac{1}{2}.f.v^2$	$\tfrac{1}{2}.f.\omega^2$	$\tfrac{1}{2}.R.i^2$	$\tfrac{1}{2}.R_H.p^2$	$\tfrac{1}{2}.r_a.P^2$

K = compliance; K = rotation compliance; M_a = acoustic inertia; C_a = acoustic capacitance; f = translation friction; f = rotational friction; R_H = hydraulically resistance; r_a = acoustic resistance; J = moment of inertia; M = momentum (of torque); v = speed; v = voltage

We provide a number of applications of table 6-1 in the figures 6-6 to 6-16.

2.2.2 Applications:

1. Frictional dissipation

a. Electrical

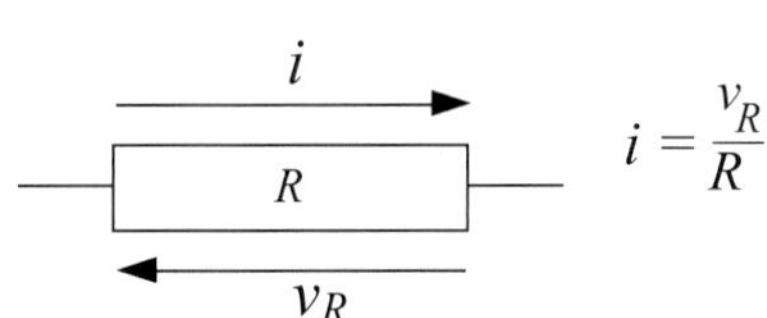

$$i = \frac{v_R}{R}$$

Fig. 6-6

b. Mechanical translation

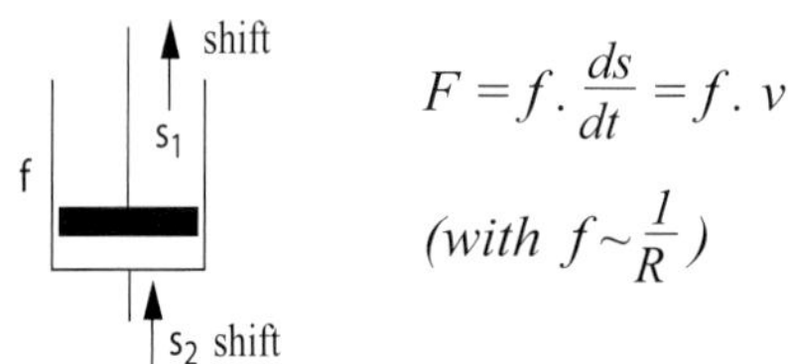

$$F = f \cdot \frac{ds}{dt} = f \cdot v$$

$$(with \ f \sim \frac{1}{R})$$

Fig. 6-7

c. Hydraulic

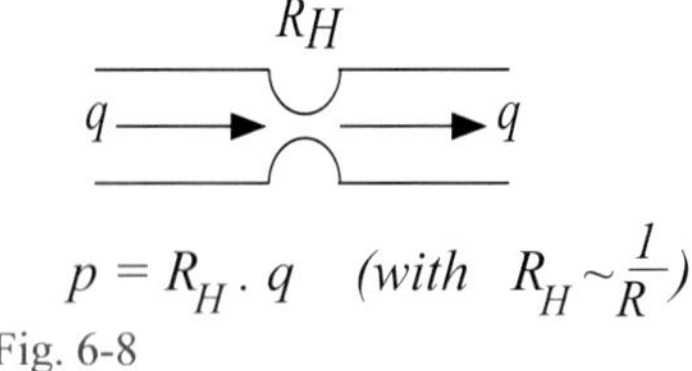

$$p = R_H \cdot q \quad (with \ R_H \sim \frac{1}{R})$$

Fig. 6-8

d. Mechanical rotation

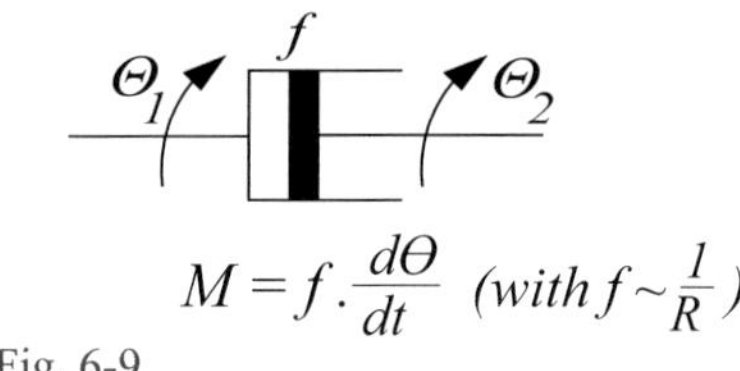

$$M = f \cdot \frac{d\Theta}{dt} \quad (with \ f \sim \frac{1}{R})$$

Fig. 6-9

2. Displacement

a. Electrical

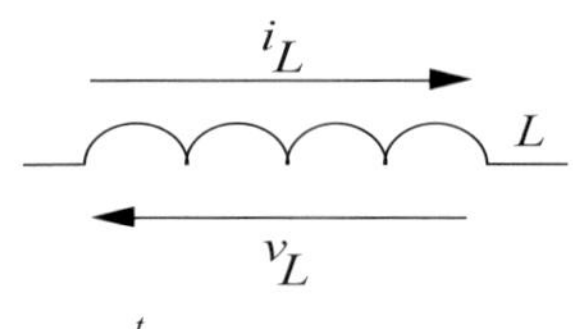

$$i_L = \frac{1}{L} \int_0^t v_L \cdot dt$$

Fig. 6-10

b. Mechanical translation

$$F = K \cdot (S_1 - S_2) \quad (with \ K \sim \frac{1}{L})$$

Fig. 6-11

c. Hydraulic

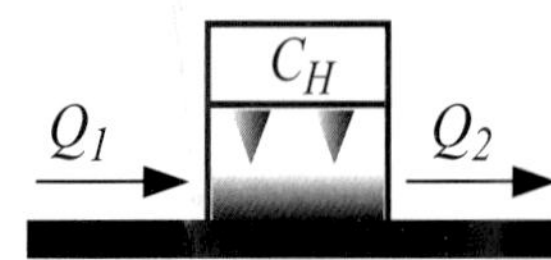

$$Q = Q_1 - Q_2$$

$$p = \frac{Q}{C_H} \quad (with \ C_H \sim \frac{1}{L})$$

Fig. 6-12

d. Mechanical rotation

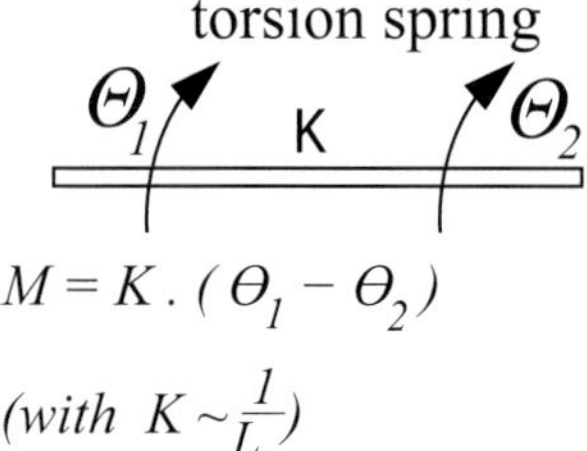

$$M = K \cdot (\Theta_1 - \Theta_2)$$

$$(with \ K \sim \frac{1}{L})$$

Fig. 6-13

3. Inertia

a. Electrical

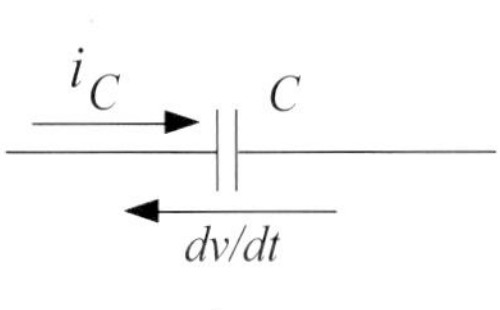

$$i_C = C \cdot \frac{dv}{dt}$$

Fig. 6-14

b. Mechanical translation

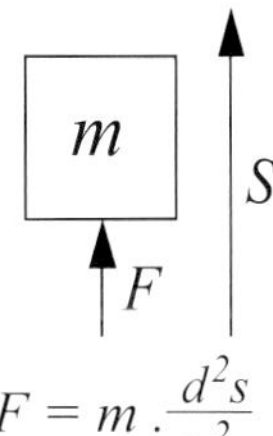

$$F = m \cdot \frac{d^2 s}{dt^2}$$

Fig. 6-15

c. Hydraulic

d. Mechanical rotation

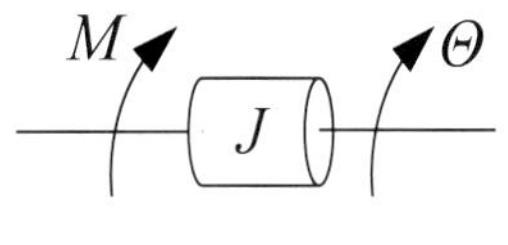

$$M = J \cdot \frac{d^2 \Theta}{dt^2}$$

Fig. 6-16

2.3 Building blocks of the analog computer

An analog computer consisted of essentially four types of parts:

- operational amplifiers switched as summing amplifiers
- operational amplifiers switched as integrators
- accurately adjustable potentiometers, the so-called coefficient potentiometers
- non-linear elements.

An example of solving a simple problem with an analog computer is now shown. Consider the equation of a simple servo system:

$$\frac{1}{\omega_0^2} \cdot \frac{d^2 y(t)}{dt^2} + \frac{2 \cdot \beta}{\omega_0} \cdot \frac{d y(t)}{dt} + y(t) = x(t) \tag{6-3}$$

Herein: $y(t)$ = output (angular position of the shaft)

        $x(t)$ = input (reference angular position)

        β = damping factor

        ω_0 = natural angular frequency

We can rewrite the differential equation (DE) : $\dfrac{1}{\omega_0^2} \cdot \ddot{y} + \dfrac{2 \cdot \beta}{\omega_0} \cdot \dot{y} + y = x(t)$

For simplicity we have written $y(t)$ as y.

Set $\omega_0 = 0.2$ rad/s and $\beta = 0.2$, then we find: $25 \cdot \ddot{y} + 2 \cdot \dot{y} + y = x(t)$

We isolate the highest derivative on the left-hand side:

$$\ddot{y} = -0.08 \cdot \dot{y} - 0.04 \cdot y + 0.04 \cdot x(t)$$

(6-4)

Fig. 6-17 shows the block diagram as implemented in an analog computer

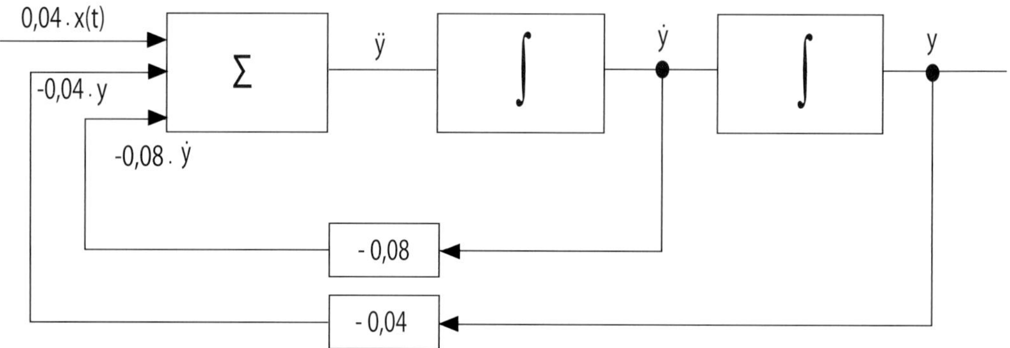

Fig. 6-17: Solution of a linear DE using an analog simulator

The name analog computer was a point of discussion in the sixties. Some authors did prefer to use the term "analog simulator".

3. SIMULATION WITH BLOCK DIAGRAMS: SIMULINK

Fifty years ago it was common to make use of block diagrams in telecommunication and in measurement and control theory. The block orientated simulation methods have been developed from simulating control systems. To simulate a switching network using block orientated methods produces problems with switching on and off of semiconductors. This is extremely difficult to describe with a set of differential equations (DE) since the structure of this set of DE's is constantly changing because of switching. In some situations it is almost impossible to describe the behaviour of a diode based on the switching state. In **Caspoc** this problem has been solved by the program as it creates the DE's based on the state of the switches.

Simulink, a powerful graphical preprocessor of **Matlab**, allows us to describe dynamic systems in the simple form of a block diagram. A graphical library is available with generators (adjustable step functions, sine functions,...) and a number of blocks in which the transfer function and its parameters can be adjusted to create a first order system, an integrator, a PID controller, etc. Comparators and the required connection lines provide the possibility to create a complete control block diagram. Useful components such as a double channel oscilloscope allow us to create a reference input step on one channel and the controlled output of the system (the response) on the other channel and these can be saved and printed.

Numeric example 6-1:

We consider here numeric example 13-1 of p. 13-7. It is an exercise with a buck converter with continuous current. The principal diagram is shown in fig. 13-4.

We redraw this diagram in fig. 6-18.

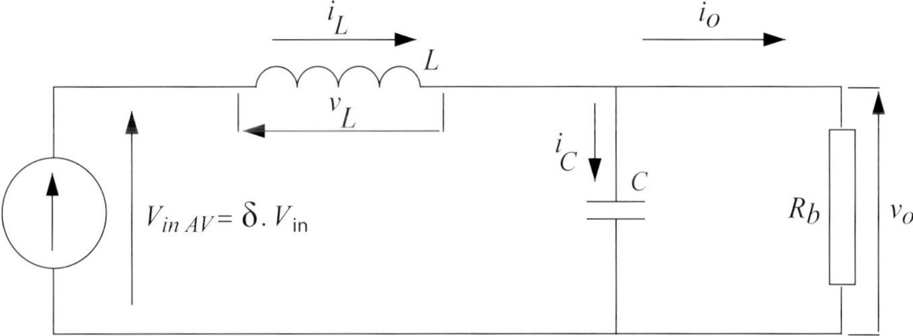

Fig. 6-18: Buck converter with continuous current

Voltage source V_i and switch S (with duty cycle δ) from fig. 13-4 are here represented by $\delta . V_i$. We can describe the following relationships:

$$v_L = \delta . V_{in} - v_0 = L . \frac{di_L}{dt} \tag{6-5}$$

$$i_L = \frac{1}{L} \int_0^t (\delta . V_{in} - v_0) . dt \tag{6-6}$$

$$v_c = v_0 = \frac{1}{C} \int_0^t i_c . dt \tag{6-7}$$

$$i_L = i_o + i_c \tag{6-8}$$

General rule: the outputs of integrating blocks are always used as the starting point for the setup of the model. We are familiar with this rule from the analog computer. With the expressions (6-6) and (6-7) we can draw two integrators in fig. 6-19. We also use expression (6-8) to complete the block diagram.

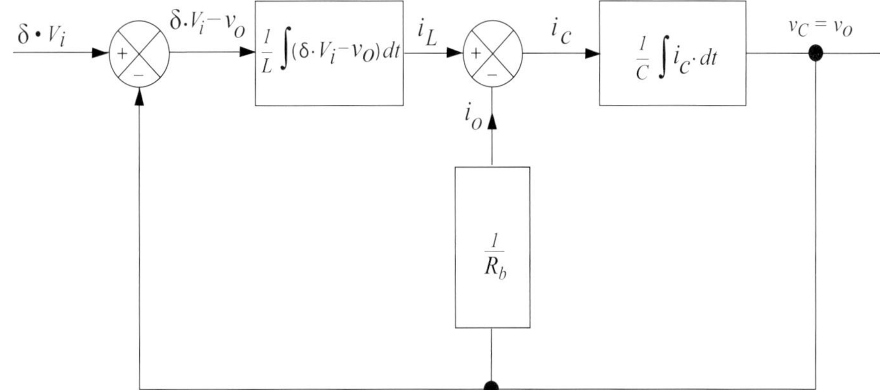

Fig. 6-19: Block diagram for fig. 6-18

Next we draw the block diagram in Simulink and we fill in the details of numeric example 13-1:
$L = 83.4\mu H$; $C = 100\mu F$; $R_b = 2.5\Omega$, so that: $\dfrac{1}{L} = 11990$; $\dfrac{1}{C} = 10^4$; $R_b = 0.4$ mho.
With for example $V_{in} = 10V$ and $V_{0\,AV} = 5V$, we find: $\delta = 0.5$.

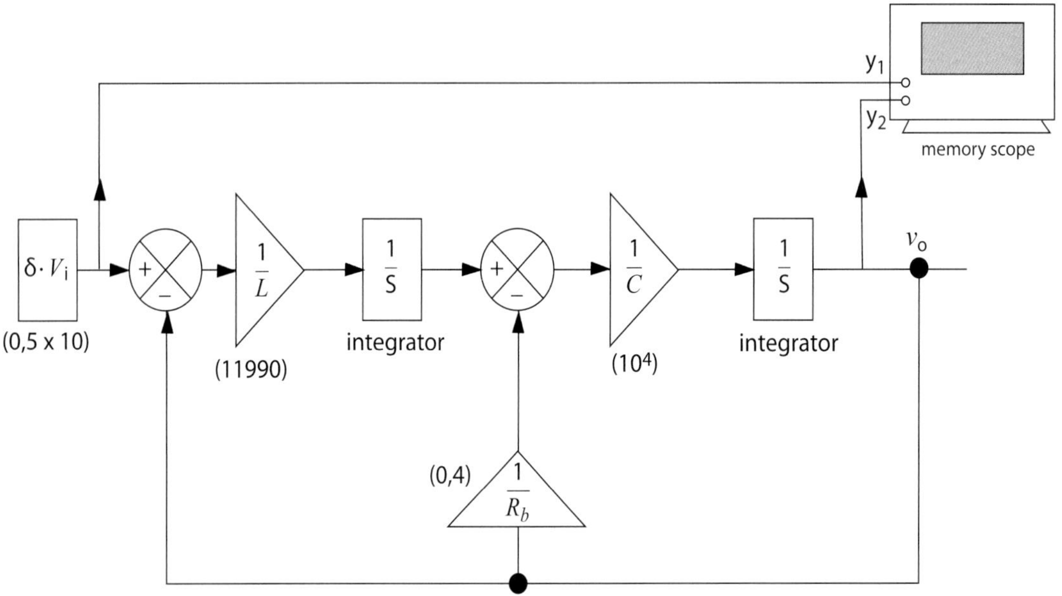

Fig. 6-20: Simulink block diagram of fig. 6-18

4. SIMULATION OF AN ELECTRICAL NETWORK – SPICE

4.1 Spice

Spice (**S**imulation **P**rogram with **I**ntegrated **C**ircuit **E**mphasis) was developed by the University of California-Berkely. A number of different commercial versions of Spice exist.

In Spice semiconductors are modelled by a set of equations, that have to be solved for each calculation step. In this process, depending upon the required accuracy, the integration step size is adjusted to obtain a convergent solution. This adjustment of the integration step size increases the calculation time. In **Caspoc** this problem is solved by including switching models, which require no adjustment of the integration step size.

In newer versions of Spice transfer functions and Laplace functions can be defined. The disadvantage of this is that these equations are added to the solution set of Spice, the result of which is that convergence is less likely as the functions become more complex. In **Caspoc** these functions are filled in using the block diagram method. Convergence is then less dependent on these functions. A digital controller in the form of an algorithm cannot be modelled in **Spice.**

4.2 Numeric example 6-2:

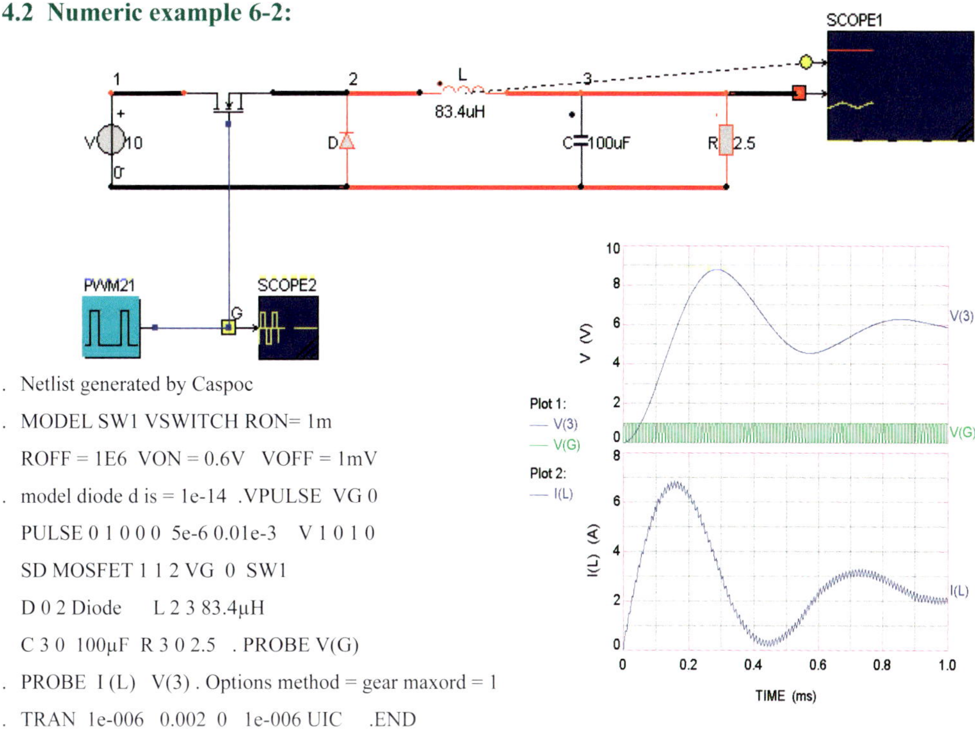

. Netlist generated by Caspoc
. MODEL SW1 VSWITCH RON= 1m
  ROFF = 1E6 VON = 0.6V VOFF = 1mV
. model diode d is = 1e-14 .VPULSE VG 0
  PULSE 0 1 0 0 0 5e-6 0.01e-3   V 1 0 1 0
  SD MOSFET 1 1 2 VG 0 SW1
  D 0 2 Diode   L 2 3 83.4µH
  C 3 0 100µF R 3 0 2.5   . PROBE V(G)
. PROBE I (L) V(3) . Options method = gear maxord = 1
. TRAN 1e-006 0.002 0 1e-006 UIC   .END

Fig. 6-21: Spice simulation of a buck converter with continuous current

5. MULTILEVEL MODEL AND SIMULATION: CASPOC

5.1 Levels in model forming

Caspoc (**C**omputer **A**ided **S**imulation for **PO**wer **C**onverters) was developed by
Dr. MSc. P. Van Duijsen from the firm Simulation Research. Caspoc is a multilevel simulation
packet. The starting point of model forming of a mechatronic system is a multilevel model of the
system. This multilevel model describes the independent components of the system, such as the
electrical, mechanical and control components and the links that are present between these independent parts. A multi level model contains the following three levels:
1. system level
2. circuit level
3. component level

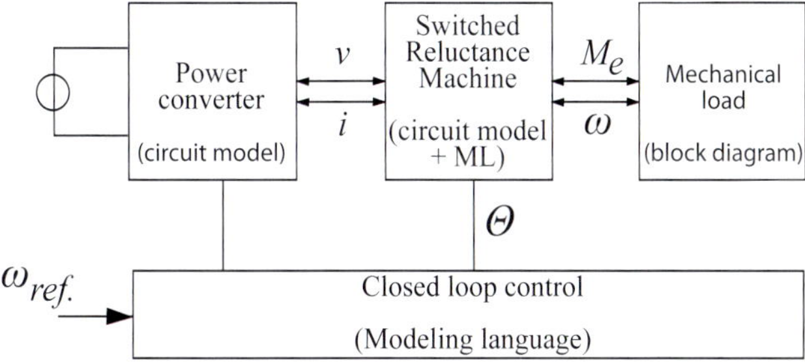

Fig. 6-22: Multilevel model of an electrical drive

In fig. 6-22 it is shown that the model of a power converter is linked with the model of an electrical machine, a switched reluctance machine. This model in turn is linked with the model of a mechanical load. Modelling of a mechatronic system is split over these three levels, see fig. 6-23.

At the lowest level, the component level, the independent components are described. Included here are all the electrical components of power electronics, the mechanical actuators such as electrical machines, the sensors for transforming the mechanical variables to electrical variables and mechanical components for driving the load such as gear boxes and belts.

In the middle level we describe the connections of the independent mechanical and electrical components that form a circuit. The most common example here is an electrical circuit. Also the mechanical components can be considered a circuit, think about the comparisons described earlier.

At the highest level, the system level, the control of the system is described. Here in addition to the internal controller for the power electronics we find the external control algorithm described. This algorithm for example could be programmed by a process operator in the factory.

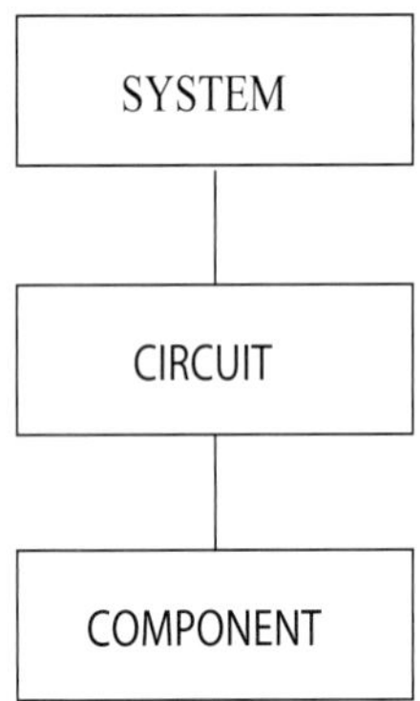

Fig. 6-23: Model levels

5.2 Different types of models

From the mechatronic viewpoint the whole system needs to be modelled. The majority of the existing modelling and simulation packets are focused on only one form of modelling. In this way we can distinguish the differences between the simulation packets for electrical circuits and dynamic systems.

1. **Electric circuit simulation:**

   A large variety of programs exist for simulating electric circuits. The most widely known are the simulation packets based on SPICE.
   - Advantage: designed for electric circuits.
   - Disadvantage: not suitable for control algorithms and dynamic non-linear mechanical systems.

2. **Dynamic systems simulation: (block diagram)**
   - Advantage: able to create models of non-linear dynamic systems.
   - Disadvantage: electrical circuits first need to be described by differential equations.

3.   **Control algorithm: (Pascal, C)**
   . Advantage: describing a control algorithm.
   . Disadvantage: no possibility for modelling dynamic systems, unless part of a complete simulation program.

A possibility to create a model of all three types is a multilevel model. This multilevel model is composed of the following minimum parts:
1. Circuit model
2. Block diagram model
3. Modelling language (Pascal, C)

The combination of the above mentioned models is a multilevel model. The connections between the circuit model, the block diagram model and the modelling language exist in the block diagram. The simulation is composed of all the models of the multilevel model. In this way it is possible to study the interaction between the different subsystems. Changing the mechanical load, such as torque pulses, have a direct effect on the electrical behaviour of the power converter.

5.3 Multilevel model forming and simulation

The three levels are divided as shown in table 6-2. Here we can see that the electric circuit is solved using "modified nodal analysis".

The components and system behaviour can be modelled in the block diagram method. The block diagram method is extended with a "modelling language". With the help of this "modelling language", system blocks with multiple inputs and outputs can be defined.

The "modelling language" in the current version of Caspoc is Pascal, in which functions in C can be added. Using this program language equations can be built, or complex algorithms can be implemented for example to model a digital controller.

Table 6-2: Implementation levels in Caspoc

System behaviour	Block diagram + Modelling Language
Electric circuit with components	Circuit analysis (MNA)
Component behaviour	Block diagram + Modelling Language

MNA = modified nodal analysis

Between the different levels a link needs to be created. These links are shown in table 6-3. As a result of all the different modelling techniques the possibilities for simulating power electronics and electronic drives is quite diverse. Simple rectifiers can be modelled including choppers for stabilising DC voltage. Additionally a voltage source converter with a three phase controlled rectifier and an induction machine with vector control can be implemented. The voltage source converter with rectifier with electrical components can be simulated using the circuit analysis method, while the induction machine and the mechanical kinematic equations can be described using system blocks. Finally the vector control can be described using the "modelling language".

As a result of this type of implementation it has become relatively simple to simulate an entire circuit. The circuit model can be quickly adjusted by the addition or removal of electrical components. Optimising the controller design can be achieved by writing a few lines of program code. This type of technique results in optimal simulation. Additionally the simulation of a system block which represents an analog process, such as integration, is carried out much quicker than a circuit with multiple components.

Table 6-3: Links between the implementation methods

Electric circuit →→ block diagram	Voltage (junction) Current (component) Condition (switch)
Block diagram →→ electric circuit	Controlled current source Controlled voltage source Switch (SCR, GTO, IGBT)
Block diagram →→ modelling language	Signal (block diagram)
Modelling language →→ block diagram	Signal (Modelling Language)

5.4 Conclusion

Using models and simulation can save time and money in the development process. In the case of a mechatronic design special requirements need to be formulated for the modelling and simulation software. As a result of the different techniques available for mechatronics, electronics, mechanics and information technology, the use of a multilevel simulation packet greatly simplifies the modelling of mechatronic systems. The multilevel model simplifies the creation of models of the independent components of a drive system.

For power electronics a circuit model is used. The equations of the electrical machine are located in the block diagram and the algorithm of the controller form part of the modelling language. The combination of the three models, the multilevel model, is simulated as one model.

5.5 Numeric example 6-3:

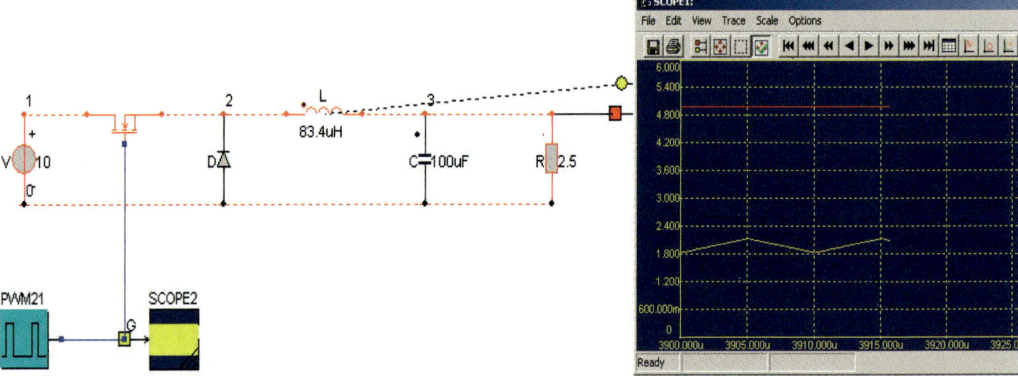

Fig. 6-24: CASPOC simulation of a buck converter with continuous current

Output voltage and current in the coil in nominal service

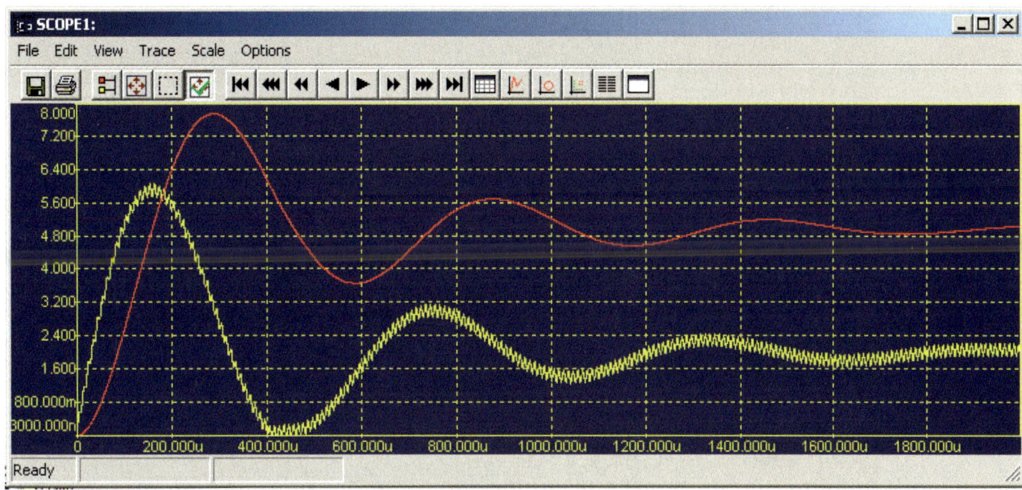

Fig. 6-25:   Output voltage and current at start-up of the circuit of fig. 6-24 (transient behaviour). CASPOC simulation

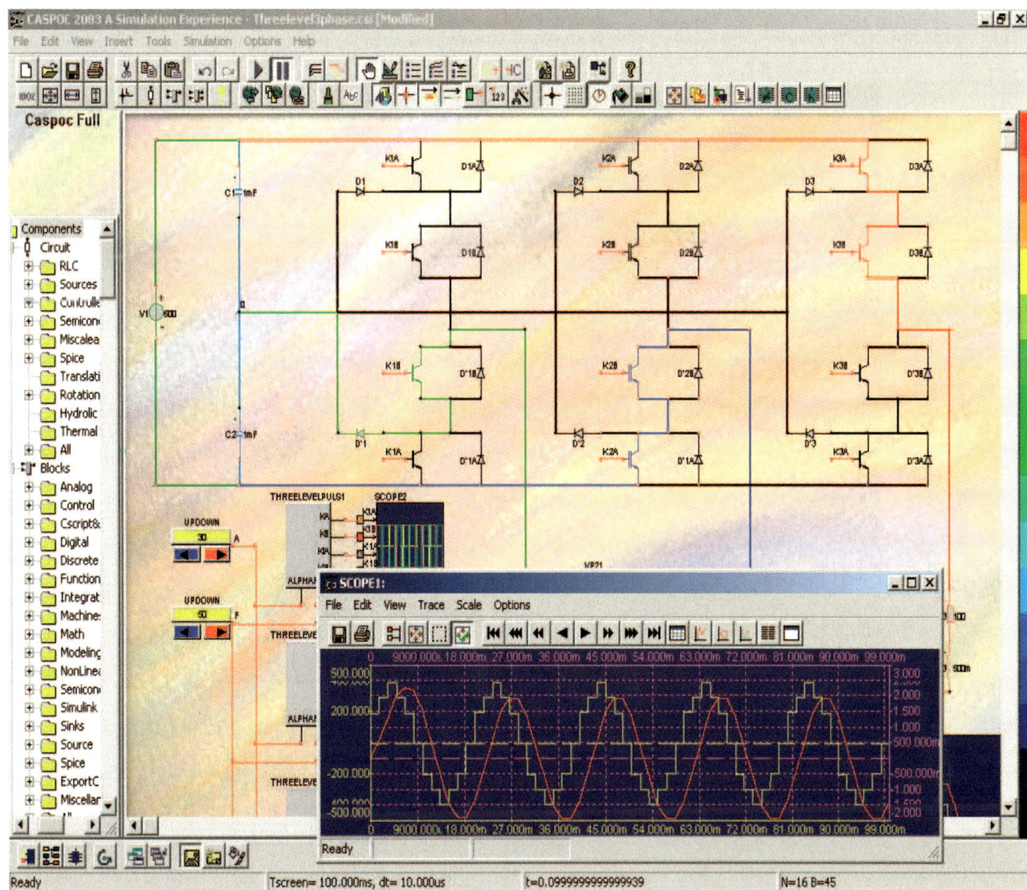

Photo Simulation Research: Simulation of a three-phase current source inverter (CSI) with the accompanying waveforms, provided by a CASPOC simulation

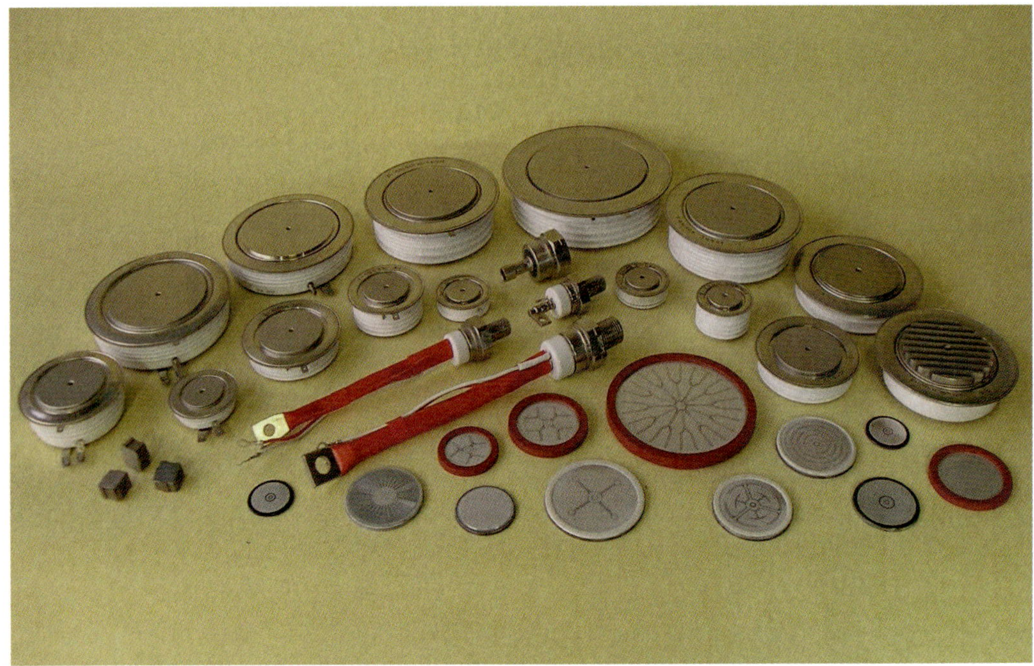

Photo Westcode (KWx): A selection of semiconductors from Westcode.

Photo Hitachi: PDD 4k5 6k5 IGBTs

PART 2

ELECTRONIC POWER CONVERTERS

7 INDUCTIVELY LOADED RECTIFIERS

CONTENTS

Uncontrolled rectifiers can be classified in various ways: according to the supply voltage, the type of load, the application, etc...

Depending on the supply voltage, we can distinguish between single phase and three-phase rectifiers.

As far as the load goes we distinguish resistive, capacitive and resistive inductive loads.

If we look at applications we distinguish two large groups:

1. Linear and switch mode power supplies : mainly used for the supply to electronic equipment and circuits.

2. Industrial applications: this is mainly inductively loaded single phase rectifiers and three-phase bridge rectifiers. This is the material discussed of this chapter.

   Examples of such industrial applications include: battery chargers, DC welding equipment, DC traction, electrical supply for electrochemical processes (electrolysis, chrome plating, nickel coating), rectifiers for automobile alternators, the supply for frequency converters, etc ...

1.   CURRENT FLOW OF AN INDUCTIVE CIRCUIT WITH A SINUSOIDAL SUPPLY

1.1 Mechanical switch

At instant t_1 we close the switch of the circuit with a resistive inductive load and a sinusoidal voltage $v = \hat{v} \cdot \sin \omega t$. This is shown in fig. 7-1. The actual applied voltage for the series circuit of R_b and L_b is shown to the right of the switch S.

Assuming $Z = \sqrt{R_b^2 + \omega^2 \cdot L_b^2}$; $\mathrm{tg}\ \Phi = \dfrac{\omega . L_b}{R_b}$; $\cos \Phi = \dfrac{R_b}{Z}$; $\tau = \dfrac{L_b}{R_b}$ and $\alpha = \omega \cdot t_1$.

Calculating leads us to the following expression for the current flow through the load

$$i(t) = \frac{\hat{v}}{Z} \cdot \left[\sin(\omega t - \phi) + \sin(\phi - \alpha) \cdot e^{\frac{(t_1 - t)}{\tau}} \right] \tag{7-1}$$

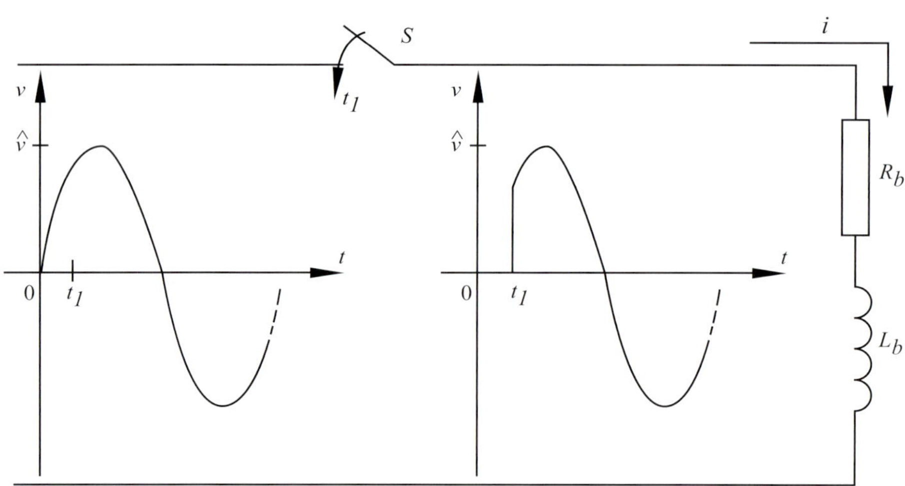

Fig. 7-1: R–L circuit connected to an AC voltage source

Remarks

1. In expression (7-1) we notice that the current has two components:

- $\dfrac{\hat{v}}{Z} \cdot \sin(\Phi - \alpha) \cdot e^{\frac{(t_1 - t)}{\tau}}$. This is an exponentially decaying term which has practically disappeared after $5.\tau$. This is the transient term.

- $\dfrac{\hat{v}}{Z} \cdot \sin(\omega t - \Phi)$. After $5.\tau$ only this term remains, this is referred to as the steady state term. This is the nominal current which we have learned about in the classical electrical science of an RL-circuit when connected to a sinusoidal voltage. In expression (7-1) we have actually taken account of what happens immediately after closing the switch when connecting a sinusoidal voltage and an RL circuit, in other words we have taken the transient behaviour into account.

2. If we close the switch in figure 7-1 at $t_1 = 0$ then expression (7-1) becomes:

$$i(t) = \frac{\hat{v}}{Z} \cdot \left[\sin(\omega t - \phi) + \sin \phi \cdot e^{-t/\tau} \right] \qquad (7\text{-}2)$$

For this special case ($t_1 = 0$) we have drawn the transient term and the nominal term together in fig. 7-2 (this is the total current $i(t)$!).
After $5.\tau$ the transient term is practically zero and $i(t)$ = steady state term.

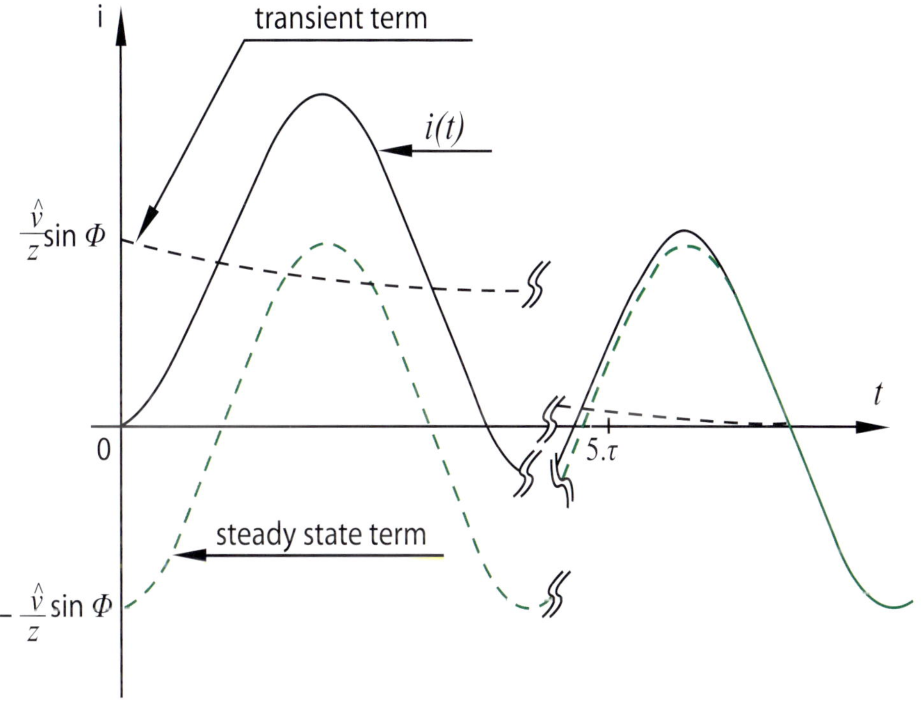

Fig. 7-2: Current flow of fig. 7-1 if $t_1 = 0$.

1.2 Numeric examples:

1. Supply voltage: 230 V – 50 Hz.   Load: 10 Ω – 100 mH.
    Switch closes at $t_1 = 0$. Determine the current flow.

Solution:

$$\tau = \frac{L_b}{R_b} = 10 \text{ ms} \; ; \; Z = 32.96 \, \Omega \; ; \; \Phi = 72°20' \; ; \; \hat{v} = 325 \text{ V} \; ;$$

$$i(t) = 9.85 \left[\sin(\omega t - 72°20') + 0.952 \cdot e^{-100.t} \right].$$

Fig. 7-3 shows the solution.

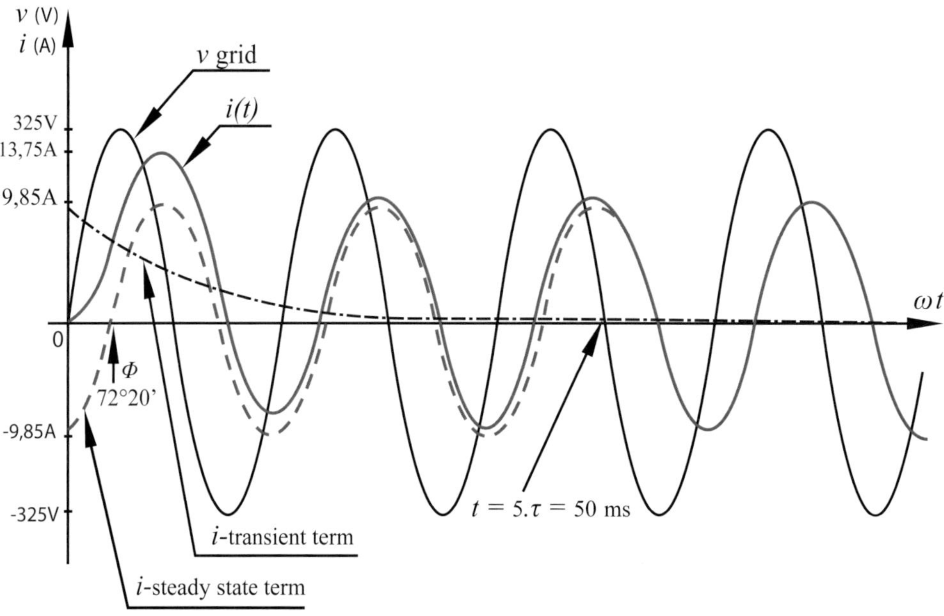

Fig. 7-3: Current flow

2. Same details but now the switch closes at $\alpha = 72°20'$

$$i(t) = 9.85 \left[\sin \left(\omega t - 72°20' \right) + 0 \right] = 9.85 \cdot \sin \left(\omega t - 72°20' \right).$$

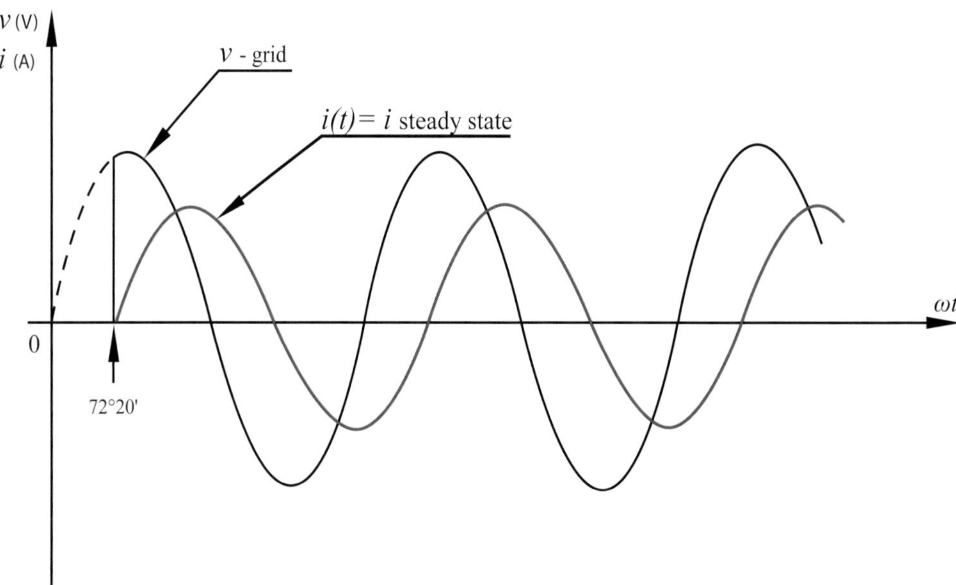

Fig. 7-4: Current flow

Notice in fig. 7-4 there is no transient term, in other words we immediately arrive at the "nominal" state. The current lags the voltage by 72°20' .

1.3 Calculating the current flow in fig. 7.1

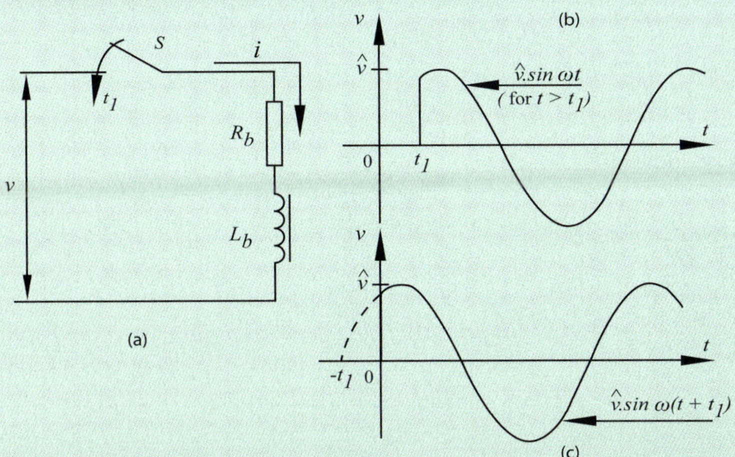

Fig. 7-5: *R-L* circuit connected to an AC voltage source.

A resistive and inductive circuit is connected via a mechanical switch S at instant t_1 to a sinusoidal voltage $v = \hat{v} . \sin \omega t$ (fig. 7-5). Our goal is to determine the current flow in the circuit.
To describe the applied voltage with Laplace notations we first consider fig. 7-5c.
Here in: $v = \hat{v} . \sin \omega (t + t_1) = \hat{v} . \sin (\omega t + \alpha)$ with $\alpha = \omega t_1$.
This voltage only exists for $t \geq 0_+$.

$$F(s) = \mathscr{L} \left[\hat{v} . \sin (\omega t + \alpha) \right] = \hat{v} . \frac{s . \sin \alpha + \omega . \cos \alpha}{s^2 + \omega^2}$$

By moving the voltage waveform of fig. 7-5c over a distance t_1 we arrive at fig. 7-5b.

Application of the time delay rule gives: $\quad F(s) = \hat{v} . e^{- t_1 s} . \dfrac{s . \sin \alpha + \omega . \cos \alpha}{s^2 + \omega^2}$

Set: $\quad \tau = \dfrac{L_b}{R_b} ; \quad Z = \sqrt{ R_b^2 + \omega^2 . L_b^2 } ; \quad tg\ \phi = \dfrac{\omega L_b}{R_b} ; \quad \sin \phi = \dfrac{\omega L_b}{Z} ; \quad \cos \phi = \dfrac{R_b}{Z}$

The current in fig. 7-5a can be written as:

$$I(s) = \frac{\hat{v} . e^{- t_1 s} . \dfrac{s . \sin \alpha + \omega \cos \alpha}{s^2 + \omega^2}}{(R_b + s . L_b)} = \hat{v} . e^{- t_1 s} \left[\frac{s . \sin \alpha + \omega . \cos \alpha}{(R_b + s . L_b) . (s^2 + \omega^2)} \right]$$

Expressions between square brackets need to be split into partial fractions:

$$\frac{a}{R_b + s . L_b} + \frac{b . s + c}{(s^2 + \omega^2)} = \frac{s . \sin \alpha + \omega . \cos \alpha}{(R_b + s . L_b) . (s^2 + \omega^2)}$$

Solving gives:

$$a = \frac{L_b}{Z} . \sin (\phi - \alpha)$$

$$b = \frac{1}{Z} . \sin (\alpha - \phi)$$

$$c = \frac{\omega}{Z} . \cos (\phi - \alpha)$$

So that:
$$I(s) = \hat{v} \cdot e^{-t_1 s} \cdot \left[\frac{\frac{L_b}{Z} \cdot \sin(\phi - \alpha)}{(R_b + s \cdot L_b)} + \frac{\frac{1}{Z} \cdot s \cdot \sin(\alpha - \phi) + \frac{\omega}{Z} \cdot \cos(\phi - \alpha)}{(s^2 + \omega^2)} \right]$$

$$I(s) = \frac{\hat{v} \cdot e^{-t_1 s}}{Z} \cdot \left[\frac{\sin(\phi - \alpha)}{(s + \frac{1}{\tau})} + \frac{s \cdot \sin(\alpha - \phi) + \omega \cdot \cos(\alpha - \phi)}{(s^2 + \omega^2)} \right]$$

Inverse Laplace transform:

$$i(t) = \frac{\hat{v}}{Z} \cdot \left[e^{-\frac{(t - t_1)}{\tau}} \cdot \sin(\phi - \alpha) + \sin(\omega t - \omega t_1 + \alpha - \phi) \right]$$

$$i(t) = \frac{\hat{v}}{Z} \cdot \left[e^{-\frac{(t - t_1)}{\tau}} \cdot \sin(\phi - \alpha) + \sin(\omega t - \phi) \right]$$

$$i(t) = \frac{\hat{v}}{Z} \cdot \left[\sin(\omega t - \phi) + \sin(\phi - \alpha) \cdot e^{\frac{(t_1 - t)}{\tau}} \right] \tag{7-1}$$

With $\alpha = \omega t_1$ and $\omega \tau = tg\, \phi$ we can re-write (7-1):

$$i(t) = \frac{\hat{v}}{Z} \cdot \left[\sin(\omega t - \phi) + \sin(\phi - \alpha) \cdot e^{(\alpha - \omega t) \cdot cotg\, \phi} \right] \tag{7-3}$$

2.   CURRENT AND VOLTAGE FLOW OF AN INDUCTIVELY LOADED SINGLE PHASE RECTIFIER

2.1 Current flow

The diode in fig. 7-6 can be compared to a switch which is closed at $t = 0$. Expression (7-2) indicates the current flow:

$$i = \frac{\hat{v}}{Z} \cdot \left[\sin\phi \cdot e^{-t/\tau} + \sin(\omega t - \phi) \right] \tag{7-2}$$

Here in: $Z = \sqrt{R_b^2 + \omega^2 L_b^2}$; $\Phi = bgtg\dfrac{\omega L_b}{R_b}$; $\tau = \dfrac{L_b}{R_b}$.

We consider the same numerical example as that on p. 7-3. The current $i(t)$ of fig. 7-3 can be redrawn as in fig. 7-7. There is however a fundamental difference, since in contrast with the mechanical switch the diode prevents the current becoming negative. The diode blocks at instant t_2, see fig. 7-7. Every period ($T,\ 2\ T...$) the transient behaviour reoccurs again.

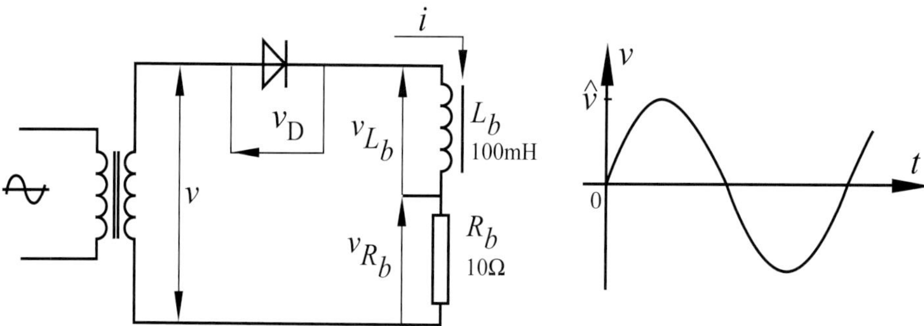

Fig. 7-6: Inductively loaded diode circuit.

The 8.65 ms and the 13.75 A in fig. 7-7 are calculated in section 2.2.

In section number 2-5 we find that $t_2 = 14.7$ ms.

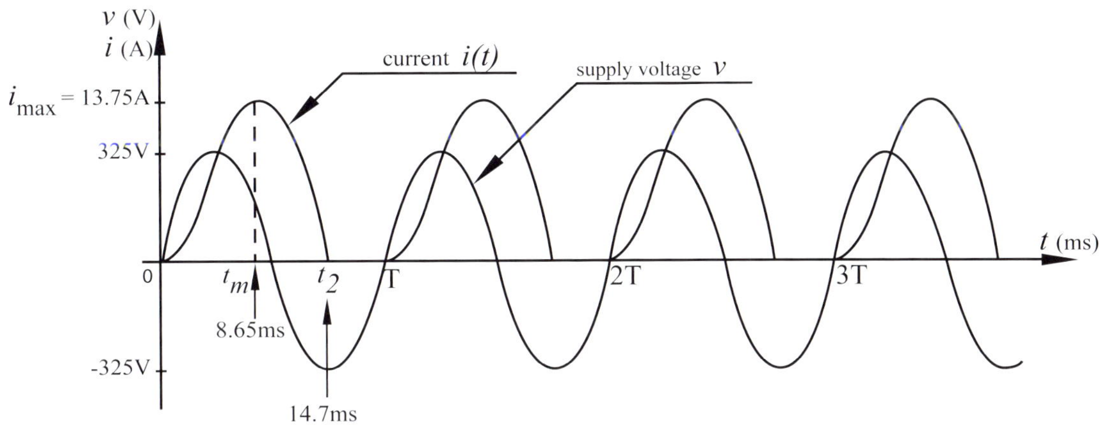

Fig. 7-7: Current flow of fig. 7-6.

2.2 Time of maximum current strength

By neglecting the voltage drop across the diode for every instant in time the voltage is:

$$u = u_{R_b} + u_{L_b} = i \cdot R_b + L_b \cdot \frac{di}{dt}$$

The instant (t_m) of maximum current strength occurs at $\frac{di}{dt} = 0$.

From (7-2): $\frac{di}{dt} = 0 = \frac{\hat{v}}{Z} \left[-\frac{\sin \phi}{\tau} \cdot e^{-t_m/\tau} + \omega \cdot \cos (\omega t_m - \phi) \right]$ or: $e^{-t_m/\tau} = \frac{\omega \tau \cos (\omega t_m - \phi)}{\sin \phi}$

so that: $$t_m = -\tau \cdot \ln \left[\frac{\omega \tau \cos (\omega t_m - \phi)}{\sin \phi} \right] \qquad (7-4)$$

Expression (7-4) may be considered as exact insofar as $v_D \ll \hat{v}$.

At the instant (t_m) of maximum current strength $\frac{di}{dt} = 0$ and it follows from

$v = i \cdot R_b + L_b \cdot \frac{di}{dt}$ that $v = i_{max} \cdot R_b$.

The time of i_{max} and also of $v_{R_b} = i_{max} \cdot R_b$ is the intersection of v_{R_b} with the v-curve. Function (7-4) is difficult to use since t_m appears in the left and the right term. The easiest method is to choose a few values for t_m in the right-hand side and calculate if we get the same values on the left-hand side.

With the data from fig. 7-6 we find that $t_m = 8.65$ ms. With $\hat{v} = 230 \times \sqrt{2} = 325.26$ V it follows from (7-2): $i_{max} = 13.75$ A. This is indicated in fig. 7-7.

2.3 Voltage waveform. Equal surface area criterion

If we neglect the voltage drop across the diode in fig. 7-6 , then we find for every instant in time:
$v = v_{L_b} + v_{R_b}$. Here in : $v_{R_b} = i . R_b$.
We reconsider the current waveform in fig. 7-8a that we already determined in fig. 7-7, but now for just one period of the supply voltage. The current reaches its maximum value i_{max} at time t_m .
In fig. 7-8b we now draw $v_{R_b} = i . R_b$, together with the supply voltage v.
At every instant $v_{L_b} = v - v_{R_b}$ is described.
From zero to t_m the coil is energised with an energy $\frac{1}{2} . L_b . i_{max}^2$.
Between t_m and t_2 the polarity of v_{L_b} reverses (see fig. 7-8b) while i remains positive: the energy $\frac{1}{2} . L_b . i_{max}^2$ flows out of the coil. At t_2 , i is zero.
The energy in the coil has also become zero.
In fig 7-8c we have only drawn v_{L_b} . The shaded surface area indicates the flux in the coil.

From zero to t_m the flux is built up $\left[\Delta \Phi = \int_{o}^{t_m} v_{L_b} . dt \right]$ (Wb) and between t_m and t_2 the flux will return to 0.
The shaded area below the t-axis is exactly equal to the shaded area above the t-axis.
The **average voltage** v_{L_b} across the coil is zero.
For an **ideal coil** this **equal surface area criterion** is always valid. Note that the (perpendicular) shaded surface area to the left of t_m in fig. 7-8b is equal to the shaded surface area to the right of t_m . These are the same surface areas as shown in fig. 7-8c!

2.4 Average output voltage of a single phase rectifier

The average output voltage in fig. 7-8d (and in fig. 7-6!) depends upon the conduction angle β .

$$V_{di} = \frac{1}{2.\pi} \int_{o}^{\beta} v . d\omega t = \frac{1}{2.\pi} \int_{o}^{\beta} \hat{v} . \sin \omega t . d\omega t \qquad \boxed{V_{di} = \frac{\hat{v}}{2.\pi} . (1 - \cos \beta)} \qquad (7-5)$$

$V_{di} :$ d = average value (= DC voltage)
      i = ideal. We consider a rectifier without losses in the transformer or the diode.

Since $v_{L_{b\,(AV)}} = 0$, the output DC current may be determined from: $I_d = \frac{V_{di}}{R_b}$.

The value of the conduction angle β will be determined in section 2.5.

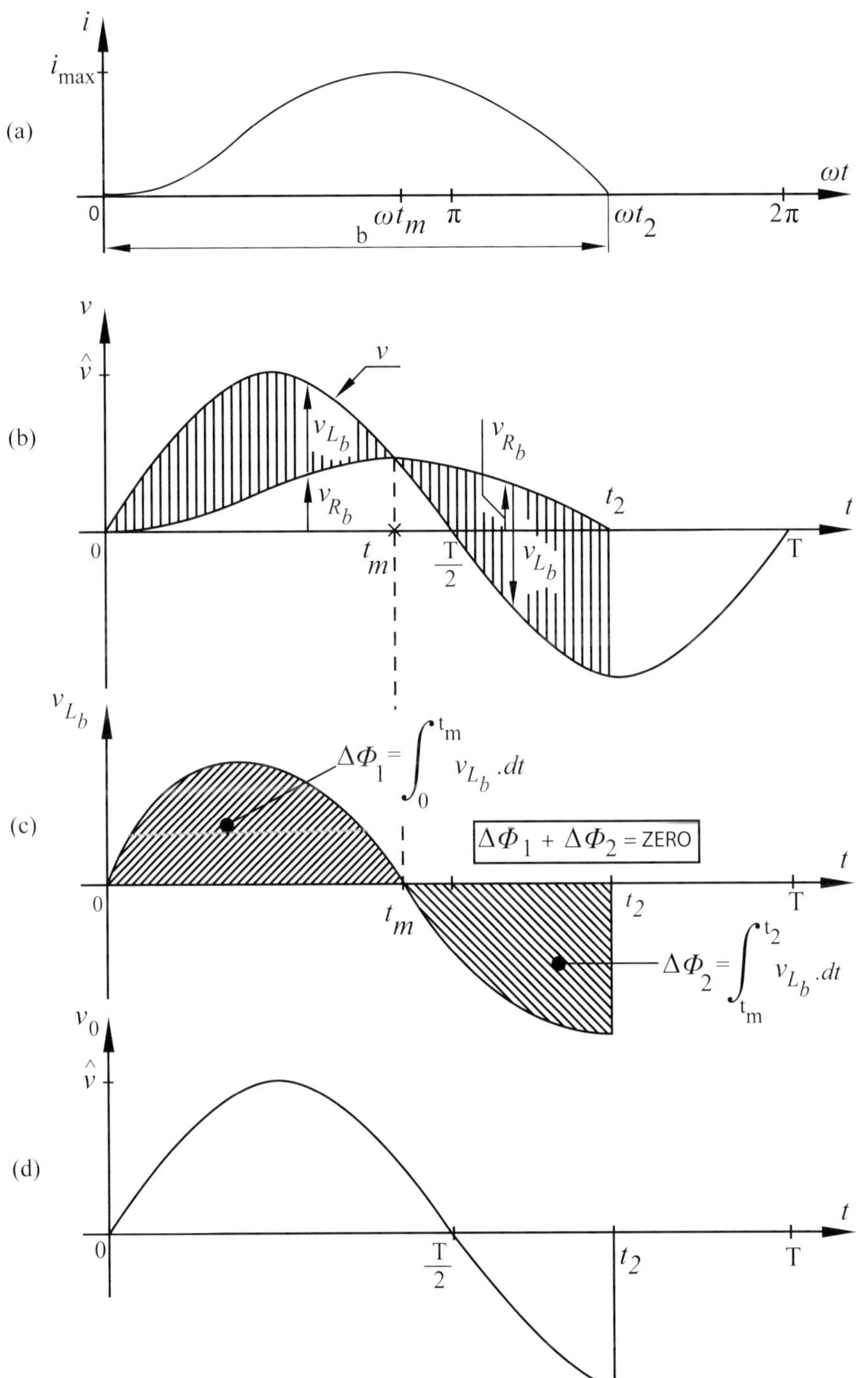

Fig. 7-8: Current and voltage waveforms from the configuration in fig. 7-6

2.5 Conduction angle

The angle $\omega t_2 = \beta$ (fig. 7-8a) is called the conduction angle. We can determine β by making the current equal to zero in expression (7-2):

$$0 = \frac{\hat{v}}{Z}\left[\,\sin\,(\,\omega t_2 - \Phi\,) + \sin\,\Phi\,.\,e^{-\,t_2/\tau}\,\right]$$

With $\omega t_2 = \beta$ and $\omega.\tau = \text{tg }\Phi$ we find: $\sin\,(\,\Phi - \beta\,) = \sin\,\Phi\,.\,e^{-\,\beta/\text{tg}\Phi}$.

The value of β as a function of Φ can only be determined by iteration. The solution is graphically presented in fig. 7-9.

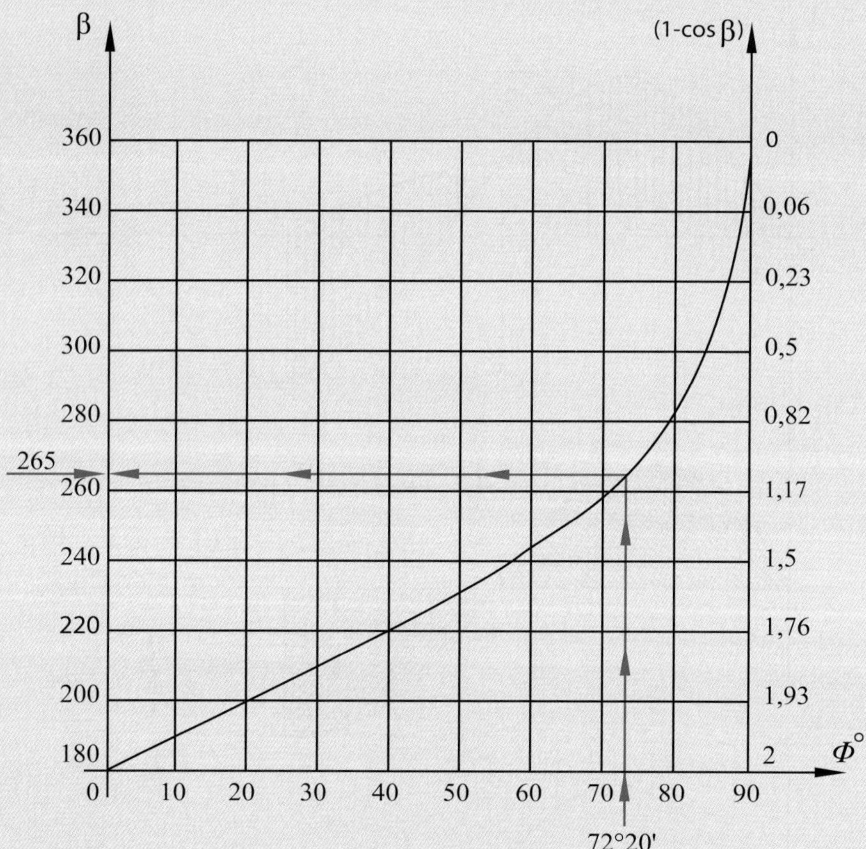

Fig. 7-9: Relationship between conduction angle and inductivity in the configuration of fig. 7-6.

With $\Phi = 72°20'$ we find: $\beta = 265°$.

From $\omega t_2 = \beta = 265°$ it follows: $t_2 = 14.7$ ms. This is shown in figure 7-7.

3. APPROXIMATIONS IN THE STUDY OF RECTIFIERS

For simplicity in the study of rectifiers a number of ideal properties are assumed, namely:
1. Switches without voltage drop or leakage current.
2. Switching occurs without time delay.
3. The supply voltage is perfectly sinusoidal.
4. There is a constant DC output current.

These ideal properties are closely approximated by industrial configurations. In the following cases we make use of these approximations.

4. HALF WAVE THREE-PHASE RECTIFIER - RESISTIVE LOAD

4.1 Operation

From t_1 (fig. 7-10) v_1 is the most positive secondary voltage and D_1 conducts. Via D_1, v_1 is applied to the cathodes of D_2 and D_3 so that these block. At t_2 , v_2 is larger than v_1 and D_2 conducts. At the cathode of D_1 , v_2 appears so that D_1 blocks. In the same fashion D_3 conducts from t_3 .

Remarks
1. The phase order of the supply determines which (diode-) switch conduct and which switches turn off (commutate). Since the supply network dictates this sequence we talk about net commutation or natural commutation. The points 1, 2, 3, 4,... (fig. 7-10) are the natural commutation points.
2. As will later become clear later, in this circuit pre-magnetisation of the transformer core occurs.

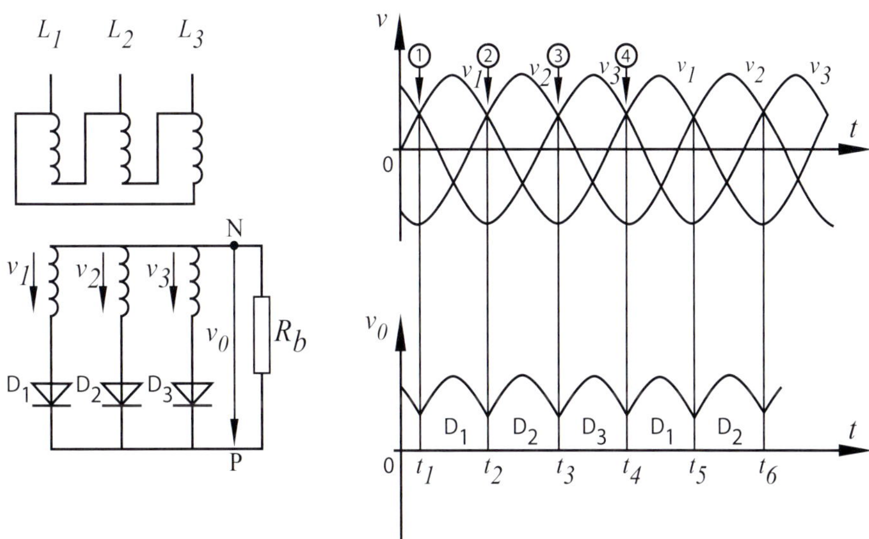

Fig. 7-10: Half wave three-phase rectifier. Configuration and voltage waveform.

4.2 Commutation

Up to now we have assumed that the diodes commutate immediately in the natural commutation points. From practical point of view that the leakage reactance of the supply transformer prevents the currents in the conducting diode from suddenly decreasing to zero at instant t_2 (fig. 7-10) D_2 conducts and D_1 cannot yet block. Fig. 7-11 indicates an equivalent circuit for this situation. The phases 1 and 2 are temporarily short-circuited. The short-circuited voltage v_K at the cathodes of D_1 and D_2 is indicated by:

$$v_K = \frac{v_1 + v_2}{2}$$

After t_2, v_2 becomes more positive and v_1 becomes less positive so that i_{d2} increases and i_{d1} decreases. The difference $v_K - v_1$ together with X_{L1} determines the time at which i_{d1} becomes zero, while $v_2 - v_K$ and X_{L2} determine the increase of i_{d2}.

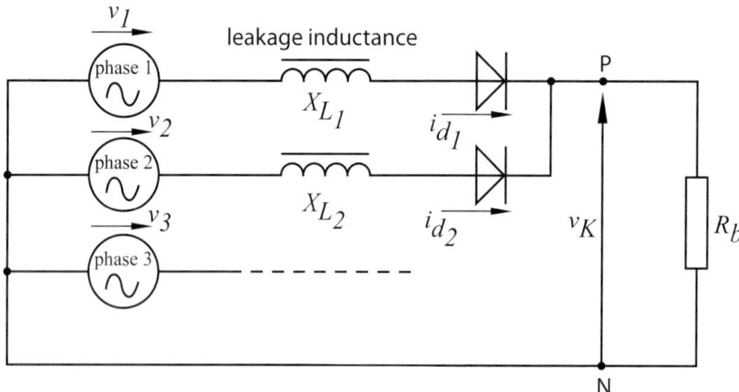

Fig. 7-11: Short-circuited voltage v_K during commutation

The commutation time can also be expressed as the overlap angle µ. The shaded surface area (fig. 12-7) shows the voltage loss as a result of commutation.

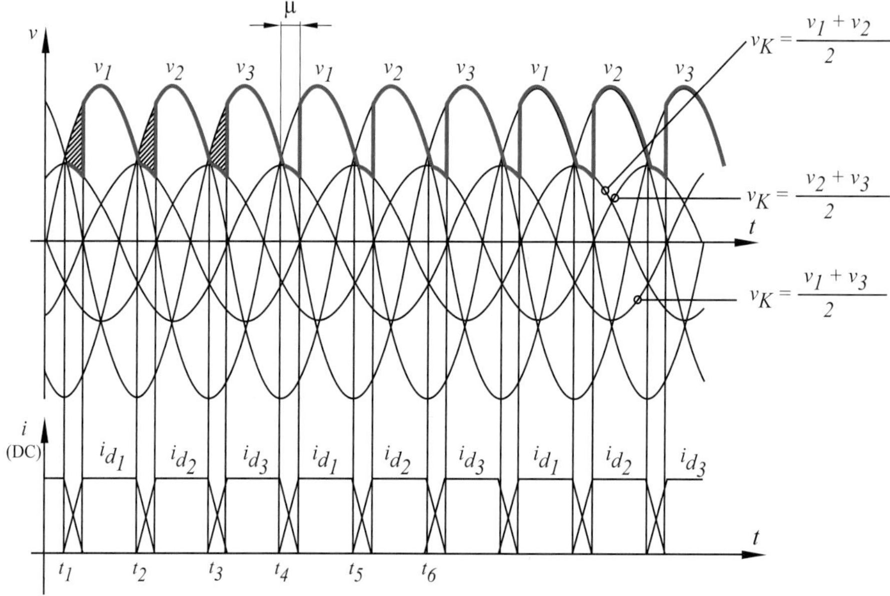

Fig. 7-12: Voltage and current waveform during commutation

The gradual slopes of the flanks during commutation (fig. 7-12) results in less harmonics in comparison to the theoretical square wave pattern of the diode current.

4.3 Average rectified voltage

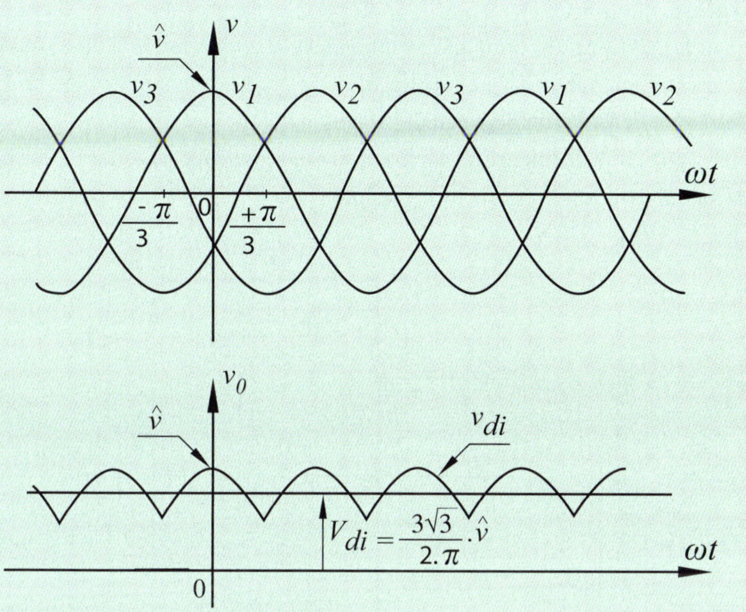

Fig. 7-13: Rectified voltage over half wave rectified

We select the zero point of ωt in fig. 7-13 in such a way that it corresponds with the maximum of the phase voltage v_1. We can then write $v_1 = \hat{v} . \cos \omega t$.

The average rectified voltage in the case of an ideal rectifier is then easy to calculate:

$$V_{di} = 3 . \left[\frac{1}{2.\pi} \int_{-\pi/3}^{+\pi/3} \hat{v} . \cos \omega t . d\omega t \right] = \frac{3}{2.\pi} . \hat{v} . \sin \omega t \Big|_{-\pi/3}^{+\pi/3} = \frac{3.\hat{v}}{2.\pi} . 2 . \sin \frac{\pi}{3} = \frac{3.\sqrt{3}}{2.\pi} . \hat{v}$$

$$V_{di} = \frac{3.\sqrt{3}}{2.\pi} . \hat{v} \qquad (7\text{-}6)$$

If V_f is the secondary voltage of the transformer, then $\hat{v} = \sqrt{2} . V_f$ and we find for V_{di}:

$$V_{di} = \frac{3.\sqrt{6}}{2.\pi} . V_f = 1.17 . V_f \qquad (7\text{-}7)$$

The key output DC current has the same time based waveform as v_{di} and : $I_d = \dfrac{V_{di}}{R_b}$

5. THREE-PHASE BRIDGE CIRCUIT

5.1 Three-phase bridge circuit. Resistive load

5.1.1 Configuration

The diode bridge of fig. 7-14 can in principle be immediately connected to a three-phase network. To adjust the value of the output voltage a transformer is placed at the input of the bridge rectifier. This provides galvanic isolation between the supply network and the DC voltage output. In contrast to normal supply transformers the star point of the secondary may not be earthed. If it is earthed then we can only work with a "floating" output.

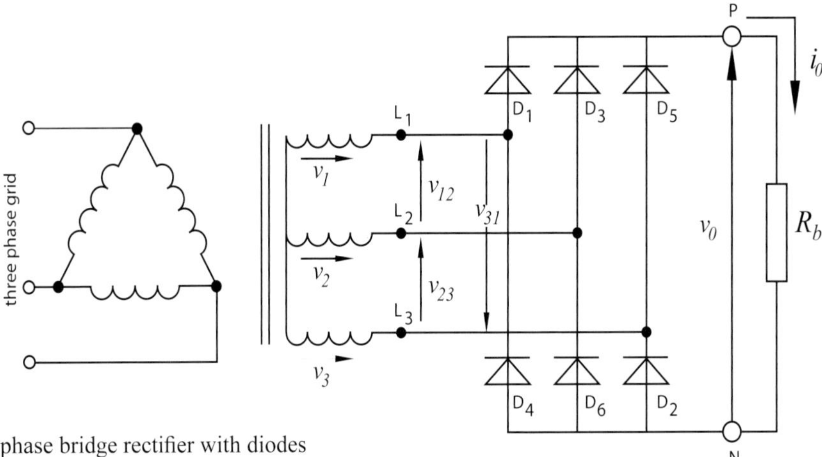

Fig. 7-14: Three-phase bridge rectifier with diodes

5.1.2 Line voltages

Between the three line conductors there are three AC voltages present. In relation to each other these voltages are separated by one third of a period. The notation used for the instantaneous line voltages in fig. 7-14 has the following meaning: the order of the subscript letters of v_{12} means that v_{12} is positive when the potential of line L_1 is higher than the potential of line L_2 .

If v_{12} is the largest instantaneous line-voltage, then current will flow from L_1 , through D_1, R_b, D_6 , to L_2 .

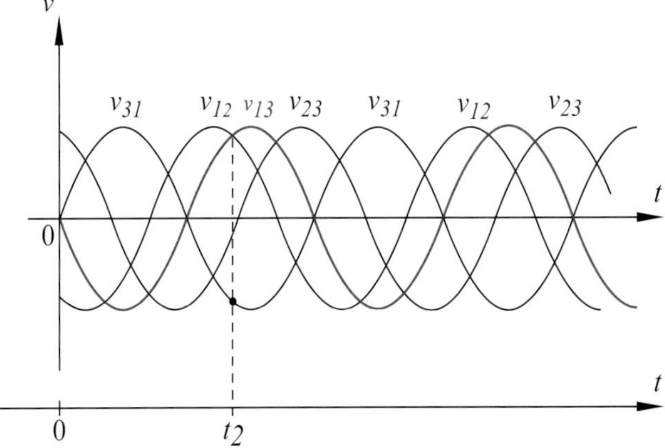

Fig. 7-15: Line voltages of the three-phase diode bridge of fig. 7-14

When the largest instantaneous voltage is v_{23} then D_3 and D_2 conduct. If v_{31} is the largest instantaneous value then there is conduction via L_3 through D_5, R_b, D_4 to L_1.

We now look at what happens when v_{31} has the largest negative value of all the line voltages. This is shown in fig. 7-15 from the instant t_2. Now L_1 is positive in relation to L_3 and a current will flow from L_1 through D_1, R_b, D_2 to L_3. Instead of working with the negative part of v_{31} it is simpler to reason with the positive part of v_{13}.

We find the same scenario with the negative parts of the other line voltages. In future we work only with the six positive voltages as shown in fig. 7-16 while realizing that in reality we have three-line voltages (v_{12}, v_{23}, v_{31}) and their inverse forms (v_{21}, v_{32}, v_{13}).

In fig. 7-16 we have only drawn the part of the line voltage that has the largest instantaneous value of all the six line voltages. The line voltage v_{13} has been drawn completely for a didactic reason that will later becomes obvious.

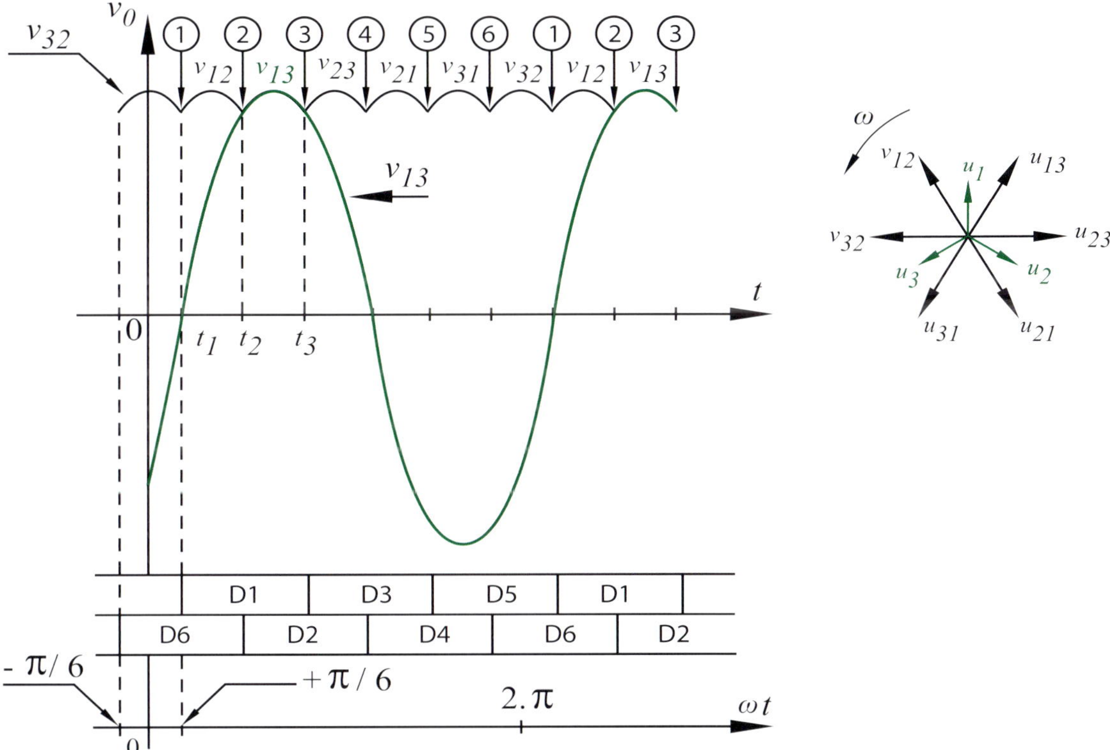

Fig. 7-16: Instantaneous value of the largest line voltage. Commutation points.

5.1.3 Natural commutation points

From instant t_1 in fig. 7-16 v_{12} becomes the largest instantaneous voltage of the three phase network. In other words L_1 is positive with respect to L_2. The current flows then from L_1, through D_1, R_b, D_6 to L_2. Diodes D_1 and D_6 therefore conduct. At instant t_2, v_{13} becomes the largest voltage and D_1 and D_2 will conduct. We say that D_2 has taken over from D_6 or that there is commutation from D_6 to D_2.

At t_3, v_{23} becomes larger than v_{13} and D_3 together with D_2 will conduct, there is commutation from D_1 to D_3. The points 1, 2, 3 etc… in fig. 7-16 are the points where the diodes commutate in a natural manner. These points are referred to as **points of natural commutation.**

We see that there is always one diode from the top half of the bridge that conducts together with one diode from the bottom half of the bridge. If we ignore the voltage loss across the conducting diode pair, then the output voltage v_o is formed by all the peak values of v_{12}, v_{13}, v_{23}, as is shown in fig. 7-16. The output voltage v_o of the configuration in fig. 7-14 is again drawn in fig. 7-17, now together with the output current i_o of the bridge rectifier.

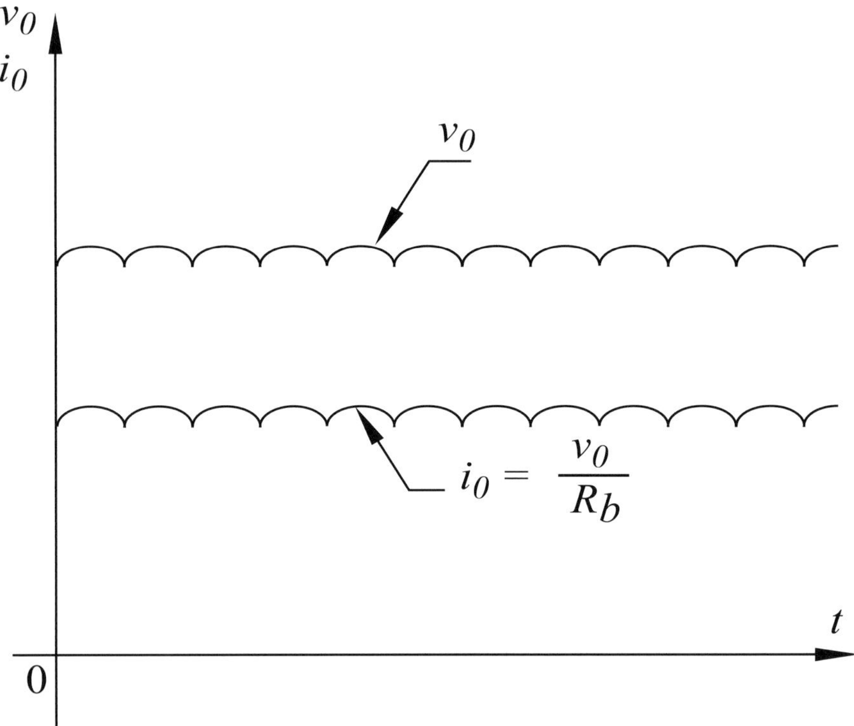

Fig. 7-17: Output voltage and output current of a bridge rectifier

5.1.4 Reference line voltage

The natural commutation points are formed at the intersection of two line voltages. The commutation point 1 is the instant that diode D_1 conducts for the first time (together with D_6 !). To electronically determine the commutation points, it is easier to use the zero crossover of a voltage rather than the intersection point of two voltages of several hundred volts. Luck is on our side because in fig. 7-16 we see that the commutation point 1 (intersection of v_{32} and v_{12}) corresponds with the zero crossover of v_{13} . We call v_{13} the reference voltage for D_1 . For every diode in the circuit shown in fig. 7-14 we can determine the reference line voltage. The result is presented in table 7-1. A diode conducts for the first time at the zero crossover of its reference line voltage.

A simple rule of thumb allows us to quickly determine the reference line voltage:

1. for a diode from the top half of the bridge we consider the line that is connected to the anode and after that the previous supply line. For example diode D_1 : anode (line 1), previous supply line (3): the zero crossover of the line voltage v_{13} is the natural commutation point of diode D_1.

2. for a diode from the bottom half of the bridge : the inverse voltage of the diode above it. Example: diode D_6 : Reference line voltage for D_3 is v_{21} so that the inverse v_{12} reference is for D_6 .

Table 7-1

reference line voltage	diode
v_{13}	D_1
v_{23}	D_2
v_{21}	D_3
v_{31}	D_4
v_{32}	D_5
v_{12}	D_6

Table 7-2

driving voltage	Pair of diodes
v_{12}	$D_6 + D_1$
v_{13}	$D_1 + D_2$
v_{23}	$D_2 + D_3$
v_{21}	$D_3 + D_4$
v_{31}	$D_4 + D_5$
v_{32}	$D_5 + D_6$

5.1.5 Driving voltage

We have seen that the diodes D_6 and D_1 conduct together with the voltage v_{12} . We refer to v_{12} as the **driving voltage** for $D_6 + D_1$. In this way in fig. 7-14 we can determine the driving voltage for every pair of diodes that conduct simultaneously. These driving voltages are indicated in table 7-2.

5.1.6 Pulse number

Important factors that determines the design of a rectifier are amongst others the value and the frequency of the ripple voltage.

We call pulse number p:

$$p = \frac{\text{ripple frequency fundamental harmonic of the output voltage}}{\text{frequency mains supply}} \qquad (7\text{-}8)$$

The majority of single phase rectifiers are two pulse and the majority of three phase bridge rectifiers are six pulse. The indexes of E_1 , B_2 , B_6 refer to the pulse number. The lower the pulse number and the larger the permitted output ripple, the cheaper the output filter will be. In fig. 7-16 we see that the output voltage has a frequency six times larger than the grid frequency: $p = 6$.

5.1.7. Average rectified voltage and current

From fig. 7-16 we can calculate V_{di} :

$$V_{di} = 6 \cdot \left[\frac{1}{2.\pi} \int_{-\pi/6}^{+\pi/6} \hat{v} \cdot cos\ \omega t \cdot d\omega t \right] = \frac{3.\hat{v}}{\pi} \cdot sin\ \omega t \Big|_{-\pi/6}^{+\pi/6} = \frac{3.\hat{v}}{\pi}$$

With $V_{line} = \dfrac{\hat{v}}{\sqrt{2}}$ becomes: $\qquad V_{di} = \dfrac{3 \cdot \hat{v}}{\pi} = \dfrac{3 \cdot \sqrt{2} \cdot V_{line}}{\pi}$ $\qquad\qquad$ (7-9)

V_{di} : $\qquad d$ = average value (= DC voltage)

$\qquad\qquad i$ = ideal rectifier (transformer and diode losses are ignored)

If the line voltage is 3 x 230V then we find V_{di} = 310 V. At 3 x 400V becomes V_{di} = 540V .
The average DC current through R_b is given by: $I_d = \dfrac{V_{di}}{R_b}$

5.2 Three-phase bridge. Resistive-inductive load

5.2.1 Voltage, current

Fig. 7-18 shows the configuration. Without the coil the current flow would be as in fig. 7-17.
With a coil L_b in series with R_b (fig. 7-18) the ripple in i_0 is even less so that as a good approxima-
tion i_0 may be considered constant (fig. 7-18). We call this constant DC current I_d .
In table 7-2 we see that v_{12} and v_{13} are driving voltages for diode D_1 while v_{21} and v_{31} are
driving voltages for diode D_4 . We can easily sketch i_{D_1} and i_{D_4} in fig. 7-18.
These currents (i_{D_1} and i_{D_4}) together form the current i_{S_1} .
In an identical manner i_{S_2} and i_{S_3} can be drawn. If the turns ratio is 1:1, then i_{P_1} (= i_{S_1}),
i_{P_2} (= i_{S_2}) and i_{P_3} (= i_{S_3}) can easily been drawn.

From the primary phase currents we can derive the line currents i_{L_1} , i_{L_2} , i_{L_3} (see evaluation
question no. 7.2).

The output voltage is the same as for a resistive load: $\qquad V_{di} = \dfrac{3 \cdot \sqrt{2} \cdot V_{line}}{\pi}$ $\qquad\qquad$ (7-9)

The larger the self inductance L_b , the more constant the output current.

With the average voltage across the coil zero, the average current is: $\quad I_d = \dfrac{V_{di}}{R_b}$ $\qquad$ (7-10)

In fig. 7-18 note that i_{S_1} is practically a block shaped current for 2/3 of period.
The effective value of the secondary line current follows from (5-6):

$$I_{RMS} = I_{line} = \sqrt{\delta} \cdot I_d = \sqrt{\frac{2}{3}} \cdot I_d \quad \text{or:} \qquad I_{line} = \sqrt{\frac{2}{3}} \cdot I_d \qquad\qquad (7-11)$$

5.2.2 Transformer power

The secondary apparent power of the transformer is: $S_{sec.} = \sqrt{3} \cdot V_{line} \cdot I_{line}$

$$S_{sec.} = \sqrt{3} \cdot \frac{\pi \cdot V_{di}}{3 \cdot \sqrt{2}} \cdot \sqrt{\frac{2}{3}} \cdot I_d = 1.05 \cdot P_{di} \qquad \boxed{S_{sec.} = 1.05 \cdot P_{di}} \qquad (7\text{-}12)$$

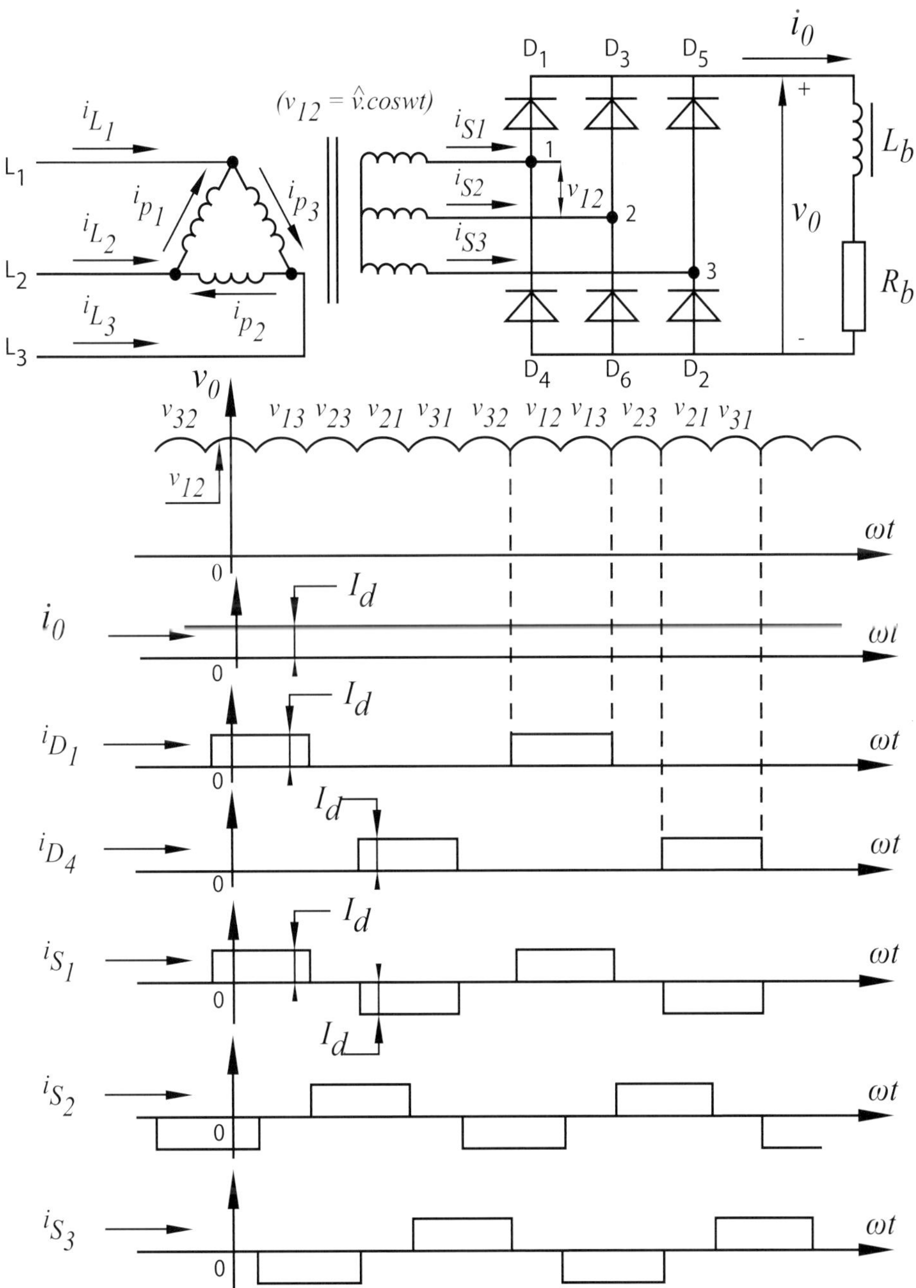

Fig. 7-18: Inductively loaded three-phase bridge

The solution of evaluation question 7.2c indicates that the apparent power of the primary side is equal to the secondary apparent power (7-12), so that: $S_{prim} = 1.05 \cdot P_{gi}$.
The apparent power of the transformer is:

$$S = \frac{1}{2} \cdot (S_{prim} + S_{sec}) = 1.05 \cdot P_{gi} \qquad \boxed{S = 1.05 \cdot P_{gi}} \qquad (7\text{-}13)$$

In German literature the apparent power is called the "Bauleistung". If we connect the bridge directly to a three phase supply obviously the value of the rectified voltage is directly related to the value of the supply voltage.

Remark
In studying rectifiers we assume ideal behaviour, this also includes ignoring the magnetising current of the transformer.

5.3 Twelve-pulse rectifier

When we connect two three-phase rectifiers which are 30° out of phase with each other in series we produce a 12-pulse rectifier (fig. 7-19). By connecting one secondary of the supply transformer in star and the other in delta we achieve the 30° phase displacement. The fifth and seventh harmonics disappear on the AC-side (see p. 8.59: 12-pulse circuit). A configuration as shown in fig. 7-19 can be used to supply a "three level" inverter. Such an inverter (NPC-inverter = neutral point clamped inverter) can be used to supply large AC-motors (> 1MW) .

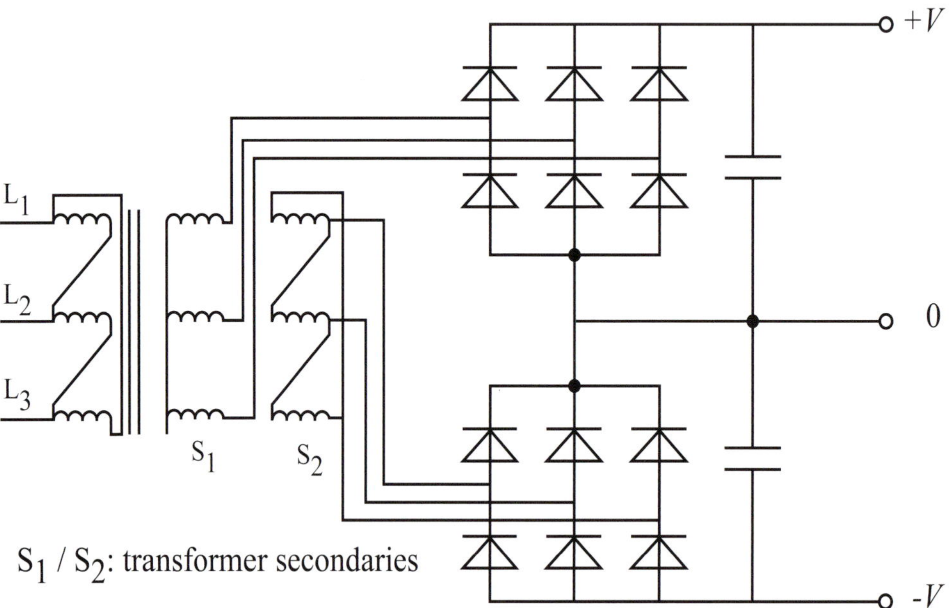

Fig. 7-19: Twelve-pulse rectifier used to supply an NPC-inverter

6. TABLE 7-3 SINGLE-PHASE RECTIFIERS

		HALF WAVE	FULL WAVE (center tap)	FULL WAVE (bridge)
V_{di} = average DC output voltage, at no load I_d = DC current, true load Ideal diode and transformer. Mains frequency f $P_{di} = V_{di} \cdot I_d$		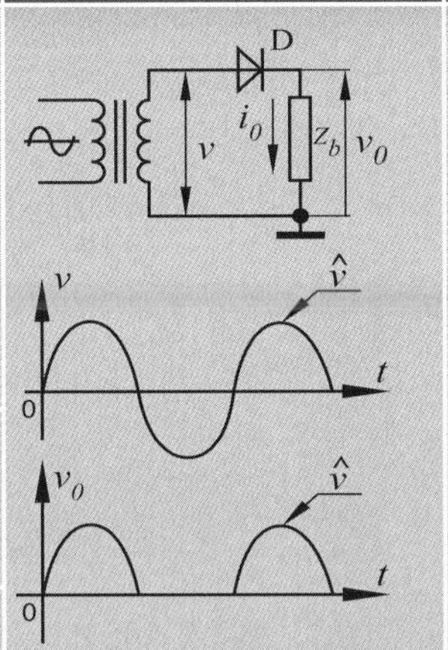 Fig. 7-20	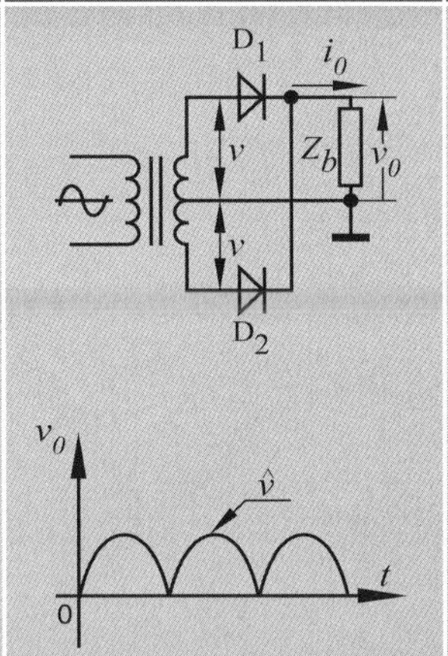 Fig. 7-21	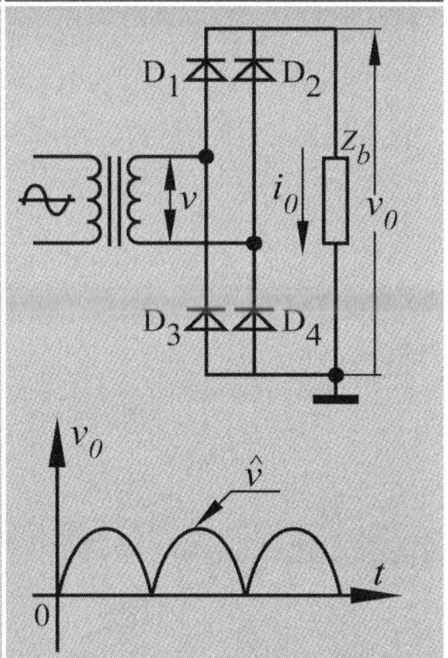 Fig. 7-22

LOAD				
	load			
I_{RMS}	R	$1.57 \cdot I_d$	$1.11 \cdot I_d$	$1.11 \cdot I_d$
V_{RMS}	R	$1.57 \cdot V_{di}$	$1.11 \cdot V_{di}$	$1.11 \cdot V_{di}$
Ripple frequency	R	f	$2 \cdot f$	$2 \cdot f$
Ripple factor r (%)	R	121	47.2	47.2
TRANSFORMER				
	load			
V_{RMS} secondary	R/L	$2.22 \cdot V_{di}$	$2.22 \cdot V_{di}$ (total)	$1.11 \cdot V_{di}$
I_{RMS} secondary	sine wave	$1.57 \cdot I_d$	$0.785 \cdot I_d$	$1.11 \cdot I_d$
	square wave	----	$0.71 \cdot I_d$	I_d
VA secondary	sine wave	$3.49 \cdot P_{di}$	$1.23 \cdot P_{di}$	$1.23 \cdot P_{di}$
	square wave	----	$1.57 \cdot P_{di}$	$1.11 \cdot P_{di}$
VA primary	sine wave	$3.49 \cdot P_{di}$	$1.23 \cdot P_{di}$	$1.23 \cdot P_{di}$
	square wave	----	$1.11 \cdot P_{di}$	$1.11 \cdot P_{di}$
PER DIODE (minimal specifications)				
	load			
Average current I_{FAV}	R/L	I_d	$0.5 \cdot I_d$	$0.5 \cdot I_d$
Effective current $I_{F(RMS)}$	R	$1.57 \cdot I_d$	$0786 \cdot I_d$	$0.786 \cdot I_d$
Peak current I_{FPM}	R	$3.14 \cdot I_d$	$1.57 \cdot I_d$	$1.57 \cdot I_d$
	L	----	I_d	I_d
Maximum reverse V_{RRM}	R/L	$3.14 \cdot V_{di}$	$3.14 \cdot V_{di}$	$1.57 \cdot V_{di}$
Form factor $a = \dfrac{I_{RMS}}{I_{AV}}$	R	1.57	1.57	1.57

7. TABLE 7-4 THREE-PHASE RECTIFIERS

		HALF WAVE	FULL WAVE (center tap)	FULL WAVE (bridge)

V_{di} = average DC output voltage, at no load

I_d = DC current, true load

Ideal diodes and transformer

Mains frequency f

$P_{di} = V_{di} \cdot I_d$

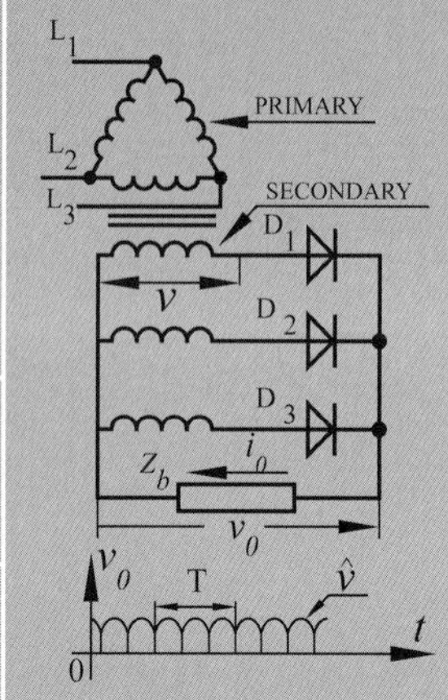

Fig. 7-23   Fig. 7-24   Fig. 7-25

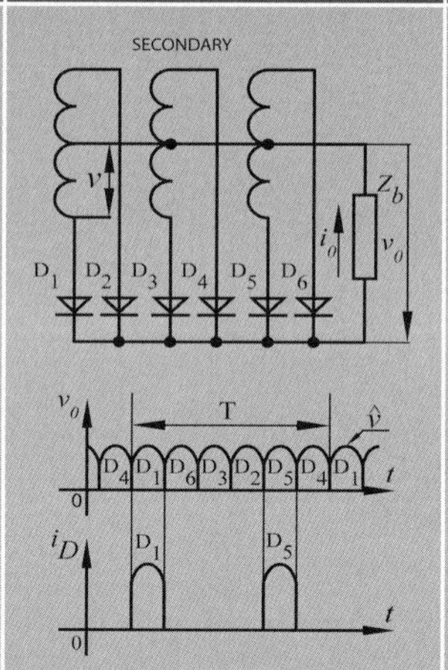

LOAD

	load	HALF WAVE	FULL WAVE (center tap)	FULL WAVE (bridge)
I_{RMS}	R	$1.02 \cdot I_d$	I_d	I_d
V_{RMS}	R	$1.02 \cdot V_{di}$	V_{di}	V_{di}
Ripple frequency	R	$3 \cdot f$	$6 \cdot f$	$6 \cdot f$
Ripple factor r (%)	R	18.3	4.3	4.3

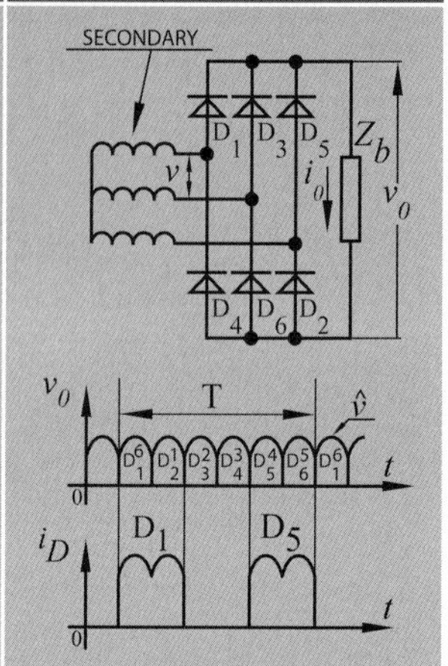

TRANSFORMER

	load	HALF WAVE	FULL WAVE (center tap)	FULL WAVE (bridge)
V_{RMS} secondary phase	sine/square wave	$0.855 \cdot V_{di}$	$0.74 \cdot V_{di}$	$0.428 \cdot V_{di}$
I_{RMS} secondary	sine/square wave	$0.58 \cdot I_d$	$0.408 \cdot I_d$	$0.816 \cdot I_d$
VA secondary	sine/square wave	$1.48 \cdot P_{di}$	$1.81 \cdot P_{di}$	$1.05 \cdot P_{di}$
VA primary	sine/square wave	$1.21 \cdot P_{di}$	$1.28 \cdot P_{di}$	$1.05 \cdot P_{di}$

PER DIODE (minimal specifications)

	load	HALF WAVE	FULL WAVE (center tap)	FULL WAVE (bridge)
Average current I_{FAV}	R/L	$0.333 \cdot I_d$	$0.167 \cdot I_d$	$0.333 \cdot I_d$
Effective current $I_{F(RMS)}$	R/L	$0.577 \cdot I_d$	$0.408 \cdot I_d$	$0.577 \cdot I_d$
Peak current I_{FPM}	R	$1.21 \cdot I_d$	$1.05 \cdot I_d$	$1.05 \cdot I_d$
Peak current I_{FPM}	L	I_d	I_d	I_d
Maximum reverse V_{RRM}	R/L	$2.09 \cdot V_{di}$	$2.09 \cdot V_{di}$	$1.05 \cdot V_{di}$
Form factor $a = \dfrac{I_{RMS}}{I_{AV}}$	R/L	1.73	2.44	1.73

8. EVALUATION

7.1   Consider a single phase rectifier loaded with a motor in series with an extra filter coil (L_m and R_m). Sketch the switching matrix (p. 1.3) for this configuration.

7.2   In the configuration shown in fig. 7-18 the transformer has a winding ratio 1:1.
a) Sketch i_{p_1} , i_{p_2} , i_{L_2} , as a function of time.
b) Calculate the effective value of i_{L_2} .
c) Determine the primary apparent power and prove these is equal to the secondary apparent power.

7.3   Consider an SKN400 (Semikron) with an I_{FAV} = 200A. The diode is used in a three-phase bridge rectifier. Calculate with expression (2-4) the dissipation per diode. The form factor a = 1.73.

7.4   Is the number of (diodes) switches in the configuration of fig. 7-14 correct according to the theory on p. 1.3.

7.5   Why may we express diode power $P_{max} = I_{FM} \cdot V_{RRM}$ in kVA?

7.6   A three-phase bridge rectifier is resistively and inductively loaded. The average output voltage V_{di} should be 100 V. For a load of 5Ω ($= R_b$) , calculate:
- diode specifications: I_{FAV} ; I_{FRMS} , V_{RRM}
- transformer specifications: V_{RMS} (secondary phase voltage) ; I_{RMS} (secondary phase current) ; apparent power

7.7   How much energy is stored in the coil shown in fig. 7-6?

7.8   In German literature what is meant by "Bauleistung" of a transformer?

7.9   What is meant by "reference line voltage" of a three-phase bridge rectifier?

7.10   Why does the exponentially decaying term of expression (7-1) practically disappear after $t = 5 . \tau$?

7.11   How large is v_D in fig. 7-6 if the diode is conducting the nominal current?

7.12   What is meant by "the equal surface area criterion"?

7.13   What is meant by "conduction angle" of a rectifier?

7.14   We sometimes refer to natural commutation points. What does this mean?

7.15   Consider the configuration shown in fig. 7-1. Supply voltage 230 V - 50 Hz. Load: 10 Ω - 100mH. The switch is closed at $t_1 = \dfrac{3 . T}{4}$. Determine the current flow.

7.16   Considered of configuration shown in fig. 7-1. Supply voltage 230 V - 50 Hz. Load: 0 Ω -100mH. Determine the current flow in the following situations:
1. $t_1 = 0$;
2. $t_1 = T/4$
3. $t_1 = T/2$.

7.17   Apparently the diodes in fig. 7-14 are numbered in an odd manner. Write down the conduction order of the diode pairs and discover for yourself the logic.

7.18 A three-phase bridge circuit with diodes is consecutively loaded as shown in fig. 7-26. Impedance (b) is highly inductive, while in configuration (c) a small inductance and a large capacitance are used. Sketch the theoretical current flow of the three load conditions.

Fig. 7-26:

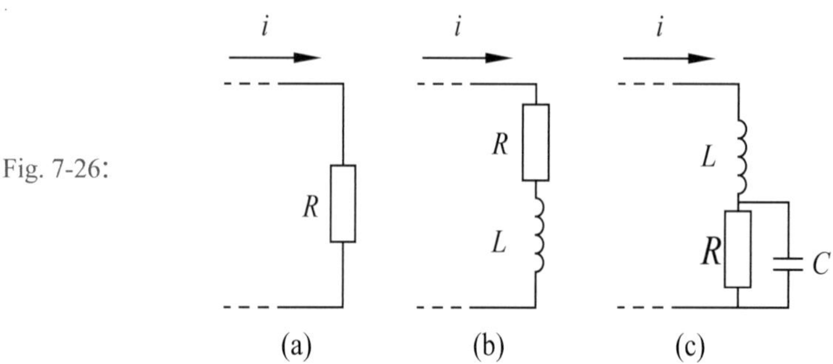

7.19 Try to sketch, starting from fig. 7-10, the configuration of a half wave three-phase rectifier resistively and inductively loaded. In addition sketch the waveform of the output voltage and current.

7.20 Fig. 7-27 shows a full wave three-phase rectifier (the so-called double star circuit). Sketch the waveform of the output voltage and current through diode D_1 and diode D_4 .

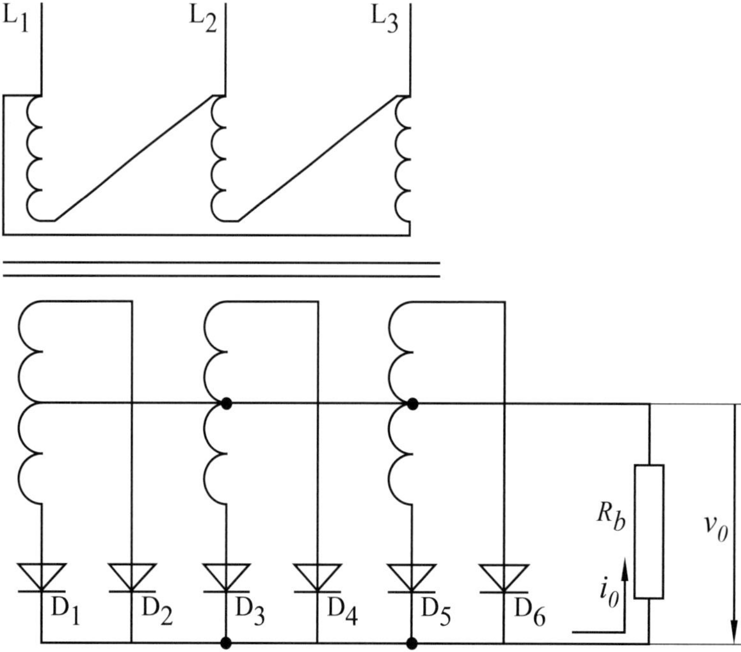

Fig. 7-27: Double star circuit (three-phase full wave rectifier)

7.21 What is the pulse number of the configuration of fig. 7-27?

7.22 What is the pulse number of the configuration in fig. 7-10?

7.23 Which ideal components would we need so that approximations 1, 2, 3, 4 (no. 3 p. 7.11) would not be approximations but exact properties?

8 CONTROLLED RECTIFIERS

CONTENTS

The term controlled rectifier is reserved for rectifiers with an electronically controlled output voltage. The switch of choice is an SCR. The reason being that in this type of circuit the (thyristor) switches commutate naturally under the influence of the supply voltage (line commutation, natural commutation). Another name for this type of circuit is a DC controller. Controlled rectifiers find application in the control of DC motors, the supply of electrochemical processes, DC welding equipment, the control of small universal motors, etc ...

1. HALF WAVE SINGLE PHASE CONTROLLED RECTIFIER (E_1) - RESISTIVE LOAD

Fig. 8-1 shows the configuration. Since only one half of the period from the single phase supply voltage is rectified, this is called a half wave single phase rectifier. The pulses v_s that come from the control circuit, can vary between zero and 180° in relation to the supply voltage v.

1.1 Current flow through the load

A thyristor can conduct under the precondition that the anode voltage is positive with respect to the cathode and at the same time there is a positive voltage present between the gate and the cathode. In fig. 8-1b and -c we see that after the angle α the conditions are met. The thyristor would begin to conduct with a current that can be described at every instance by: $i_o = \dfrac{v_o}{R_b}$.

If the current falls below the holding current I_H then the thyristor switches off. Every positive halve period of the supply voltage this pattern repeats itself.

1.2 Voltage across the load

In the case of a resistive load the output voltage v_o scaled to R_b has the same form as the current i_o. Compare the form of fig. 8-1b with that of fig. 8-1d!

1.3 Firing angle α

The angle α that is generated in the control circuit determines the instant of switch-on of the SCR. This angle α is referred to as the firing angle of the thyristor circuit.

1.4 Conduction angle β

The angular difference between the moment of switch on and reaching the holding current is called the conduction angle β. Since the holding current is small, I_H is almost the 180° point or the zero crossover point:

$$\alpha + \beta \approx 180° \qquad (8\text{-}1)$$

Remarks

1. The unidirectional character of the thyristor results in half wave rectification.
   At α = 0°, the current is maximum half of a sine wave. By varying α the resulting DC current can be controlled.
   This explains the name controlled rectifier or silicon controlled rectifier, or in short SCR.
2. Since the firing angle α is changed, this control principle is sometimes called phase control.
3. The thyristor has minimal power losses.

4. Disadvantages of phase control:
- poor power factor;
- thyristor has to deal with a large *di/dt* ;
- results in harmonics causing radio and TV disturbance.

1.5 Voltage form across the thyristor

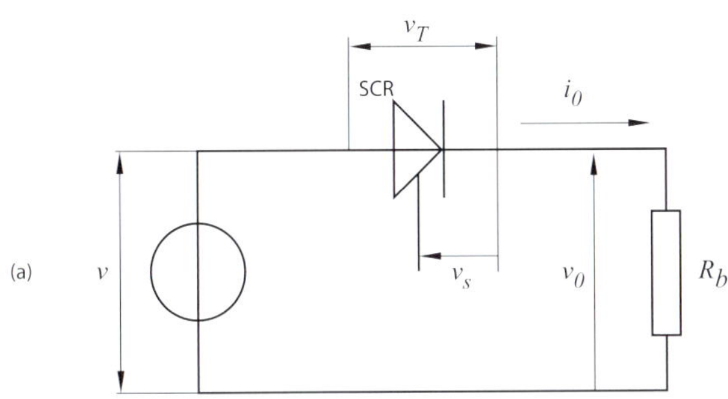

As long as the thyristor is not fired (between $0°$ and α) we can compare its behaviour to an open switch. The voltage across its terminals follows faithfully the supply voltage. See fig. 8-1e. The forward voltage drop across a conducting SCR is approximately 1.6 V. After the thyristor switches off (is extinguished) because the i_o has fallen below I_H , the voltage again follows the supply voltage.

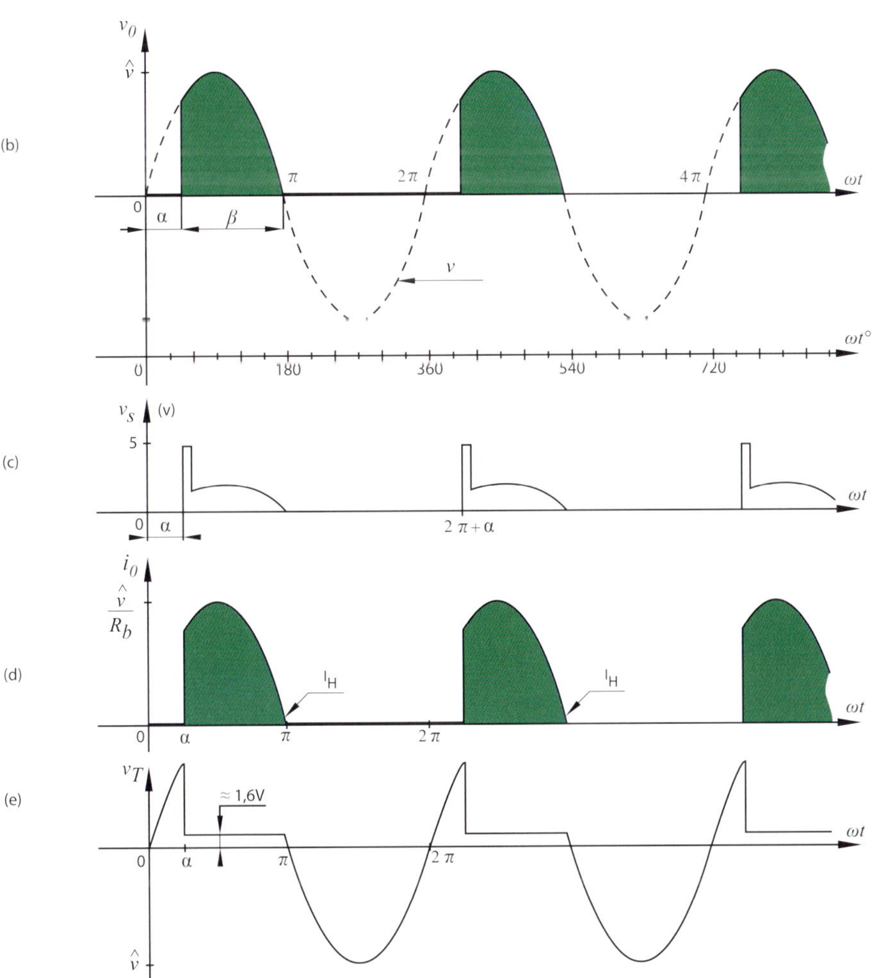

Fig. 8-1: Circuit and waveforms of a resistive loaded E_1 - controller.

1.6 Values of DC voltage and effective voltage

We call V_{dia} the average value or the DC voltage across the resistor R_b. This average value will decrease the more the firing angle α increases. The relationship between V_{dia} and α is determined in the following expression:

$$V_{dia} = \frac{\hat{v}}{2.\pi} (1 + \cos \alpha)$$

(8-2)

For the RMS value of the voltage across R_b we find:

$$V_{RMS} = \frac{\hat{v}}{2} \sqrt{\frac{180° - \alpha}{180°} + \frac{\sin 2.\alpha}{2.\pi}}$$

(8-3)

$\hat{v}$ is the amplitude of the supply voltage and α is given in degrees.

Remarks

1. For an explanation of the subscripts used with V_{dia}, see p. 8-5.
2. Expressions (8-2) and (8-3) are valid for an ideal transformer and an ideal SCR. The real values will be somewhat smaller.
3. To quickly calculate V_{dia} and V_{RMS} for random supply voltages we have published table 8.1. on p. 8-10.
4. The form factor is the ratio $a = \dfrac{V_{RMS}}{V_{dia}}$

Numeric example 8-1:

Required:

1) what is the value of α in fig. 8-1?
2) determine for this value of α the DC value and the RMS value of the current through $R_b = 24\,\Omega$ when the supply voltage is 230 V - 50 Hz.

Solution:

1) $\alpha = 45°$
2) we first determine the DC- and the RMS value of the voltage across the load R_b.

$$\text{a) (8-2)} \rightarrow\rightarrow\rightarrow\rightarrow \quad V_{dia} = \frac{\hat{v}}{2.\pi} (1+ \cos \alpha) = \frac{\sqrt{2} \cdot 230}{2.\pi} \cdot (1 + \cos 45°) = 88.37 \text{ V}$$

$$I_d = \frac{88.37}{24} = 3.68 \text{ A}$$

$$\text{(8-3)} \rightarrow\rightarrow\rightarrow\rightarrow \quad V_{RMS} = \frac{\hat{v}}{2} \sqrt{\frac{180° - \alpha}{180°} + \frac{\sin 2.\alpha}{2.\pi}} = \frac{\sqrt{2} \cdot 230}{2} \sqrt{\frac{135°}{180°} + \frac{\sin 90°}{2.\pi}} = 155 \text{V}$$

$$I_{RMS} = \frac{155}{24} = 6.46 \text{ A}$$

b) Fig. 8-2: $\quad V_{dia} = 89 \text{ V} \rightarrow\rightarrow I_d = 3.7 \text{ A} \;;\; V_{RMS} = 154 \text{ V} \rightarrow\rightarrow I_{RMS} = 6.41 \text{ A}$

c) Table 8-1: $\quad V_{dia} = 0.3842 \times 230 = 88.37 \text{ V} \rightarrow\rightarrow I_d = 3.68 \text{ A} \;;$
$\quad V_{RMS} = 0.6742 \times 230 = 155 \text{ V} \rightarrow\rightarrow I_d = 6.46 \text{ A}.$

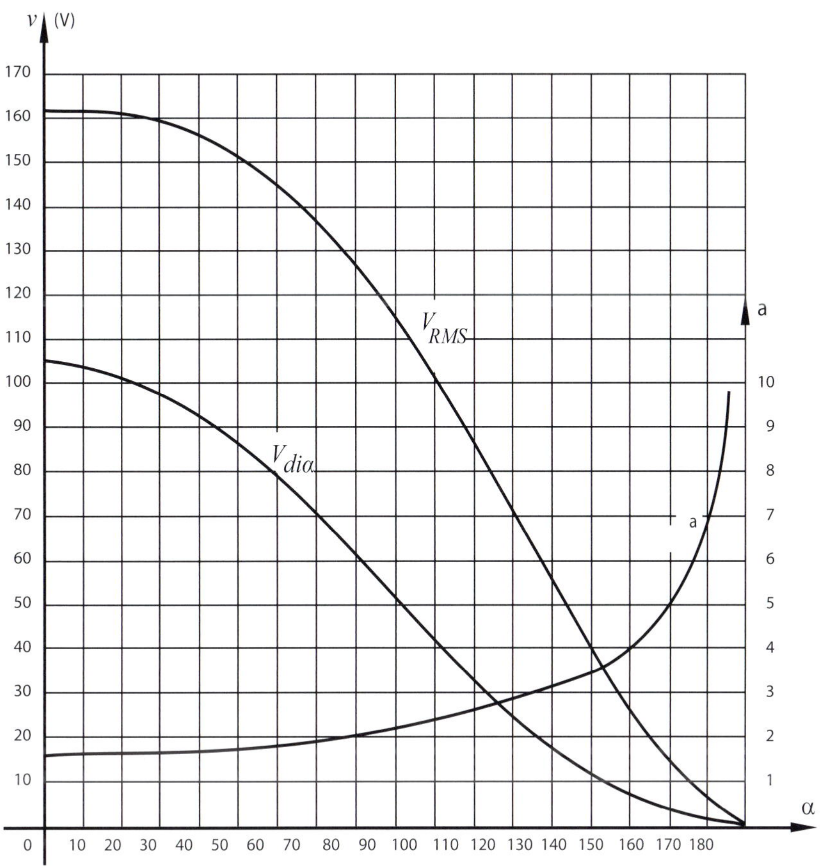

Fig. 8-2: DC voltage, RMS voltage and form factor of a controlled single phase rectifier
(RMS supply voltage: 230V), α is given in degrees

1.7 Formulas

Fig. 8-1 shows once again the voltage form across the load with a firing angle α when the supply voltage $v = \hat{v} \cdot \sin \omega t$. We can determine V_{dia} and V_{RMS} across the load.

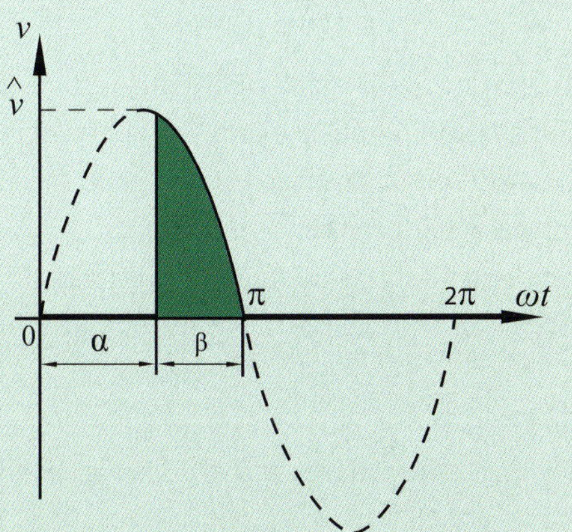

V_{dia} :

$d =$ average value ($=$ DC voltage)

$i =$ ideal. This means that we ignore losses in the rectifier (transformer and semiconductors)

$\alpha =$ firing angle

Fig. 8-1b : Voltage form of a single phase
half wave controlled rectifier

$$1. \quad V_{di\alpha} = V_{AV} = \frac{1}{2.\pi} \int_{\alpha}^{\pi} v \cdot d\omega t = \frac{1}{2.\pi} \int_{\alpha}^{\pi} \hat{v} \cdot \sin \omega t \cdot d\omega t = \frac{\hat{v}}{2.\pi} \cdot (- \cos \omega t) \Big|_{\alpha}^{\pi}$$

$$V_{di\alpha} = \frac{\hat{v}}{2.\pi} \cdot (1 + \cos \alpha) \qquad (8\text{-}2)$$

$$2. \quad V_{RMS} = \sqrt{ \frac{1}{2.\pi} \int_{\alpha}^{\pi} v^2 \cdot d\omega t }$$

$$\int_{\alpha}^{\pi} v^2 \cdot d\omega t = \int_{\alpha}^{\pi} \hat{v}^2 \, \sin^2 \omega t \cdot d\omega t = \frac{\hat{v}^2}{2} \int_{\alpha}^{\pi} (1 - \cos 2\omega t) \, d\omega t = \frac{\hat{v}^2}{2} (\pi - \alpha) + \frac{\hat{v}^2}{4} \sin 2.\alpha$$

So that:
$$V_{RMS} = \frac{\hat{v}}{2} \sqrt{ \frac{180° - \alpha}{180°} + \frac{\sin 2.\alpha}{2.\pi} } \qquad (8\text{-}3)$$

2. RESISTIVE AND INDUCTIVE LOADED E_1-CONTROLLER

2.1 Voltage form

After a time t_1 or an angle $\omega t_1 = \alpha$ the SCR fires (fig. 8-3). The SCR starts to conduct and by neglecting the approximately 1.6 V voltage drop (v_T) the output voltage v_o is practically equal to the supply voltage. This is shown in fig. 8-3b.

2.2 Current flow

After the firing angle α the thyristor starts to conduct. The current can only gradually increase because of the inductive nature of the load. With a current i_o through a coil, there is a corresponding magnetic energy $\frac{L_b \cdot i_o^2}{2}$.

Together with a rising i_o , $v_{R_b} = i_o \cdot R_b$ also increases.

In addition, at every moment in time $v_o = v_{R_b} + v_{L_b}$.

At the instant that $v_{R_b} = v_o$, v_{L_b} becomes zero, see fig. 8-3d to the right. This instant corresponds to the end of the magnetic charging of the coil, in other words the current i_o has reached its maximum. From this instant the polarity of v_{L_b} reverses. At this point the magnetic energy starts to discharge and the current starts to decrease.

At the instant $t_2 = \delta/\omega$ the current has reached the holding value I_H and the thyristor switches off. We call δ the extinction angle. Between 180° and $\omega t_2 = \delta$ the output voltage of this circuit is negative !

The exact form of i_o is provided by expression (7-1).

The average value of v_{L_b} is zero (equal surface area criterion, see p. 7-8). This is shown in fig. 8-3d (green areas!!)

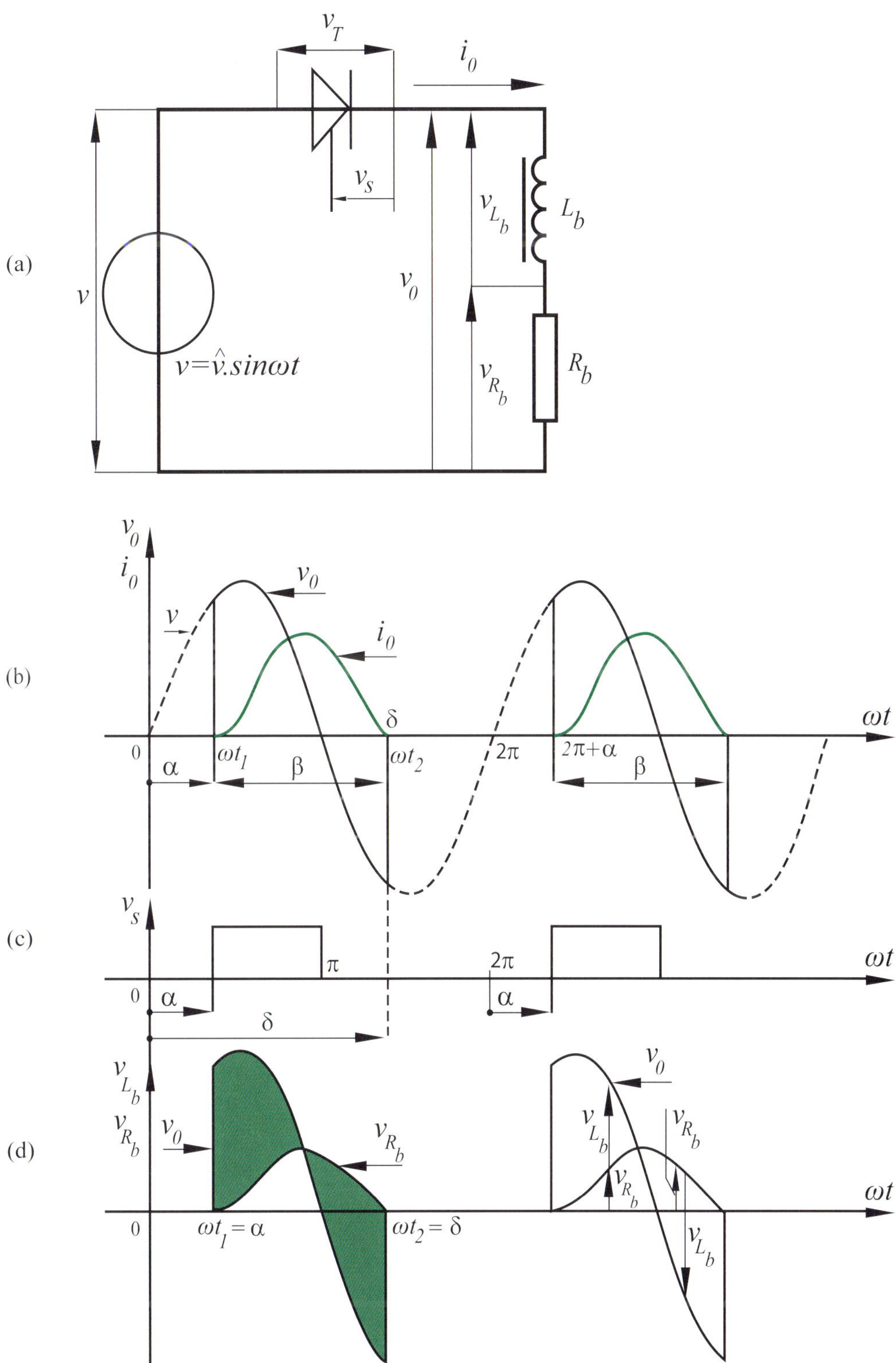

Fig. 8-3: Configuration and waveforms of inductively loaded E_1 - controller

2.3 Determining the extinction angle

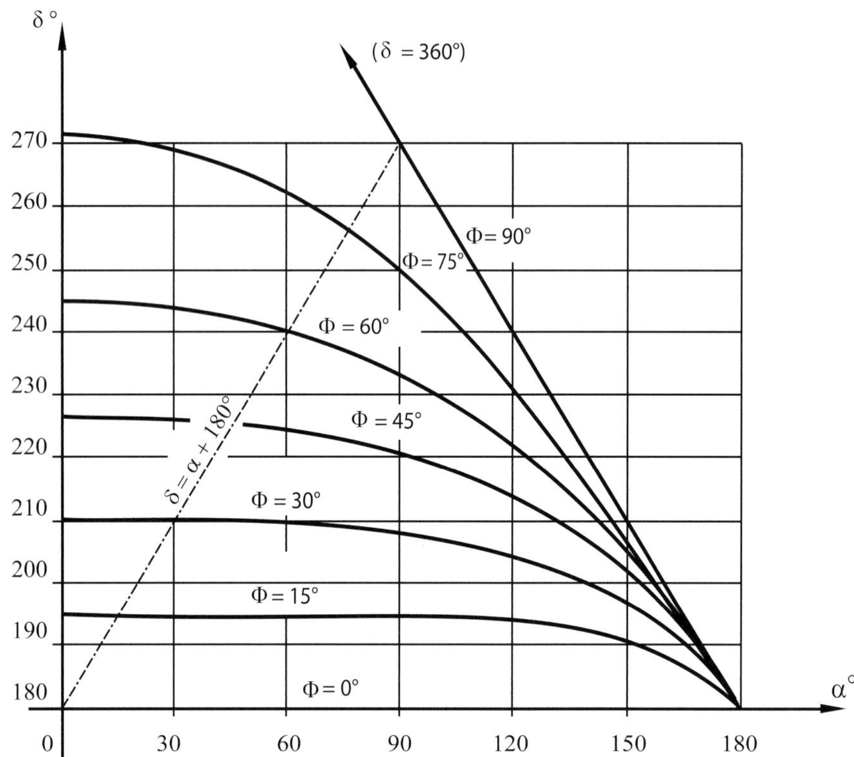

Fig. 8-4: Extinction angle with inductively loaded controlled rectifier

Fig. 8-4 allows us to calculate the extinction angle δ for a resistive inductive loaded half wave controlled rectifier.

The angle Φ is found from $tg\ \Phi = \dfrac{\omega \cdot L_b}{R_b}$. The average output voltage can be found with:

$$V_{di\alpha} = \frac{\hat{v}}{2.\pi}\ (\ \cos \alpha - \cos \delta\ )$$

(8-5)

Numeric example 8-2:

Given: In fig. 8-3 the load consist of 10 Ω and 100 mH. The firing angle is α = 90°.
Supply: 230V - 50Hz

Required: Determine the conduction angle β of the SCR and the average output voltage.

Solution: $tg\ \Phi = \dfrac{\omega \cdot L_b}{R_b} = \dfrac{2.\pi.50.10^{-1}}{10} = 3.1416 \longrightarrow\longrightarrow\ \Phi = 72°20'$

Fig. 8-4: α = 90° and $\Phi = 72°20' \longrightarrow\longrightarrow\ \delta \approx 247°$
The conduction angle: β = δ − α = 157°.
Average output voltage: (8-5) $\longrightarrow\longrightarrow V_{di\alpha} = \dfrac{\sqrt{2} \cdot 230}{2.\pi} \cdot (\ \cos 90° - \cos 247°\ ) = 20.23V$

2.4 Extinction angle δ

Expression (7-1) provides the current wave form for an *R-L* circuit where the mechanical switch is closed at time t_1 connecting it to a sinusoidal voltage.

We reconsider (7-1):
$$i(t) = \frac{\hat{v}}{Z}\left[\sin(\omega t - \Phi) + \sin(\Phi - \alpha)\,.\,e^{(\alpha - \omega t).\text{cotg}\,\Phi}\right] \qquad (7\text{-}1)$$

Here in: $\quad Z = \sqrt{R_b^2 + \omega^2.\,L_b^2}\;\;;\;\; tg\,\Phi = \dfrac{\omega.L_b}{R_b}\;\;;\;\; \alpha = \omega t_1\,.$

An important difference with the mechanical switch in this case is the fact that the current through the SCR can never be negative and so the current will cease to flow at t_2 $(= \delta/\omega)$. Here in δ is the extinction angle.

When $\omega t_2 = \delta$ the thyristor switches off $(i = 0)$.

From (7-1): $\quad 0 = \dfrac{\hat{v}}{Z}\left[\sin(\delta - \Phi) + \sin(\Phi - \alpha).\,e^{(\alpha - \delta).\text{cotg}\,\Phi}\right]$

or :
$$\sin(\delta - \Phi) = \sin(\alpha - \Phi).\,e^{(\alpha - \delta).\text{cotg}\,\Phi} \qquad (8\text{-}4)$$

If we wish to find δ for a specific firing angle α with a specific load ($\Phi = bg\,tg\,\omega.L_b/R_B$) then quite some calculation (using iteration) is required to solve (8-4). Fig. 8-4 however provides an elegant solution for this problem.

2.5 Average output voltage

The average output voltage is easily determined:

$$V_{dia} = \frac{1}{2.\pi}\int_{\alpha}^{\delta} \hat{v}\,.\sin \omega t\,.d\omega t = \frac{\hat{v}}{2.\pi}(\cos\alpha - \cos\delta) \quad \text{or:}\quad V_{dia} = \frac{\hat{v}}{2.\pi}(\cos\alpha - \cos\delta) \qquad (8\text{-}5)$$

3. TABLE 8-1 SINGLE PHASE POWER CONTROL

SINGLE PHASE DC- and AC-POWER CONTROL OHMIC LOAD (R_b / L_b load + freewheel diode at DC-control)							
	DC controller						AC controller
	HALF WAVE			FULL WAVE			
$\alpha°$	$V_{di\alpha}$ (V)	V_{RMS} (V)	a	$V_{di\alpha}$ (V)	V_{RMS} (V)	a	V_{RMS} (V)
0	0.4500	0.7071	1.5713	0.9003	1.0000	1.1107	1.0000
5	0.4493	0.7071	1.5737	0.8986	0.9999	1.1127	0.9999
10	0.4467	0.7067	1.5820	0.8935	0.9994	1.1185	0.9994
15	0.4425	0.7058	1.5950	0.8850	0.9981	1.1278	0.9981
20	0.4366	0.7040	1.6124	0.8732	0.9956	1.1401	0.9956
25	0.4291	0.7011	1.6338	0.8581	0.9915	1.1554	0.9915
30	0.4200	0.6968	1.6590	0.8400	0.9855	1.1732	0.9855
35	0.4095	0.6911	1.6876	0.8189	0.9773	1.1934	0.9773
40	0.3975	0.6836	1.7197	0.7950	0.9667	1.2159	0.9667
45	0.3842	0.6742	1.7548	0.7685	0.9535	1.2407	0.9535
50	0.3698	0.6629	1.7925	0.7395	0.9375	1.2677	0.9375
55	0.3542	0.6496	1.8339	0.7084	0.9187	1.2968	0.9187
60	0.3376	0.6342	1.8785	0.6752	0.8969	1.3283	0.8969
65	0.3202	0.6168	1.9263	0.6404	0.8722	1.3619	0.8722
70	0.3021	0.5972	1.9768	0.6041	0.8446	1.3981	0.8446
75	0.2833	0.5757	2.0321	0.5667	0.8142	1.4367	0.8142
80	0.2642	0.5523	2.0904	0.5283	0.7810	1.4783	0.7810
85	0.2447	0.5270	2.1536	0.4894	0.7453	1.5229	0.7453
90	0.2251	0.5000	2.2212	0.4502	0.7071	1.5706	0.7071
95	0.2055	0.4715	2.2944	0.4109	0.6668	1.6228	0.6668
100	0.1860	0.4416	2.3742	0.3720	0.6245	1.6787	0.6245
105	0.1668	0.4105	2.4610	0.3336	0.5806	1.7404	0.5806
110	0.1481	0.3785	2.5557	0.2962	0.5353	1.8072	0.5353
115	0.1300	0.3458	2.6600	0.2599	0.4891	1.8818	0.4891
120	0.1125	0.3127	2.7795	0.2251	0.4422	1.9644	0.4422
125	0.0960	0.2793	2.9094	0.1920	0.3950	2.0572	0.3950
130	0.0804	0.2460	3.0597	0.1608	0.3479	2.1635	0.3479
135	0.0659	0.2131	3.2336	0.1318	0.3014	2.2868	0.3014
140	0.0527	0.1809	3.4326	0.1053	0.2559	2.4302	0.2559
145	0.0407	0.1498	3.6806	0.0814	0.2119	2.6032	0.2119
150	0.0302	0.1201	3.9768	0.0603	0.1698	2.8159	0.1698
155	0.0211	0.0921	4.3649	0.0422	0.1303	3.0876	0.1303
160	0.0136	0.0664	4.8823	0.0271	0.0939	3.4650	0.0939
165	0.0077	0.0433	5.6234	0.0153	0.0613	4.0065	0.0613
170	0.0034	0.0237	6.9706	0.0068	0.0335	4.9264	0.0335
175	0.0009	0.0084	9.3333	0.0017	0.0119	7.0000	0.0119
180	0.0000	0.0000		0.0000	0.0000		0.0000

Remarks

1) RMS supply voltage $= 1$V .

2) $V_{di\alpha}$ and V_{RMS} for the full wave rectifier follow from the formulas on p. 5.8 fig. 5-26 .

3) V_{RMS} of an AC-controller follows from (9-1) on p. 9.3

4.   HARMONICS IN THE CASE OF PHASE-CONTROLLED SINE WAVE

Fig. 8-5 shows the voltage and current in the case of a controlled single phase rectifier with resistive load. We can decompose the voltage waveform in a Fourier series, calculation using the current would follow a similar pattern.

The voltage $v = \hat{v} \cdot \sin \omega t$ exists in the interval corresponding with the angles α to π (fig. 8-5) .

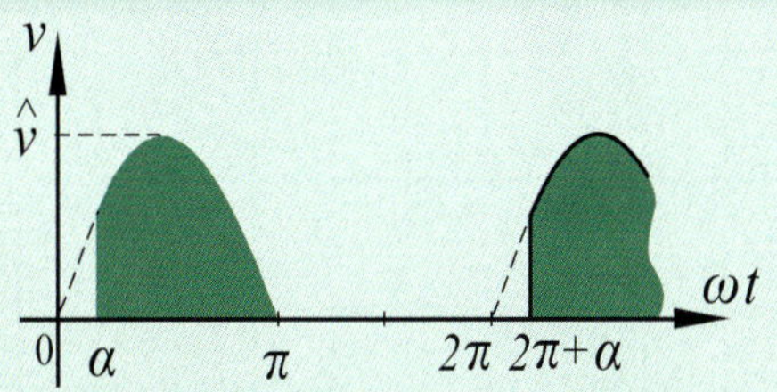

There is no symmetry with respect to the origin or the ordinate axis and the function is not symmetrical with the x-axis: therefore no simplification is possible. There is a DC component together with a series of sine and cosine terms present.

Fig. 8-5: Phase control of single phase AC voltage

We calculate the Fourier terms:

Constant term (= DC term):

$$A_0 = \frac{1}{2.\pi} \int_{\alpha}^{\pi} v \cdot d\omega t = \frac{1}{2.\pi} \int_{\alpha}^{\pi} \hat{v} \cdot \sin \omega t \cdot d\omega t = \frac{\hat{v}}{2.\pi} \left(1 + \cos \alpha \right) = V_{dia}$$

First harmonic:

$$A_1 = \frac{1}{\pi} \int_{\alpha}^{\pi} \hat{v} \cdot \sin \omega t \cdot \sin \omega t \cdot d\omega t = \frac{\hat{v}}{2.\pi} \left[(\pi - \alpha) + \frac{\sin 2.\alpha}{2} \right]$$

$$B_1 = \frac{1}{\pi} \int_{\alpha}^{\pi} \hat{v} \cdot \sin \omega t \cdot \cos \omega t \cdot d\omega t = \frac{\hat{v}}{2.\pi} \left[\frac{(\cos 2.\alpha - 1)}{2} \right]$$

k = 2 to ∞ :

$$A_k = \frac{1}{\pi} \int_{\alpha}^{\pi} \hat{v} \cdot \sin \omega t \cdot \sin k\omega t \cdot d\omega t = \frac{\hat{v}}{2.\pi} \left[\frac{\sin (1+k).\alpha}{(1+k)} - \frac{\sin (k-1).\alpha}{(k-1)} \right]$$

$$B_k = \frac{1}{\pi} \int_{\alpha}^{\pi} \hat{v} \cdot \sin \omega t \cdot \cos k\omega t \cdot d\omega t$$

$$B_k = \frac{\hat{v}}{2.\pi} \left[\frac{\cos (1+k).\alpha - \cos (1+k).\pi}{(1+k)} + \frac{\cos (k-1).\pi - \cos (k-1).\alpha}{(k-1)} \right]$$

Amplitude: $X_k = \sqrt{A_k^2 + B_k^2}$.   Fase: $\quad \Phi_k = bg\ tg\ \dfrac{B_k}{A_k}$ if $A_k > 0$

$$\Phi_k = bg\ tg\ \frac{B_k}{A_k} + \pi \ \text{if}\ A_k < 0$$

We calculate the first ten harmonics and draw these (together with A_0) as a function of α in fig. 8-6.

We set for example $\alpha = 90°$ and measure the amplitude of the Fourier components in fig. 8-6. Now it is possible to draw the frequency spectrum of fig. 8-7.

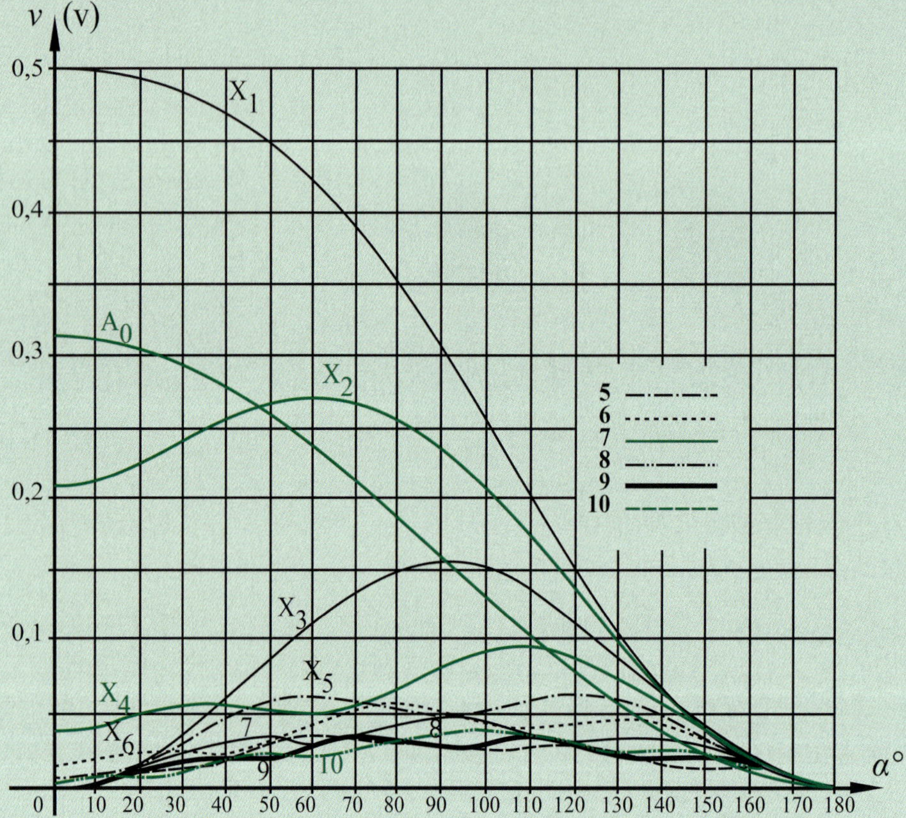

Fig. 8-6: DC component (A_0) and the first ten harmonics (X_1 to X_{10}) of the chopped sine wave from fig. 8-5. Here in is $\hat{v} = 1V$.

Take note that we have only calculated the amplitudes. To recreate the original waveform from these harmonics we would also need to calculate phase delay (Φ_k) per harmonic in order to establish the positions of each harmonic in relation to each other (at $t = 0$!).

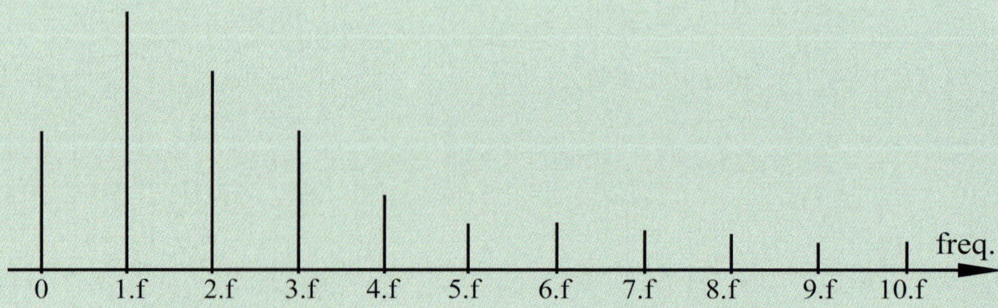

Fig. 8-7: Frequency spectrum of a chopped sine wave (fig. 8-6) in the case of $\alpha = 90°$.

5. DISTORTION POWER

From AC theory we are familiar with the principles for determining power and power factor:

Power: • active ($P = V . I . \cos \varphi$)

           • reactive ($Q = V . I . \sin \varphi$)

           • apparent ($S = V . I$)

Power factor: $\lambda = \dfrac{P}{S}$

These principles apply as long as we are dealing with sinusoidal voltage and current. Very often in power electronics we have to deal with non-sinusoidal waveforms. The input voltage to a rectifier may be considered sinusoidal if this comes from a strong supply network but the current may be square waved. We can establish some definitions for dealing with power calculations in the case of a sinusoidal voltage and non-sinusoidal current. In general any periodic, non-sinusoidal current, can be decomposed into a Fourier series.

The effective value of this current is:

$$I = \sqrt{I_1^2 + I_2^2 + I_3^2 + \ldots + I_n^2} \qquad I^2 = I_1^2 + \sum_{n=2}^{\infty} I_n^2$$

I = RMS (effective) value of the periodic non-sinusoidal current

I_1 = RMS value of the first (fundamental) harmonic

I_n = RMS value of the higher harmonics

To calculate the higher harmonics review p. 5.9.

In the case where the first harmonic is out of phase with respect to the sinusoidal voltage we can write: $I^2 = I_{1P}^2 + I_{1Q}^2$

I_{1P} = active component, in phase with the voltage $V \longrightarrow\longrightarrow I_{1P} = I_1 . \cos \varphi_1$ (zie fig. 8-8b)

I_{1Q} = reactive component, 90° out of phase with respect to $V \longrightarrow\longrightarrow I_{1Q} = I_1 . \sin \varphi_1$

So that: $I^2 = I_{1P}^2 + I_{1Q}^2 + \sum_{n=2}^{\infty} I_n^2$ (8-6)

Multiplying the left and right term in (8-6) with the square of the RMS voltage V gives:

$$(V . I)^2 = (V . I_{1P})^2 + (V . I_{1Q})^2 + \sum_{n=2}^{\infty} (V . I_n)^2$$

S = $V . I$ = total apparent power (VA)

P = $V . I_{1P}$ = active power (only current and voltage of the same frequency produce active power), (W)

Q_1 = $V . I_{1Q}$ = reactive power (var)

D = $\sqrt{\sum_{n=2}^{\infty} (V . I_n)^2}$ = distortion power. This power is the result of the distortion of the non-sinusoidal current. As the voltage (V) and the current (I_n , with $2 \leq n < \infty$) have different frequencies this is a "wattles" power.

So that: $S^2 = P^2 + Q_1^2 + D^2$ (8-7)

Expression (8-7) is three dimensionally presented in fig. 8-8a. From this graphic we have extracted the power triangles shown in fig. 8-8b/c.

Based on this we can define a displacement power factor and a distortion factor:

$$\cos \varphi_1 = \frac{P}{S_1} = \text{displacement power factor;} \qquad \varphi_1 = \text{displacement (shift) angle}$$

$$\cos \delta = \frac{S_1}{S} = \frac{I_1}{I} = \text{distortion factor ;} \qquad \delta = \text{distortion angle}$$

$$\lambda = \cos \varphi_1 \cdot \cos \delta = \frac{P}{S_1} \cdot \frac{S_1}{S} = \frac{P}{S} = \text{power factor}$$

These definitions also remain valid for three phase systems:

$$\text{Displacement factor} = \cos \varphi_1 = \frac{P}{S_1} \qquad\qquad (8\text{-}8)$$

$$\text{Distortion factor} = \cos \delta = \frac{I_1}{I} = \frac{S_1}{S} \qquad\qquad (8\text{-}9)$$

$$\text{Power factor} = \lambda = \cos \varphi_1 \cdot \cos \delta = \frac{P}{S} \qquad\qquad (8\text{-}10)$$

Remarks

1. Do not mix up the terms displacement factor ($\cos \varphi_1$) and power factor ($\cos \varphi_1 \cdot \cos \delta$). If the current becomes sinusoidal, then is $\delta = 0$ and $\cos \delta = 1$, so that only in this case the power factor is equal to $\cos \varphi_1$.

2. In determining the RMS value of the current (p. 8.13) we have ignored the DC component I_0 since on the supply side of the transformer no I_0 is present. If I_0 has to be calculated, then this is included in the distortion power as:

$$D = \sqrt{(V \cdot I_0)^2 + \sum_{n=2}^{\infty} (V \cdot I_n)^2}$$

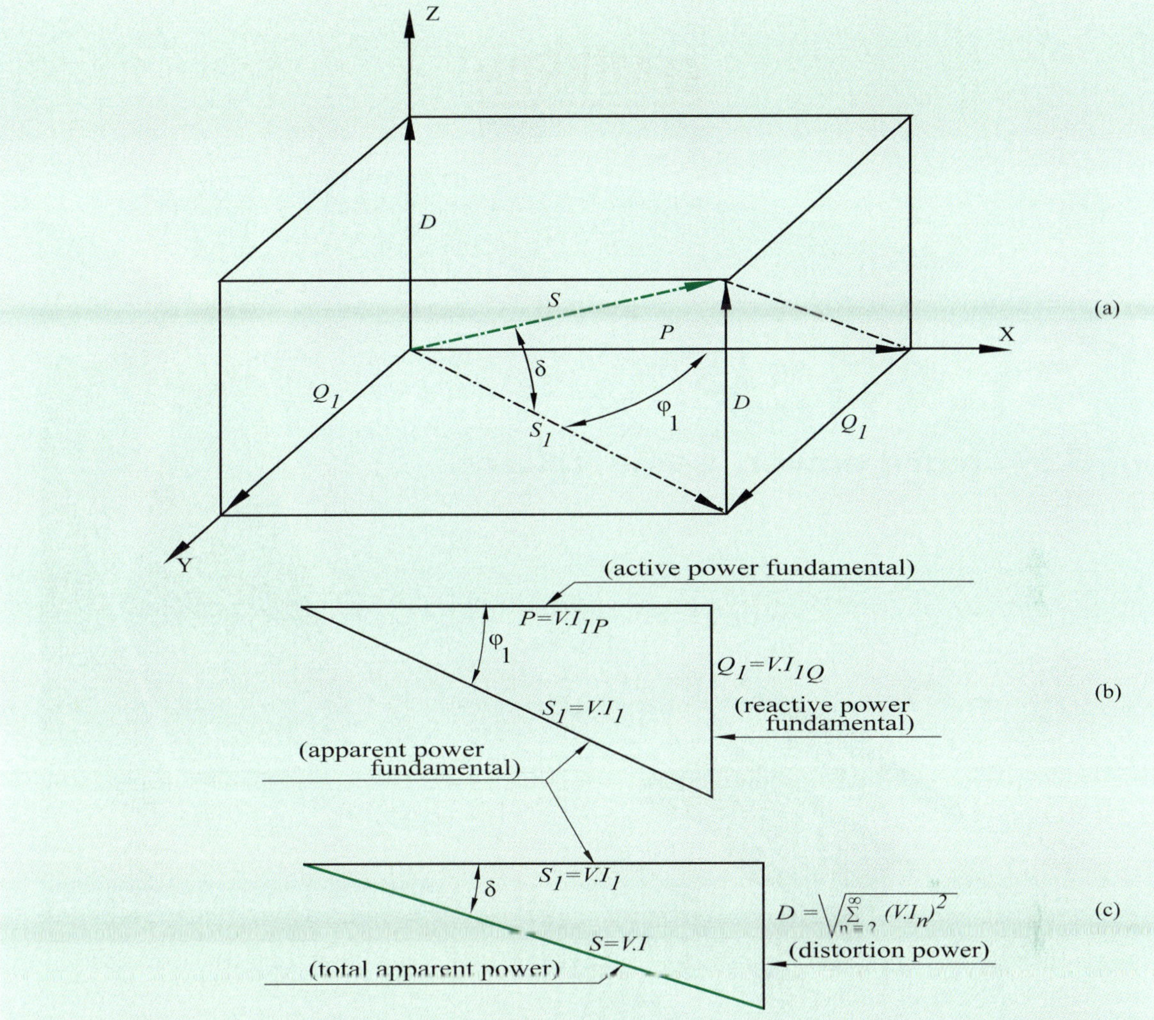

Fig. 8-8: Power triangles with sinusoidal voltage and non-sinusoidal current

Numeric example 8-3:

Given:

In fig. 8-1 we replace the thyristor with a diode.

The supply voltage can be written as $v = \hat{v} . \sin \omega t$. The diode may be considered as ideal.

Required:

Determine the power factor of this half-wave single phase rectifier.

Solution:

In table 5-1 (p. 5.7) we can determine that for such a rectifier the RMS value of the voltage over the load is : $V_{RMS} = \hat{v}/2$

From this : $I_{RMS} = \dfrac{\hat{v}}{2 . R_b}$ and $P = V_{RMS} \cdot I_{RMS} = \dfrac{\hat{v}^2}{4 . R_b}$

The apparent power of this circuit is: $S = V_{line} \cdot I_{RMS} = \dfrac{\hat{v}}{\sqrt{2}} \cdot \dfrac{\hat{v}}{2 . R_b} = \dfrac{\hat{v}^2}{2 \sqrt{2} . R_b}$

Power factor: $\lambda = P/S = \dfrac{\hat{v}^2 . 2 . \sqrt{2} . R_b}{4 . R_b . \hat{v}^2} = 0.707$

Numeric example 8-4:
Given:
Fig. 8-1 with $R_b = 10\ \Omega$; $V_{line} = 230$ V - 50 Hz ; $\alpha = 90°$; $V_T = 0$.

Required:
Determine P, S, displacement factor, distortion factor, power factor, Q_1, D.

Solution:

- Fundamental harmonic of current
  p. 8-11: $A_1 = 81.32$ V ; $B_1 = -51.77$ V ; $X_1 = 96.4$ V ; $V_1 = 68.16$ V

  $I_1 = 6.816$ A ; $\varphi_1 = bgtg\ \dfrac{B_1}{A_1} = -32°48'$; $I_{1P} = I_1 \cdot \cos \varphi_1 = 5.75$ A

- RMS value of current
  Table 8-1: $V_{RMS} = 0.5 \times 230 = 115$ V ; $I_{RMS} = \dfrac{115}{R_b} = 11.5$ A

- Active power:
  $P = V_{RMS} \cdot I_{RMS} = 115 \times 11.5 = 1322.5$ W

  or: $P = V_{line} \cdot I_{1P} = 230 \times 5.75 = 1322.5$ W

- Total apparent power
  $S = V \cdot I = 230 \times 11.5 = 2645$ VA

- Apparent power of fundamental harmonic
  $S_1 = V \cdot I_1 = 230 \times 6.816 = 1567.68$ VA

- Displacement factor
  $\cos \varphi_1 = \dfrac{P}{S_1} = \dfrac{1322.5}{1567.68} = 0.8436$

- Distortion factor
  $\cos \delta = \dfrac{S_1}{S} = \dfrac{1567.68}{2645} = 0.5926$

- Power factor
  $\lambda = \dfrac{P}{S} = \dfrac{1322.5}{2645} = 0.5$

- Reactive power of fundamental harmonic
  $Q_1 = \sqrt{S_1^2 - P^2} = 841.8$ var

- Distortion power
  $D = \sqrt{S^2 - S_1^2} = 2130.35$ var

6.   FULL WAVE B_2 -CONTROLLED RECTIFIER, RESISTIVE INDUCTIVE LOAD

In fig. 8-9a we have replaced the four diodes in the classic rectifier bridge with thyristors. This bridge is known as a single phase full wave controlled B_2 - rectifier.

6.1 Thyristor control

The supply voltage is considered positive if terminal A in fig. 8-9a is positive with respect to terminal B. The thyristor pair Th_1-Th_4 is simultaneously controlled during the positive half period of the supply voltage. After that the thyristor pair Th_2-Th_3 is fired during the negative half of supply voltage period. Just as was the case with the half wave rectifier the firing angle of the thyristors is calculated with respect to the zero crossover of the supply voltage.
For the thyristor pair Th_1-Th_3 this is with respect to 0, $2.\pi$, ...(radians) and for Th_2-Th_3 with respect to π, $3.\pi$, ...(radians).

6.2 Discontinuous current ($\alpha > \Phi$)

Consider the case where the firing angle α is greater than the phase delay Φ of the load impedance Z. This is studied in fig. 8-9b.
At a time that corresponds to the firing angle Th_1-Th_4 conduct and i_o begins to flow and increase as studied in fig. 8-3. The extinction angle can be determined from fig. 8-4, just as we have done in numeric example 8-2.
During the negative half cycles of the supply voltage the thyristors Th_2-Th_3 conduct after $\pi + \alpha$ and we have the current flow through the load as previously. See fig. 8-9b.
The load current does not flow continuously. This is referred to as **discontinuous current**.
Notice that we have drawn the voltage v_{R_b} over the resistor in fig. 8-9b. This waveform drawn to scale R_b is exactly the same form as the current waveform.

6.3 Continuous current ($\alpha \leq \Phi$)

If $\alpha = \Phi$ then we see in fig. 8-4 that $\delta = \alpha + \pi$. The current stops flowing through Th_1-Th_4 at the instant that Th_2-Th_3 is triggered. This is the border of discontinuous current.
The current goes to zero but immediately increases as shown in fig. 8-9c.

If $\alpha < \Phi$ then the current through Th_1-Th_4 has not yet become zero when Th_2 - Th_3 is triggered. The current switches or commutates from Th_1-Th_4 to Th_2-Th_3 etc. This is shown in fig. 8-9d.

When a large inductance is present in the load the ripple in the output current will be small resulting in an almost flat current form. The current through a thyristor (pair) has a quasi block form.

6.4 Equal surface area criterion.

The criterion of **equal voltage time–surface areas** is also valid here. This is obvious from the green surface areas in the figures 8-9 b/c/d.

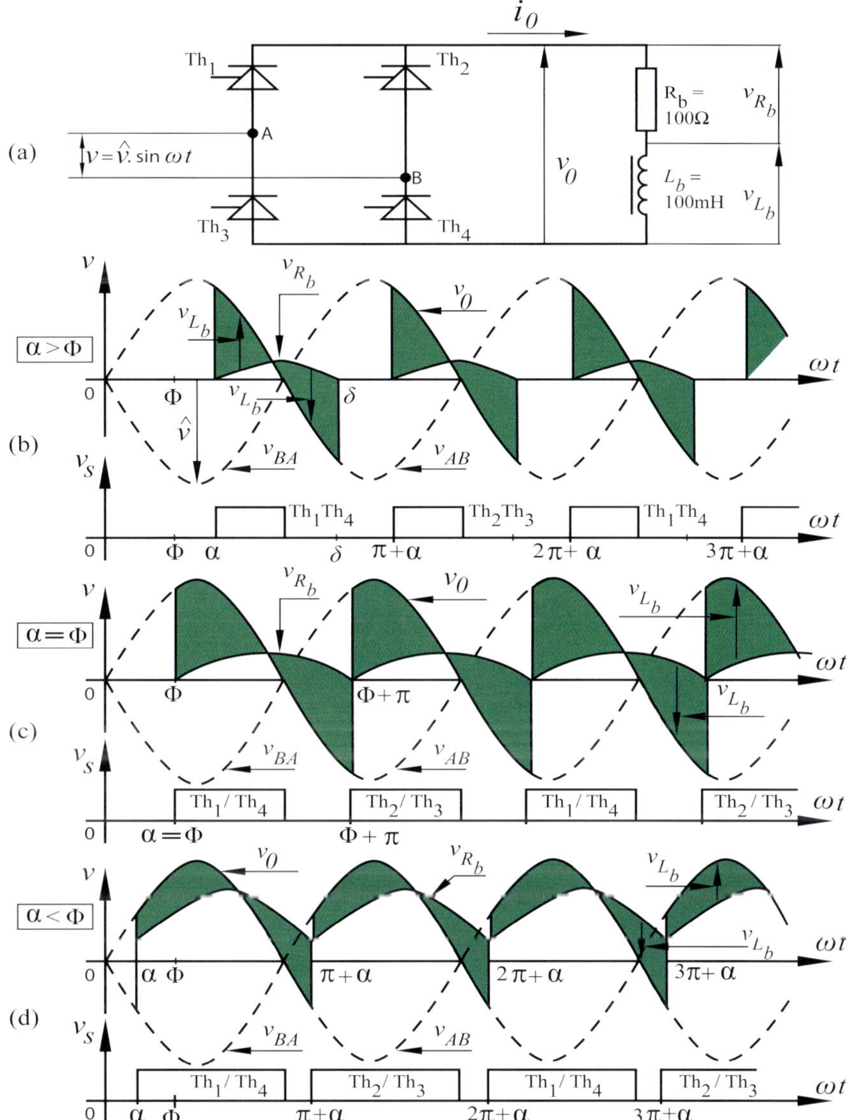

Fig. 8-9: A full controlled bridge. Resistive and inductively loaded

Numeric example 8-5:

Required: for the configuration shown in fig. 8-9:

    1. how large is Φ

    2. if $\alpha = 120°$, for how long is the current discontinuous if the supply frequency is 50Hz

Solution:

1. $tg\,\Phi = \dfrac{\omega.L_b}{R_b} = \dfrac{2.\pi.50.10^{-1}}{10} = 3.1416 \rightarrow\rightarrow\rightarrow \Phi = 72°20'$

2.   Fig. 8-4 $\rightarrow \delta = 229°$; discontinuous current $= (\,\alpha + 180°\,) - 229° = 71°$;

    $t = \dfrac{71}{360}$ x 20 x $10^{-3} = 3.95$ ms

6.5 Current waveforms in fig. 8-9

We proceed with fig. 8-9 on p. 8.21 to facilitate the reasoning on the present page.

If a resistive and inductive load is connected to a sinusoidal voltage $v = \hat{v} \cdot \sin \omega t$ at a time $t_1 = \frac{\alpha}{\omega}$ after the zero cross over, then the current form is given by expressions 7.1 and 7.3.

The current is composed of a steady state term $i_1 = \frac{\hat{v}}{Z} \cdot \sin (\omega t - \Phi)$ and a transient term $i_2 = \frac{\hat{v}}{Z} \cdot \sin (\Phi - \alpha) \cdot e^{(\alpha - \omega t).cotg\ \Phi}$. If $\alpha > \Phi$ then the positive part of $(i_1 + i_2)$ flows and $v_{R_b} = R_b \cdot i_o$ looks like that shown in fig. 8-9b.

Here in is: $i_o = \frac{\hat{v}}{Z} \left[\sin (\omega t - \Phi) + \sin (\Phi - \alpha) \cdot e^{(\alpha - \omega t).cotg\ \Phi} \right]$. The current is discontinuous.

In the case where the firing angle $\alpha = \Phi = bgtg\ \dfrac{\omega.L_b}{R_b}$ the exponential transient term $i_2 = 0$ and only the steady state term $i_1 = \frac{\hat{v}}{Z} \cdot \sin (\omega t - \Phi)$ will flow through the RL-load.

As a result of the semiconductors only the positive half of this i_1 current can flow so that $v_{R_b} = i_o \cdot R_b$ consists of a full wave rectified sinusoid (fig. 8-9c).

With $\alpha < \Phi$ then $\delta > \alpha + 180°$ (see fig. 8-4) and thyristor pair Th_1-Th_4 do not conduct at the instant that the other pair are fired ($\omega t = \pi + \alpha$ in fig. 8-9d) so that the current commutates from Th_1-Th_4 to Th_2-Th_3, etc...

6.6 Average output voltage

We calculate $V_{di\alpha}$ for the three cases shown in fig. 8-9.

1. $\boxed{\alpha > \Phi}$ $V_{di\alpha} = 2 \cdot \left[\dfrac{1}{2.\pi} \int_{\alpha}^{\delta} v \cdot d\omega t \right] = \dfrac{1}{\pi} \int_{\alpha}^{\delta} \hat{v} \cdot \sin \omega t \cdot d\omega t$

$$V_{di\alpha} = \frac{\hat{v}}{\pi} \cdot (\cos \alpha - \cos \delta) \tag{8-11}$$

This output voltage is obviously double with respect to the voltage in fig. 8-3 (expression 8-5). We can find the extinction angle δ in fig. 8-4.

2. $\boxed{\alpha = \Phi}$ $V_{di\alpha} = 2 \cdot \left[\dfrac{1}{2.\pi} \int_{\alpha}^{\alpha + \pi} \hat{v} \cdot \sin \omega t \cdot d\omega t \right] = \dfrac{\hat{v}}{\pi} \cdot (- \cos \omega t) \Big|_{\alpha}^{\alpha + \pi}$

$$V_{di\alpha} = \frac{2 \cdot \hat{v}}{\pi} \cdot \cos \alpha \tag{8-12}$$

3. $\boxed{\alpha < \Phi}$ $V_{di\alpha} = 2 \cdot \left[\dfrac{1}{2.\pi} \int_{\alpha}^{\alpha + \pi} \hat{v} \cdot \sin \omega t \cdot d\omega t \right]$; $\quad V_{di\alpha} = \dfrac{2 \cdot \hat{v}}{\pi} \cdot \cos \alpha$ (8-13)

We find the same expression as for $\alpha = \Phi$.

For $\alpha \leq \Phi$ the output voltage is determined by the firing angle α and the voltage $\hat{v}$ and is independent of the load impedance. Notice that the green surface areas in fig. 8-9 demonstrate the equal surface area criterion: $v_{L_b} = 0$

6.7 DC current

As $v_{L_b} = 0$ then: $\boxed{I_{d\alpha} = \dfrac{V_{di\alpha}}{R_b}}$ (8-14)

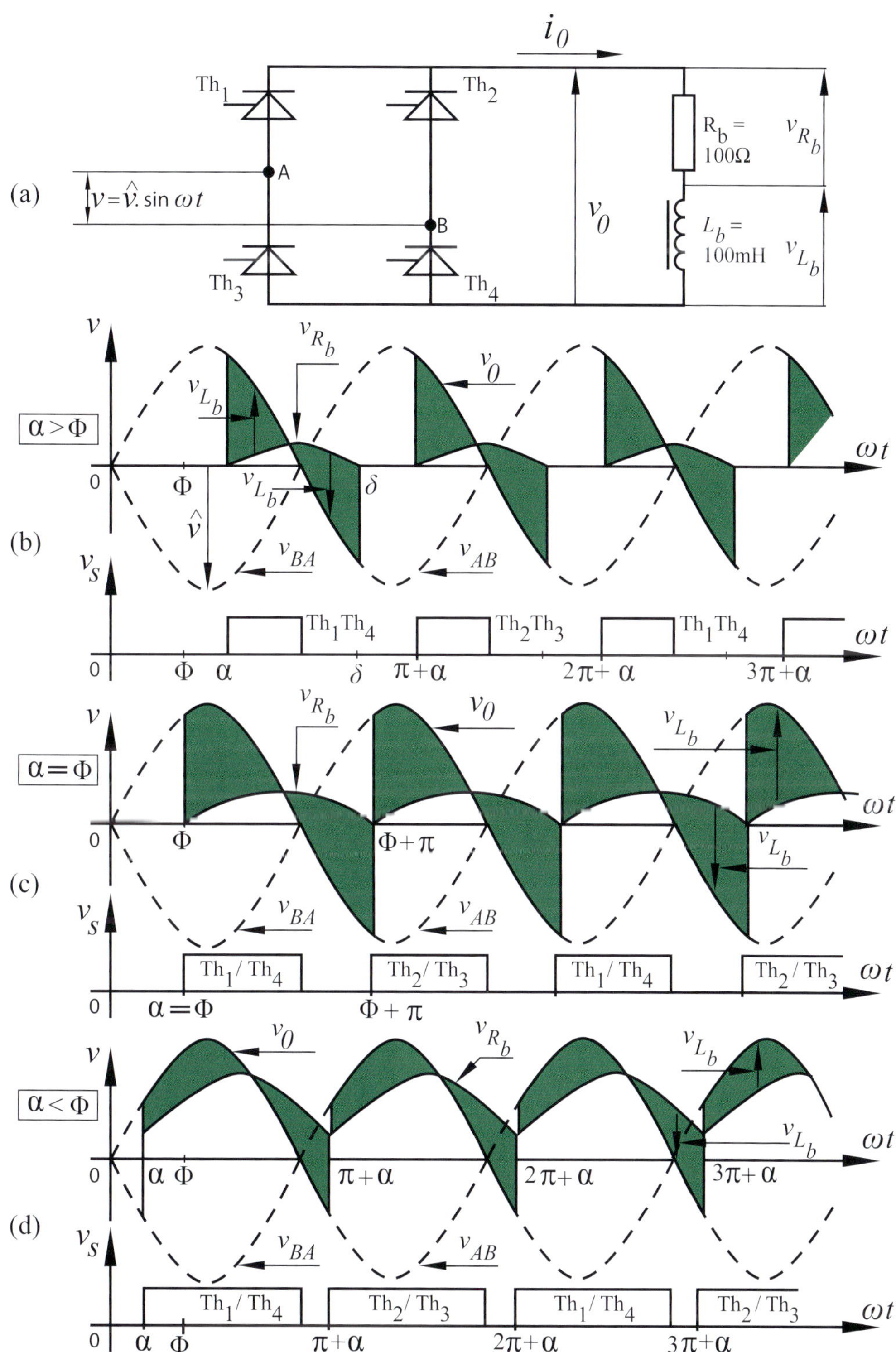

Fig. 8-9: Full wave B_2 -controlled rectifier, resistive inductive load

7.   FULLY CONTROLLED B_6-RECTIFIER. RESISTIVE LOAD.

7.1 The circuit

By replacing the diodes with thyristors in the circuit shown in fig. 7-14 a controlled rectifier is obtained. See fig. 8-10. The numbering of the semiconductors and the coding of the line voltages are maintained as in fig. 7-14.

In fig. 8-11a the secondary phase voltages of the transformer are drawn and shows the relationship with the secondary line voltages. These line voltages are the supply for the bridge shown in fig. 8-10. We see in fig. 8-11a the familiar $30°$ phase shift of a three-phase system. This has already been covered during the theory of electrical science.

7.2 Natural commutation points, driving voltage, reference voltage

Consider the case where all thyristors are controlled by a DC voltage between their respective gates and cathodes. The bridge behaves as a normal diode bridge. To determine which SCR's are conducting we can use table 8-3. This table is identical to table 7-2, only now the diodes have been replaced by thyristors.

Table 8-2   **Table 8-3**

reference line voltage	thyristor
v_{13}	Th_1
v_{23}	Th_2
v_{21}	Th_3
v_{31}	Th_4
v_{32}	Th_5
v_{12}	Th_6

driving voltage	thyristor pair
v_{12}	$Th_6 + Th_1$
v_{13}	$Th_1 + Th_2$
v_{23}	$Th_2 + Th_3$
v_{21}	$Th_3 + Th_4$
v_{31}	$Th_4 + Th_5$
v_{32}	$Th_5 + Th_6$

The points 1, 2, 3… in fig. 8-11b are called the natural commutation points as was the case in fig. 7-16 for the diode bridge of fig. 7-14. In table 7-1 the reference voltage for each diode was given. This information is again presented in fig. 8-2 with the exception that the diodes are replaced by thyristors. Instead of steady DC voltage we now use pulse control to the gates of the SCR's. The instantaneous time that the gate pulse is applied to a particular thyristor influences the form and the magnitude of the output voltage v_o of the thyristor bridge. This will be studied in the following text.

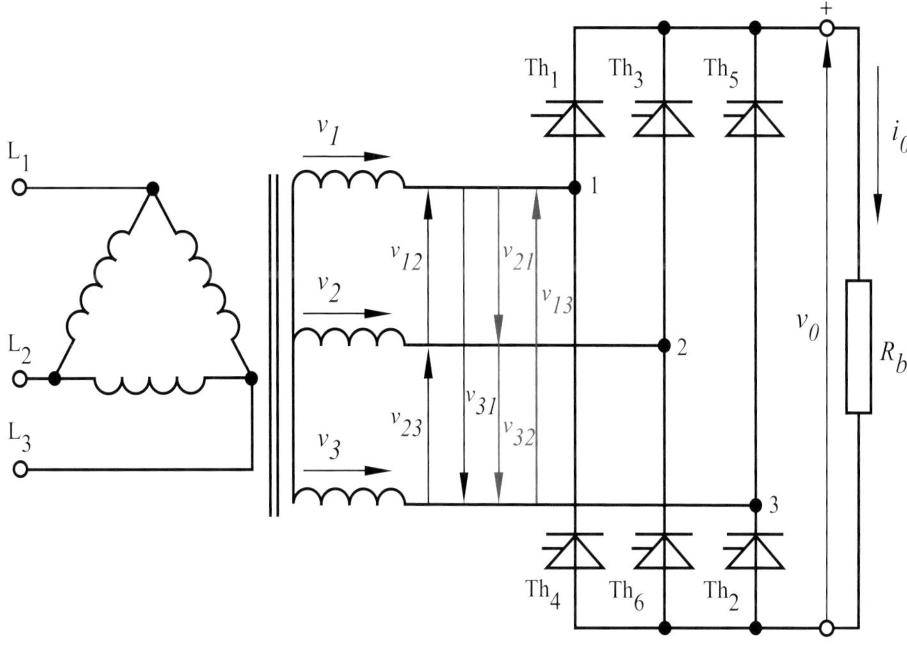

Fig. 8-10: Full controlled B_6 bridge with resistive load

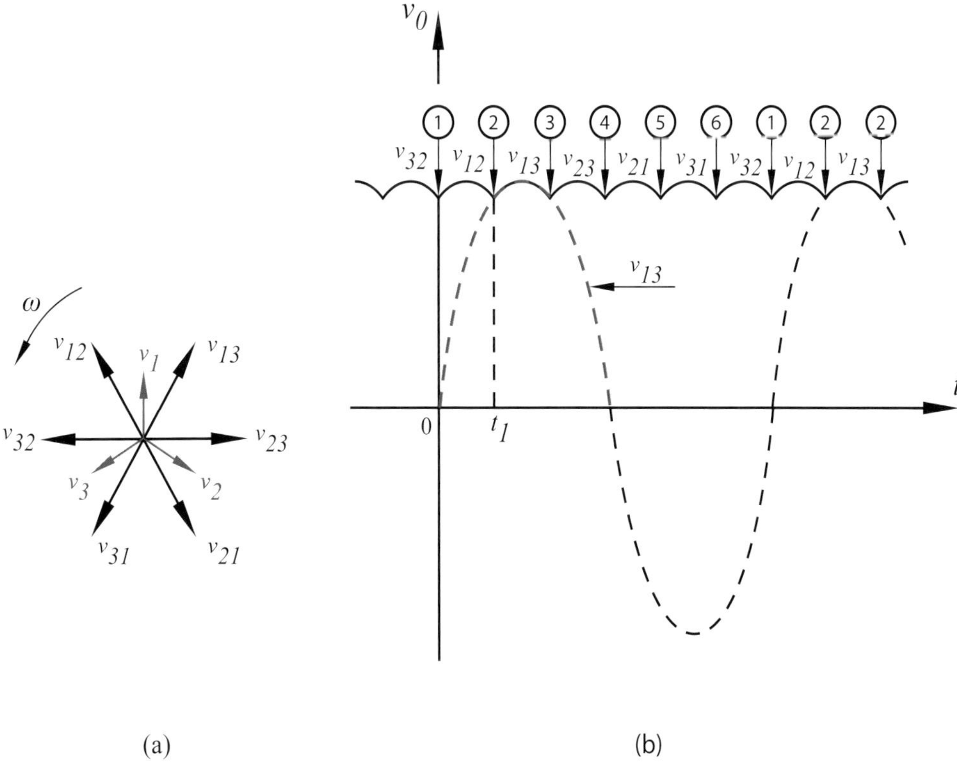

(a)   (b)

Fig. 8-11:   a. Supply voltage in fig. 8-10

            b. Natural commutation points and reference voltage

7.3 Thyristor bridge operation when α = 0°

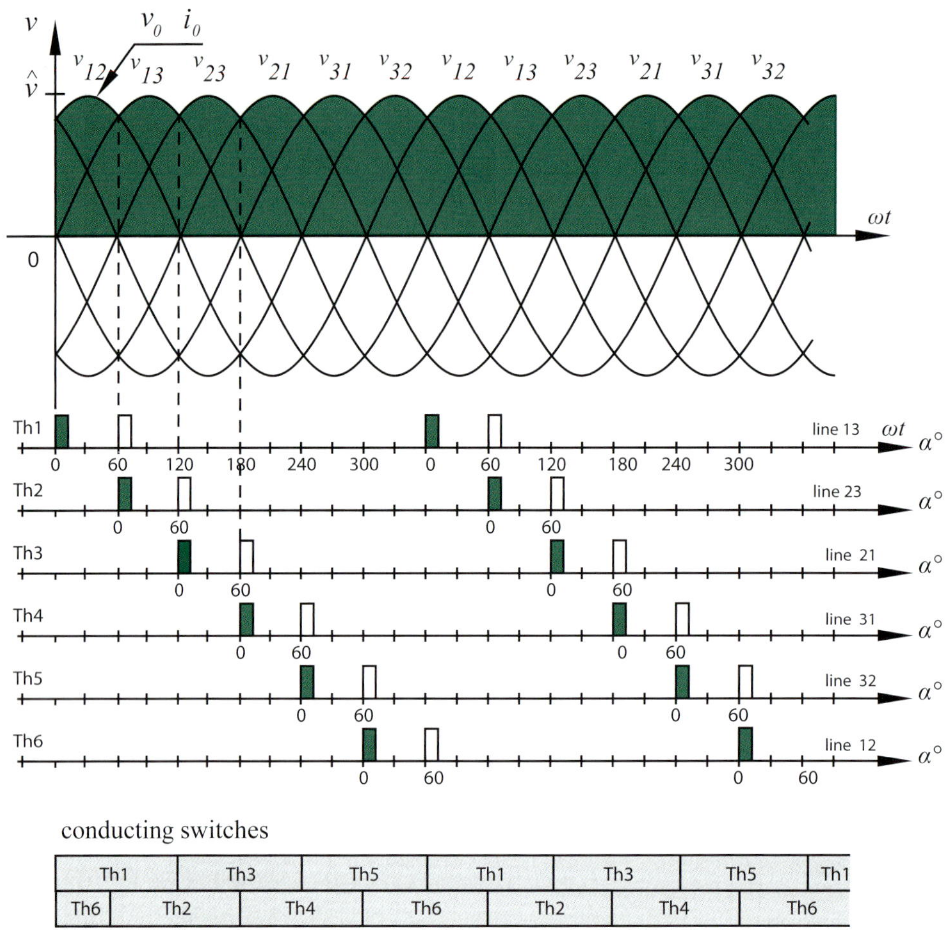

Fig. 8-12: Voltage waveform of fig. 8-10 when α = 0°

As in the case of a diode bridge we let the zero crossover of the *t*-axis coincide with the zero cross-over of v_{13} . The reference firing angle (α = 0°) of Th$_1$ is exactly the zero crossover of v_{13} . As a result in the following diagrams the α -axis of Th$_1$ corresponds with the ωt-axis. Consider the case where the firing angle for Th$_1$ is 0° . A gate pulse at ωt = 0° can cause Th$_1$ to conduct since v_{12} is the most positive voltage at that instant. To complete the circuit a thyristor in the bottom half of the bridge needs to conduct. According to Table 8-3 this is Th$_6$. To initiate conduction of the bridge we therefore need to simultaneously pulse Th$_1$ and Th$_6$. Up to now we have only fired Th$_1$ and Th$_6$ at ωt = 0°. If we now deliver a pulse at ωt = 60° to Th$_2$ then, since $v_{13} > v_{12}$, Th$_2$ will take over conducting from Th$_6$ while Th$_1$ remains conducting. This is further developed in fig. 8-12.

Remarks

1. In the previous discussion we fired Th$_1$ at ωt = 0° . A sixth of a period later Th$_2$ is fired , 60° later Th$_3$, etc.… . This explains the location of the first pulse for every thyristor in fig. 8-12. The second pulse is needed to start the conduction of the bridge. An SCR from the top half of the bridge must conduct with one from the bottom half of the bridge at all times.

Two control pulses per thyristor is referred to as a double pulse. Instead of working with two pulses we could also work with one pulse that is at least 60° wide.

2. Normally we have just three line voltages v_{12}, v_{23}, v_{31}. In table 8-3 we see that v_{12} is the driving voltage for Th_1-Th_6 and that v_{21} serves the thyristor pair Th_3-Th_4. In this way we have six driving voltages which are simply the three line voltages and there inverses.

7.4 Thyristor bridge operation when α = 30°

The first pulse is applied to Th_1 at $\omega t = 30°$. Since v_{12} is the dominant line voltage at that instant we also provide a control pulse to Th_6 to start the bridge.

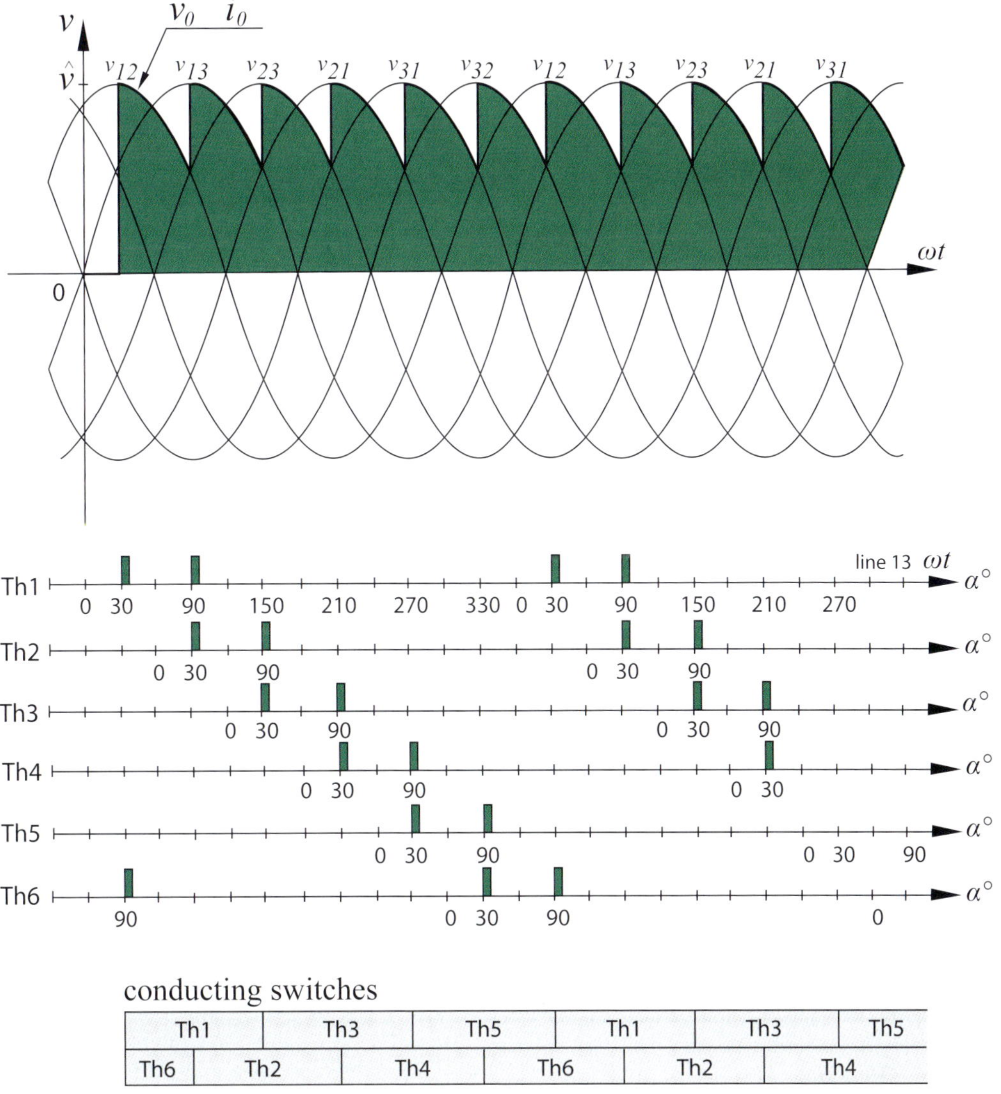

conducting switches

Th1		Th3		Th5		Th1		Th3		Th5

Th6		Th2		Th4		Th6		Th2		Th4

Fig. 8-13: Voltage form for fig. 8-10 with α = 30°

Up to now we have only caused Th_1 and Th_6 to conduct at $\omega t = 30°$. At $\omega t = 90°$ we now apply a control pulse to Th_2. Since $v_{13} > v_{12}$, Th_2 will take over from Th_6 while Th_1 remains conducting. This is further developed in fig. 8-13.

7.5 Thyristor bridge operation when $\alpha = 60°$

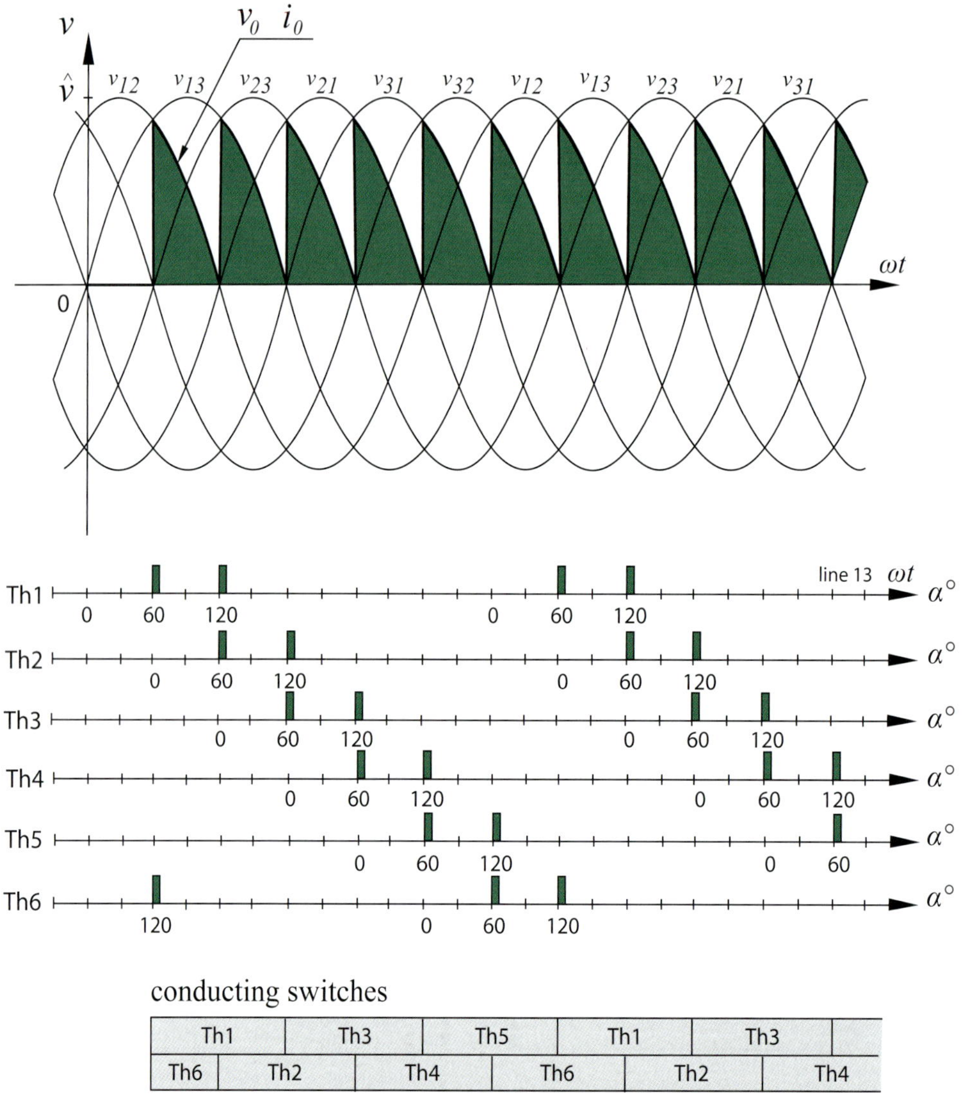

Fig. 8-14: Voltage form for fig. 8-10 with $\alpha = 60°$

In fig. 8-14 is $\alpha = 60°$. The output voltage (and current) are continuous. This is the border with discontinuous current.

Remark

Notice that α for each thyristor is the same so that we talk about the firing angle α of the bridge.

7.6 Thyristor bridge operation when α = 90°

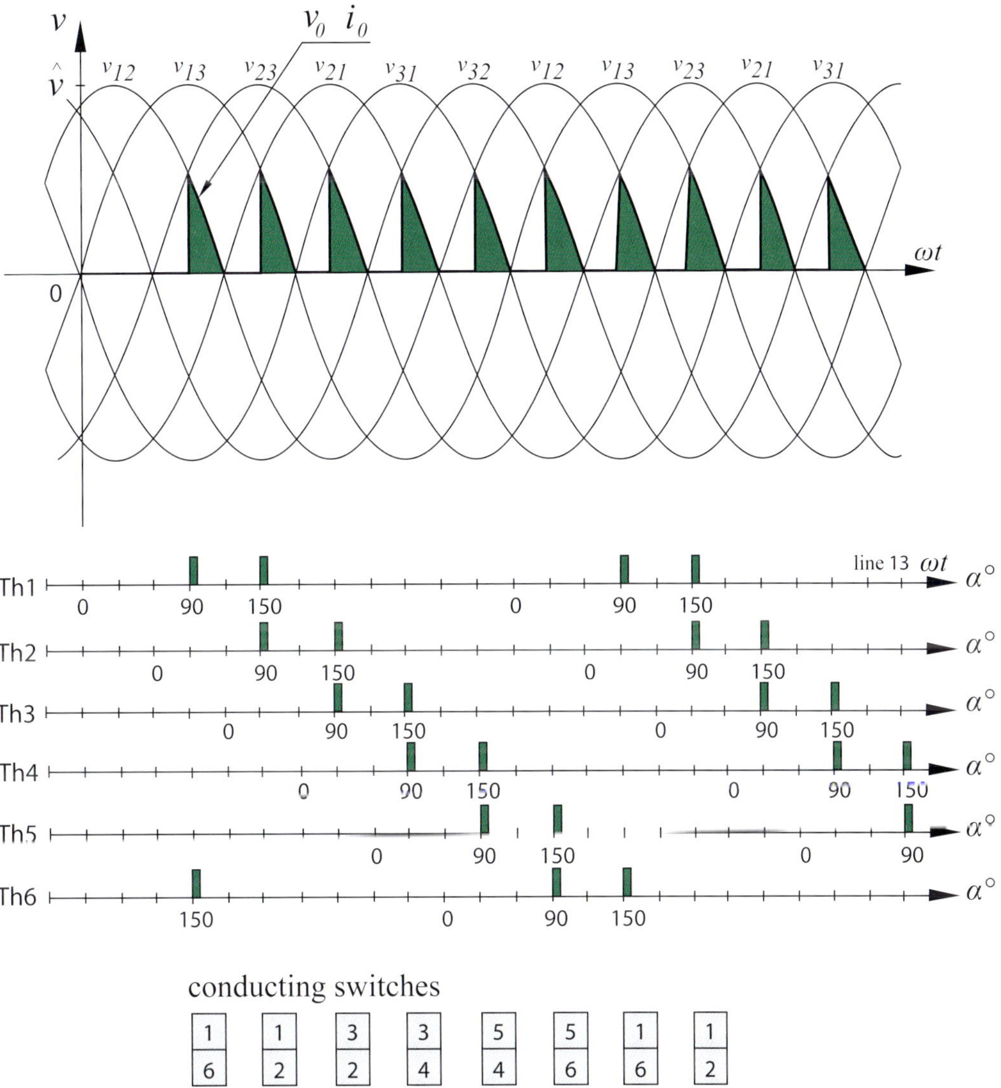

Fig. 8-15: Voltage form for fig. 8-10 with α = 90°

As mentioned previously for α > 60° with a resistive load we have discontinuous current and voltage. This is clearly visible in the above figure where the firing angle is 90°.

Notice that the bridge has to be continually started so that **the double pulse is not only needed to start the bridge but also to keep it conducting.**

7.7 Thyristor bridge operation when α = 120°

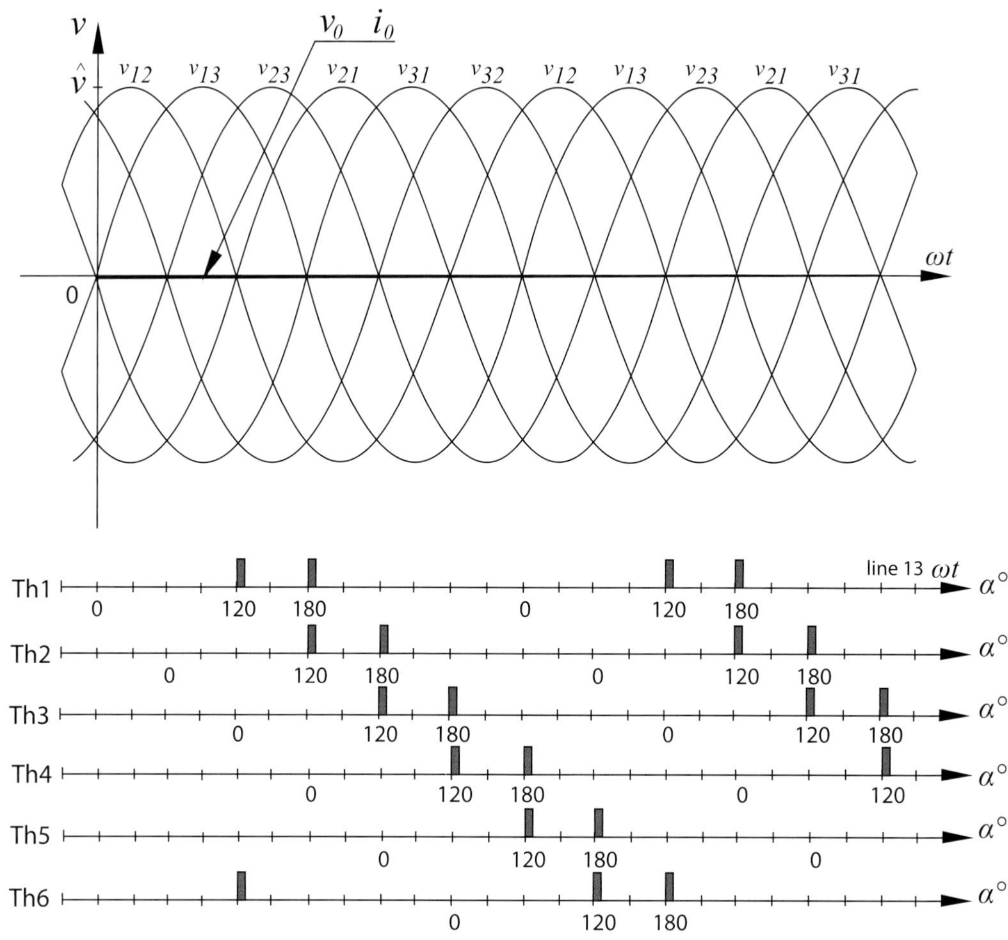

Fig. 8-15: Voltage form for fig. 8-10 with α = 120°

At α = 120° (fig. 8-16) the output voltage is zero. Th_1 together with Th_2 can conduct with v_{13}, but the SCR that receives the pulse at that instant is Th_6 instead of Th_2. Since the supply voltage v_{12} is zero for Th_1 and Th_6 the bridge will not start.

7.8 Control characteristic

From what we have discussed we can make the following conclusions:
1. The DC voltage is positive for 0° < α < 120°.
2. The output voltage and current for the load resistor are in phase and have the same form.
3. From α > 60° there is discontinuous service. It is therefore logical to implement the double pulse continuously, it is not only necessary for starting the bridge but also required in the case of discontinuous service.

4. The definition of pulse number of a rectifier is (see p. 7.17):

$$p = \frac{ripple\ frequency\ fundamental\ harmonic}{frequency\ mains} \tag{8-15}$$

5. Output voltage:

In determining the output voltage we distinguish two areas:

- continuous service ($0° < \alpha \leq 60°$):

$$V_{di\alpha} = \frac{3.\hat{v}}{\pi} \cos \alpha \tag{8-16}$$

- discontinuous service ($60° < \alpha \leq 120°$):

$$V_{di\alpha} = \frac{3.\hat{v}}{\pi} \left[1 + \frac{\cos \alpha}{2} - \frac{\sqrt{3}}{2}.\sin \alpha \right] \tag{8-17}$$

Here in $\hat{v}$ is the amplitude of the line voltage.

If $V_{di} = \frac{3.\hat{v}}{\pi}$ then (8-16) becomes: $V_{di\alpha} = V_{di}.\cos \alpha$ or: $\frac{V_{di\alpha}}{V_{di}} = \cos \alpha$

If the firing angle is greater than $60°$, then it follows from (8-17):

$$\frac{V_{di\alpha}}{V_{di}} = 1 + \frac{\cos \alpha}{2} - \frac{\sqrt{3}}{2}.\sin \alpha \tag{8-18}$$

We are now able to calculate $\frac{V_{di\alpha}}{V_{di}}$ as a function of α and to graph the results in fig. 8-17. The curve in fig. 8-17 is the **control characteristic** of the resistively loaded B_6-controller.

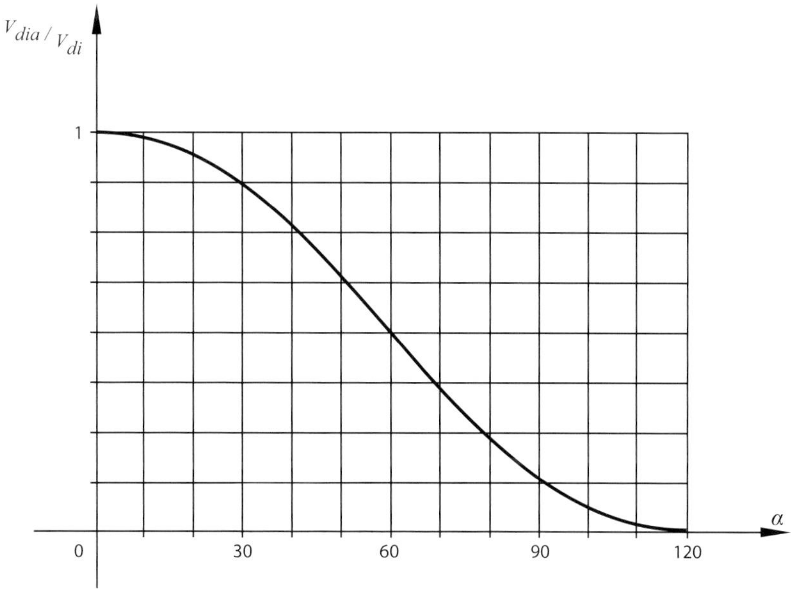

Fig. 8-17: Control characteristic resistively loaded B_6 -controller.

7.9 Calculating the output voltage

In calculating the output voltage we distinguish between two areas:

- continuous service ($0° < \alpha \leq 60°$)
- discontinuous service ($60° < \alpha < 120°$)

7.9.1 $0° < \alpha \leq 60°$

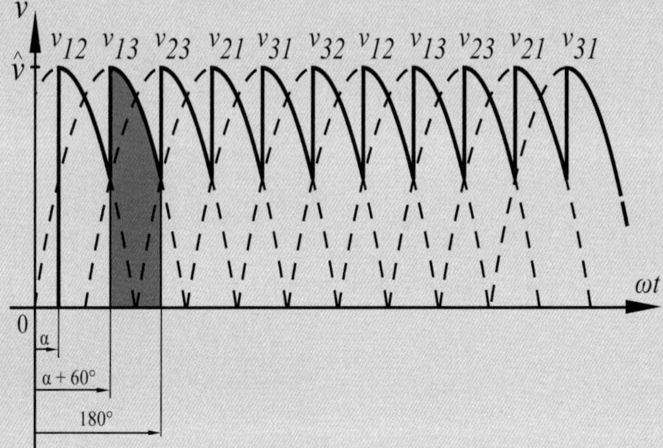

Fig. 8-18: Voltage form fully controlled bridge at $\alpha = 30°$

$$V_{di\alpha} = V_{AV} = 6 \cdot \frac{1}{2.\pi} \int\limits_{\alpha+60°}^{\alpha+120°} \hat{v} \cdot \sin \omega t \cdot d\omega t = \frac{3 \cdot \hat{v}}{\pi} \left(- \cos \omega t \right) \Big|_{\alpha+60°}^{\alpha+120°}$$

$$V_{di\alpha} = \frac{3.\hat{v}}{\pi} \cos \alpha \qquad\qquad (8\text{-}16)$$

7.9.2 $60° < \alpha \leq 120°$

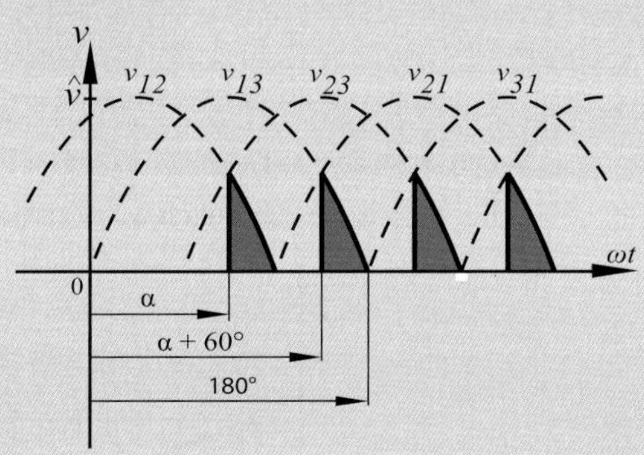

Fig. 8-19: Voltage form fully controlled bridge at $\alpha = 90°$

$$V_{di\alpha} = 6 \cdot \frac{1}{2.\pi} \int\limits_{\alpha+60°}^{180°} \hat{v} \cdot \sin \omega t \cdot d\omega t = \frac{3 \cdot \hat{v}}{\pi} \left(- \cos \omega t \right) \Big|_{\alpha+60°}^{180°}$$

$$V_{di\alpha} = \frac{3.\hat{v}}{\pi} \left[1 + \frac{\cos \alpha}{2} - \frac{\sqrt{3}}{2} \cdot \sin \alpha \right] \qquad\qquad (8\text{-}17)$$

Numeric example 8-6:

Given: The configuration in fig. 8-10. $R_b = 31\Omega$. Line voltage 3 x 230V.

Required: Determine the output voltage and current for $\alpha = 0°$, 90° and 120°.

Solution:

1) $\alpha = 0°$

    (8-16): $V_{di\alpha} = \dfrac{3.\hat{v}}{\pi} \cos \alpha = \dfrac{3 . \sqrt{2} . 230}{\pi} . \cos 0° = 310.6 \text{ V}$

    $I_{d\alpha} = \dfrac{V_{di\alpha}}{R_b} = \dfrac{310.6}{31} = 10 \text{ A}$

    We write: $V_{di\alpha} = V_{di\,0°} = V_{di} = \dfrac{3.\hat{v}}{\pi}$

    In general: $V_{di\alpha} = V_{di} . \cos \alpha$ also: $I_{d\alpha} = I_{d\,0°} = I_d$

2) $\alpha = 90°$

    (8-17): $V_{di\,90°} = \dfrac{3.\hat{v}}{\pi}\left[1 + \dfrac{\cos \alpha}{2} - \dfrac{\sqrt{3}}{2} . \sin \alpha\right] = \dfrac{3.\sqrt{2}.230}{\pi}\left[1 + 0 - \dfrac{\sqrt{3}}{2}\right] = 41.61 \text{ V}$

    $I_{di\,90°} = \dfrac{V_{di\alpha}}{R_b} = \dfrac{41.61}{31} = 1.34 \text{ A}$

3) $\alpha = 120°$

    (8-17): $V_{di\,120°} = \dfrac{3.\hat{v}}{\pi}\left[1 + \dfrac{\cos \alpha}{2} - \dfrac{\sqrt{3}}{2} .\sin \alpha\right] = \dfrac{3 . \sqrt{2} . 230}{\pi} .[1 - 0.25 - 0.75] = 0 \text{ V}$

    $I_{d\,120°} = 0 \text{ A}$

8.   FULL CONTROLLED B_6 -RECTIFIER- RESISTIVE AND INDUCTIVE LOAD

8.1 Rectifier operation of thyristor bridge

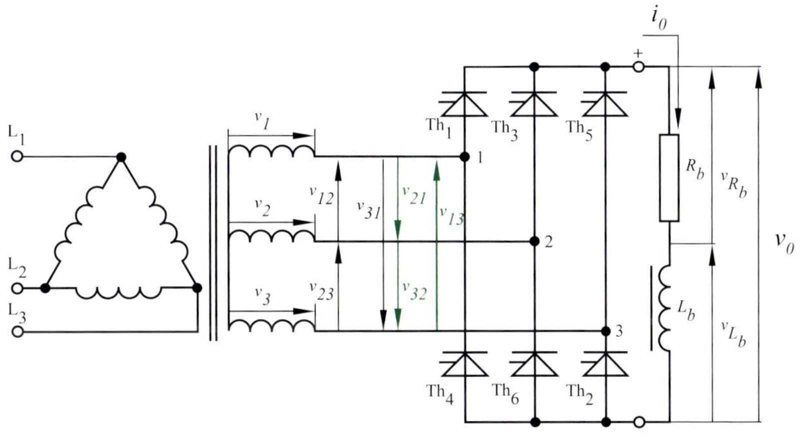

Fig. 8-20: Fully controlled B_6 -rectifier with resistive and inductive load

With a firing angle up to 60° we encounter the same voltage form as with a resistive load. The current form depends on how inductive the load is. (L_b/R_b). Fig. 8-21 shows the current and voltage waveform at $\alpha = 60°$.

The bridge is started at t_0 by applying control pulses to the gates of Th_1 and Th_6.

Clearly the voltage form is the same as already discussed in fig. 8-14.

Between t_0 and t_1 the current increases and the magnetic energy in the coil increases. At time t_1 the rectified voltage v_o is equal to $i_o . R_b$, in other words the voltage across the coil is zero. From that instant the magnetic energy in the coil starts to decay and the current begins to drop. Since an SCR stops conducting when the anode current falls below the holding current, Th_1 and Th_6 remain conducting. At instant t_2, Th_2 receives a control pulse and since $v_{13} > v_{12}$, Th_2 will take over from Th_6. The voltage v_{13} is now dominant and the current again increases until at the intersection of v_{R_b} with v_{13} when it will once again decrease, etc.

This pattern continues to repeat itself and after a time a steady state is reached. In fig. 8-22 this steady state is redrawn in detail. The steady state condition is determined by the equal surface area criterion ($A_1 = A_2$).

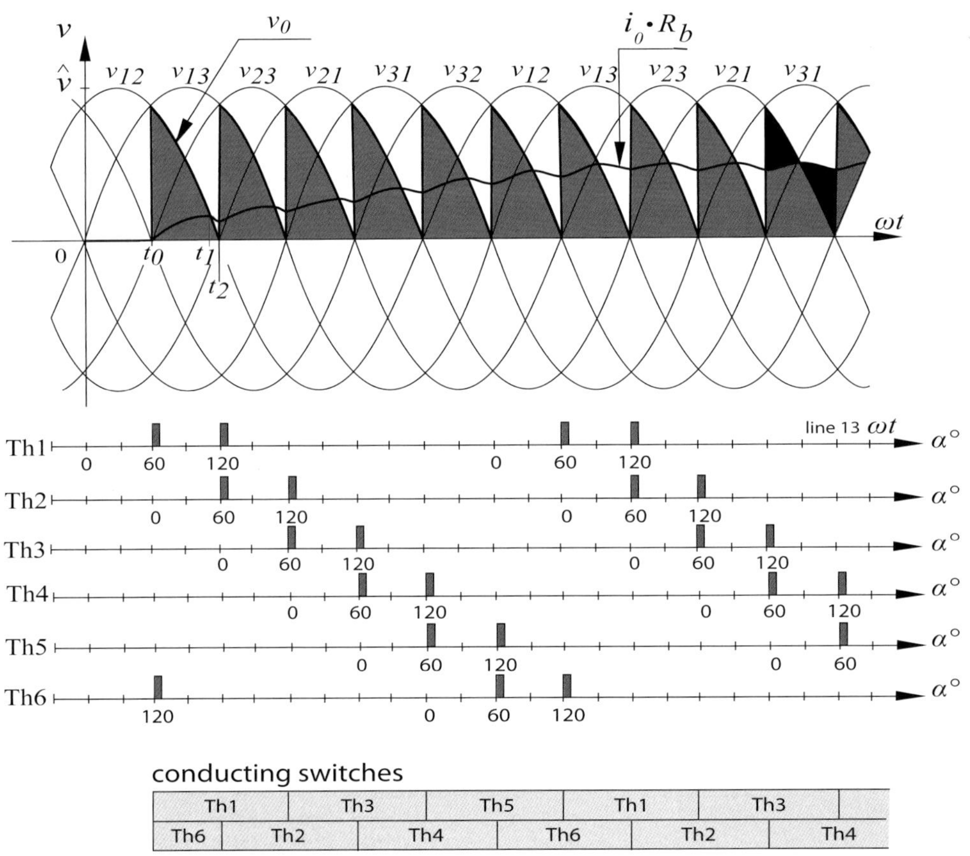

Fig. 8-21: Current and voltage waveform fig. 8-20 for $\alpha = 60°$

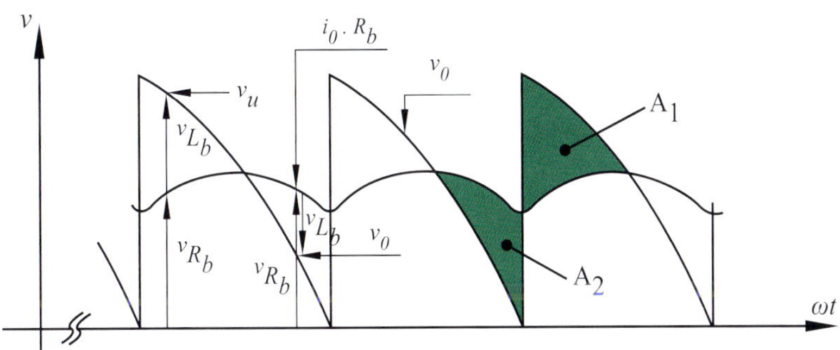

Fig. 8-22: Steady state condition fig. 8-20 for $\alpha = 60°$

In fig. 8-23 we consider a firing angle of $90°$. Between t_0 and t_1 the current increases the energy stored in the coil and at t_2 this energy is depleted (equal surface area criterion!). A small V_{dia} is the result, see the black areas at the extreme right of fig. 8-23.

For practical values of R_b we make only a small error when we assume that V_{dia} is practically zero when $\alpha = 90°$. In practice there will always be discontinuous current (see remark at the bottom of p. 8.34!)

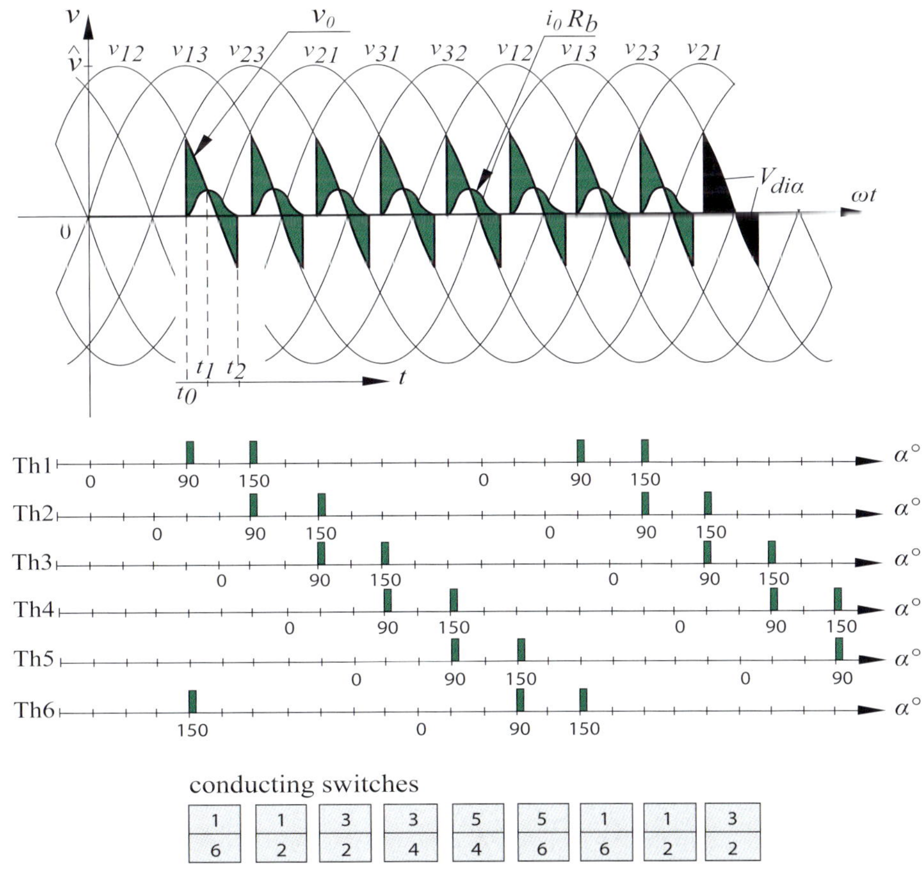

Fig. 8-23: Current and voltage waveform fig. 8-20 for $\alpha = 90°$

Numeric example 8-7:

Given: Fig. 8-20 with $R_b = 10\ \Omega$ and $L_b = 55$ mH
Secondary phase voltage 3 x 230 V - 50 Hz
Firing angle $\alpha = 90°$ (fig. 8-23).

Required:
1.   Determine the conduction angle per thyristor pair
2.   Average output voltage $V_{di\alpha}$
3.   The maximum energy stored in the coil

Solution:
1.   Conduction angle thyristors

$$\Phi = bgtg\ \frac{\omega.L_b}{R_b} = bgtg\ \frac{2.\pi.50.55.10^{-3}}{10} = 60°$$

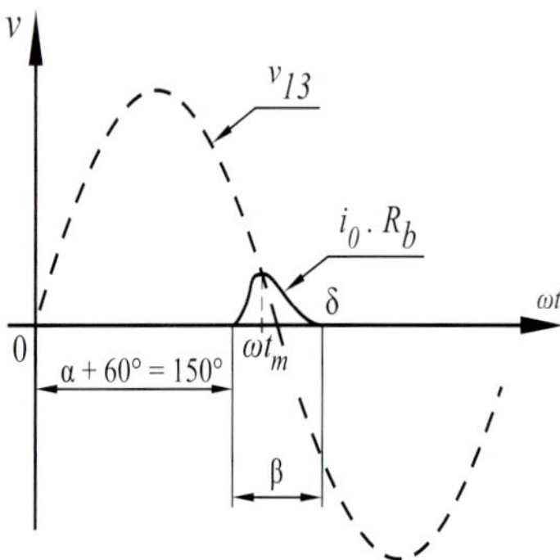

Fig. 8-4: $\alpha = 150°$; $\Phi = 60°$ gives: $\delta = 205°$
Conduction angle: $\beta = \delta - 150° = 55°$

Fig. 8-23 bis

2.   Output voltage
a)   Single phase rectifier:

$$(8\text{-}5):\ V_{di\alpha} = \frac{\hat{v}}{2.\pi}\ (\cos\alpha - \cos\delta) = \frac{\sqrt{3}\ .\ \sqrt{2}\ .\ 230}{2\ .\ \pi}\ (\cos 150° - \cos 205°) = 3.61\ V$$

b)   Three-phase bridge: $V_{di\alpha} = 6 \times 3.61 = 21.66$ V

3.   Maximum energy in the coil: $W = \dfrac{L\ .\ i_{max}^2}{2}$

$$(7\text{-}3):\ i(t) = \frac{\hat{v}}{Z}\ .\ \left[\sin(\omega t - \Phi) + \sin(\Phi - \alpha)\ .\ e^{(\alpha - \omega t)\ .\ cotg\ \Phi}\right]$$

Determine $\omega.t_m$:

$$\frac{di}{dt} = 0 \rightarrow \frac{\hat{v}}{Z}\left[\omega\ .(\cos\omega\ t_m - \Phi) + \sin(\Phi - \alpha)\ .(-\ \omega.cotg\ \Phi)\ .e^{(\alpha - \omega\ t_m)\ .cotg\ \Phi}\right] = 0$$

$$\cos(\omega\ t_m - \Phi) = \sin(\Phi - \alpha)\ .\ cotg\ \Phi\ .\ e^{(\alpha - \omega\ t_m).cotg\ \Phi}$$

with $\Phi = 60°$ and $\alpha = 150°$ we find, after iteration : $\omega.t_m = 176°25'$

$$\hat{v} = \sqrt{3}\ .\ \sqrt{2}\ .\ 230\ ;\ Z = \sqrt{R_b^2 + \omega^2\ L_b^2} = 19.96\ \Omega\ ;\ \omega.t_m = 176°25'$$

$(7\text{-}3) \rightarrow i_{max} = 3.65$ A

$$W = \frac{1}{2}\ .\ L\ .\ i_{max}^2 = \frac{1}{2} \times 55 \times 10^{-3} \times 3.65^2 = 0.366\ J$$

Remark

In fig. 8-23 note that Th_1 and Th_2 conduct simultaneously with v_{13} at $\omega t = 150°$. This is redrawn
in fig. 8.23bis. A firing angle of 150° means that $\delta < 210°$ (fig. 8-4!) since $\Phi = 90°$ is only theoretically possible. There will **always** be **discontinuous current** in the case of fig. 8-23

Photo CT-concept: IGBT with driver. This driver is suitable for ABB-IGBTs of 4500V and 6500 V

Photo ABB (KWx): Large power thyristors. Types up to 8.5kV - 1200A and 1800V - 6.1kA

8.2 Inverter service of a fully controlled three phase bridge

A fully controlled three-phase bridge can regulate energy flow from the load to the supply under certain conditions:

- the firing angle has to be greater than 90°
- a co-operating emf has to be present in the load circuit
- to ensure continuous service, the load needs to be sufficiently inductive.

Fig. 8-24 shows a configuration whereby the three conditions are met. In this figure for every instant in time: $v_o = -E + v_{R_b} + v_{L_b}$.

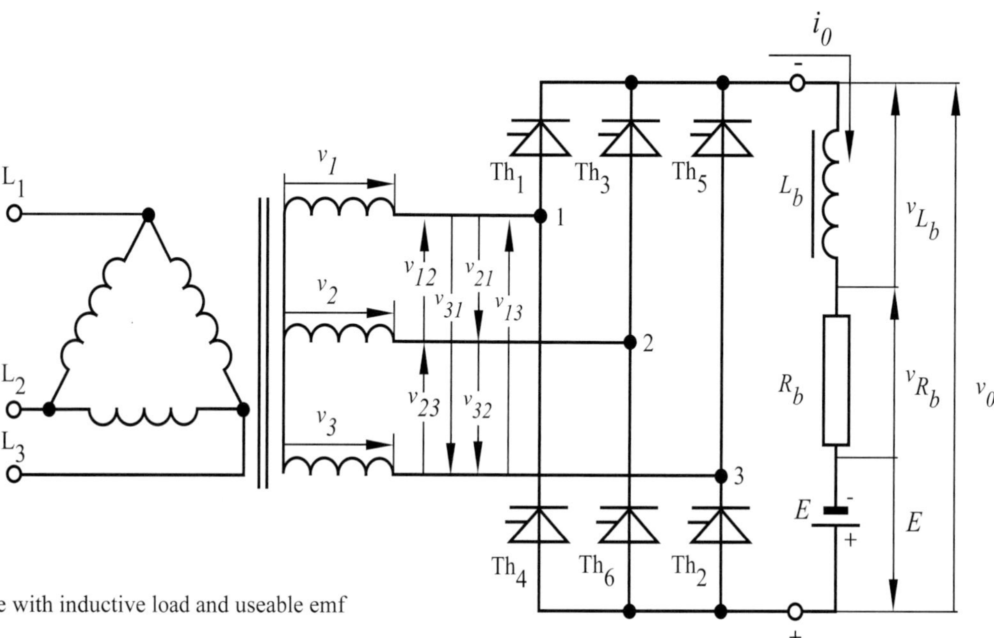

Fig. 8-24: Bridge with inductive load and useable emf

Consider when $\alpha = 120°$. Fig. 8-25c shows the output voltage waveform of the configuration in fig. 8-24. At t_0 (fig. 8-25a) control pulses are applied to Th_1 and Th_6. Before this instant the negative voltage E was across the output, in other words the cathode of Th_1 was E volt negative with regard to the anode of Th_6. It is clear that at t_1 the SCR's Th_1 and Th_6 will start to conduct with the driving voltage E since $v_{12} = 0$. The output voltage v_o will follow the voltage v_{12} .

Between t_0 and t_1 the coil stores energy. At time t_1 , v_{L_b} becomes zero and $v_o = -E + v_{R_b}$. The energy stored in the coil will begin to dissipate from t_1 and if this energy is sufficiently high, or in other words if the circuit is sufficiently inductive, then Th_1 and Th_6 will conduct until t_2 .

At t_2 , Th_2 and Th_1 will receive a control pulse so that Th_2 will take over from Th_6 since the voltage v_{13} is less negative than v_{12} and at the same time $v_{13} > E$. Via the thyristors Th_1 and Th_2 the output voltage of the bridge will follow v_{13} , etc ...

The equal surface area criterion is also valid here, this is shown in fig. 8-25b. Due to the presence of the unidirectional thyristors the current direction remains the same as for rectifier service. The source with an emf E delivers energy back to the grid. This energy recuperation from DC current source to the AC network is known as **inverter service**.

Remark

Depending on α , E , L_b , discontinuous service may or not occur.

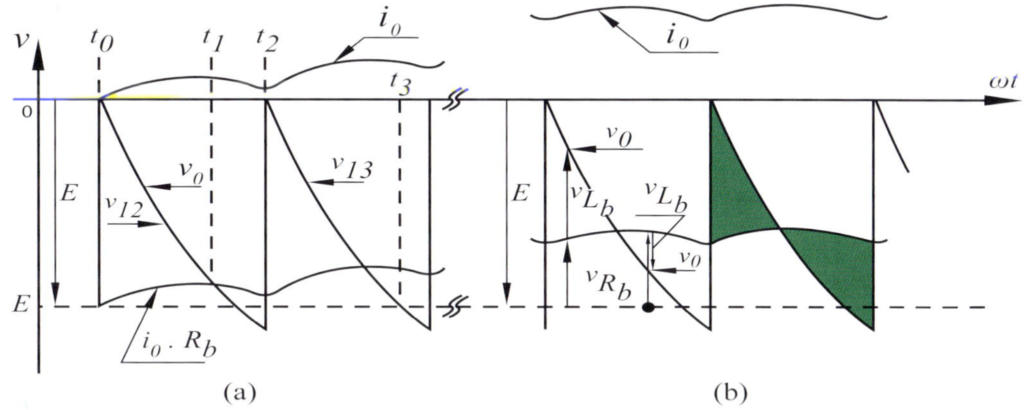

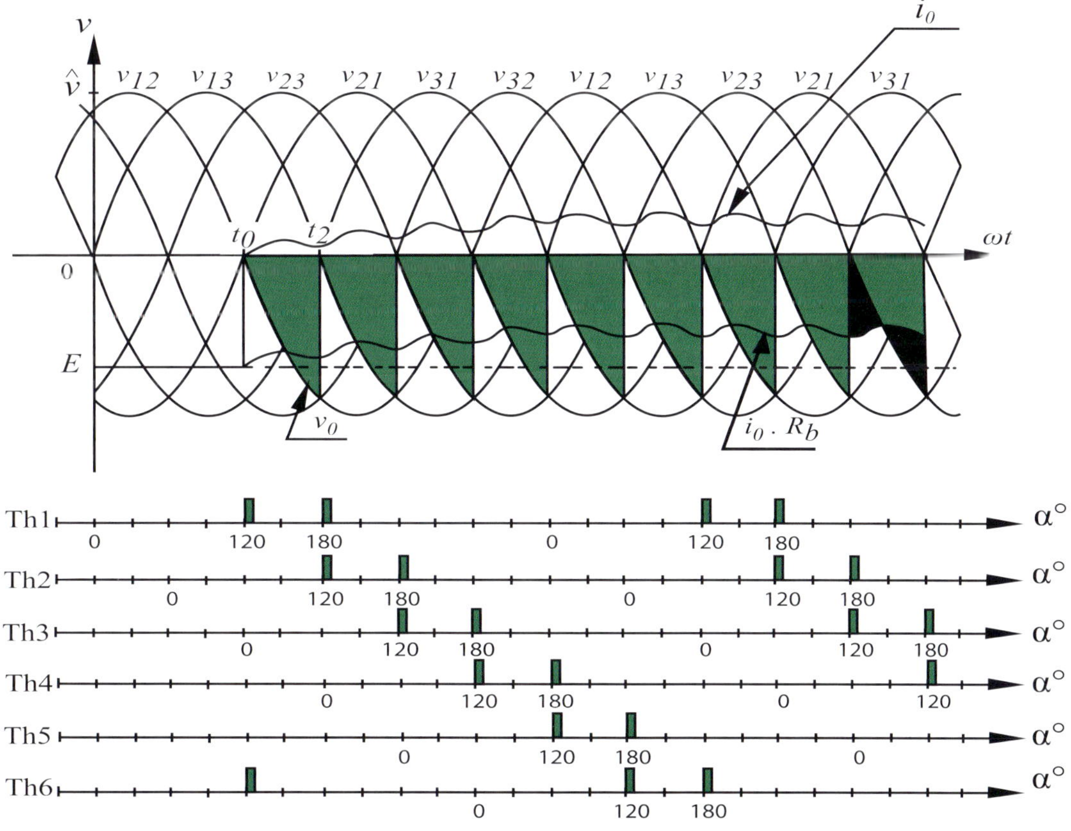

Fig. 8-25: Output current and voltage waveform fig. 8-24 for $\alpha + 120°$. We assume continuous current service

8.3 Firing failure of inverter

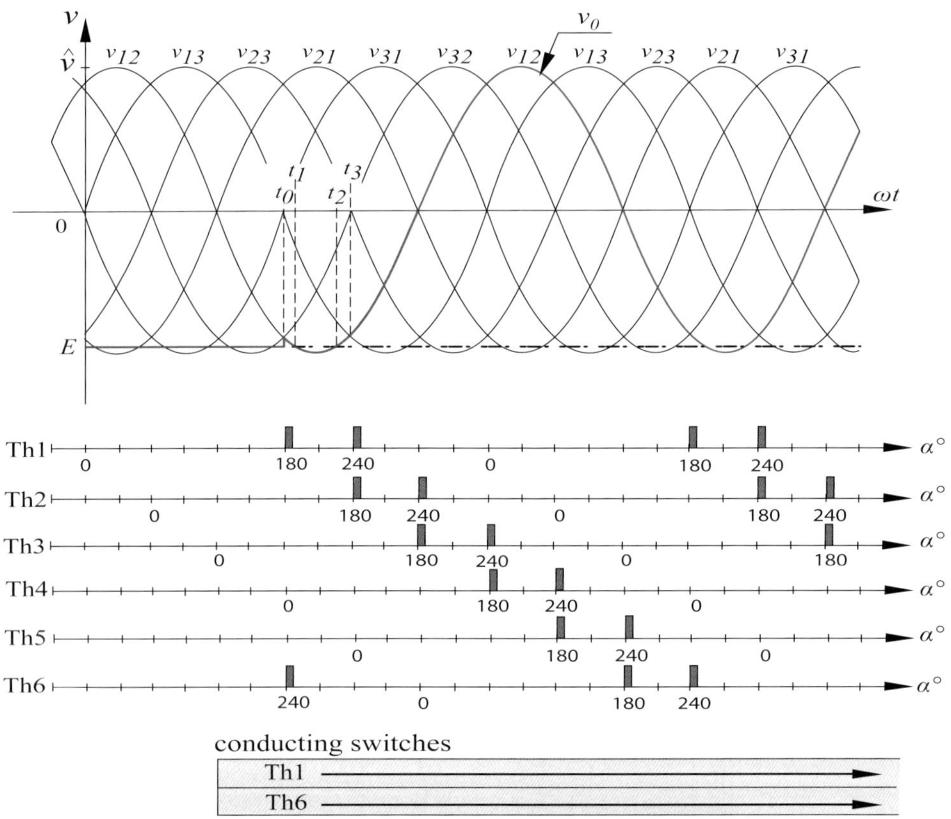

Fig. 8-26: Output voltage fig. 8-24 with $\alpha = 180°$

Consider the case where the firing angle in fig. 8-24 is 180°. This situation is shown in fig. 8-26. At the instant t_0, Th_1 and Th_6 conduct and the output voltage v_o follows the line voltage v_{12}. Between t_0 and t_1 the stored energy in the coil increases. At t_1 becomes $v_{12} = E$ and with a resistive load the thyristors can switch off, but first the magnetic energy in the coil has to dissipate.

In addition thyristors do not switch off instantly, they have a switch off time. With practical values of inductance in the circuit combined with the switch off time it can easily happen that Th_1 and Th_6 remain conducting between t_1 and t_2. After t_2, E and v_{12} remain the driving voltages. At instant t_3, Th_1 and Th_2 receive a control pulse. Thyristor Th_6 does not switch off instantly (turn-off time). In addition the voltage difference between E and v_{12} increases immediately, while the voltage difference between E and v_{13} decreases. It is clear that Th_6 together with Th_1 remain conducting. The output voltage therefore follows v_{12} (fig. 8-26). The current through the load is large and can only be interrupted by the SCR fuses. This mechanism is referred to as firing failure of the inverter . To avoid this the firing angle is limited to less than 180°, a practical value is for example $\alpha_{max} = 150°$.

8.4 Control characteristic B_6-controller

The expressions for the average output voltage of a B_6-controller are:

1. With a resistive load:

$$0° < \alpha \leq 60° : \qquad V_{dia} = \frac{3 \cdot \hat{v}}{\pi} \cdot \cos \alpha \qquad\qquad (8\text{-}16)$$

$$60° < \alpha \leq 120° : \qquad V_{dia} = \frac{3 \cdot \hat{v}}{\pi} \left(1 + \frac{\cos \alpha}{2} - \frac{\sqrt{3}}{2} \cdot \sin \alpha\right) \qquad (8\text{-}17)$$

2. Bridge with a large inductive load (without discontinuous current) is calculated in no. 8-5:

$$V_{dia} = \frac{3 \cdot \hat{v}}{\pi} \cos \alpha = V_{di} \cdot \cos \alpha \qquad\qquad (8\text{-}19)$$

Here in : V_{dia} = average output voltage of bridge (V)

$\hat{v}$ = amplitude of the supply voltage (V)

α = firing angle (°)

If we express the ratio $\dfrac{V_{dia}}{V_{di}}$ in terms of α then we obtain the control characteristic in fig. 8-27.

Up to 60° the control characteristic of a resistive load and a resistive inductive load are the same. In the case of a resistive load the output voltage remains positive. When $\alpha = 120°$ it becomes zero. If the load is resistive inductive, then the rectifier can have a negative output voltage under the precondition that there is a significant voltage source present on the load side, see "inverter service".

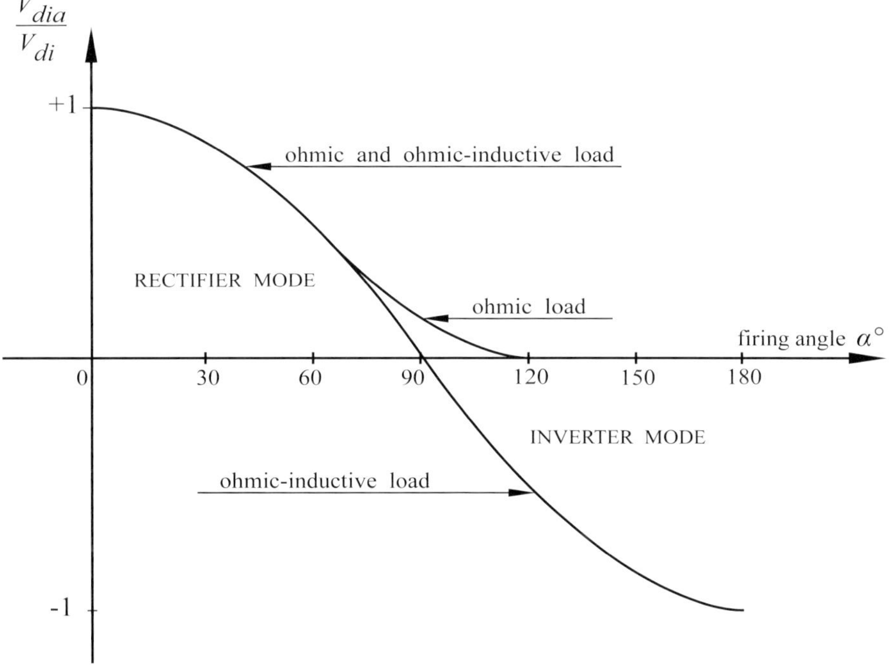

Fig. 8-27: Control characteristic B_6-controller

Numeric example 8-8:

Given:

In the configuration of fig. 8-24 the secondary voltage of the transformer is 3 x 400V, the firing angle $\alpha = 120°$, the load is heavily inductive so that discontinuous current does not occur. The emf E is 350V and $R_b = 10\Omega$. Transformer and thyristors are ideal.

Question:

What is the average value of the output current.

Solution:

$$V_{dia} = \frac{3 . \hat{v}}{\pi} . \cos \alpha = \frac{3 . \sqrt{2} . 400}{\pi} . \cos 120° = -270V$$

$v_o = -E + v_{R_b} + v_{L_b}$. With average values: $V_{dia} = -E + V_{R_b}$ since $V_{L_b} = 0$;

$$-270 = -350 + V_{R_b} \rightarrow\rightarrow V_{R_b} = 80V ; \qquad I_d = \frac{V_{R_b}}{R_b} = \frac{80}{10} = 8A.$$

8.5 Calculating the output voltage

When $\alpha < 60°$ the voltage form of v_o is the same as for a resistive load and $V_{dia} = \frac{3.\hat{v}}{\pi} . \cos \alpha$.
If $60° < \alpha \leq 90°$ and we assume the circuit is sufficiently inductive to maintain continuous current then we find the v_o voltage form of fig. 8-28a.

$$V_{dia} = 6 . \frac{1}{2.\pi} \int_{\alpha + 60°}^{\alpha + 120°} \hat{v} . \sin \omega t . d\omega t = \frac{3 . \hat{v}}{\pi} (- \cos \omega t) \Big|_{\alpha + 60°}^{\alpha + 120°} = \frac{3 . \hat{v}}{\pi} . \cos \alpha$$

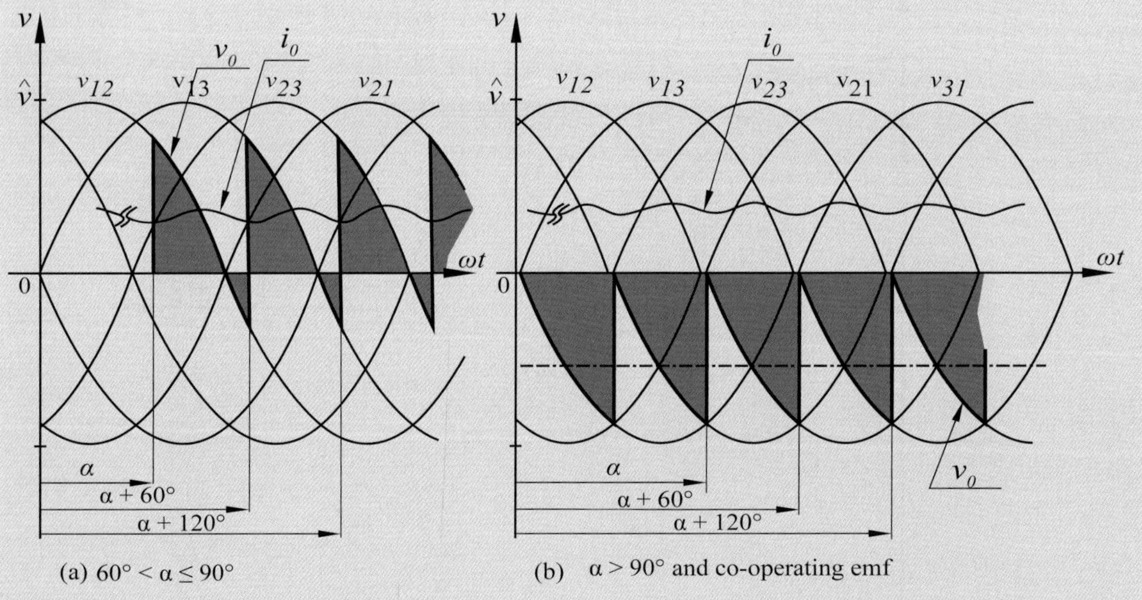

(a) $60° < \alpha \leq 90°$ (b) $\alpha > 90°$ and co-operating emf

Fig. 8-28: Voltage form with an inductively loaded full controlled SCR-bridge ($\alpha > 60°$)

With $\alpha > 90°$, a co-operating emf and sufficient inductance, then the voltage waveform is as in fig. 8-28b. The calculation of the output voltage is identical with that of fig. 8-28a, so that:

$V_{di\alpha} = \dfrac{3.\hat{v}}{\pi} \cdot \cos \alpha$. For $\alpha > 90°$ becomes $V_{di\alpha}$ negative.

In the case of continuous current the rectified (ideal) output voltage of a full controlled three phase bridge is:

$$V_{di\alpha} = \frac{3.\hat{v}}{\pi} \cdot \cos \alpha = V_{di} \cdot \cos \alpha \tag{8-19}$$

When $\alpha = 0°$ it is: $V_{di} = \dfrac{3.\hat{v}}{\pi}$ $\tag{8-20}$

For the case where $\alpha = 90°$ there is discontinuous current, see fig. 8-23. Expression (8-19) is not applicable. In numeric example 8-7 we calculate $V_{di\alpha}$.

We incur a small error when we make $V_{di90°} = 0$ as in numeric example 8-7.

8.6 Displacement factor, distortion factor, power factor with a full controlled three-phase bridge

We limit our study to the practical situation that the bridge has a large inductance as load, in other words continuous current mode with a low ripple current. The current form in the supply lines is approximated in fig. 8-29.

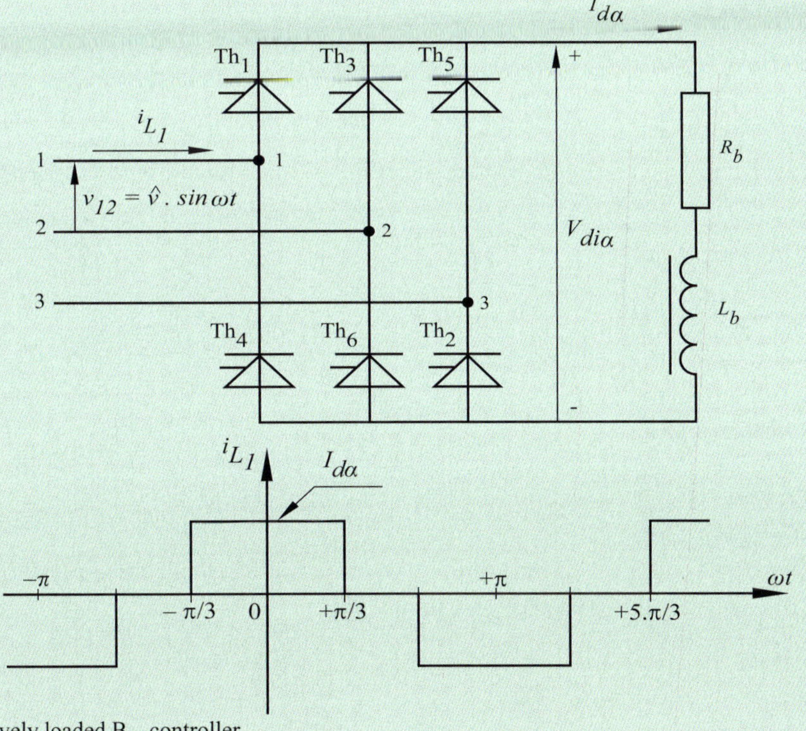

Fig. 8-29: Inductively loaded B$_6$ -controller

8.6.1 Displacement factor

With an ideal controller the active power P is equal to the DC power.

$$P = V_{di\alpha} \cdot I_{d\alpha} = \sqrt{3} \cdot V \cdot I_1 \cdot \cos \varphi_1$$

With $V_{di\alpha} = \dfrac{3.\hat{v}}{\pi} \cdot \cos \alpha = \dfrac{3\sqrt{2} \cdot V}{\pi} \cdot \cos \alpha$ this becomes : $V = \dfrac{\pi \cdot V_{di\alpha}}{3\sqrt{2} \cdot \cos \alpha}$

From (8-30) on p. 8-53 we find for the RMS-value of the fundamental harmonic:

$$I_1 = \frac{\sqrt{6}}{\pi} \cdot I_d \cdot \cos \alpha = \frac{\sqrt{6}}{\pi} \cdot I_{d\alpha}$$

So that: $V_{di\alpha} \cdot I_{d\alpha} = \sqrt{3}\ \dfrac{V_{di\alpha} \cdot \pi}{3\sqrt{2} \cdot \cos \alpha} \cdot \dfrac{\sqrt{6}}{\pi} \cdot I_{d\alpha} \cdot \cos \varphi_1$

Displacement factor: $\boxed{\cos \varphi_1 = \cos \alpha}$ (8-21)

We demonstrate this graphically in fig. 8-33 (p. 8.47)

Remark

Expression (8-21) is only valid in the case of continuous current with a practically constant DC current (in other words with a large inductive load!). The displacement factor is independent of the load impedance and depends only on the firing angle.

8.6.2 Distortion factor

$\cos \delta = \dfrac{I_1}{I}$. We previously found $I_1 = \dfrac{\sqrt{6}}{\pi} \cdot I_{d\alpha}$.

If we apply the definition of the RMS current (fig. 5-21 in table 5-1 on p. 5.8) to fig. 8-29:

$$\rightarrow\rightarrow I = \sqrt{\frac{2}{3}} \cdot I_{d\alpha}$$

So that: $\cos \delta = \dfrac{I_1}{I} = \dfrac{\sqrt{6} \cdot I_{d\alpha}}{\pi \cdot \sqrt{\dfrac{2}{3}} \cdot I_{d\alpha}} = \dfrac{3}{\pi}$ $\boxed{\cos \delta = \dfrac{3}{\pi}}$ (8-22)

8.6.3 Power factor

$\lambda = \cos \varphi_1 \cdot \cos \delta = \dfrac{3}{\pi} \cdot \cos \alpha$ $\boxed{\lambda = \dfrac{3}{\pi} \cdot \cos \alpha}$ (8-23)

With $\alpha = 0° \rightarrow\rightarrow \lambda = \dfrac{3}{\pi} = 0.955$. This is also the power factor of a three-phase diode bridge (with a large inductive load and continuous current).

8.7 Reactive power curve

With a firing angle α we find for the fundamental harmonic:

$$Q_{1\alpha}^2 + P_{d\alpha}^2 = S_{1\alpha}^2 = \left(\frac{P_{d\alpha}}{\cos\varphi_1}\right)^2 = \left(\frac{V_{di}\cdot\cos\alpha\cdot I_{d\alpha}}{\cos\varphi_1}\right)^2 = (V_{di}\cdot I_{d\alpha})^2$$

$$\rightarrow\rightarrow\rightarrow\quad \left(\frac{Q_{1\alpha}}{V_{di}\cdot I_{d\alpha}}\right)^2 + \left(\frac{P_{d\alpha}}{V_{di}\cdot I_{d\alpha}}\right)^2 = 1$$

If for example in the case of a motor control the controller maintains a constant DC current ($=$ armature current) I_d, then $V_{di}\cdot I_{d\alpha} = V_{di}\cdot I_d = P_{d\,\alpha\,=\,0°}$

and :

$$\left(\frac{Q_{1\alpha}}{P_{d\,\alpha\,=\,0°}}\right)^2 + \left(\frac{V_{di\alpha}}{V_{di}}\right)^2 = 1 \qquad\qquad (8\text{-}24)$$

Expression (8-24) is the equation of a circle with radius of one (fig. 8-30). The graphic in fig. 8-30 is known as the reactive power curve of a controlled rectifier.
Notice that the polarity of $Q_{1\alpha}$ does not change.
$Q_{1\alpha} = S_1\cdot\sin\varphi_1 = S_1\cdot\sin\alpha$.   For $0 < \alpha < 180°$ is $Q_{1\alpha} > 0$.
This means that for inverter and rectifier mode reactive power is drawn from the supply.
It is obvious that with a firing angle in the region of 90° maximum reactive power is drawn from the supply.

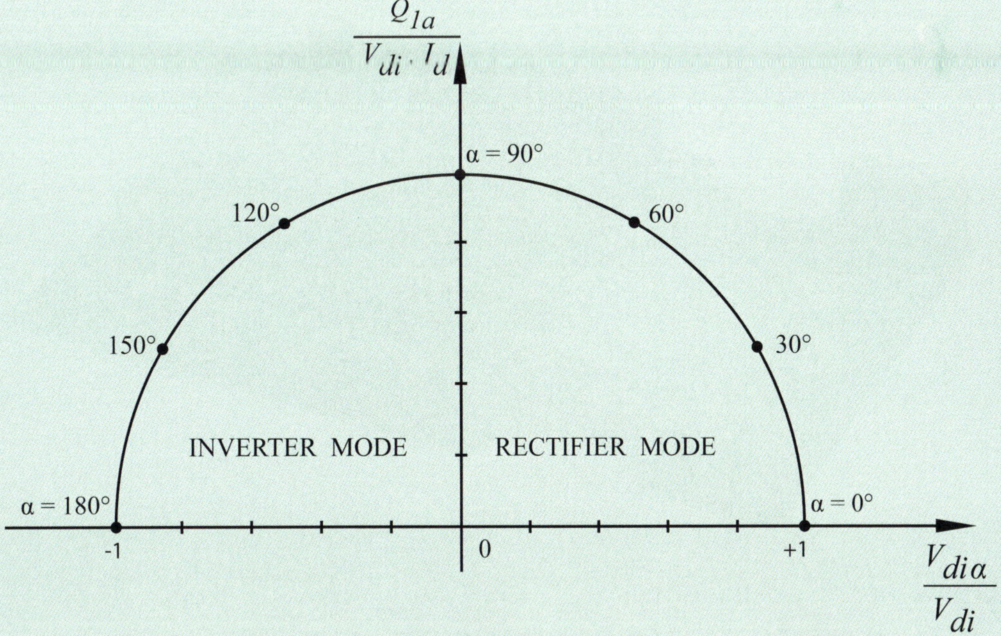

Fig. 8-30: Reactive power curve of a controlled rectifier

Photo ABB (KWx): IGCT (type 5SHY55L4500). Asymmetrical IGCT (V_{DRM} = 4500 V ; V_{RRM} = 17 V) . I_{TAVM} = 1860 A. V_T = 2.8 V .

Photo ABB (KWx) : IGCT (type 5SHX14H4510). V_{DRM} = 4500 V . I_{TAVM} = 420 A . V_{T0} = 1.65 V.
Reverse conducting IGCT with monolythic integrated freewheel diode

9. FULL CONTROLLED B_6 -RECTIFIER

9.1 Dead time of a B_6-controller

We consider a full controlled B_6 rectifier as shown in fig. 8-32. If the thyristors fire at $\alpha = 0°$ for example and immediately thereafter the controller gives the command for the firing angle to be $60°$ then it takes $60°$ until the output voltage of the bridge follows this command, see fig. 8-31a.

We refer to the 3.3ms that corresponds to this delay as the dead time of the controller.

If the firing angle of the previous thyristor is $60°$ and now the controller determines that $\alpha = 0°$ then the following thyristor fires almost immediately after the new command: the dead time is practically zero, see fig. 8-31b.

From these two extreme examples it follows that statistically the dead time may be considered to be 1.66ms. This may be rounded of to 1.7ms.

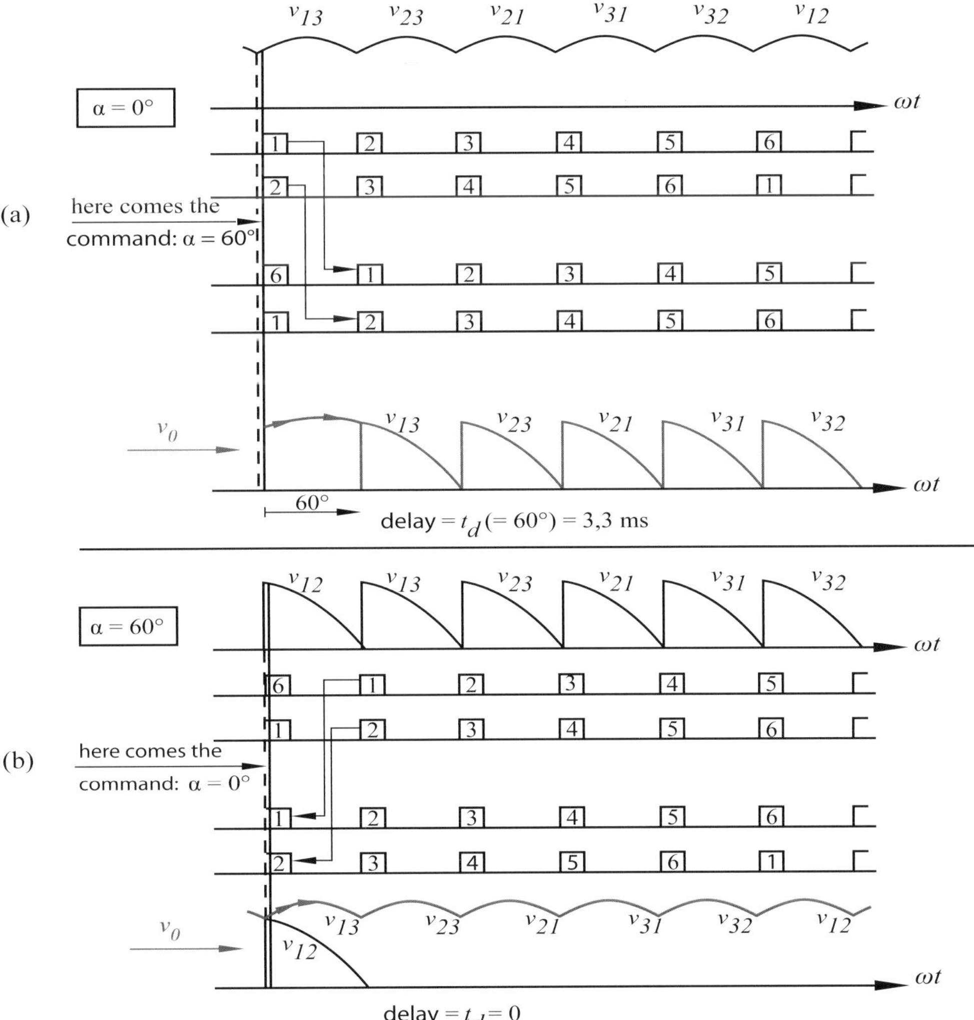

Fig. 8-31

9.2 Current waveform of a B_6-controller with *R-L* load

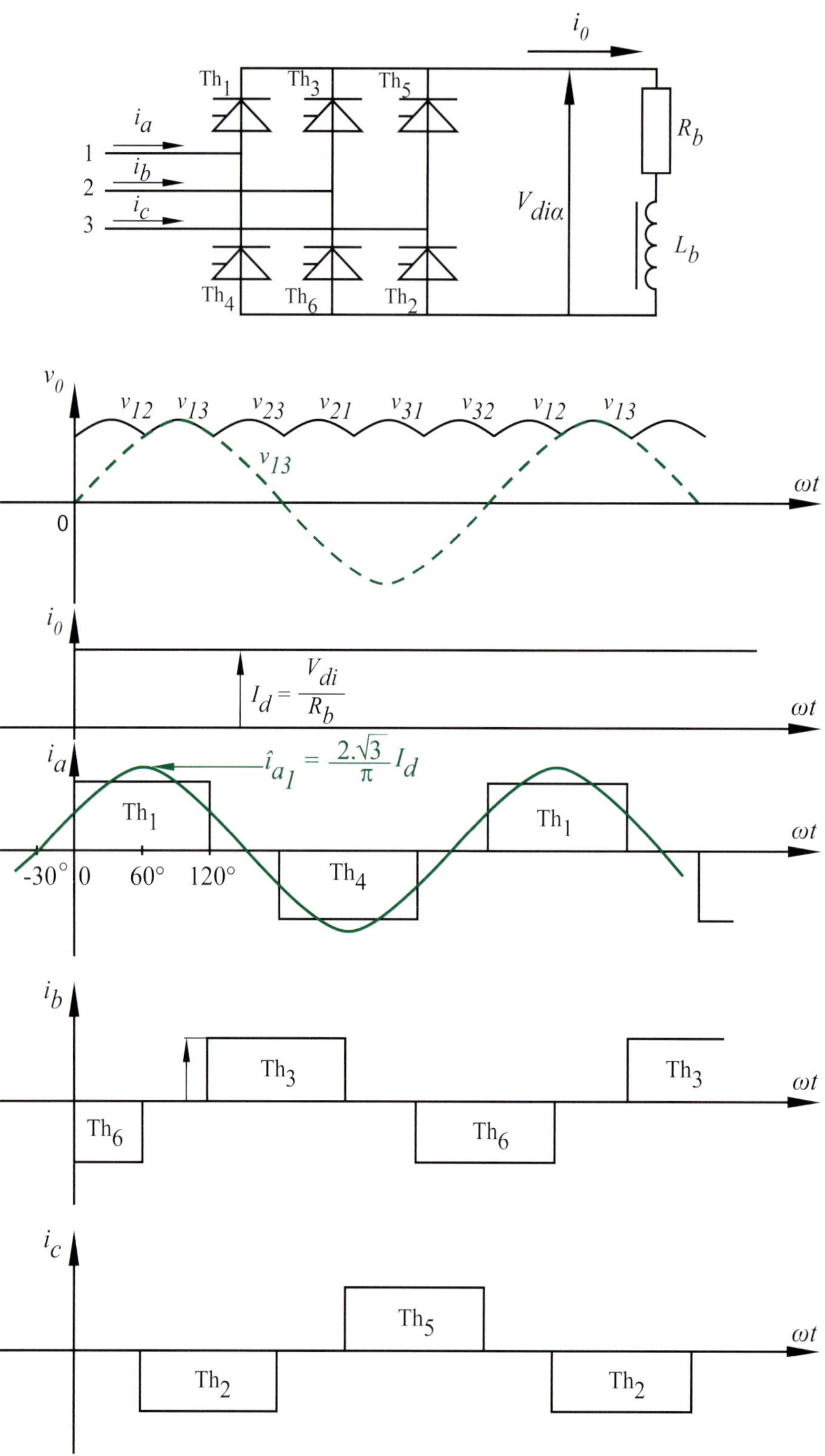

Fig. 8-32: Current waveforms B_6-controller for $\alpha = 0°$

If the load in fig. 8-32 is sufficiently inductive, a current with low ripple flows with the value

given by $I_{da} = \dfrac{V_{dia}}{R_b}$. Here in $V_{dia} = V_{di} \cdot \cos \alpha = \dfrac{3 \cdot \hat{v}}{\pi} \cdot \cos \alpha$.

At $\alpha = 0°$ the maximum DC current occurs: $I_d = \dfrac{V_{di}}{R_b}$.

At $\alpha = 60°$ we find: $I_{d60°} = \dfrac{V_{di60°}}{R_b} = \dfrac{V_{di} \cdot \cos \alpha}{R_b} = \dfrac{I_d}{2}$, in other words the DC current is halved.

We now draw the current and voltage waveforms for $\alpha = 0°$ (fig. 8-32) and for $\alpha = 60°$ (fig. 8-33)

in the case that $v_{13} = \hat{v} \cdot \sin \omega t$.

On p. 8-53 we find the amplitude of the fundamental harmonic of the line current:

$$\hat{i}_{a1} = \frac{2 \cdot \sqrt{3}}{\pi} \cdot I_{da} .$$

In fig. 8.11a we see that v_1 leads 30° on v_{13} so that i_{a1} in fig. 8-32 is in phase with v_1 or: $\varphi_1 = 0°$.

Hereby is φ_1 the phase angle between phase voltage and phase current.

From fig. 8-33 it follows that i_{a1} 30° lags v_{13} and therefore 60° with respect to v_1 so that $\varphi_1 = 60°$.
$(= \alpha$, see expression 8-21) .

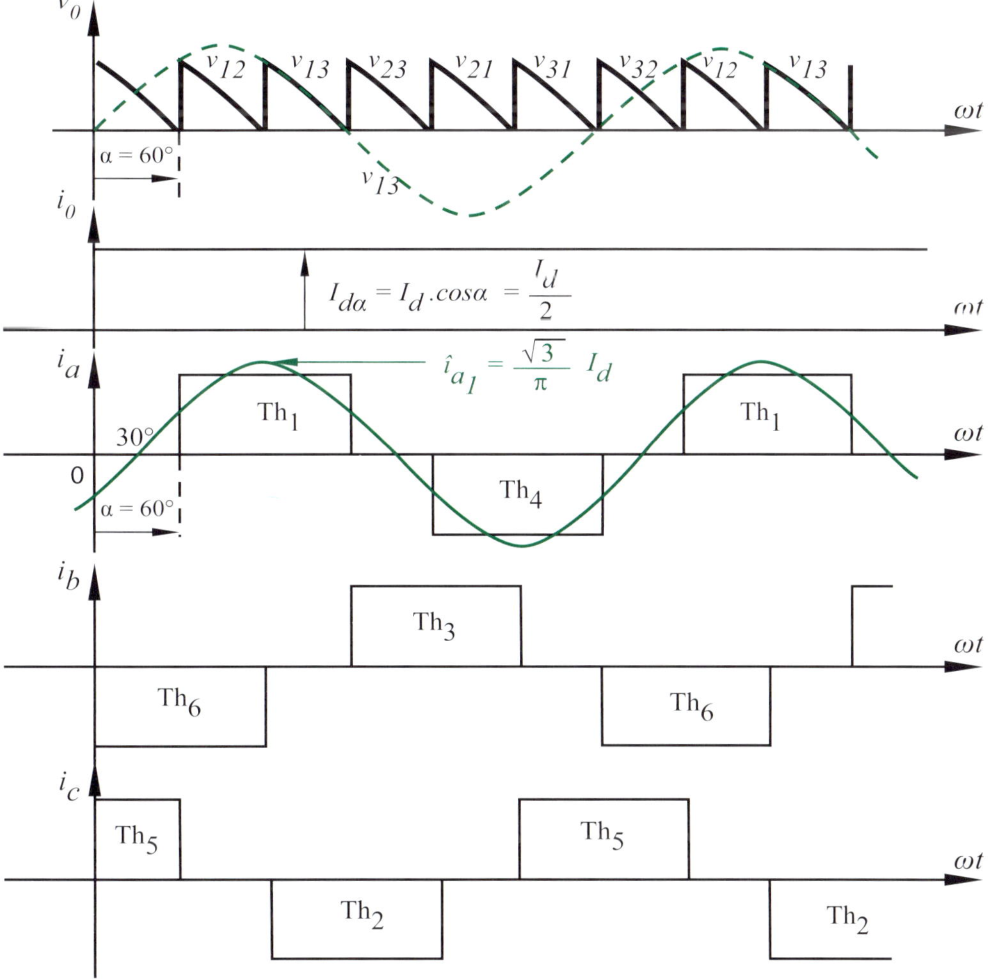

Fig. 8-33: Current waveforms B_6 -controller at $\alpha = 60°$

9.3 Harmonics at the output of a B_6-controller with a large inductive load

Fig. 8-34 shows the voltage waveform at the output of a full controlled three-phase bridge (full controlled B_6-rectifier)

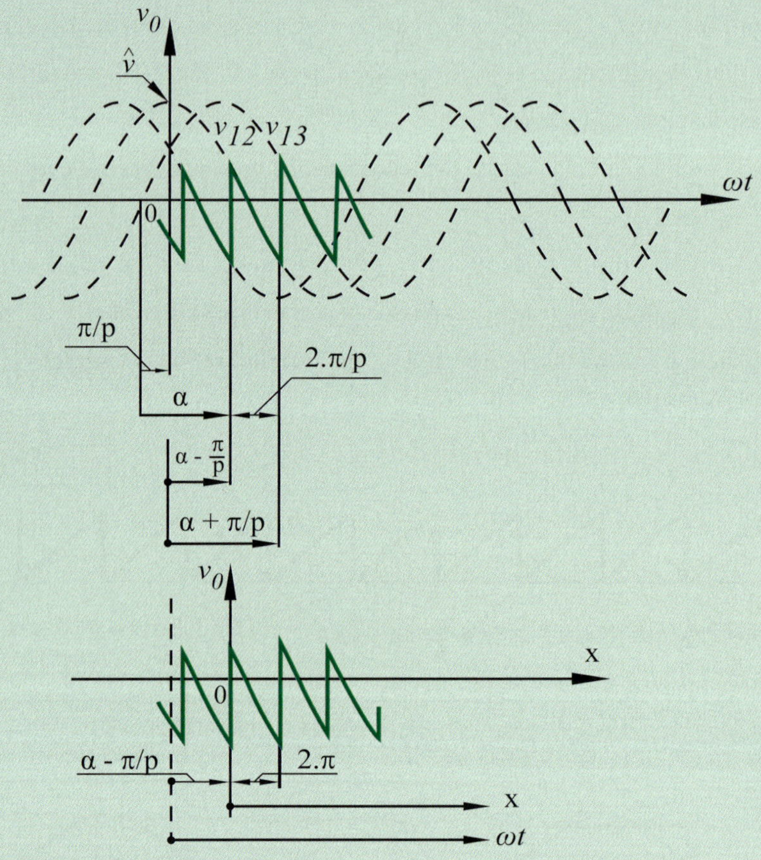

Fig. 8-34: Output voltage full controlled three-phase bridge with large inductive load

The output voltage waveform is a periodic function with period $= \frac{2 \cdot \pi}{p}$. We consider that portion of the output voltage where Th_1 and Th_6 conduct, in other words when v_o follows the voltage v_{12}. As we know the reference zero crossover for the firing angle of Th_1 is the zero crossover of v_{13}. In the lower part of fig. 8-34 we only draw the output voltage and in such a way that one period of v_o lies exactly between 0 and $2.\pi$ of the new x-axis.

Note that the value $\omega t = \frac{2 \cdot \pi}{p}$ in the upper part of the figure corresponds with $x = 2 \cdot \pi$ in the lower portion of the figure, so that: $\omega t = \frac{x}{p}$.

In addition the zero crossover of the x-axis is displaced by $(\alpha - \frac{\pi}{p})$ with respect to the ωt-axis, so that in reality: $\omega t = \frac{x}{p} + \alpha - \frac{\pi}{p}$.

The mathematical form of the v_o curve is given by: $v_{12} = v_O(t) = \hat{v} \cdot \cos \omega t$ so that:

$$v_o(x) = \hat{v} \cdot \cos\left(\frac{x}{p} + \alpha - \frac{\pi}{p}\right).$$

Here $\hat{v}$ is the amplitude of the input AC voltage of the bridge. In the case of supply via a transformer this is the amplitude of the secondary line voltage. The Fourier series for the voltage v_o in the v_o - x -axis is given by:

$$V_0(x) = A_0 + \sum_{n=1}^{n=\infty} \left[A_n . \sin(n.x) + B_n . \cos(n.x) \right]$$

- DC-component A_0 :

$$A_0 = \text{DC-component} = V_{di\alpha} = V_{di} . \cos\alpha$$

$$V_{di\alpha} = \frac{1}{2.\pi} \int_0^{2.\pi} \hat{v} . \cos\left(\frac{x}{p} + \alpha - \frac{\pi}{p}\right).dx = \frac{\hat{v}.p}{2.\pi} \int_0^{2.\pi} \cos\left(\frac{x}{p} + \alpha - \frac{\pi}{p}\right) . d\left(\frac{x}{p}\right)$$

$$= \frac{\hat{v}.p}{2.\pi} \left[\sin\left(\frac{2.\pi}{p} + \alpha - \frac{\pi}{p}\right) - \sin\left(\alpha - \frac{\pi}{p}\right) \right]$$

Solving gives: $V_{di\alpha} = \hat{v} . \frac{p}{\pi} . \sin\frac{\pi}{p} . \cos\alpha$

$$\alpha = 0° \longrightarrow\longrightarrow\longrightarrow \qquad \boxed{V_{di} = \hat{v} . \frac{p}{\pi} . \sin\frac{\pi}{p}} \qquad\qquad (8\text{-}25)$$

- Sinus components A_n :

$$A_n = \frac{1}{\pi} \int_0^{2.\pi} v_0(x) . \sin(n.x) . dx = \frac{\hat{v}}{\pi} \int_0^{2.\pi} \cos\left(\frac{x}{p} + \alpha - \frac{\pi}{p}\right) . \sin(n.x).dx$$

$$= \frac{\hat{v}}{2.\pi} \int_0^{2.\pi} \sin\left(\frac{x}{p} + \alpha - \frac{\pi}{p} + n.x\right) . dx - \frac{\hat{v}}{2.\pi} \int_0^{2.\pi} \sin\left(\frac{x}{p} + \alpha - \frac{\pi}{p} - n.x\right) . dx$$

$$A_n = \frac{-\hat{v}}{2.\pi.\left(n + \frac{1}{p}\right)} . \cos\left[\left(n + \frac{1}{p}\right) . x + \alpha - \frac{\pi}{p}\right]\Big|_0^{2.\pi}$$

$$+ \frac{\hat{v}}{2.\pi.\left(\frac{1}{p} - n\right)} . \cos\left[\left(\frac{1}{p} - n\right) . x + \alpha - \frac{\pi}{p}\right]\Big|_0^{2.\pi}$$

$$= - \frac{\hat{v}}{2.\pi.\left(n + \frac{1}{p}\right)} \left\{ \cos\left[\left(n + \frac{1}{p}\right) . 2.\pi + \left(\alpha - \frac{\pi}{p}\right)\right] - \cos\left(\alpha - \frac{\pi}{p}\right) \right\}$$

$$+ \frac{\hat{v}}{2.\pi.\left(\frac{1}{p} - n\right)} \left\{ \cos\left[\left(\frac{1}{p} - n\right) . 2.\pi + \left(\alpha - \frac{\pi}{p}\right)\right] - \cos\left(\alpha - \frac{\pi}{p}\right) \right\}$$

Since n is an integer, we get:

$$\cos\left[\left(n + \frac{1}{p}\right) . 2.\pi + \left(\alpha - \frac{\pi}{p}\right)\right] = \cos\left(\alpha + \frac{\pi}{p}\right)$$

and: $\qquad \cos\left[\left(\frac{1}{p} - n\right) . 2.\pi + \left(\alpha - \frac{\pi}{p}\right)\right] = \cos\left(\alpha + \frac{\pi}{p}\right)$

So that:

$$A_n = \frac{-\hat{v}}{2.\pi.\left(n + \frac{1}{p}\right)} \cdot \left[\cos\left(\alpha + \frac{\pi}{p}\right) - \cos\left(\alpha - \frac{\pi}{p}\right)\right]$$

$$+ \frac{\hat{v}}{2.\pi.\left(\frac{1}{p} - n\right).} \cdot \left[\cos\left(\alpha + \frac{p}{\pi}\right) - \cos\left(\alpha - \frac{\pi}{p}\right)\right]$$

$$A_n = \frac{\hat{v}}{2.\pi}\left[\frac{1}{\left(n + \frac{1}{p}\right)} + \frac{1}{\left(n - \frac{1}{p}\right)}\right] \cdot \left[\cos\left(\alpha - \frac{\pi}{p}\right) - \cos\left(\alpha + \frac{\pi}{p}\right)\right]$$

$$= \frac{\hat{v}}{2.\pi} \frac{\left(n - \frac{1}{p}\right) + \left(n + \frac{1}{p}\right)}{\left(n + \frac{1}{p}\right).\left(n - \frac{1}{p}\right)} \cdot \left(2.\sin\alpha.\sin\frac{\pi}{p}\right) = \hat{v} \cdot \frac{2.n.p}{\left(n^2.p^2 - 1\right)} \cdot \frac{p}{\pi} \cdot \sin\frac{\pi}{p} \cdot \sin\alpha$$

Replacing (8-25) in this expression leads to: $A_n = V_{di} \cdot \dfrac{2.n.p}{\left(n^2.p^2 - 1\right)} \cdot \sin\alpha$

● Cosine components B_n :

$$B_n = \frac{\hat{v}}{\pi} \int_{o}^{2.\pi} \cos\left(\frac{x}{p} + \alpha - \frac{\pi}{p}\right) . \cos(n.x) . dx$$

A calculation similar to the A_n - term gives: $B_n = - V_{di} \cdot \dfrac{2}{\left(n^2.p^2 - 1\right)} \cdot \cos\alpha$

We set $n.p = k$, whereby:

$p =$ pulse number of the circuit

$n = $ **rank of the harmonic** (1, 2, 3, ...)

$k = $ **order of the harmonic** = ratio between harmonic and supply frequency: $f_k = k . f_{grid}$

In relation to the fundamental harmonic of the output voltage of the rectifier, we have harmonics of rank n (1, 2, 3...) but in relation to the supply frequency there are only harmonics of order k present. With an M_3 controller in addition to the fundamental harmonic of 150Hz the following harmonics of 300 Hz, 450Hz, 600Hz,...will be present and with a B_6 controller we find in addition to the fundamental harmonic of 300Hz also 600Hz, 900 Hz, 1200 Hz, …components in the output signal.

As far as the output signal is concerned we only have to consider the sine and cosine components of the order $k = n.p$ in the Fourier series.

The amplitudes of the sine and cosine terms are then:

$$A_{k\alpha} = V_{di} \cdot \frac{2.k}{k^2 - 1} \cdot \sin\alpha \quad \text{and:} \quad B_{k\alpha} = - V_{di} \cdot \frac{2}{k^2 - 1} \cdot \cos\alpha$$

Since the amplitude of the harmonics is also a function of the firing angle α we have added the subscript α so that instead of A_k and B_k we write $A_{k\alpha}$ and $B_{k\alpha}$.

The RMS value of the k-th harmonic is : $\quad V_{k\alpha} = \dfrac{\sqrt{A_{k\alpha}^2 + B_{k\alpha}^2}}{\sqrt{2}}$.

$$V_{k\alpha} = \frac{V_{di}}{\sqrt{2}} \sqrt{\frac{4 \cdot k^2 \cdot \sin^2 \alpha}{(k^2 - 1)^2} + \frac{4 \cdot \cos^2 \alpha}{(k^2 - 1)^2}} = V_{di} \cdot \frac{\sqrt{2}}{(k^2 - 1)} \sqrt{k^2 \cdot \sin^2 \alpha + \cos^2 \alpha}$$

$$= \frac{V_{di} \cdot \sqrt{2}}{(k^2 - 1)} \cdot \sqrt{k^2 \cdot (1 - \cos^2 \alpha) + \cos^2 \alpha}$$

$$\frac{V_{k\alpha}}{V_{di}} = \frac{\sqrt{2}}{(k^2 - 1)} \cdot \sqrt{k^2 - (k^2 - 1) \cdot \cos^2 \alpha} \tag{8-26}$$

If the load is a DC motor these harmonics will give rise to:
- eddy current losses in the metal of the machine
- corresponding Joule losses in the armature conductors
- deterioration of the commutation of the machine.

These extra losses may be reduced by including an inductive coil (choke) in the DC side or by removing the harmonics. This last point can be accomplished for example by using a twelve pulse rectifier (see further). With the help of expression (8-26) it is possible to calculate the percentage of the RMS value of the harmonics in relation to the maximum DC voltage for different firing angles, see table 8-4.

- $p = 2$ (single phase bridge)
- $p = 3$ (half wave controlled three-phase bridge)
- $p = 6$ (full controlled three-phase bridge).

Since $n = 1, 2, 3...$ we find for $k\ (= n.p)$ only multiples of 2, 3 and 6.
In addition we specify that $f_{grid} = 50$Hz.

Table 8-4

harmonics		$V_{k\alpha} / V_{di}$ in %			
order (k)	frequency (Hz)	$\alpha = 0°$	$\alpha = 30°$	$\alpha = 60°$	$\alpha = 90°$
2	100	47.14	62.36	84.98	94.28
3	150	17.50	30.62	46.77	53
4	200	9.42	20.54	32.99	37.71
6	300	4.04	12.61	21	24.24
8	400	2.24	9.18	15.59	17.96
9	450	1.76	8.1	13.80	15.90
10	500	1.43	7.25	12.39	14.28
12	600	0.98	5.99	10.29	11.86
14	700	0.72	5.11	8.8	10.15
15	750	0.63	4.76	8.2	9.47

9.4 Current harmonics on the AC-side

By loading the supply with a controlled rectifier little changes as far as the sinusoidal supply voltage is concerned if this is considered a strong net. The current drawn by the controller is however not sinusoidal, which results in current harmonics. Together with the net impedance and its inductive, capacitive and resistive components this current harmonics result in the superposition of unwanted voltage forms on the sinusoidal supply voltage. Resonant effects may even occur.
We consider the case with continuous current, which occurs when the controller load is sufficiently inductive. With a full controlled three phase bridge (B_6 controller) we may obtain line currents on the input as shown in fig. 8-35.

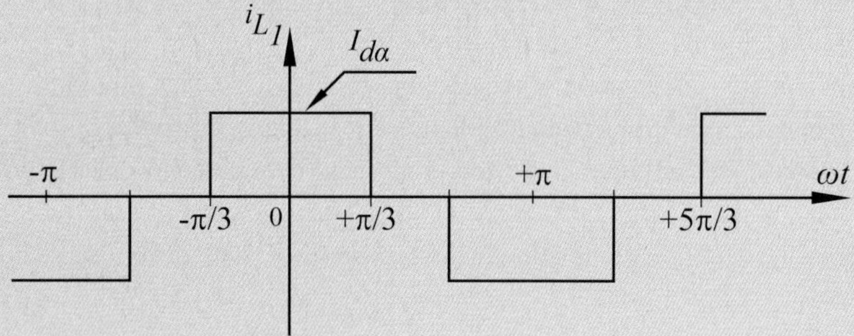

Fig. 8-35: Line currents of a full controlled three phase bridge

If we decompose the current form into a Fourier series, then it is clear that only cosine terms are present since:
1) the function is symmetrical in relation to the ordinate-axis, there are therefore no sine terms
2) the function is reversible per half period in relation to the t-axis resulting in odd numbered harmonics.

$$B_n = \frac{2}{\pi} \int_0^{\pi} f(t) \cdot \cos n\omega t \cdot d\omega t \qquad \text{Herein is } f(t) \quad \left| \begin{array}{l} = \; I_{d\alpha} \text{ between } 0 \text{ and } \pi/3 \\[2mm] = \; 0 \text{ between } \pi/3 \text{ and } 2.\pi/3 \\[2mm] = \; -I_{d\alpha} \text{ between } 2.\pi/3 \text{ and } \pi \end{array} \right.$$

$$B_n = \frac{2 \cdot I_{d\alpha}}{\pi} \int_0^{\pi/3} \cos n\omega t \cdot d\omega t \; - \; \frac{2 \cdot I_{d\alpha}}{\pi} \int_{2.\pi/3}^{\pi} \cos n\omega t \cdot d\omega t$$

$$B_n = \frac{2 \cdot I_{d\alpha}}{\pi \cdot n} \left(\sin n\omega t \right) \Big|_0^{\pi/3} - \frac{2 \cdot I_{d\alpha}}{\pi \cdot n} \left(\sin n\omega t \right) \Big|_{2.\pi/3}^{\pi} = \frac{2 \cdot I_{d\alpha}}{\pi \cdot n} \left(\sin n\frac{\pi}{3} + \sin n\frac{2 \cdot \pi}{3} \right)$$

The expression between brackets is zero for all values of n which are multiples of 2 and 3.
Except for the fundamental harmonic (with the supply frequency) we only have harmonics of the following order $k = 6 \cdot n \pm 1$, with $n = 1, 2, 3\ldots$
In general on the input side of a rectifier in addition to the fundamental harmonic the only harmonics present will be of order:

$$k = n \cdot p \pm 1 \tag{8-27}$$

In the expression $B_n = \dfrac{2 \cdot I_{d\alpha}}{\pi \cdot n} \left(\sin n \cdot \dfrac{\pi}{3} + \sin n \cdot \dfrac{2 \cdot \pi}{3} \right)$, the term between brackets

becomes $+ 2 \cdot \sin \dfrac{\pi}{3} = \sqrt{3}$ or $- 2 \cdot \sin \dfrac{\pi}{3} = - \sqrt{3}$ for the existing harmonics:

$$B_k = \pm \dfrac{2 \cdot \sqrt{3} \cdot I_{d\alpha}}{\pi \cdot k} \longrightarrow \left|\begin{array}{l} + \text{ for } k = 1, 7, 13, ... \\ - \text{ for } k = 5, 11, 17, ... \end{array}\right.$$

The harmonic analysis of the line current then becomes:

$$i(t) = \dfrac{2 \cdot \sqrt{3}\ I_{d\alpha}}{\pi} \left(\cos \omega t - \dfrac{\cos 5\ \omega t}{5} + \dfrac{\cos 7\ \omega t}{7} - \dfrac{\cos 11\ \omega t}{11} + \dfrac{\cos 13\ \omega t}{13} ... \right) \qquad (8\text{-}28)$$

The RMS value of the k-th harmonic compared to the RMS value of the fundamental harmonic and independent of the firing angle is:

$$I_k = \dfrac{1}{k} \cdot I_1 \qquad (8\text{-}29)$$

From (8-28) it follows that the amplitude of the fundamental harmonic: $\hat{i}_1 = \dfrac{2 \cdot \sqrt{3}}{\pi} \cdot I_{d\alpha}$

The RMS-value is then: $I_1 = \dfrac{2 \cdot \sqrt{3}}{\sqrt{2} \cdot \pi} \cdot I_{d\alpha} = \dfrac{\sqrt{6}}{\pi} \cdot I_{d\alpha}$

$$I_1 = \dfrac{\sqrt{6}}{\pi} \cdot I_d \cdot \cos \alpha \qquad (8\text{-}30)$$

The maximum RMS value of the fundamental harmonic of the line currents occurs when $\alpha = 0°$ · $I_{1\,max} = \dfrac{\sqrt{6}}{\pi} \cdot I_d$. Fig. 8-36 shows the frequency spectrum of the line current of fig. 8-35, whereby the relative amplitude with respect to the fundamental harmonic is determined.

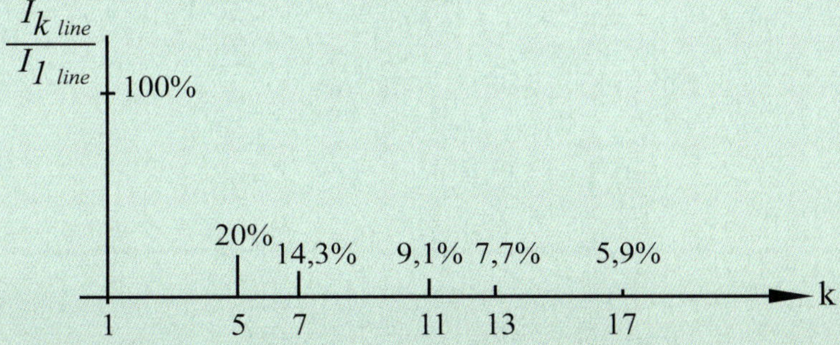

Fig. 8-36: Frequency spectrum of the current waveform in fig. 8-35

On p. 8.50 we found that on the DC side we only have voltage harmonics of the order $k = n.p$. Now we note that on the AC side we only have current harmonics of the order $n.p \pm 1$. In other words for every voltage harmonic of the order k in the output of the controller there are two current harmonics in the input with respective orders $(k + 1)$ and $(k - 1)$. Table 8-5 (p. 8.55) shows the ratio of the harmonic current strength with respect to the ground harmonic on the AC side for a few pulse numbers p.

Remarks

1. **Multi-pulse circuits**
   From table 8-5 it appears that a three pulse circuit has relatively large second and fourth harmonic supply side currents. That is the reason that for large loads a six pulse circuit is used instead of a three pulse. With a six pulse circuit the largest harmonics (in amplitude!) are the fifth and seventh. A twelve pulse circuit is an interesting option for even larger loads since then the fifth and seventh harmonic will no longer appear in the supply side current.

2. **Commutation**
   An assumed square shaped current waveform in fig. 8-35 will in practice have a form as in fig. 8-37 as a result of commutation time (this is not infinitely fast). The amplitudes of the current harmonics will in reality be lower that the theoretically calculated values.

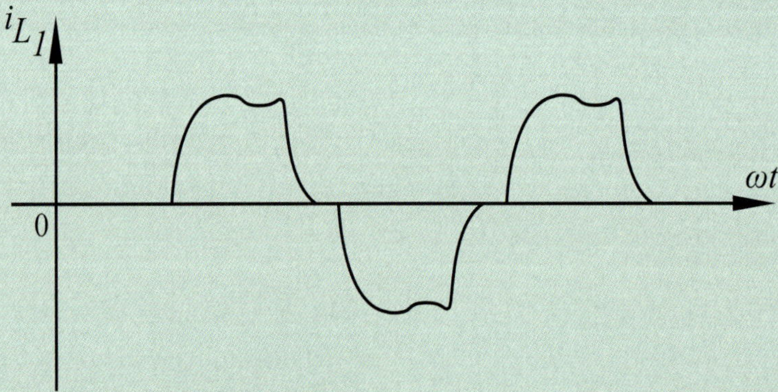

Fig. 8-37: Line currents full controlled bridge.

Table 8-5 Currents harmonics on the AC side

$p \rightarrow\rightarrow$	3	6	12
k	I_k/I_1 (%)		
2	50.00	------	------
3	------	------	------
4	25.00	------	------
5	20.00	20.00	------
6	------	------	------
7	14.28	14.28	------
8	12.50	------	------
9	------	------	------
10	10.00	------	------
11	9.09	9.09	9.09
12	------	------	------
13	7.69	7.69	7.69
14	7.14	------	------
15	------	------	------
16	6.25	------	------
17	5.88	5.88	------
18	------	------	------
19	5.26	5.26	------
20	5.00	------	------
21	------	------	------
22	4.55	------	------
23	4.35	4.35	4.35
24	------	------	------
25	4.00	4.00	4.00

10. TWELVE PULSE CONTROLLERS

By connecting two six pulse controllers together in series or in parallel when the AC supplies are 30° out of phase with respect to each other we end up with a twelve pulse controller.

The 30° phase displacement is easy to realise by connecting the primary of one supply transformer in star and the other supply transformer in delta (fig. 8-38).

In the case of two bridges (fig. 8-39) we could for example also connect the secondaries of a single transformer one in star and the other in delta. On the AC side we will have harmonics of the order $k = p.n \pm 1$.

With $p = 12$ and $n = 1, 2, 3$…this gives : $k = 11, 13, 23, 25$…

The fifth and seventh harmonics of the supply frequency that were present in the six pulse configuration no longer occur in a twelve pulse circuit.

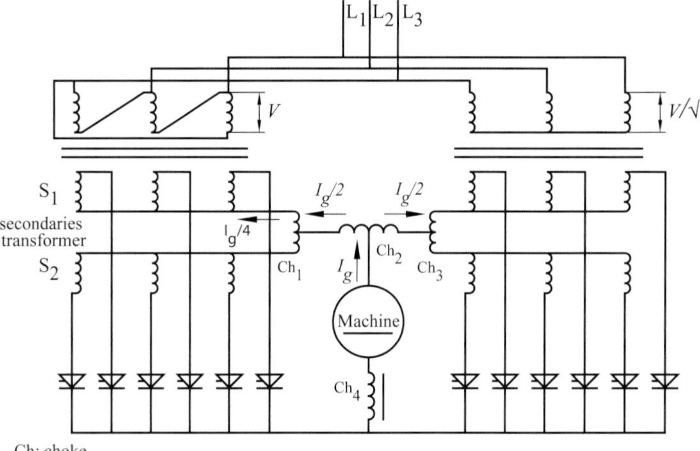

Fig. 8-38: Twelve pulse circuit created by connecting two six pulse controllers in parallel.

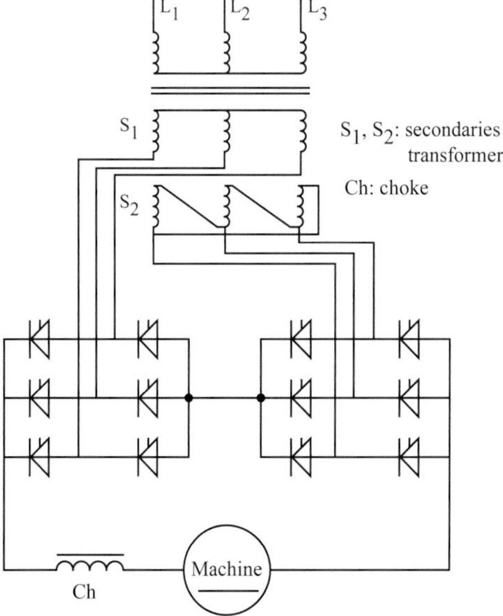

Fig. 8-39: Two AC bridge controllers connected in series supplied with voltage with a 30° phase displacement also creates a twelve pulse circuit.

11. EVALUATION

8.1   In which specific situation does an SCR require a double pulse? Why?

8.2   What are the disadvantages of using phase control with thyristors?

8.3   For an SKT600 (Semikron) we assume that the maximum ambient temperature is 50°C and that $R_{th\,j\text{-}a}$ = 0.11°C/W. What is the maximum permissible $I_{T(AV)}$? What is then the power dissipation in the thyristor? The SKT600 data can be found on p. 4.15 and p. 4.24/25.

8.4   Why are some power electronic circuits called natural commutating circuits?

8.5   What is the maximum RMS anode voltage that is allowable with an SKT10/08D configured as a controlled rectifier (E_1 -controller)?

8.6   In the configuration shown in fig. 8-10 V_{line} = 400V.What is the average rectified output voltage when α = 135° and for α = 100°.

8.7   What is meant by "miss-firing" of an inverter? How can this be prevented?

8.8   What does the expression "distortion power" means?

8.9   What is the maximum power factor of a full controlled bridge?

8.10 In fig. 8-20 the following information is provided: R_b = 10Ω , L_b = 150mH , transformer and thyristor are ideal, line voltage 230V - 50Hz. Determine with a firing angle of α = 0° and 60°: the output DC voltage; output current, power factor, displacement factor, distortion factor and reactive power.

8.11 Which specifications should the thyristors in fig. 8-20 have given the information in the previous question.

8.12 Determine the maximum inverse voltage of the semiconductors in the situations shown in fig. 8-19 and 8-21.

8.13 Determine the DC output voltage for the circuit shown in fig. 8-9 if the supply voltage is 230V - 50Hz and the firing angle is respectively 60° and 105°.

8.14 In fig. 8-3 the supply voltage is 230V - 50Hz and the load is comprised of 7.255Ω and 40mH. Determine the average output voltage if the firing angle is 60°. In addition determine the specifications the thyristor needs to comply with.

8.15 The full controlled B_2 -controller of fig. 8-9 is connected to a 230V - 50Hz supply. The load is comprised of 7.255Ω and 40mH. Calculate the average output voltage if the firing angle is respectively 30°, 60°, 90°.

8.16 Determine the largest inverse voltage across the thyristor in fig. 8-1. What is the largest possible blocking voltage across the SCR? For which value of α does this occur?

8.17 Set α = 120° in the configuration of fig. 8-20. Can the bridge be brought into conduction? Confirm this with a graphic such as fig. 8-21 or 8-23.

8.18 Consider the speed control of an independently excited 100kW motor. The current protection is set at 1.5x de nominal armature current. How much reactive power (kvars) does this motor draw from the net at start-up?

8.19 We consider a half wave and full wave controlled rectifier, both with resistive load. Let α = 30° and 150° respectively. Determine for the rectifiers the output power with respect to the maximum possible output power (when α = 0°). Use table 8-1.

8.20 Determine for the data used in question 8.10 with $\alpha = 60°$: the RMS value of the fundamental harmonic of the input line current of the bridge, RMS values of the fifth and seventh harmonic current components of the line current, RMS values of the 300Hz- and 600Hz- voltage components of the output.

8.21 An E_1 -controller connected to a 230V - 50Hz has a resistive load of $R_b = 100\Omega$. Calculate the amplitude of the 150Hz current in the load when the firing angle is 60° and 90° respectively.

8.22 A full controlled B_6 -controller is connected to a three phase supply of 3x400V - 50Hz with a highly inductive load of 5Ω - 280mH. Calculate the RMS value of the 350Hz current in the supply line when the firing angle is 30°.

8.23 Sketch the output voltage waveform with a firing angle α for a highly inductive load in the following situations:
a) full wave full controlled single phase bridge
b) half wave full controlled three phase bridge (M_3, see fig. 8-40 lower down).
Sketch in a similar fashion to fig. 8-34, a periodic voltage $v_o = f(x)$ and derive an expression for $\omega t = f(x)$.

8-24 Determine the maximum output voltages for a B_2- controller, M_3- and B_6 -controller. The starting point is expression (8-25).

8.25 Given a full control highly inductive loaded B_6 -controller. Calculate the ratio of the RMS values of the harmonics in the output when $\alpha = 90°$ in relation to $\alpha = 0°$. Check your answer with the values from table 8-4.

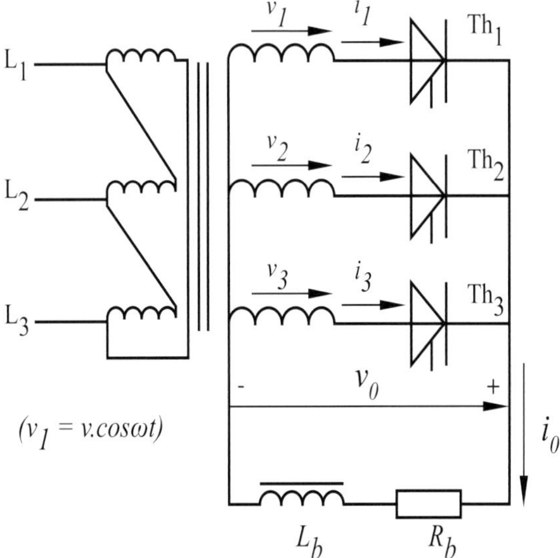

Fig. 8-40: M_3 -controller (half wave full controlled three phase bridge)

9 AC-CONTROLLERS

CONTENTS

1. AC-controller with phase control
2. AC-controller with integral cycle control.
3. Turn-off snubber for thyristors
4. Solid state relays (SSR)
5. Radio interference suppression (RFI) of thyristors
6. Evaluation

With the term AC-controller we mean an AC-AC converter which does not introduce a frequency change between input and output. The standard semiconductor used is the triac, or for larger power levels anti-parallel SCR's. Since the switches are predominantly thyristors they switch off due to the influence of the supply voltage. AC-controllers therefore belong to the family of net or natural commutation circuits just like controlled rectifiers. In terms of absolute numbers AC controllers are perhaps the most wide spread thyristor applications.

Applications include:

. electronic lighting control;

. motor control of domestic appliances;

. temperature control of ovens and water heaters;

. AC welding equipment, etc…

A special application of the AC-controller is the controlled switch. In this case the current is not regulated but the triacs and anti parallel thyristors fulfil the same role as a (mechanical) switch. An example of this is a solid state relay.

1. AC-CONTROLLER WITH PHASE CONTROL

1.1 Single phase controller with resistive load

1.1.1 Single phase AC-controller using a triac

The triac may be considered is as bi-directional switch since the current can flow in both directions. By applying a gate pulse v_c the triac conducts as shown in fig. 9.1. It switches off when below the holding current I_H.

For a resistive load this coincides with the zero crossover of the supply voltage.

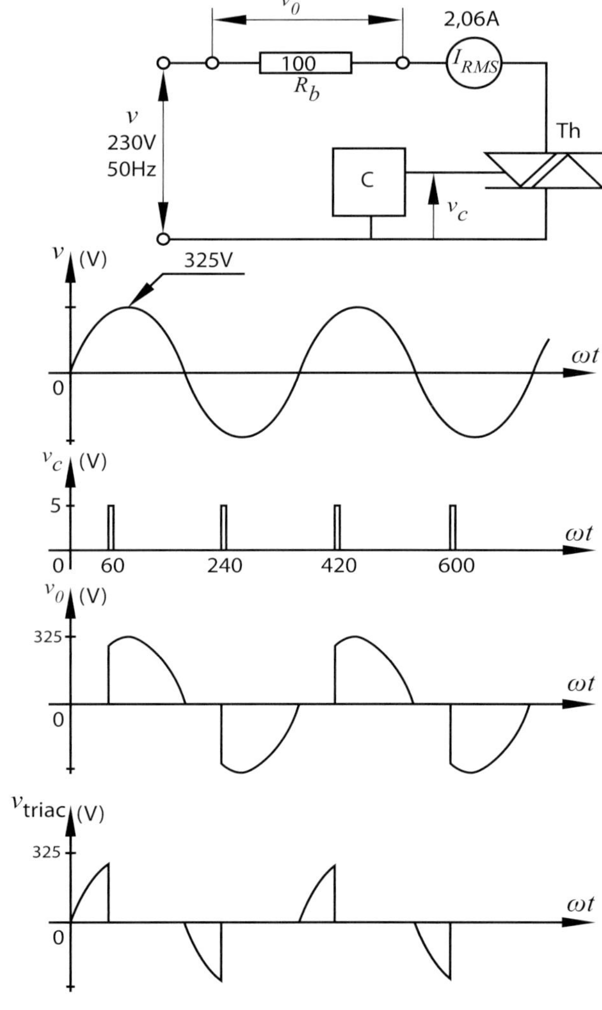

Fig. 9-1: Triac circuit with phase control

When the control pulses have a firing angle of $\alpha = 60°$ then the triac conducts from $\omega t = 60°$ / $240°$ / $420°$ / ..., as shown in fig. 9-1. When $\alpha = 0°$ practically the full supply voltage is across the load, with the exception of the voltage drop across the triac of 1.5V.

In fig. 9-1 we can see that by varying α we vary v_o and also the current through the triac and load is varied from zero to maximum. Calculation of the RMS value of the voltage across the load gives:

$$V_{RMS} = \frac{\hat{v}}{\sqrt{2}} \sqrt{\frac{180° - \alpha}{180°} + \frac{\sin 2\alpha}{2.\pi}} \qquad (9\text{-}1)$$

Here in:

V_{RMS} = RMS voltage across load

$\hat{v}$ = amplitude of the supply voltage

α = firing angle (degrees)

The voltage drop of about 1.5V across the triac is neglected. In table 8-1 the RMS value of the load voltage has been calculated for a range of α with a supply voltage of 1V (RMS).

1.1.2 Determining the RMS output voltage

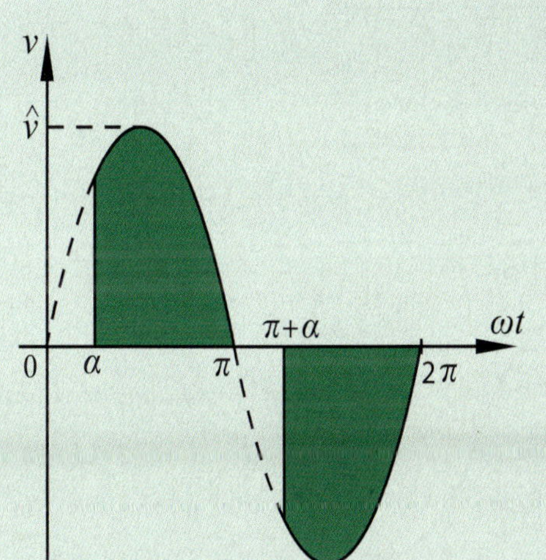

Fig. 9-2: Output voltage of an AC-controller

$$V_{RMS} = \sqrt{2 \cdot \left[\frac{1}{2.\pi} \int_{\alpha}^{\pi} v^2 . d\omega t \right]}$$

Evaluating the integral:

$$\int_{\alpha}^{\pi} v^2 . d\omega t = \int_{\alpha}^{\pi} \hat{v}^2 . \sin^2 \omega t . d\omega t$$

$$= \frac{\hat{v}^2}{2} \int_{\alpha}^{\pi} (1 - \cos 2\omega t) . d\omega t$$

$$- \frac{\hat{v}^2}{2} . (\pi - \alpha) + \frac{\hat{v}^2}{4} . \sin 2.\alpha$$

so that: $V_{RMS} = \frac{\hat{v}}{\sqrt{2}} . \sqrt{\frac{(180° - \alpha)}{180°} + \frac{\sin 2\alpha}{2\pi}} \qquad (9\text{-}1)$

Numeric example 9-1:

Given: fig. 9-1. Firing angle 30°, 60°, 150°.

Required: RMS current through the triac for the three firing delays.

Solution:

a)　α = 30° :　● (9-1) →→→ $V_{RMS} = \frac{230 . \sqrt{2}}{\sqrt{2}} . \sqrt{\frac{150°}{180°} + \frac{\sin 60°}{2.\pi}} = 226.67$ V

$$I_{RMS} = \frac{V_{RMS}}{R_b} = \frac{226.67}{100} = 2.27 \text{ A}$$

　　　● table 8-1:　$V_{RMS} = 0.9855 \times 230 = 226.66$ V ;　$I_{RMS} = 2.27$ A

b)　α = 60° :　● (9-1) →→→ $V_{RMS} = 206.29$ V →→→ $I_{RMS} = \frac{V_{RMS}}{R_b} = 2.06$ A

c)　α = 150° : ● (9-1) →→→ $V_{RMS} = 39$ V →→→ $I_{RMS} = 0.39$ A

1.1.3 Single phase AC-controller using SCR's

Triacs are constructed to handle a maximum current of about 100A. To control larger AC currents we can use two SCR's connected anti parallel as shown in fig. 9-3.

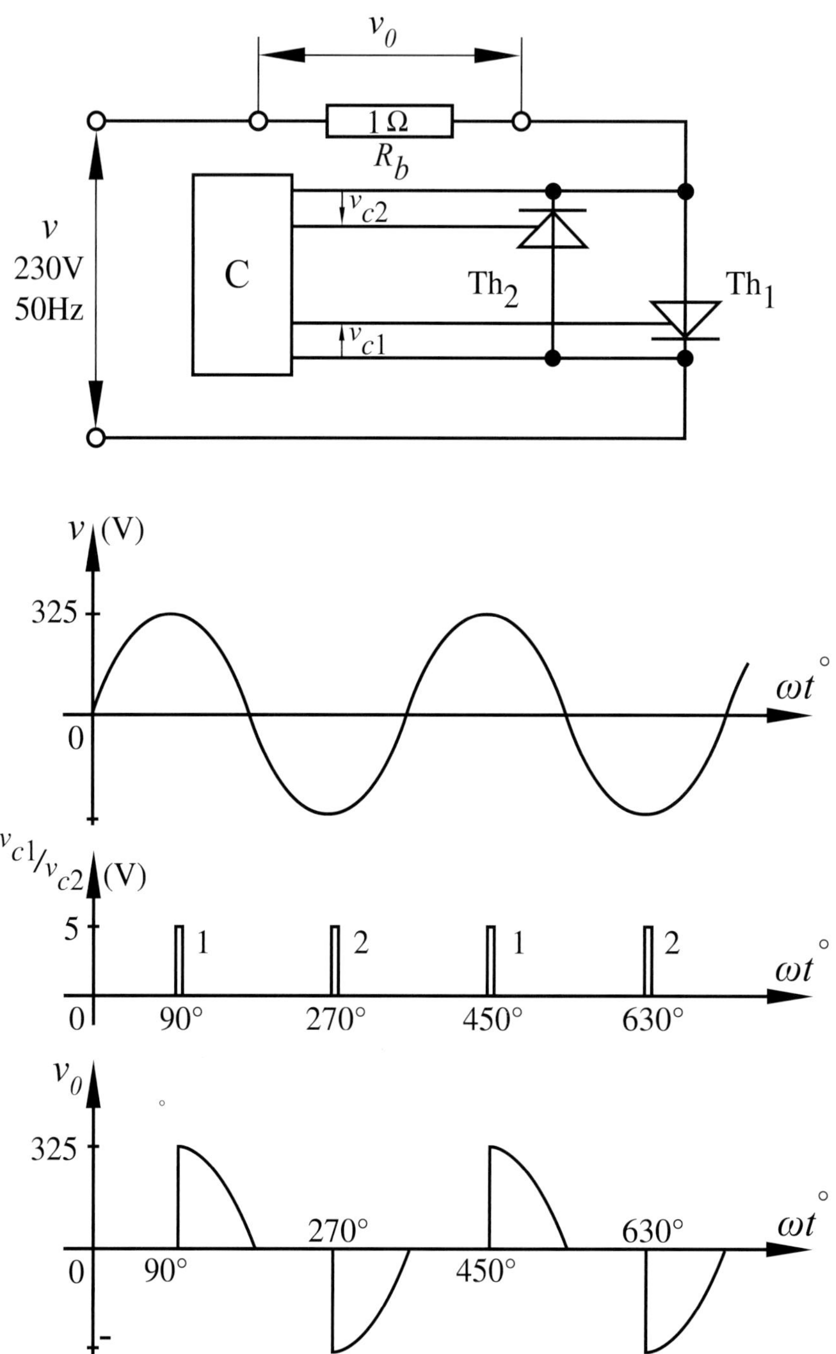

Fig. 9-3: Anti parallel configuration of SCR's.

The SK120KQ (see p. 9.6) is just such an anti-parallel configuration of SCR's. We recognise the data as for an SCR. Obviously $I_{T(AV)}$ is not provided since we are not using DC current in the power circuit but rather AC current, which explains I_{RMS} and I_{TSM} for the current under study.

Numeric example 9-2:

Required:

1)   Calculate the current strength in fig. 9-3 using expression (9-1).
2)   Determine $I_{eff.}$ in fig. 9-3 using table 8-1 on p. 8-10.
3)   What is the maximum reverse voltage across one SCR in the configuration of fig. 9-3 when $\alpha = 90°$?
4)   Determine the average and RMS current through one SCR in fig. 9-3.

Solution:

1)   (9-1) $\rightarrow\rightarrow V_{RMS} = \dfrac{230 . \sqrt{2}}{\sqrt{2}} . \sqrt{\dfrac{180° - 90°}{180°} + \dfrac{\sin 180°}{2.\pi}} = 162.63\,\text{V}$

$\qquad\rightarrow\rightarrow I_{RMS} = \dfrac{V_{RMS}}{R_b} = 162.63\,\text{A}$

2)   Table (8-1) $\rightarrow\rightarrow V_{RMS} = 0.7071 \times 230 = 162.63\ \text{V} \rightarrow\rightarrow I_{RMS} = 162.63\ \text{A}$

3)   If neither thyristor is conducting the supply voltage is across the anode cathode connections. (for one SCR this is an inverse voltage and for the other this is a blocking voltage). In fig. 9-3 one of the thyristors fires at the instant ($\omega t = 90° / 270° / ...$) that the amplitude of the supply voltage is across the thyristor, therefore: v_{max} (inverse) $= 325\text{V}$.

4)   One SCR behaves as an E_1-contoller and therefore the formulas (8-2) and (8-3) may be used, or table 8-1 with $\alpha = 90°$ as is shown in fig. 9-3 .

$\qquad$ • $V_{dia} = 0.2251 \times 230 = 51.77\ \text{V}$

Table (8-1) $\rightarrow\rightarrow \alpha = 90°$ $\qquad$ • $I_{da} = \dfrac{V_{dia}}{R_b} = \dfrac{51.77}{1} = 51.77\ \text{A}$

$\qquad$ • $V_{RMS} = 0.5 \times 230 = 115\ \text{V}$

$\qquad$ • $I_{RMS} = \dfrac{V_{RMS}}{R_b} = \dfrac{115}{1} = 115\ \text{A}$

SK 120 KQ

SEMITOP® 2

Antiparallel Thyristor Module

SK 120 KQ

Preliminary Data

Features
- Compact Design
- One screw mounting
- Heat transfer and isolation through direct copper bonded aluminium oxide ceramic (DBC)
- Glass passived thyristor chips
- Up to 1600V reverse voltage
- UL recognized, file no. E 63 532

Typical Applications
- Soft starters
- Light control (studios, theaters...)
- Temperature control

V_{RSM} V	V_{RRM}, V_{DRM} V	I_{RMS} = 134 A (full conduction) (T_s = 85 °C)
900	800	SK 120 KQ 08
1300	1200	SK 120 KQ 12
1700	1600	SK 120 KQ 16

Symbol	Conditions	Values	Units
I_{RMS}	W1C ; sin. 180° ; T_s = 100°C	94	A
	W1C ; sin. 180° ; T_s = 85°C	134	A
I_{TSM}	T_{vj} = 25 °C ; 10 ms	2000	A
	T_{vj} = 125 °C ; 10 ms	1800	A
i^2t	T_{vj} = 25 °C ; 8,3...10 ms	20000	A²s
	T_{vj} = 125 °C ; 8,3...10 ms	16200	A²s
V_T	T_{vj} = 25 °C, I_T = 300 A	max. 1,85	V
$V_{T(TO)}$	T_{vj} = 125 °C	max. 0,9	V
r_T	T_{vj} = 125 °C	max. 3,5	mΩ
$I_{DD}; I_{RD}$	T_{vj} = 25 °C, $V_{RD}=V_{RRM}$	max. 1	mA
	T_{vj} = 125 °C, $V_{RD}=V_{RRM}$	max. 20	mA
t_{gd}	T_{vj} = 25 °C, I_G = 1 A; di_G/dt= 1 A/µs	1	µs
t_{gr}	V_D = 0,67 *V_{DRM}	2	µs
$(dv/dt)_{cr}$	T_{vj} = 125 °C	1000	V/µs
$(di/dt)_{cr}$	T_{vj} = 125 °C; f= 50...60 Hz	100	A/µs
t_q	T_{vj} = 125 °C; typ.	80	µs
I_H	T_{vj} = 25 °C; typ. / max.	100 / 200	mA
I_L	T_{vj} = 25 °C; R_G = 33 Ω; typ. / max.	200 / 500	mA
V_{GT}	T_{vj} = 25 °C; d.c.	min. 2	V
I_{GT}	T_{vj} = 25 °C; d.c.	min. 100	mA
V_{GD}	T_{vj} = 125 °C; d.c.	max. 0,25	V
I_{GD}	T_{vj} = 125 °C; d.c.	max. 5	mA
$R_{th(j-s)}$	cont. per thyristor	0,45	K/W
	sin 180° per thyristor	0,47	K/W
$R_{th(j-s)}$	cont. per W1C	0,225	K/W
	sin 180° per W1C	0,235	K/W
T_{vj}		-40 ... +125	°C
T_{stg}		-40 ... +125	°C
T_{solder}	terminals, 10s	260	°C
V_{isol}	a. c. 50 Hz; r.m.s.; 1 s / 1 min.	3000 / 2500	V~
M_s	Mounting torque to heatsink	2,0	Nm
M_t			Nm
a			m/s²
m		19	g
Case	SEMITOP® 2	T 2	

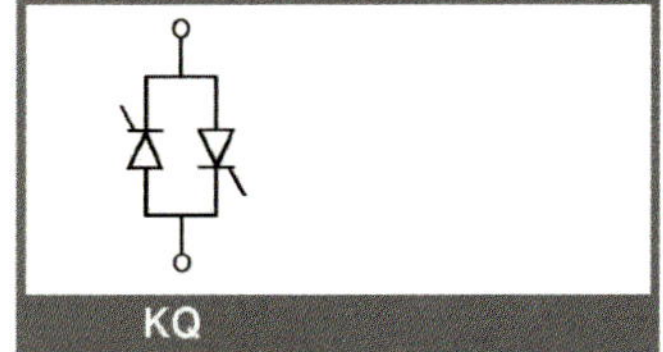

1.2 Resistive-inductive load of a single phase AC-controller

On p. 7.5/6 we determined the current flow through an RL-circuit which was connected at time t_1 to a sinusoidal voltage $v = \hat{v} . \sin \omega t$. We reproduce the result:

$$i(t) = \frac{\hat{v}}{Z} . \left[\sin (\omega t - \Phi) + \sin (\Phi - \alpha) . e^{(\alpha - \omega t).\cotg \Phi} \right] \qquad (9\text{-}2)$$

With: $Z = \sqrt{R_b^2 + \omega^2.L_b^2}$; $\Phi = bgtg \dfrac{\omega.L_b}{R_b}$; $\alpha = \omega.t_1$

The circuit behaviour depends upon the relationship between firing angle α and phase displacement Φ .

a)   First case: $\alpha = \Phi$

From (9-2): $i_o = \dfrac{\hat{v}}{Z} . \sin (\omega t - \Phi)$. This is the current which lags the voltage by the angle Φ. This is referred to as the steady state current. This is the current which flows (steady state) when the R_b - L_b load is connected to the supply **without** triac.

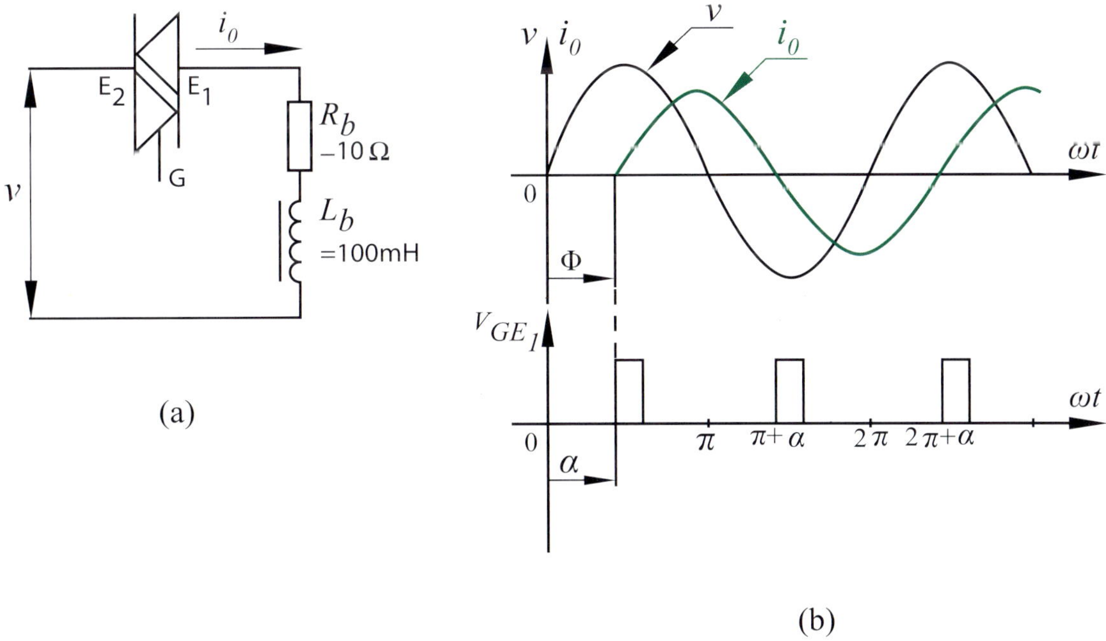

(a)

(b)

Fig. 9-4: Resistive–inductive loaded AC-controller

It is clear that with $\alpha = \Phi$ the AC-controller is not controlled, the triac operates as a mechanical switch. This is referred to as a **solid state switch** rather than an AC-controller.

b. Second situation: $\alpha < \Phi$

When we first switch on we obtain the situation as is shown in fig. 8-3, in which expression (7-1) is valid. From fig. 8-4 it follows that when $\alpha < \Phi$ the extinction angle is $\delta > \alpha + \pi$.

Notice in fig. 9-5 that control pulse number 2 has no effect on the triac since this is already conducting. After the extinction angle δ the triac is blocking and will switch on only after pulse 3. The result is a series of DC current pulses, in both the load and in the supply net.

Using a triac in the primary side of an unloaded transformer can cause problems. Indeed, in the case where Φ approaches 90° it is quite possible that $\alpha < \Phi$. If this transformer is constructed of high quality iron, therefore with a low no load current, then the DC current or start current can saturate the transformer. This results in **a very large DC current** in the transformer and triac!

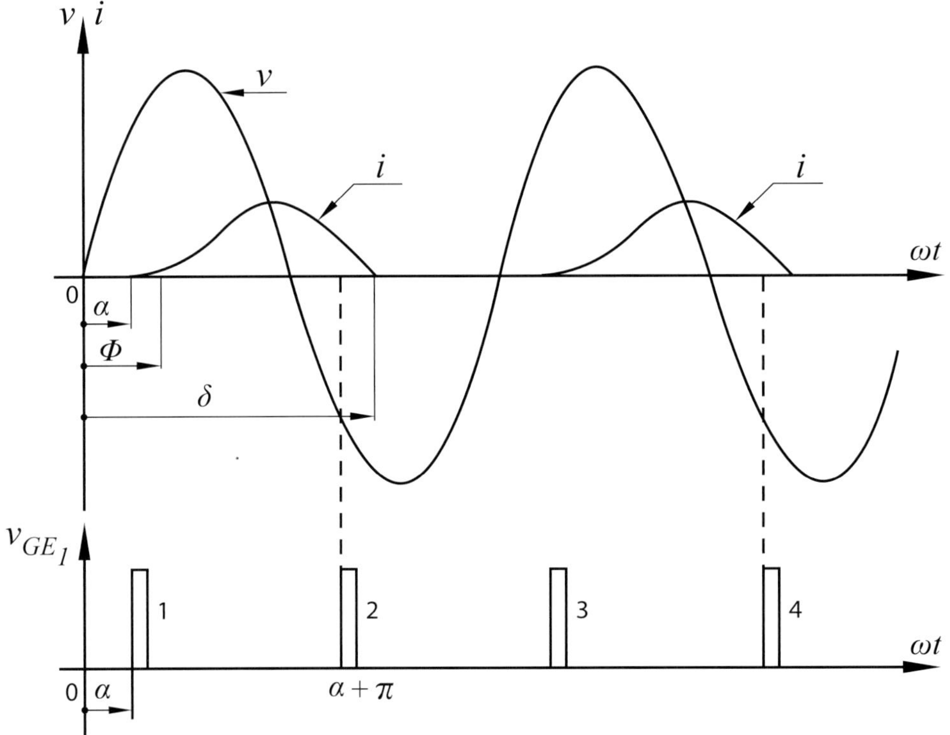

Fig. 9-5: DC current pulses when $\alpha < \Phi$

The DC current pulses in fig. 9-5 can be avoided by replacing the short control pulses with a longer pulse of width π-α (fig. 9-6). In this way there is always a control pulse present at the instant the triac switches off. The transient term will fade away and the steady state is reached after a few periods: $i = \frac{\hat{v}}{Z} \cdot \sin{(\omega t - \Phi)}$. As shown in fig. 9-6 the triac has assumed the role of a mechanical switch. The power in the load is exactly as in fig. 9-4, non controlled.

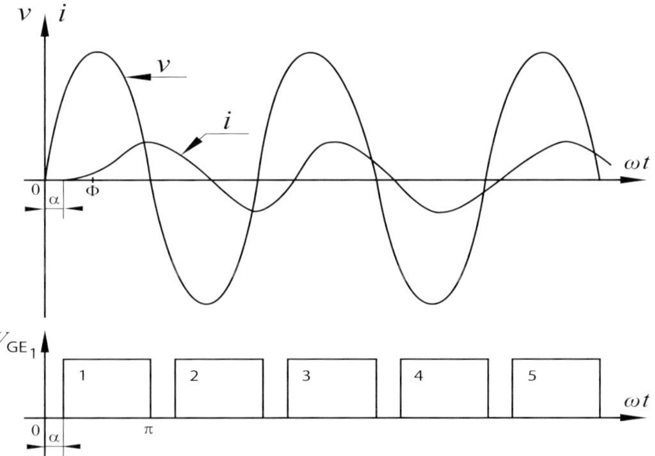

Fig. 9-6: Current flow when α < Φ and the width of the control pulse = π − α

c. Third situation: **α > Φ**

In this case the configuration of fig. 9-4a will function as an AC - controller. Pulse number 1 switches the triac on with delay α . During the interval α < ωt < δ (fig. 9-7) the current flow is described by expression (9-2). The current is zero before ωt = π + α so that at the second pulse the triac can switch on with a negative going current which also is described by expression (9-2).

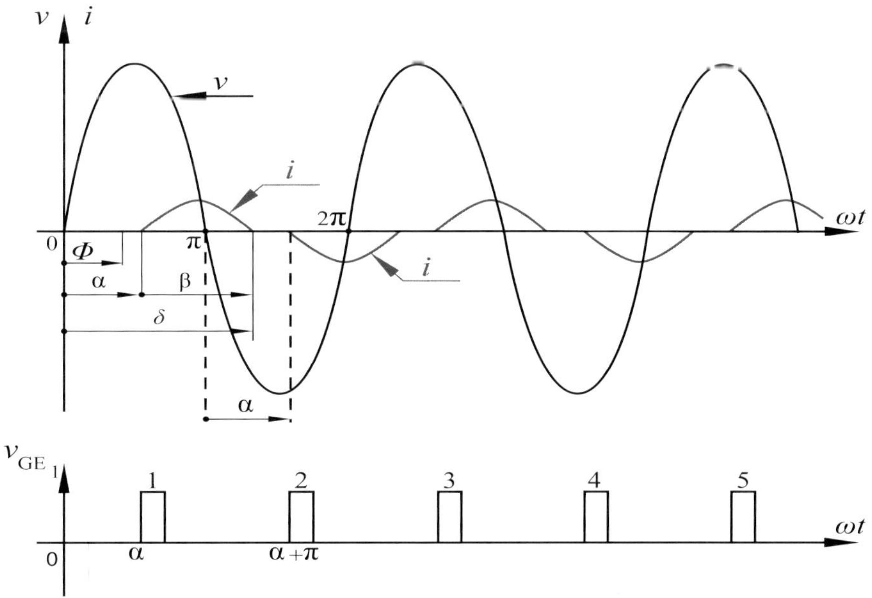

Fig. 9-7: Current waveform for α > Φ

We see in fig. 9-7 that by changing α we regulate the current through the load.
This is an AC-controller. An AC-controller is only possible when α > Φ .

1.3 Resistive-capacitive loaded single phase AC-controller

An electronic transformer has a capacitive input impedance. If such a transformer is operated via a dimmer switch, then the AC-controller has a capacitive load. Fig. 9-8a shows the setup of a resistive-capacitive loaded AC-controller. Consider continuous conduction of the circuit, in other words the AC-controller functions as a switch. In steady state the current through the switch is given by:

$$i = \frac{\hat{v}}{\sqrt{R^2 + \left(\frac{1}{\omega.C}\right)^2}} \cdot \sin\left(\omega t + \Phi\right). \quad \text{Here in is } \Phi = bgtg\,\frac{1}{\omega CR}$$

If we wish to switch on without transient behaviour present, then the capacitor should have zero energy at the moment of triggering, this is at $v_C = 0$.

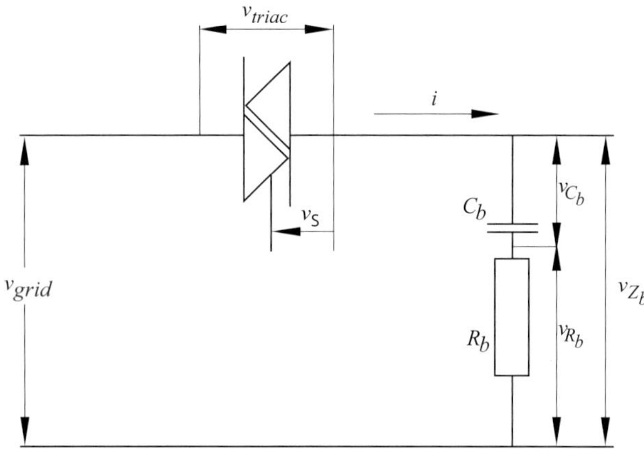

(a) Configuration

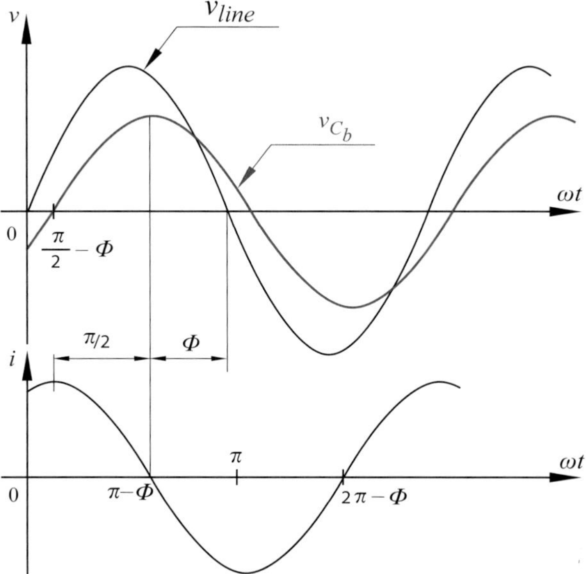

(b) Triac triggered without transient behaviour

(Fig. 9-8)

If the triac is triggered initially at $v_C = 0$ then we see in fig. 9-8b that the firing angle should be: $\alpha = \pi - (\Phi + \frac{\pi}{2}) = \frac{\pi}{2} - \Phi$. There after the triac triggers at the zero crossover of the current, that is at $(\pi - \Phi)$, $(2.\pi - \Phi)$, etc.

As was the case with the resistive inductive load we can develop an expression for the current flow (including the transient term) and then we will find that the AC-controller functions as a controller when $\alpha > \pi - \Phi$. Controlling the current is achieved with a firing angle between π and $(\pi - \Phi)$!

Fig. 9-8c shows that the firing pulses, for practical reasons, best occur from π.

In order to control between π and $\pi - \Phi$. This circuit is sometimes called **phase-drop off** rather than phase control (as was the case with resistive and resistive inductive loads).

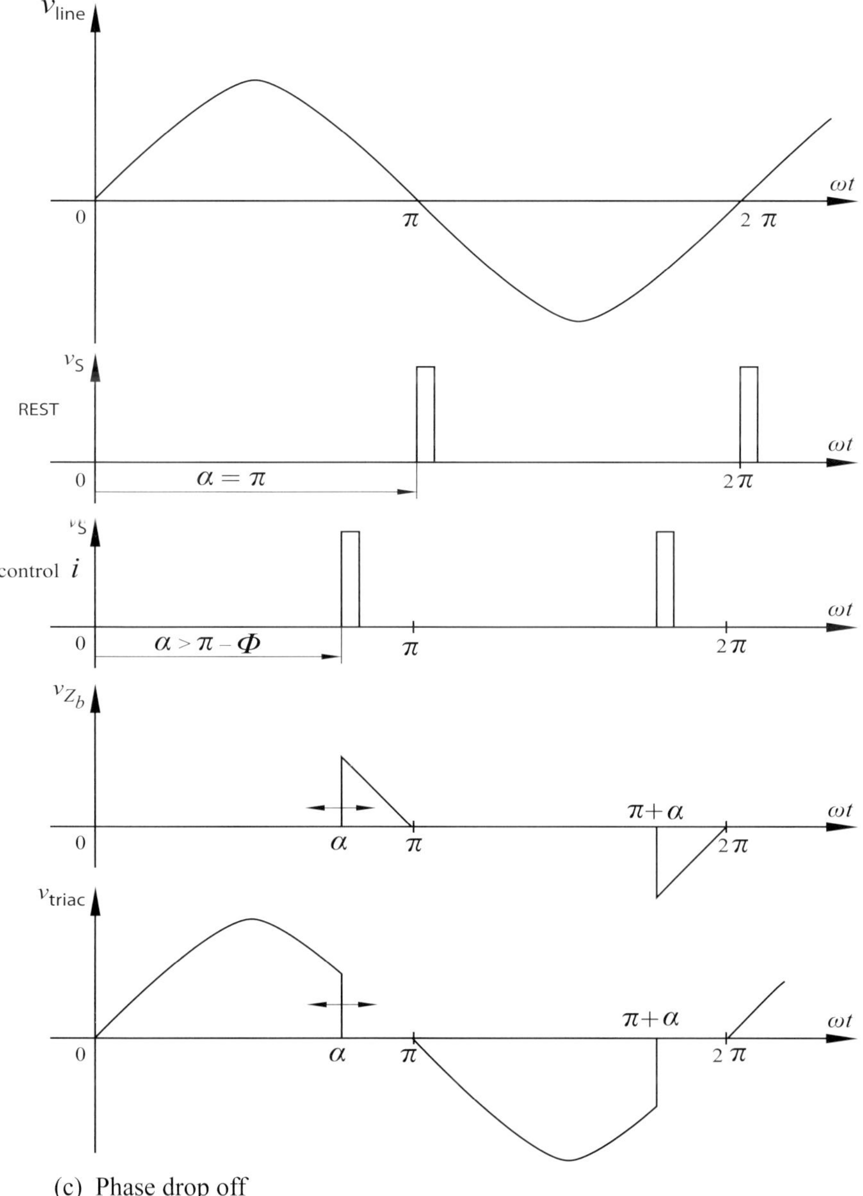

(c) Phase drop off

Fig. 9-8: Resistive-capacitive loaded AC-controller

1.4 Three-Phase AC-controllers

Fig. 9-9 shows how we can apply power control with a three phase consumer. Every triac is obviously interchangeable with a set of anti-parallel SCR's.

Without special precautions the circuit shown in fig. 9-9a will not always work since the triacs cannot switch on. They only have a voltage supply from one side! In order to work with this type of configuration some extra circuitry is required. If the three phases are available then a delta configuration as shown in fig. 9-9b will work without any problem.

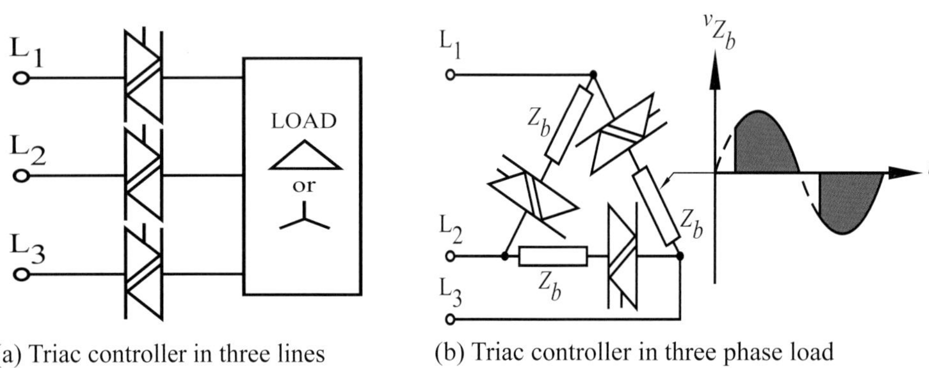

(a) Triac controller in three lines   (b) Triac controller in three phase load

Fig. 9-9: Three phase AC-controller

2. AC - CONTROLLER WITH INTEGRAL CYCLE CONTROL

2.1 Period control with a resistive load

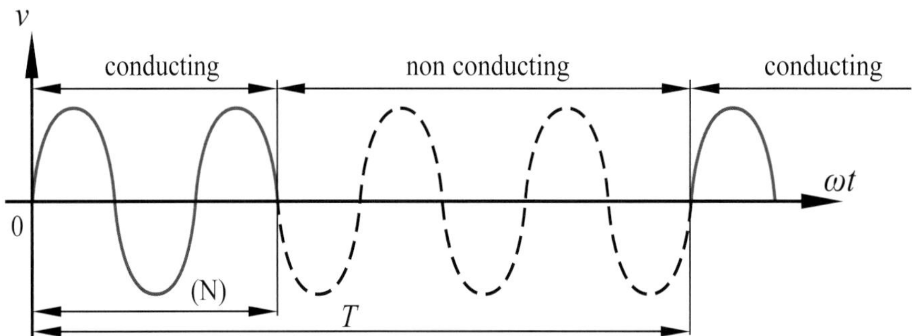

Fig. 9-10: Time ratio control

In a configuration as shown in fig. 9-1 (or 9-3) a triac (or two anti-parallel SCR's) is controlled in such a manner that there is conduction and non-conduction over a number of full half periods (fig. 9-10). This enables us to talk about period control or time ratio control (TRC).

Sometimes this is called "pulse burst modulation" or "burst firing".

Period control is an ON-OFF control strategy for an AC-controller.

We call T the period of a full ON-OFF cycle. This can be expressed in time (in fig. 9-10: $T = 80$ms from a 50Hz supply!). We could also set $T = 4$ in fig. 9-10: this is the number of complete supply voltage cycles in one ON-OFF cycle.

Suppose N = number of periods that the AC-controller is conducting , then the duty cycle is $\delta = \frac{N}{T}$. Calculation reveals that the RMS voltage across the load R_b is : $V_{R_b} = V \cdot \sqrt{\delta}$.   (9-3)

Here in is V the RMS supply voltage.

The power in R_b is found with : $P = \dfrac{V^2}{R_b} \cdot \delta$ (9-4)

The power factor λ is found with : $\lambda = \sqrt{\delta}$ (9-5)

Since the semiconductors effectively conduct from the zero crossover of the voltage this is referred to as **zero voltage switching**. Zero voltage switching occurs not only in control circuitry but is also implemented with contactors. These are simply ON-OFF circuits (see fig. 9-14 and 9-15).

2.2 Period control with resistive inductive load

Fig. 9-11 shows period control with an inductive load. At switch off ($i = I_H$) we see in fig. 9-11 that the triac has to deal with a large dv/dt. The triac needs to be rated for this large dv/dt. It may be necessary to implement a turn off snubber network (see p. 9.15).

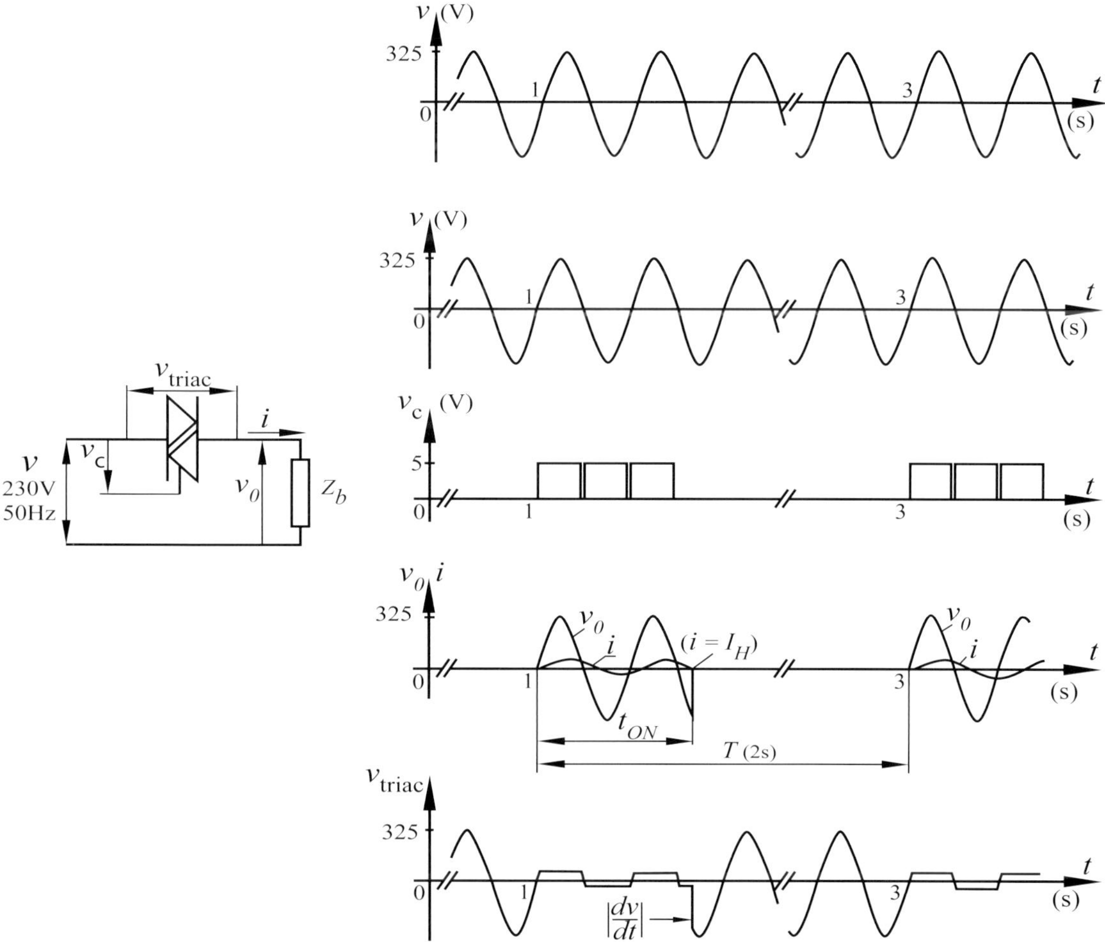

Fig. 9-11: Period control with an inductive load.

2.3 Period "Time ratio Control"

Table 9-2 indicates the standard as produced by Cenelec (European commission for electric standards).This standard indicates the repeat period (T) in which domestic temperature controllers may switch on and off.

Table 9-2

Power	Repeat period T (s)		
(W)	220 V	240 V	380 V
600	0.2	0.2	-----
800	0.8	0.3	0.1
1000	2.0	1.0	0.2
1200	4.6	2.0	0.2
1400	7.0	4.3	0.2
1600	10	6.3	0.3
1800	16	8.9	0.5
2000	24	13	0.9
2200	32	17	1.3
2400	40	24	1.9
2600	-----	31	2.6
2800	-----	-----	3.6

(Cenelec-standard for domestic temperature control indicating ON-OFF switching ratios)

It should be noted that T has to be small in relation to the time constant of the load or the system being controlled. Practical values for T are between 0.2 and 40s.

2.4 Advantages of period control

1. power factor is better than phase control.
2. by switching at the zero cross over EMI is kept to a minimum.

2.5 Applications

Temperature control is an interesting application for period control. It is not suitable for lighting applications since it causes flicker with a frequency of $1/T$. This is inconvenient since we are talking about a few Hertz and a flicker of 30 Hz or less is noticeable and irritating.

3.   TURN-OFF SNUBBER FOR THYRISTORS

RC-snubbers are often connected in parallel with SCR and triac circuits. The goal is to prevent the thyristors from firing accidentally due to a high rate of rise of voltage (dv/dt). The self inductance *L* of the main circuit is needed to determine the dimensions of the snubber components (fig. 9-12). This inductance *L* can be the inductance of the supply transformer, the inductance of the circuitry, an added coil etc...

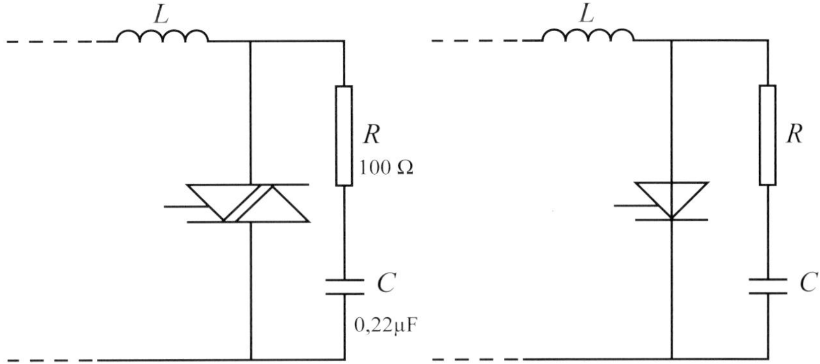

Fig. 9-12: Thyristor turn off snubber

Practical values of the snubber elements are for example: 100Ω and 0.22µF.

4.   SOLID-STATE RELAYS (SSR)

The term solid-state has its origin in the early days of semiconductor technology, which referred to the control of electrons within the material (e.g. Si). This is in contrast to controlling the electrons outside the material (in the electron tube between cathode and anode!).
Solid-state relays therefore use semiconductors (usually triacs). They have no stir contacts in contrast to electro mechanical relays.

4.1 Construction

Normally the SCR's are single pole types, since the most expensive parts are in the output circuit. Single pole switches are often referred to as SPST type (single-pole single-throw).
Three-phase SCR's also exist.

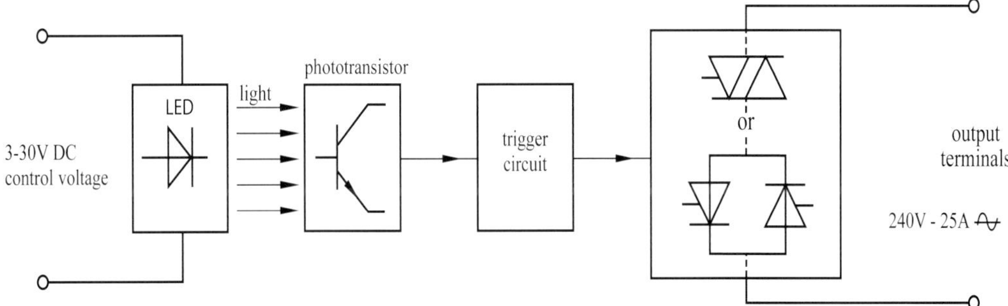

Fig. 9-13: Block diagram of an SSR

4.2 Operation

With electromechanical relays the output is controlled via a magnetic flux. In contrast, an SSR uses light flow as indicated in fig. 9-13. The firing circuit is connected via an opto-coupler which in turn controls triacs or two anti-parallel thyristors. In most cases, zero voltage switching is used. It is important to reduce EMI-RFI with this relay in applications such as computer and process control which may be sensitive to electrical noise.

Some types have " random turn on". Certain manufacturers include internal snubbers in the SSR. In connections with external snubbers it is worth noting that some manufacturers produce snubbers with the snubber R and C in one casing.

Photo Siemens: Solid state relays

4.3 Properties of an SSR

Important specifications include the nominal supply voltage and the RMS current for which the SSR is suited, as well as the control voltage. There are DC controlled SSR's and AC controlled SCR's. Other distinctions can be made between zero voltage switches and random on switches. Worth noting are the Crydom SSR's. These "solid-state relays" can be used as a DC contactor. The power switch is now an IGBT. This SSC-series can be used for DC loads up to 25 A (75 A peak) and 1000V.

4.4 Applications with an SSR

Fig. 9-14 and 9-15 show typical contactor applications using an SSR.

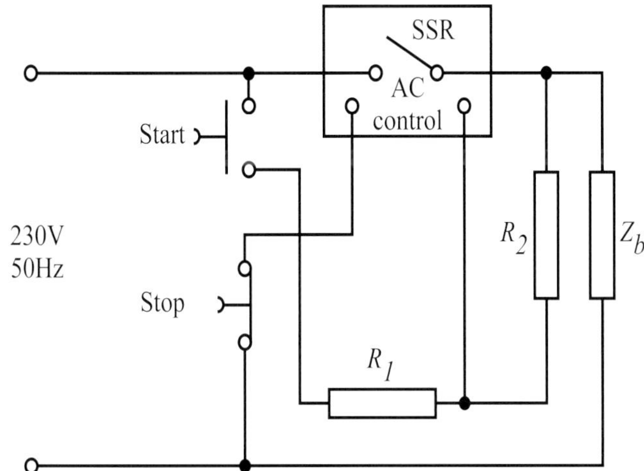

Fig. 9-14: Start-stop contactor using an SSR.

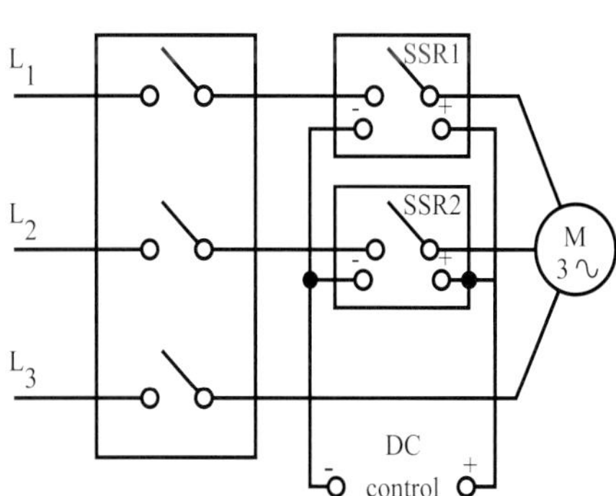

Fig. 9-15: Starting an AC motor using an SSR.

In fig. 9-14 we see an SSR being used as with classic push button control. After pressing "start" the SSR output is closed. R_1 is a limiting resistor for the input circuit. The voltage across Z_b together with R_2 and the stop push button function as a hold on until the stop button is pressed. Note that here an SCR with an AC input is needed.

For the three phase circuit shown in fig. 9-15 only two SCR's are used. If there was an SSR in each line it would be difficult to switch since the output thyristors of the SSR would only be fed from one side, in other words there would not be any voltage across the thyristor. In fig. 9-15 a mechanical three phase switch is needed to ensure the motor and relay are voltage free.

5. RADIO INTERFERENCE SUPPRESSION (RFI) OF THYRISTORS

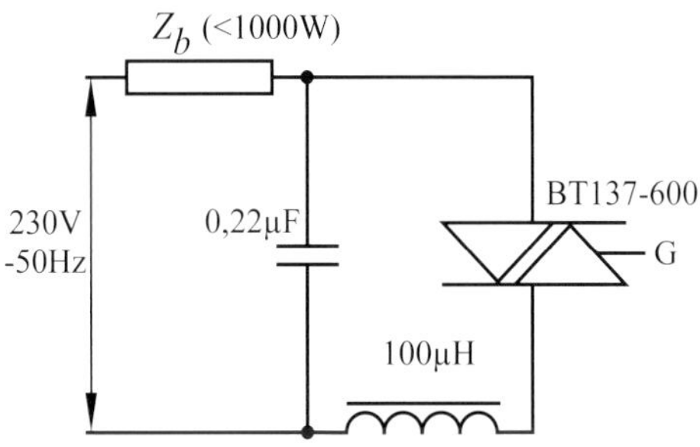

Fig. 9-16: Input filter of a thyristor circuit.

The currents associated with turn-on and turn-off of thyristors contain harmonics up to the MHz-region. This can cause disturbance in the medium wave band. This EMI primarily reaches other equipment via the supply cables. To limit the effect of this electrical pollution an LC-filter is sufficient in most cases as shown in fig. 9-16. The coil has a coefficient of self inductance between 100µH and 1mH, and has to conduct the load current. The capacitor usually has a value between 0.1 and 0.22µF. In fact this LC-filter is the input filter of our switching matrix from chapter 1.

6. EVALUATION

9-1 In the circuit shown in fig. 9-1 determine the maximum voltage allowed across the triac.

9-2 In the circuit shown in fig. 9-4a what is the RMS value of the current if the supply voltage is 230V-50Hz?

9-3 Why does the current continue to flow after the zero crossover of the voltage in the load shown in fig. 9-11.

9-4 We want to control the temperature of a 240V-2kW oven by means of period control. What is the minimum period of this TRC?

9-5 What do EMI and RFI stands for?

9-6 In fig. 9-1 $\alpha = 90°$. Determine I_{RMS} and V_{RMS} across the load of 100Ω . The triac may be considered ideal.

9-7 State 4 applications of AC controllers.

9-8 We want to use a triac to control an AC current of 12 A when the voltage is 230V. Determine the most important specifications for the triac if we are using a voltage safety margin of 2.3.

9-9 What do the following mean: TRC, zero voltage switching, SSR?

9-10 Which natural commutation circuits do you know? What is a synonym for natural commutation?

9-11 In the circuit shown in fig. 9-4 the load is 5Ω - 200mH. The circuit is connected to a 230V-50Hz supply. What happens if we employ a firing angle of 60° and:
a) the control pulses have a duration of 200µs
b) the control pulses last until the end of the half period.

9-12 What are the advantages of period control with respect to phase control.

10 CYCLOCONVERTERS

CONTENTS

1. Continuous cycloconverter
2. Trapezium cycloconverter

The cycloconverter is a single stage AC-AC converter, sometimes called a direct converter. We can distinguish two types of cycloconverters: continuous cycloconverters and trapezium converters. The preferred switch is an SCR. As will become obvious in the study of these direct converters the area of application is limited to very large slow motors in the range between 4 to 12 MW.

1. CONTINUOUS CYCLOCONVERTER

With this type of converter the output frequency may be continuously varied. In fig. 10-1 such a three phase converter is drawn. The load consists of the stator windings of a synchronous motor. Per phase the switching matrix is drawn which consists of two anti parallel bridges.

We now consider one phase (motor winding U_1U_2). By varying the firing angle α of the bridge we can arrange for an average voltage and sinusoidal form as shown in fig. 10-2. In this example the output voltage has a frequency four times lower than the supply frequency. Due to the inductive nature of the load the current ($i_{U_1U_2}$) lags the voltage ($v_{U_1U_2}$). If $v_{U_1U_2}$ and $i_{U_1U_2}$ have the same polarity this corresponds to rectifier operation of bridge A_1 or A_2 and energy flows from the supply to the machine. If $v_{U_1U_2}$ and $i_{U_1U_2}$ have opposite polarity then the instantaneous product is negative, which means that energy flows from the machine to the supply network. In this case the bridge (A_1 or A_2) operate as an inverter. To prevent a short circuit between A_1 and A_2 a pause is introduced when switching between the bridges. These short pauses are insignificant due to the low frequency of the bridge and it is possible to operate without circulating current.
The output frequency depends upon the supply frequency. With an infinite pulse number theoretically $f_{out} = f_{in}$.
With a six pulse converter there is a practical maximum of about 40% of the supply frequency. For a 50Hz supply this means a continuously controllable output frequency between 0 and 20 Hz. A cycloconverter is interesting for slow moving AC machines driving a mechanical load without a gearing mechanism.

To create a three phase voltage the bridges B_1/B_2 and C_1/C_2 need to be fired 120° and 240° later than bridges A_1/A_2 .

Due to the large amount of thyristors with accompanying control electronics the cycloconverter is only interesting when the power level is very large. At the same time the output frequency is low so that predominantly slow machines are suitable for this control method. In practice this means that cycloconverters drive synchronous machines in steel mills, cement factories, etc ... in the power range from 4 to 12 MW.

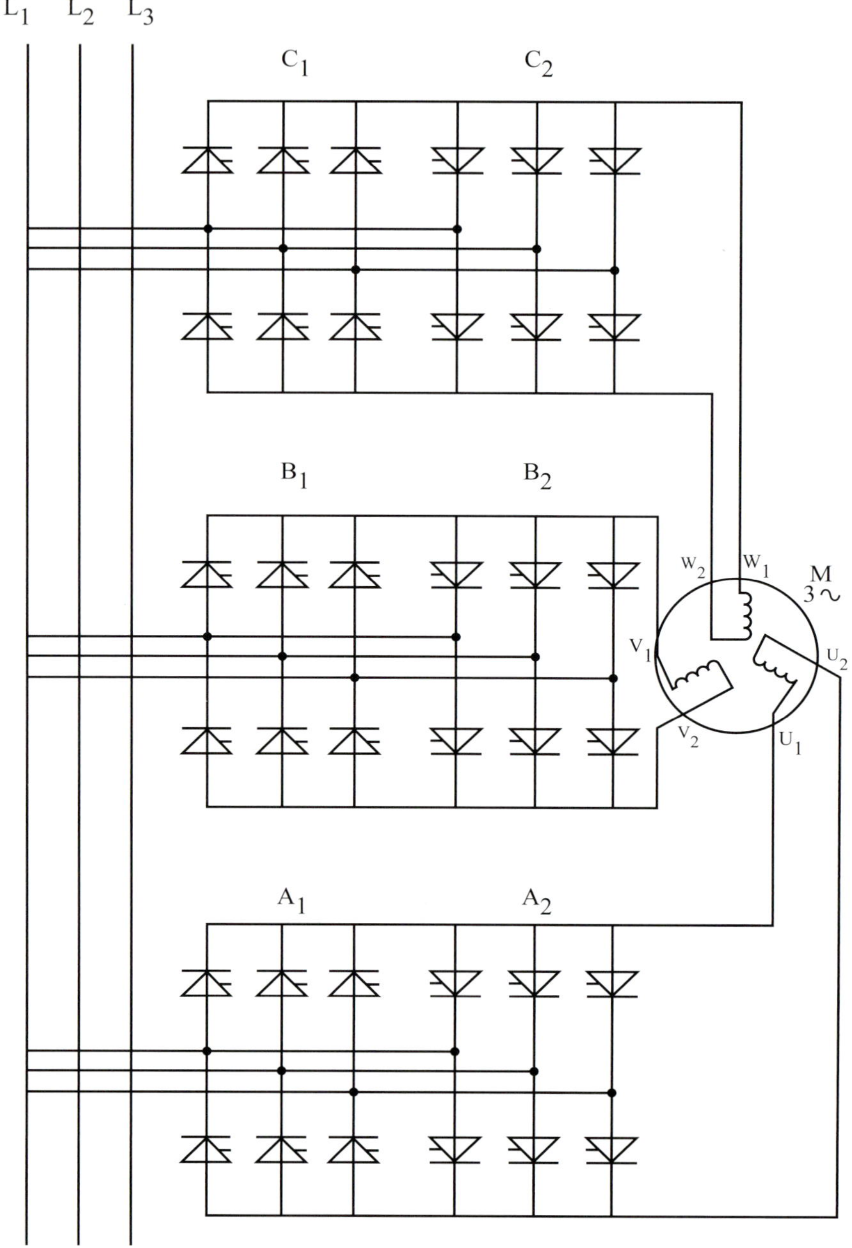

Fig. 10-1: Continuous cycloconverter used to control a three-phase motor.

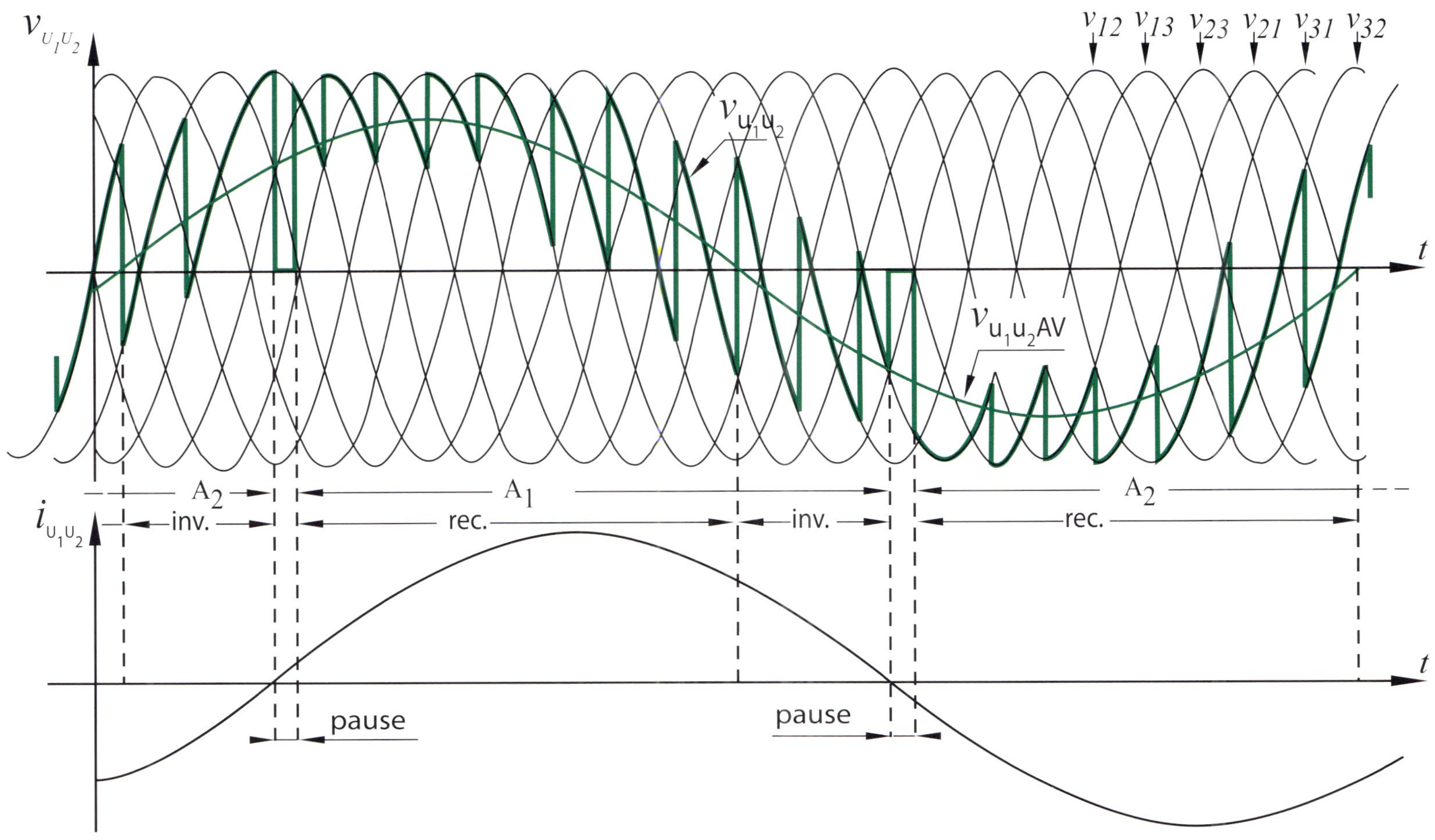

Fig. 10-2: Output voltage and current per phase of the cycloconverter shown in fig. 10-1 where $f_{out} = f_{supply}/4$

2. TRAPEZIUM CYCLOCONVERTER

The basic principle of this converter is composed of thyristor bridges in which the SCR's conduct in a certain pattern such that the load current is AC with a frequency less than the supply frequency. Fig. 10.3 shows the basic setup to obtain $f/2$ and $f/3$. It is clear that $f/4$ ... are also possible.

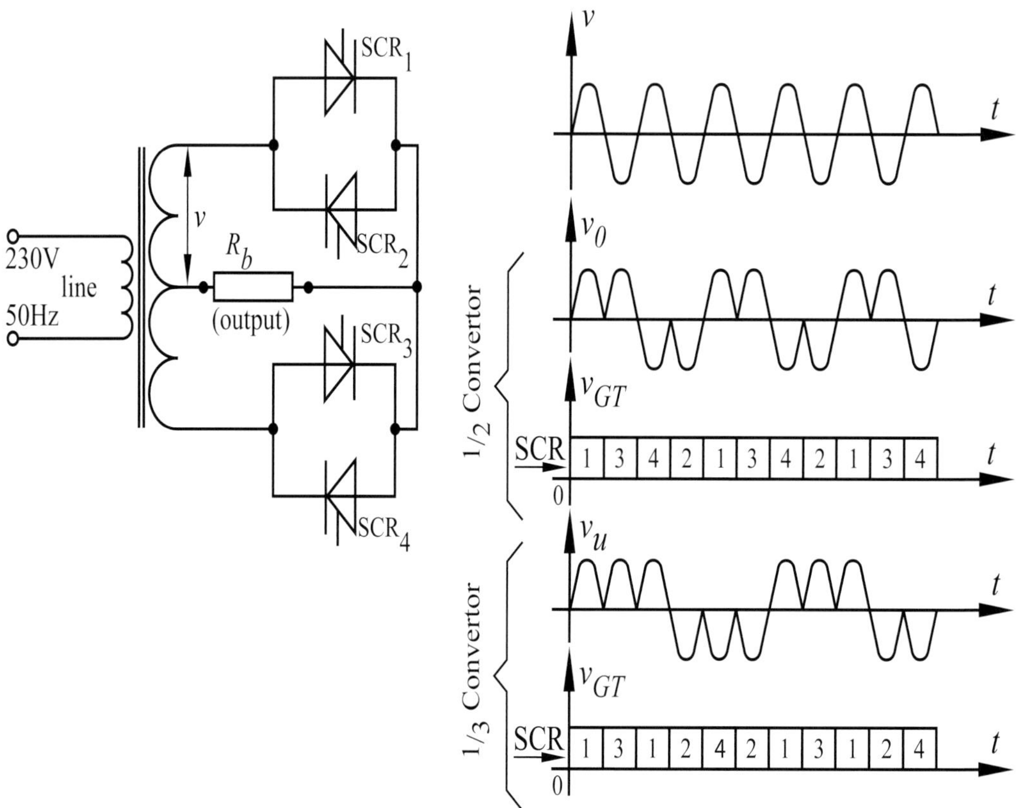

Fig. 10-3: Cycloconverter with pulse patterns for f/2 and f/3

Starting with a 50Hz supply we are able to obtain output frequencies in controlled steps from for example 5Hz to 25Hz.

The configuration in fig. 10-3 can be extended for three phase supply resulting in a three phase output. On the other hand starting with a three phase supply a single phase output can be realized with a very large current. For this two anti parallel three-phase thyristor bridges A_1 and A_2 are used. In the rectifier quadrant the firing angle $\alpha = 0°$ and in the inverter quadrant $\alpha = 150°$ (on the border of miss-firing).

Fig. 10-4 shows what this involves for a resistive inductive load. The average output voltage is trapezium shaped hence the name trapezium converter. The disadvantage of this type of converter is that the frequency is only variable in steps.

As a result of phase chopping a continuous converter draws more reactive power from the supply than a trapezium converter. On the other hand (with almost sinusoidal current) a continuous converter has much lower harmonic distortion than a trapezium converter.

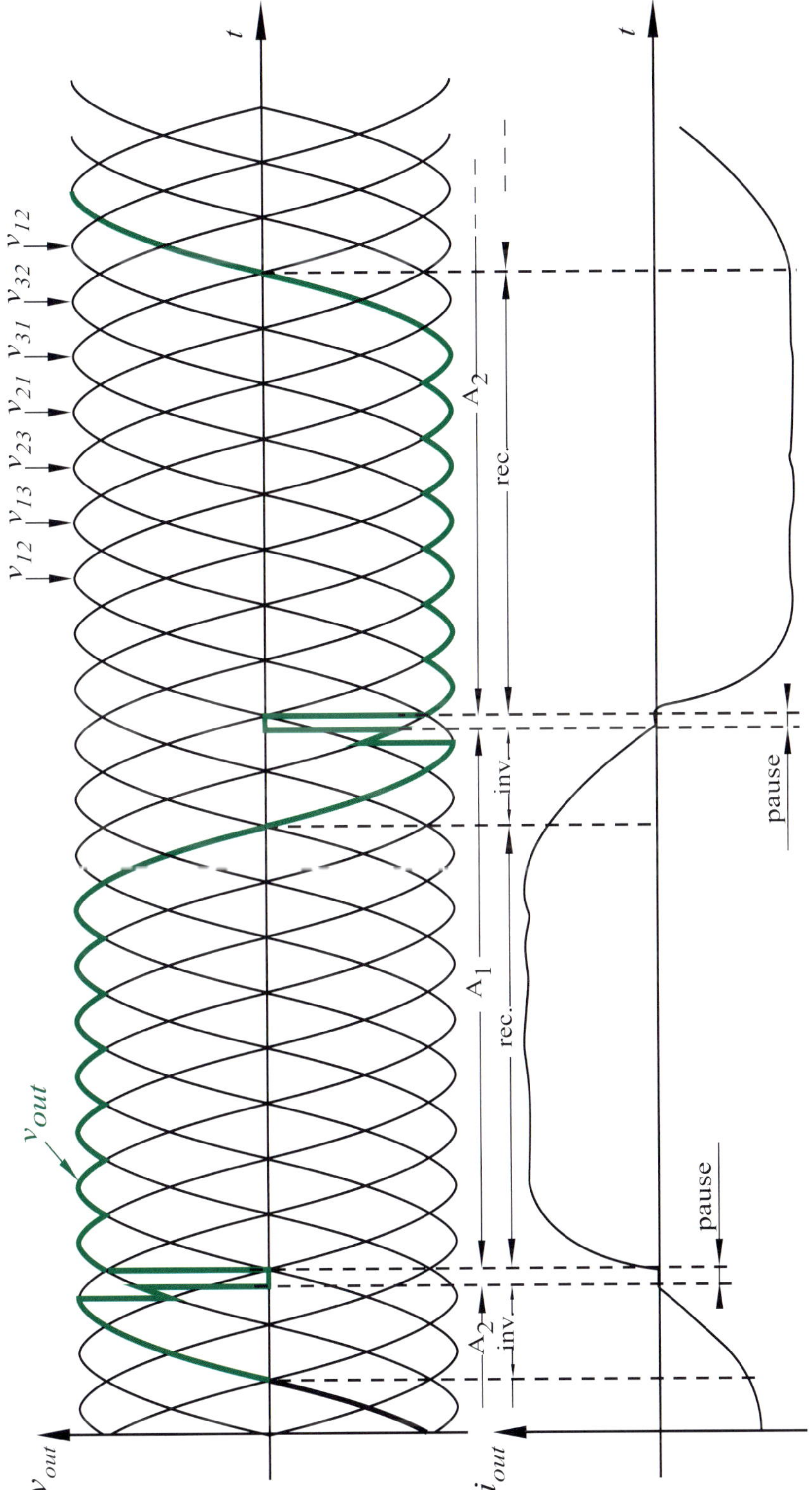

Fig 10-4: Output waveform of a trapezium converter which converts a three phase voltage to a single phase voltage with lower frequency

Photo Ixys (KWx):   Collection of Ixys-semiconductors

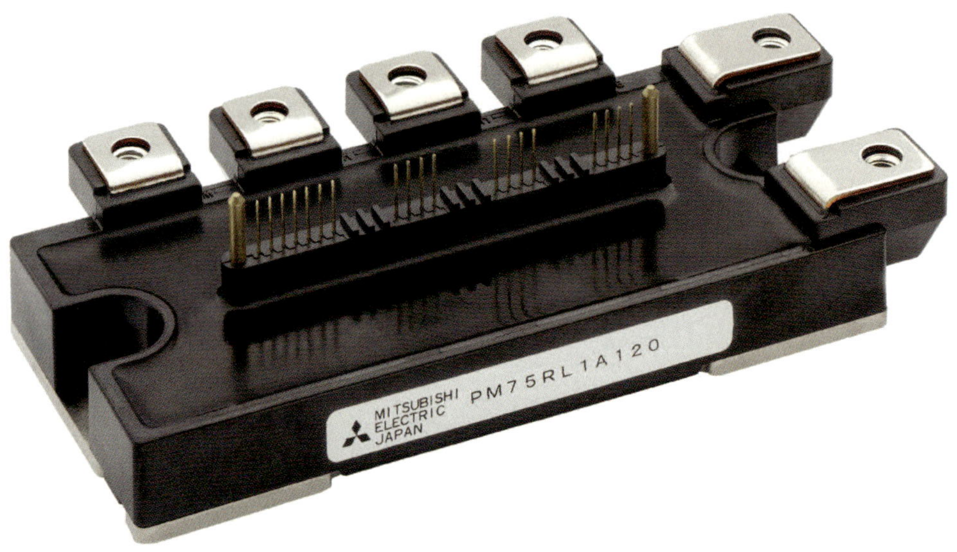

Photo Mitsubishi: PM75L1A120 This is a so called intelligent power module (IPM). It is a complete three phase inverter with 6 IGBT's with control circuitry in a single casing.
The inverter operates at 1200V-75A. Measurements 35x120mm.

11 CONTROL OF THYRISTORS

CONTENTS

1. Firing pulses
2. Pulse transformers
3. Control IC for SCR and triac
4. Triac control with a diac.
5. Evaluation

Control circuits for thyristors will mostly generate pulses. Except in the case of GTO's constructing control circuits using discrete components is a thing of the distant past.

SCR and triac are normally controlled via an integrated control circuit. An exception to this is the simple diac used to control a triac.

Depending upon the application more often than not the control function is implemented in a microprocessor or microcontroller.

The galvanic separation between control circuit and thyristor occurs via pulse transformers or opto couplers. Opto couplers are most frequently used with microprocessor control. For large powers and high voltages the galvanic separation via fiberglass connections has the added advantage that the control circuits and the power circuits are on different racks often separated by tens of centimeters.

1.   FIRING PULSES

In fig. 4-27 examples of firing pulses for thyristors were shown.

Fig. 11-1 shows the current form for a pair of resistive loads (25W and 1000W) connected to a 230V-50Hz supply. To reach the thyristor latching current of for example 24mA it takes 497µs for a 25W load and 12.4µs for a 1000W load. It is clear that the necessary time span of the firing pulses is in part dependent upon the load power. If the load is inductive then the time span increases since the latching current has to be reached first.

Fig. 11-2 shows the different pulses for zero voltage switching of a triac. The triac is consecutively a resistive load, an inductive load and then a highly inductive load. To limit the gating power use is made of a pulse train with a frequency of 5 to 7kHz rather than a wide pulse. This pulse train may also be necessary when using a pulse transformer since this cannot transmit a wide pulse.

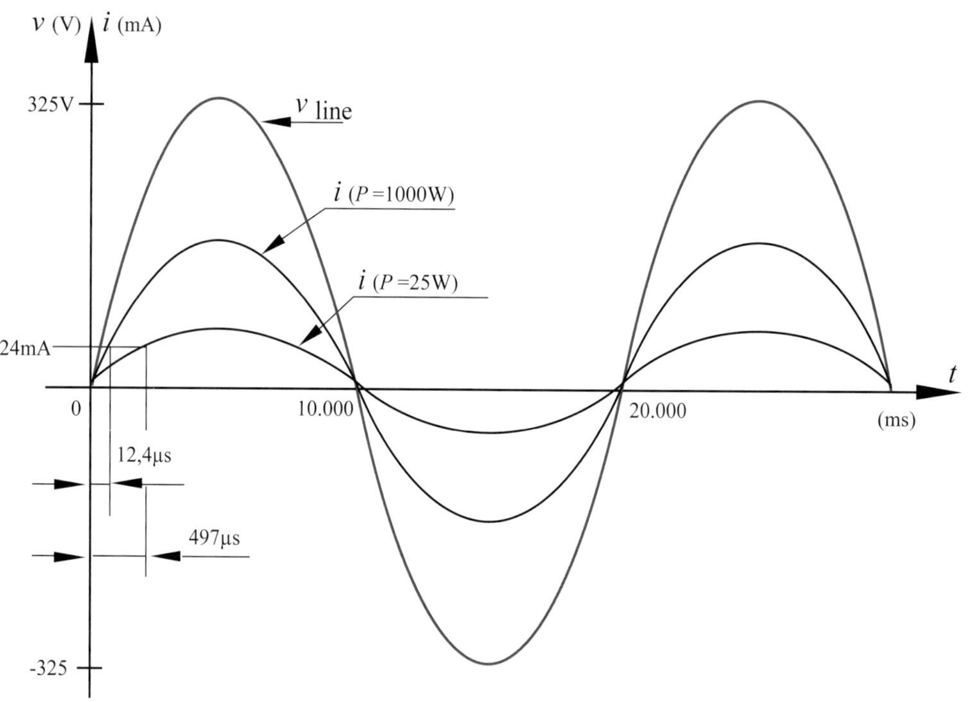

Fig. 11-1: Time required to reach a certain current

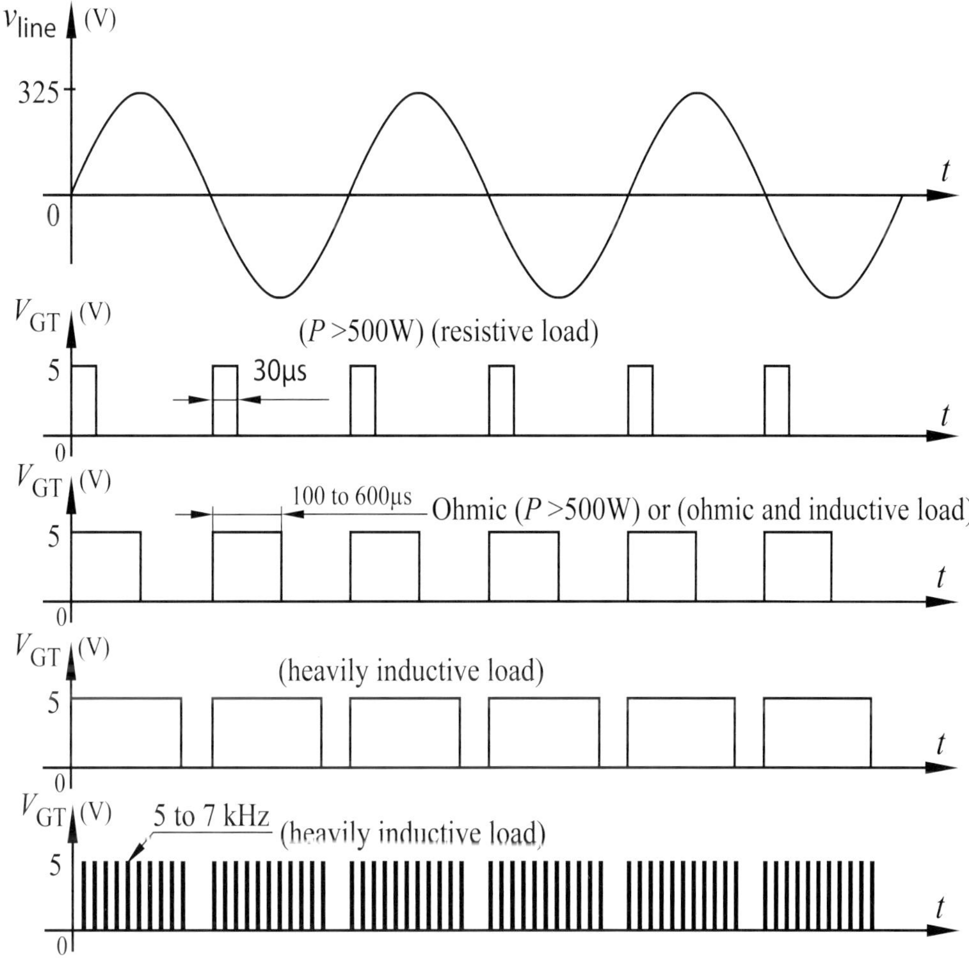

Fig. 11-2: Pulses for zero voltage switching of a triac with various loads

2. PULSE TRANSFORMERS

2.1 Properties of control pulses

The most important advantage of using pulse transformers in the control circuitry is the galvanic separation it provides. In addition, especially in three phase thyristor bridges the control circuitry is simplified by the use of pulse transformers. Depending upon the application the control pulse for a thyristor should have the following properties

- open circuit emf of 7 to 10V
- I_{GT} and V_{GT} : see thyristor datasheets (20 to 500mA)
- rise time: 0.3 to 3µs
- repeat frequency: 50Hz up to 7kHz.
- pulse duration: 20µs to 2 ms.

2.2 Pulse transfer

The core of a transformer reaches the saturation flux quicker with a higher voltage than with a lower voltage. The product $V . t = \Phi_S$ is constant (fig. 11-3b).
By applying a square wave pulse to the primary side of the pulse transformer a secondary voltage is obtained as shown in fig. 11-4.
With an undamped secondary the output voltage may look like that in fig. 11-4a.
Notice that, due to an almost always present load resistance on the secondary, the square wave input appears deformed (damped) on the secondary side (see fig. 11-4b):

- the rising flank has a rise time t_r.
- the transformer is saturated after a time t

On p. 11-6 an extract is provided from a datasheet for Schaffner- pulse transformers.
Hereby:

$V_o . t$ =   time voltage integral: the product of the pulse height and width (at half height): see fig. 11-4b

t_r =   rise time with a given load impedance R_b and at 70% of the pulse height

R_b =   load impedance used for measuring of t_r

R_p / R_S =   DC resistor primary and secondary

L_p =   primary self inductance at 1kHz, with open secondary

C_k =   capacitive coupling between primary and secondary

V_p =   RMS test voltage allowed for 1 minute between primary and secondary

$V_{nom.}$ =   max. allowed (RMS) nominal voltage.

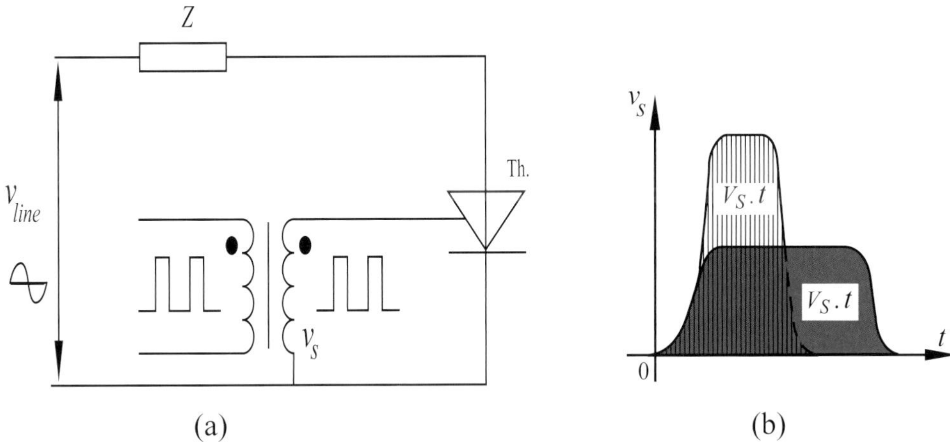

Fig. 11-3: Circuit with pulse transformer

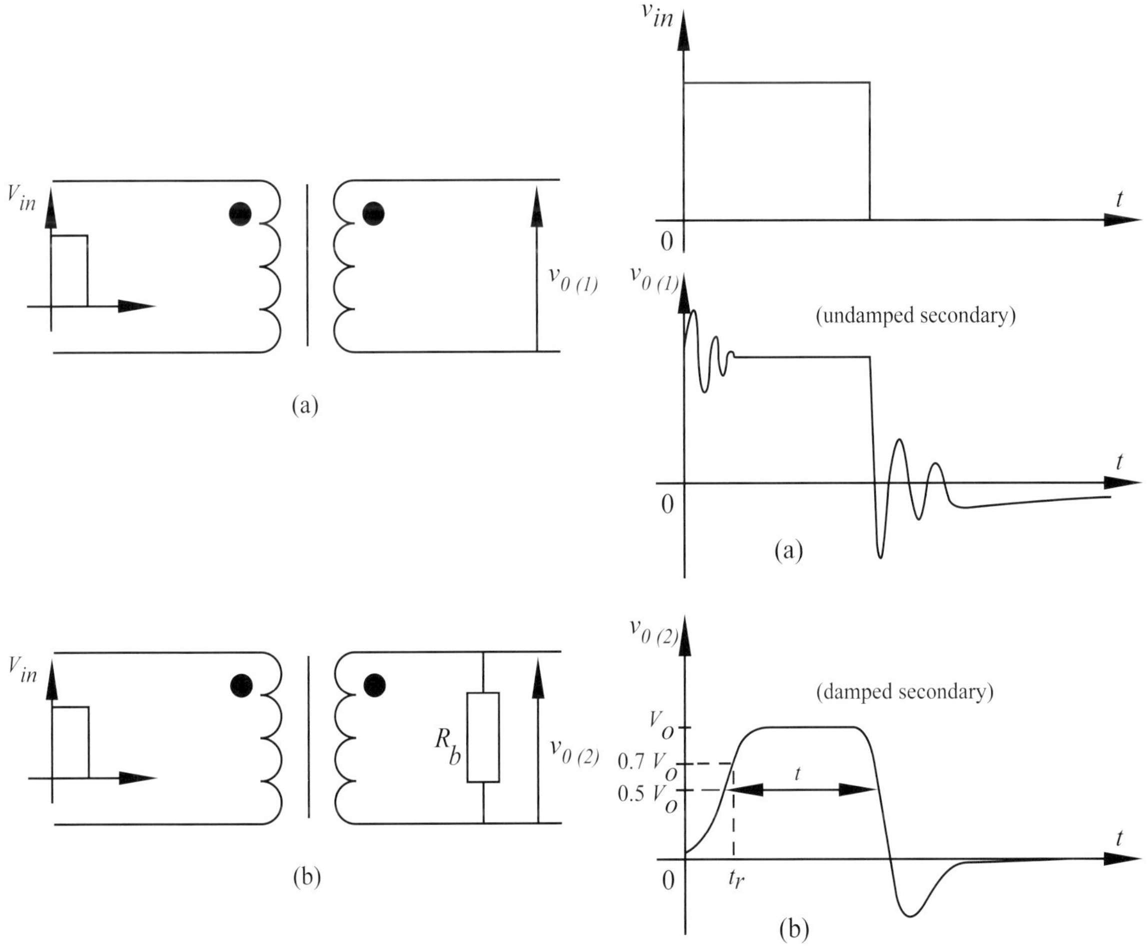

Fig. 11-4: Pulse transformer: pulse forms

2.3 Table 11-1

A selection of pulse transformers from SCHAFFNER

1. Pulse transformers with a single secondary

Type	Turns ratio	Ignition current I_{ign} [A]	Voltage U_{nom} [V]	Voltage U_p [kV]	Voltage-time area $V_o t$ [V µs]	Rise-time t_r [µs]	Inductance L_p [mH]	Inductance L_{str} [µH]	Resistance R_p [Ω]	Resistance R_s [Ω]	Coupling capacitance C_k [pF]	Weight [g]
IT 155		0.1	500	4	500	1	5	85	1		6	13
IT 245		0.1	750	4	500	1.2	8	100	1.4		10	6
IT 237		0.25	500	2.5	1'100	1	25	35	1.8	2.2	50	14
IT 239		0.25	1'000	6	350	2.3	3	80	0.8		5	13
IT 255	1:1	0.25	750	4	250	1.1	2.2	40	0.7		8	6
IT 258		1	750	3.2	250	0.25	2.5	3	0.6	0.7	80	6
IT 370		1	1'000	5	4'000	0.5	0.3	6	0.16	0.17	40	71
IT 364*		3	3'000	8	5'000	1.7	1.5	10	0.16	0.14	35	220
IT 246		0.1	750	4	200	0.4	7	35	1.8	1	7	6
IT 248	2:1	0.25	750	3.2	350	1.8	17	80	3	1.5	9	6
IT 362*		3	1'000	5	3'500	0.4	3	25	2.4	0.3	20	360
IT 260	3:1	0.1	500	3.2	200	0.3	12	30	1.8	0.7	8	6
IT 332		1	380	2.5	1'200	1.2	200	60	6	0.5	65	33

*not suitable for PCB-mounting

2. Pulse transformers with two secondaries

Type	Turns ratio	Ignition current I_{ign} [A]	Voltage U_{nom} [V]	Voltage U_p [kV]	Voltage-time area $V_o t$ [V µs]	Rise-time t_r [µs]	Inductance L_p [mH]	Inductance L_{str} [µH]	Resistance R_p [Ω]	Resistance R_s [Ω]	Coupling capacitance C_k [pF]	Weight [g]
IT 143		0.025	500	4	800	0.6	15	200	3		10	14
IT 153		0.1	500	4	600	1.4	9	120	1.4		10	14
IT 242		0.1	500	3.2	250	0.9	2.5	75	0.6		7	6
IT 243		0.1	500	3.2	250	1	2.5	85	0.7		7	6
IT 213	1:1:1	0.25	380	2.5	450	0.4	6.5	20	1.4	1.5	40	9
IT 233		0.25	500	4	300	1.3	3	45	0.75		7	13
IT 253		0.25	500	3.2	180	1.3	1.1	45	0.5		6	6
IT 312		0.25	380	2.5	1'200	1	21	35	2.4	2.7	30	24
IT 313		1	380	2.5	450	0.6	3	6	0.32	0.37	27	24
IT 249	2:1:1	0.25	500	3.2	330	3.3	17	140	2.7	1.3	9	6
IT 154		0.1	500	4	600	1.3	75	180	7	2	9	14
IT 244	3:1:1	0.1	500	3.2	200	0.7	15	70	2.6	0.8	9	6
IT 234		0.25	500	4	300	1	17	40	2	0.6	9	13
IT 314		1	380	2.5	500	1	35	20	1.5	0.65	30	25

3. CONTROL IC FOR SCR AND TRIAC

A number of companies sell linear IC's that can form part of a control module for thyristors. It is possible to build a temperature controller for an oven, a speed controller for a motor, etc...with a minimum number of components.

We consider just such a control IC, the industry standard TCA785 from Infineon.

3.1 AC controller using a TCA785

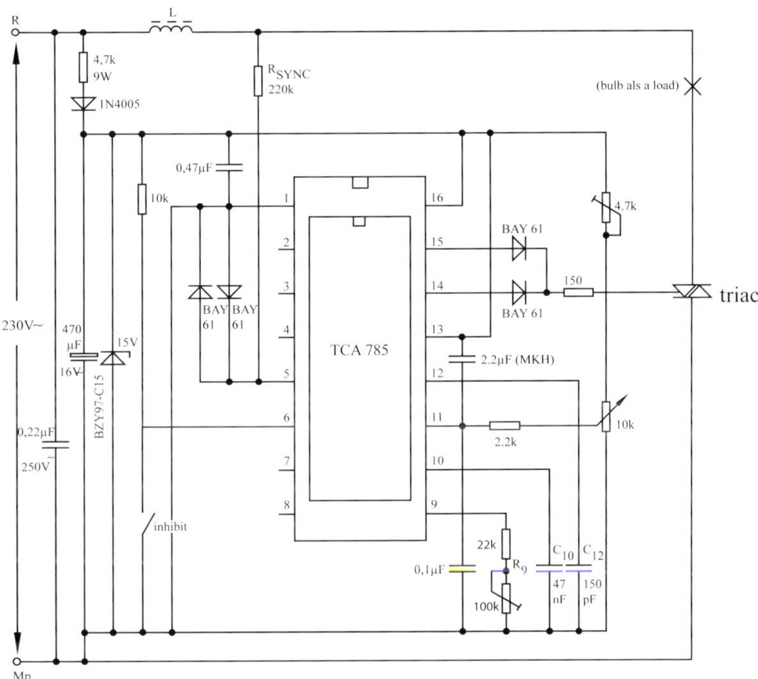

Fig. 11-5: Phase control with a TCA785 for triacs with a gate current up to 50mA

The operation of the TCA785 is explained under heading 3.2.

The IC has 16 pins. The IC is supplied with 15V DC via pin 16 and pin 1.

In fig. 11-5 the DC supply is created with rectification (1N4005), filtering (470µF) and zener stabilisation (BZY97-C15).

To detect zero crossover of the supply, pin 5 is connected to the supply voltage via a resistor $R_{sync} = 220\text{k}\Omega$.

The value of R_9 (resistor connected between pin 9 and ground!) determines the charging current I_{10} that linearly charges capacitor C_{10}. At pin 10 we have a ramp voltage in which the slope is determined by R_9 and C_{10}.

The voltage at pin 11 is regulated with a potmeter of 10kΩ and as a result the firing angle α can vary between 0° and 180°.

Pin 15 delivers a firing pulse during the positive half period of the supply voltage and pin 14 does the same during the negative half period.

Bypass capacitors are connected to pins 1, 11 and 13. In this circuit an AC current is phase controlled as shown in fig. 9-1.

3.2 Operation TCA785

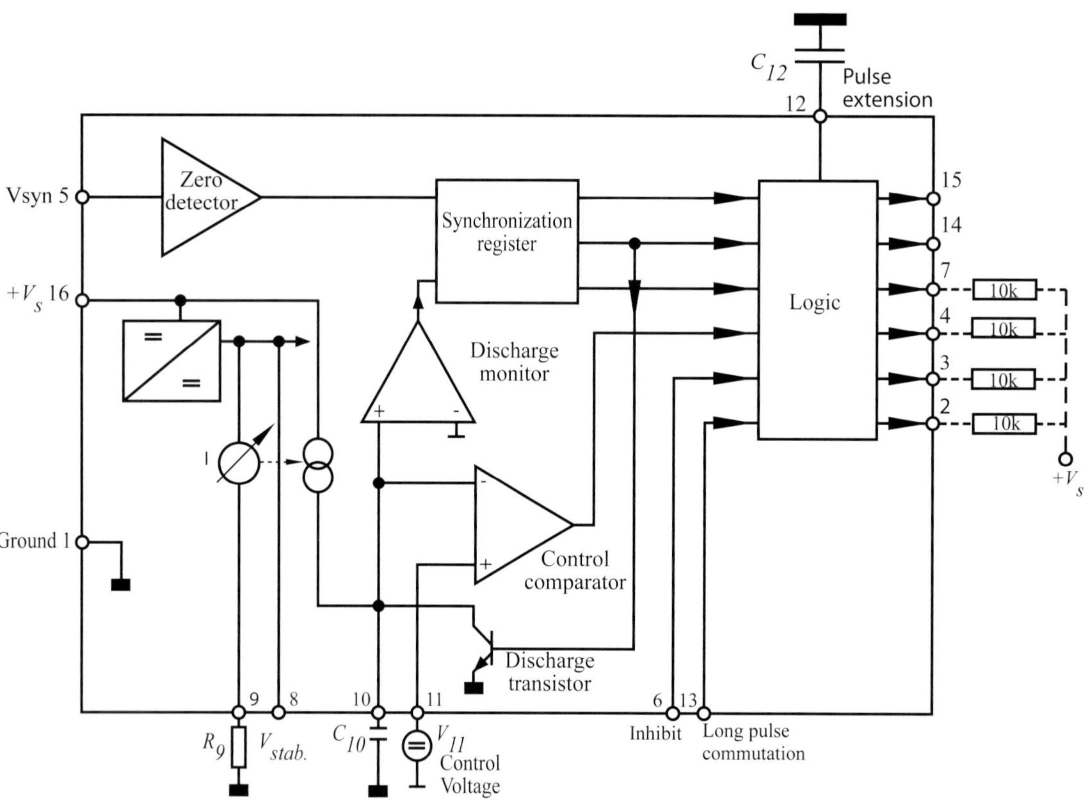

Fig.11-6: Block diagram of a TCA785

Fig. 11-6 shows a block diagram of the IC under consideration. The DC supply of 15V is connected via pin 16. Since the control pulses for the SCR's or triacs should be synchronized with the AC supply voltage, the zero cross over is detected with the zero detector. For this purpose pin 5 is connected via R_{sync} = 220kΩ with the 230V supply.

The resistor R_9 determines the value of the constant current (- generator). This current linearly charges the capacitor C_{10}. The rate of rise of the ramp voltage V_{10} is determined by C_{10} and R_9. With every zero crossover of the supply the discharge transistor conducts and C_{10} discharges, where upon the voltage ramp begins again.

If V_{10} is larger than the reference voltage V_{11} then the control comparator flips over and a pulse appears at pin 15 (during the positive half period) or on pin 14 (during the negative half cycle of the supply voltage). The pulse is approximately 30µs wide.

A pulse is also present on pins 7, 4, 3 and 2.

Fig. 11-7 shows the main wave forms on the pins of the TCA785.

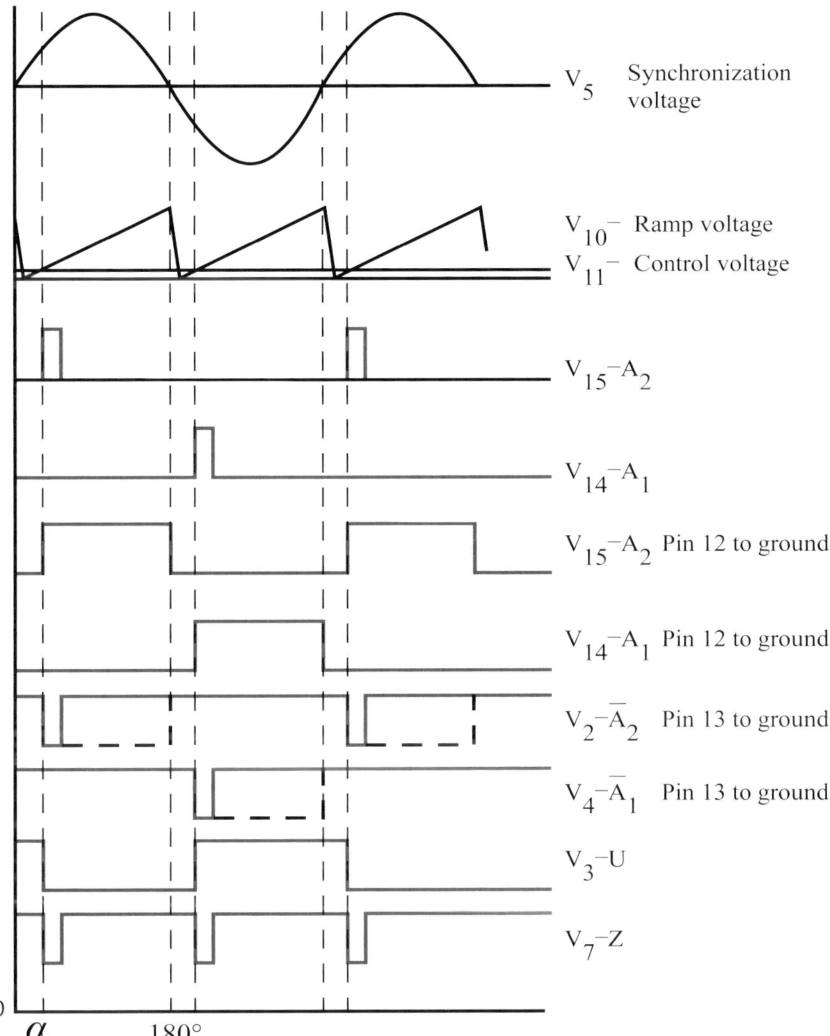

Fig. 11-7: Wave forms of a TCA785

The intersection of V_{10} and V_{11} in fig. 11-7 creates a phase delay α. This α is variable between 0° and 180°. In the characteristics we find that switch in a C_{12} = 1nF generates a pulse of approximately 620µs. Short circuiting pin 12 to ground generates a pulse to the end of the half period, as indicated in fig. 11-7.

The inverse pulses of pins 14 and 15 are generated on pins 4 and 2.

Pin 3 generates a pulse of α + 180°, while pin 7 is a NOR output of the pulses on pins 14 and 15. With a voltage V_6 < 2.5V with respect to the negative supply terminal of the IC produces an inhibit of the output pulses (pins 2/4/14/15).

If $V_6 \geq 4$V , then the pulses appear on the output of the IC. In fig. 11-5 for example pin 6 is connected to +15V. Closing switch S provides the previously mentioned inhibit.

If $V_{13} \geq 3.5$V then short pulses of 30µs are generated on pins 2 and 4 (can be widened with C_{12}). With a voltage on pin 3 less than +2V then long pulses appear on pins 2 and 4.

3.3 TCA785 (Infineon) specifications for "Phase control"

Phase control, intended to control thyristors, triacs and transistors.

The output current is 250mA.

Typical applications includes: • DC controllers

                                     • AC controllers (single phase, threephase)

- Reliable recognition of zero passage
- Large application field
- May be used as zero point switch
- Threephase operation possible (3 ICs)

Type	Ordering code	Package outlines
TCA	Q67000-A2321	DIP 16

Maximum ratings			
Supply voltage	V_s	18	V
Max. output current pin 14, pin 15	I_Q	400	mA
Inhibit voltage	V_6	V_s	V
Control voltage	V_{11}	V_s	V
Pulse control voltage	V_{13}	V_s	V
Synchronisation input curr.	I_5	200	μA
Max. outp. volt. pin 14, 15	V_Q	V_s	V
Junction temperature	T_j	125	°C
Storage temperat. range	V_s	-55 to 125	°C
Thermal resistance, system-ambient air	$R_{th\,s\text{-}a}$	80	K/W
Range of operation			
Supply voltage	V_s	8 to 18	V
Operating frequency	f	10 to 500	Hz
Ambient temperature	T_{amb}	-25 to 85	°C

Characteristics		min.	typ.	max.	
V_s = 8 à 18V. T_{amb} = –25 à 85°C. f = 50Hz.					
Open loop supply current consumption	I_s	4,5	6,5	10	mA

Synchronisation pin 5					
Input current	I_5	30		200	μA
Offset voltage	ΔV_5		30	75	mV

Control input pin 11					
Control voltage range	V_{11}	+0,2		U_{10max}	V
Input resistance	R_i		15		KΩ
Firing point	t_{fir}		$\dfrac{V_{11} \cdot R_9 \cdot C_{10}}{V_{ref.}K}$		(1)

Characteristics TCA785 (continued)

Ramp generator		min.	typ.	max.	
Max. load current	I_{10}	10		1000	µA
Load current	I_{10}		$\dfrac{V_{ref.} \cdot K}{R_9}$		(1)
Ramp voltage	V_{10}		$\dfrac{V_{ref.} \cdot K \cdot t}{R_9 \cdot C_{10}}$		(1)
Max ramp voltage	V_{10}			V_s-2	V
Saturation voltage at the capacitor	V_{10}	100	225	350	mV
Ramp resistance	R_9	3		300	kΩ
Ramp capacitance	C_{10}	500pF		1	µF (2)
Sawtooth return time	t_f		80		µs
Inhibit pin 6					
Outputs inhibited	V_{6L}		3,3	2,5	V
Outputs enabled	V_{6H}	4	3,3		V
Signal transition time	t_r	1		5	µs
Input current with $V_6 = 8V$	I_{6H}		500	800	µA
Input current with $V_6 = 1,7V$	$-I_{6L}$	80	150	200	µA
Pulse control (180° = α) pin 13					
Short pulse at Q	V_{13H}	3,5	2,5		V
Long pulse at Q	V_{13L}		2,5	2	V
Input current when $V_{13} = 8V$	I_{13H}			10	µA
Input current when $V_{13} = 1,7V$	$-I_{13L}$	45	65	100	µA
Outputs pin 2, 3, 4, 7					
Reverse current when $V_Q = V_s = 15V$	I_{CEO}			10	µA
Saturation voltage when $I_Q = 2mA$	V_{sat}	0,1	0,4	2	V
Outputs pin 14, 15					
H-output voltage when $-I_Q = 250mA$	$V_{14/15H}$	V_s-3	V_s-2,5	V_s-1	V
L-output voltage when $I_Q = 2mA$	$V_{14/15L}$	0,3	0,8	2	V
Pulse width (short pulse) without C_{12}	t_p	20	30	40	µs
Pulse width (short pulse) with C_{12}	t_p	530	620	760	µs/nF
$V_s = 8$ à $18V$. $T_{amb} = 25$ à $85°C$. $f = 50Hz$.					
Internal voltage regulator					
Reference voltage	V_{ref}	2.8	3.1	3.4	V
Switching of 10 ICs in parallel feasible.					
TC of reference voltage			2.10^{-4}	5.10^{-4}	1/K

(1) K = 1,10 ± 20%

(2) return times are to be taken into account

3.4 Large power AC controller

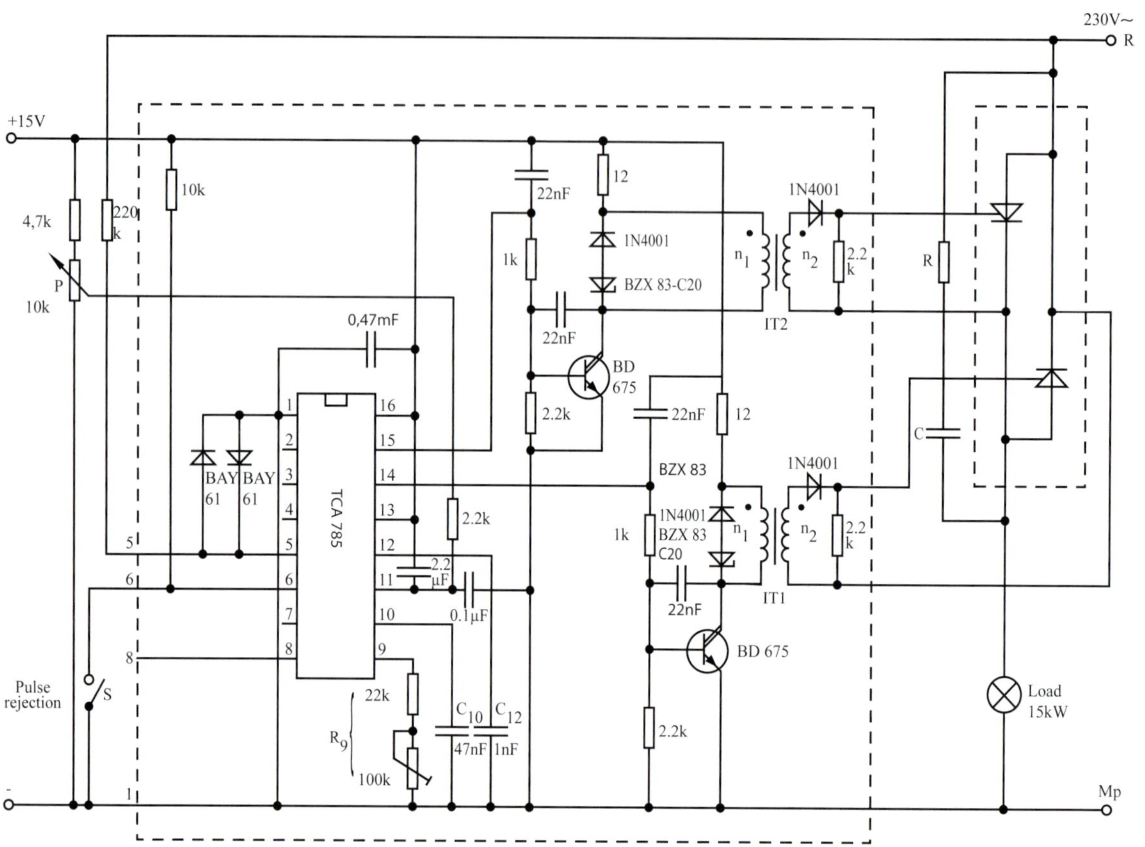

Fig. 11-8: Large power AC controller. Here two anti parallel SCR's are used instead of a triac

Numeric example 11-1:

1. We want a saw tooth voltage of 10V amplitude on pin 10. Which value should the potmeter R have (in series with 22k on pin 9 in fig. 11-8) ? Supply frequency is 50 Hz.

2. How wide is the control pulse in fig. 11-5?

Solution:

1. From the specifications of the TCA785 we find: $V_{10} = \dfrac{V_{ref.} \cdot K \cdot t}{R_9 \cdot C_{10}}$

    Since we want the saw tooth to repeat every half period of the 50Hz supply then $t = 10$ms,

    so that: $V_{10} = \dfrac{3.1 \times 1.1 \times 10 \times 10^{-3}}{R_9 \times 47 \times 10^{-9}} = 10\ \text{V}$.

    This gives $R_9 = 72.55\ \text{k}\Omega$ so that: $R = 50.50\ \text{k}\Omega$

2. From the specifications if follows (typically) : $t_p = 620\mu\text{s/nF}$ with $C_{12} = 150$pF $\rightarrow\rightarrow\rightarrow$

    $t_p \approx 620 \times 0.15 = 93\mu\text{s}$.

4. TRIAC CONTROL WITH A DIAC

Consider (fig. 11-9) the half period of the supply voltage where by A is positive with respect to B. The capacitor C_2 is charged via R_3 so that the upper plate is positive. If v_{C2} assumes a value of about 30V, then diac D conducts and triac T switches on.

As a result of the diac characteristic the voltage across its terminals remains at 10V so that C_2 can not discharge from this 10V level. Consider a resistive load connected with the triac, then the triac switches off just before the zero crossover of the voltage.

The next half period B is positive with respect A. The remaining voltage (10V) on C_2 will quickly discharge via D_4 and R_2 . The capacitor will now up to 30V charge via R_3 with the lower plate being positive. At this instant the diac is triggered and consequently also the triac. At the end of the negative half period the triac switches off and once again there is a voltage of 10V remaining on the capacitor C_2 . After this, A becomes positive with respect to B and the remaining voltage quickly discharges via R_1 and D_2 until D_1 conducts. C_2 can begin with charging almost immediately to 30V via R_3 .

The firing delay α of the AC controller is determined by time constant $R_3.C_2$. Varying R_3 allows us to vary α .

In parallel with the triac we notice a snubber network (120Ω / 0.1μF) and a filter on the input (1mH-0.22μF).

This circuit can also work without the four diode network, and without R_1 and R_2 . The remaining charge voltage of 10V on C_2 first needs to go to zero via R_3 . This results in an hysteresis effect in the circuit. As a result we prefer to include the four diode network, since the extra cost is minimal and it ensures a correct operation of this circuit.

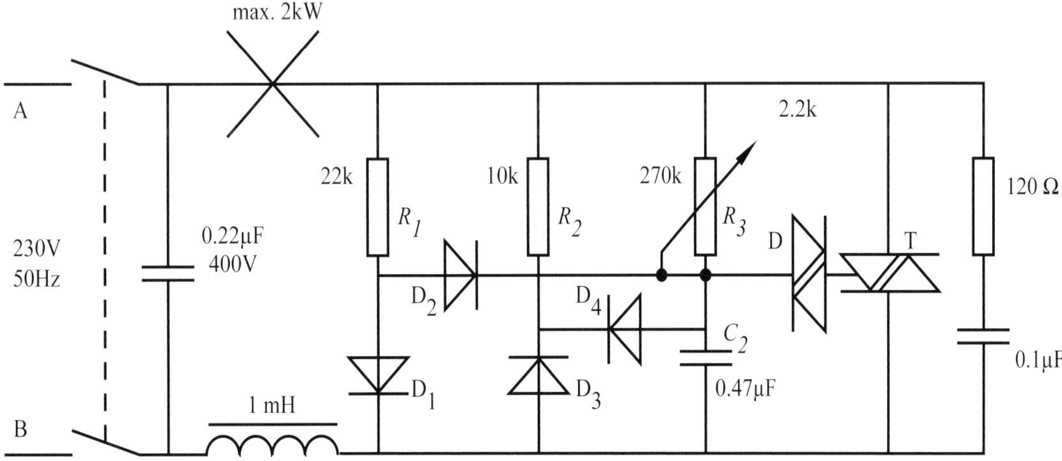

D$_1$...D$_4$: 1N4007; T: BT138-600; D: B R100-03

Fig. 11-9: Diac control of a triac.

5. EVALUATION

11.1   What is the purpose of the BAY61 diodes between pins 1 and 5 of the TCA785 shown in fig. 11-5 and 11-8?

11.2   What is the purpose of the BD675 transistors in fig. 11-8? What current and voltage should these transistors be able to handle?

11.3   Why do the gate circuits of the SCR's in fig. 11-8 need to be galvanically separated?

11.4   What is the reason that the diodes 1N4001 and BZX83C20 are used in the primary circuit of the pulse transformer shown in fig. 11-8?

11.5   In fig. 11-8 we also notice a diode 1N4001 and a 2.2kΩ resistor in the secondary side of the pulse transformer. What is the reason for this?

11.6   What is the purpose of the switch S in fig. 11-8?

11.7   Why are R and C connected in parallel with the SCR module?

11.8   What is the significance of the following terms: pulse extension, long pulse commutation, TC of reference voltage, sawtooth return time, open loop supply current consumption, transition time, firing point, saturation voltage, zero point switch?

11.9   How do we conclude that it takes 497μs to reach the holding current of 24 mA of a thyristor with a 25W load as shown in fig. 11-1?

11.10 Explain the following: $V_o.t$ and V_p for a pulse transformer.

Photo SCHAFFNER: Pulse transformers

An IT235 measures 27x22.5x13.7 mm and an IT244: 17.5x16.5x10.5 mm

12   CHOPPERS

CONTENTS

1. Operating principle of a chopper
2. Control methods
3. Resistive and resistive-inductive loaded choppers
4. Chopped resistor
5. Chopper control IC's
6. Evaluation

A constant DC voltage can be converted to another value using a chopper. A chopper is therefore a DC-DC converter.

Since there is no zero crossover in the supply voltage natural commutation is not possible. In the literature a chopper is sometimes referred to as a forced commutation circuit. Since ever form of commutation is in fact forced a better term would be artificial commutation. Due to the nature of the supply voltage in this case we need switches which can switch on and off on command. Transistors and GTO's are suited for such an application.

Choppers are used to control DC motors in the power range from a few watts to hundreds of kilowatts, in switch mode power supplies, to stabilize low DC voltages in measurement circuits, etc. Important applications include:

1   Converting an available DC voltage to a lower or a higher value. This application is especially common in switch mode power supplies.
2.   To control a DC motor where the motor supply is a DC source.

1. OPERATING PRINCIPLE OF A CHOPPER

Fig. 12-1a shows a chopper circuit with resistive load. The switch may be a thyristor or a transistor. Here after when we refer to a switch, we mean the semiconductor and its control circuitry.

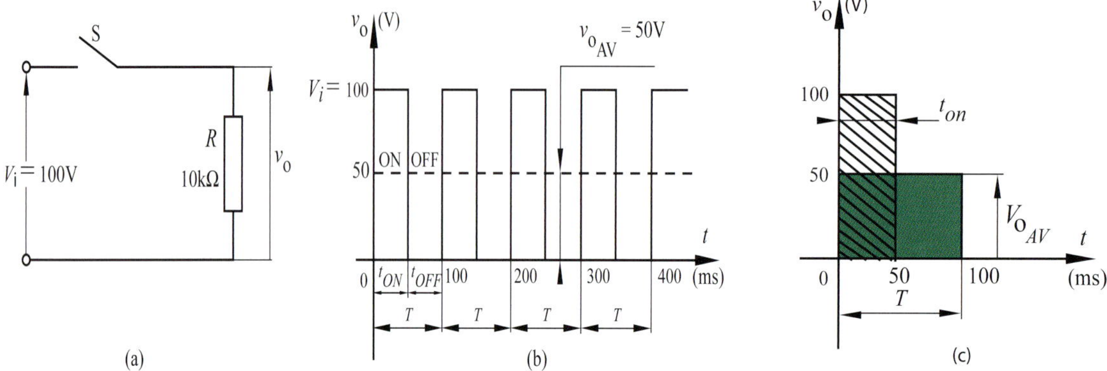

Fig. 12-1 Operating principle of a chopper

By consecutively open and closing the switch we obtain a pulsating DC voltage as shown in fig. 12-1b. From fig. 12-1c it follows that $V_{oAV} = 50V$.

We refer to the relative switch on ratio or duty cycle as :

$$\delta = \frac{t_{ON}}{T} \qquad (12\text{-}1)$$

The average output voltage is found by distributing the surface area $V_i \cdot t_{ON}$ over the time T:

$$V_{o\,AV} \cdot T = V_i \cdot t_{ON} \quad / \text{From which: } V_{o\,AV} = \frac{t_{ON}}{T} \cdot V_i$$

With $\delta = \dfrac{t_{ON}}{T}$ this becomes:

$$V_{o\,AV} = \delta \cdot V_i \qquad (12\text{-}2)$$

Numeric example 12-1:

A certain chopper has a supply voltage of 24V, the pulse frequency is 20kHz and the required output voltage is 5V. Calculate the duty cycle and the ON-time of the switch.

Solution:

$V_{oAV} = \delta \cdot V_i$ so that: $\delta = \dfrac{5}{24} = 0.2083$.

$$T = 1/f = \frac{1}{20.10^3} = 50\mu s$$

From $\delta = \dfrac{t_{ON}}{T}$ it follows that the ON-time of the switch is: $t_{ON} = 0.2083 \times 50.10^{-6} = 10.41\mu s$

Remark

The frequency used in example 12-1 is commonly used for switch mode power supplies. Normally a transistor is used as the switch.

2. CONTROL METHODS

From $\delta = \dfrac{t_{ON}}{T}$ we see that δ can be altered in two fundamental ways.

1. **Pulse width** control, in which the t_{ON} is controlled and T remains constant, see fig. 12-2.
2. **Pulse frequency** control in which T is controlled and t_{ON} remains constant, see fig. 12-3.

If the control follows a pattern we talk of modulation.
We refer to:
- PWM: pulse width modulation
- PFM: pulse frequency modulation.

A combination of pulse width modulation and pulse frequency modulation is also possible.

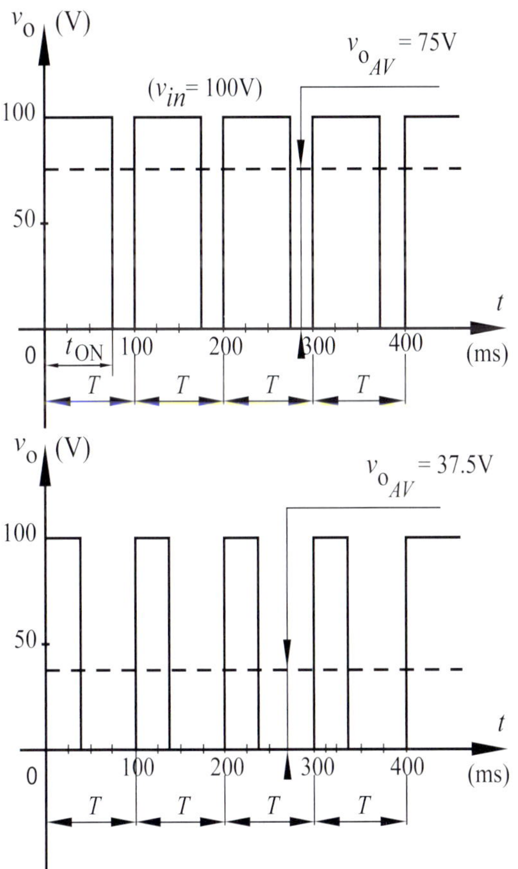

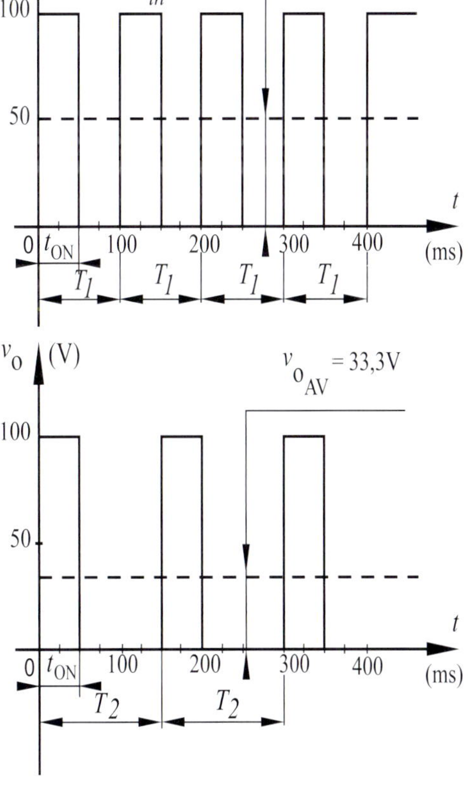

Fig. 12-2: Pulse width modulation

Fig 12-3: Pulse frequency modulation

3. RESISTIVE AND RESISTIVE-INDUCTIVE LOADED CHOPPERS

3.1 Resistive load

Fig. 12-1 shows the output voltage in the case of a resistive load. The current form of the output has the same shape as a function of time as the voltage v_o . We find:

$$V_{oAV} = \delta \cdot V_i \quad \text{and} \quad I_{AV} = \frac{V_i}{R_b} \cdot \delta \tag{12-3}$$

3.2 Resistive inductive load with $\tau \ll T$

Fig. 12-4 shows the basic configuration. Depending on the time constant $\tau = \dfrac{L_b}{R_b}$ and the chopper period T we can distinguish two practical situations:

- $\tau \ll T$
- $\tau \gg T$

In this paragraph we will study $\tau \ll T$. This is the case when the induction is low.
Fig. 12-5 shows the waveforms.
With $L_b = 35\text{mH}$, $R_b = 5\Omega$, $T = 200\text{ms}$, we find $\tau = \dfrac{L_b}{R_b} = \dfrac{35 \cdot 10^{-3}}{5} = 7\text{ms} \ll T$.

The voltage waveform at the output is identical to that of a resistive load since v_o is determined by the status of the MOSFET switch. For a conducting MOSFET is $v_o = v_i$ and for a blocking MOSFET we find $v_o = 0$.
During conduction by the MOSFET, the coil is charged with energy $W_i = \dfrac{L_b \cdot i_o^2}{2}$
When the MOSFET opens this energy $L_b \cdot i_o^2 / 2$ that is stored at that instant in the coil, will discharge via the freewheel diode D and the resistor R_b . After $5.\tau = 35\text{ms}$ this energy is as good as dissipated and converted to heat in the resistor R_b .

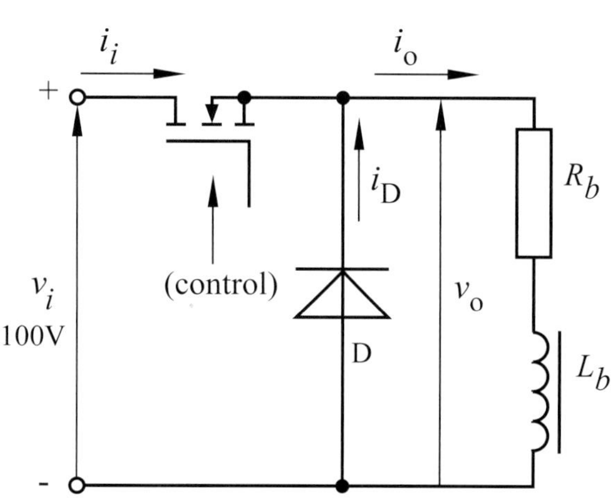

Fig. 12-4: Inductively loaded chopper with free wheel diode

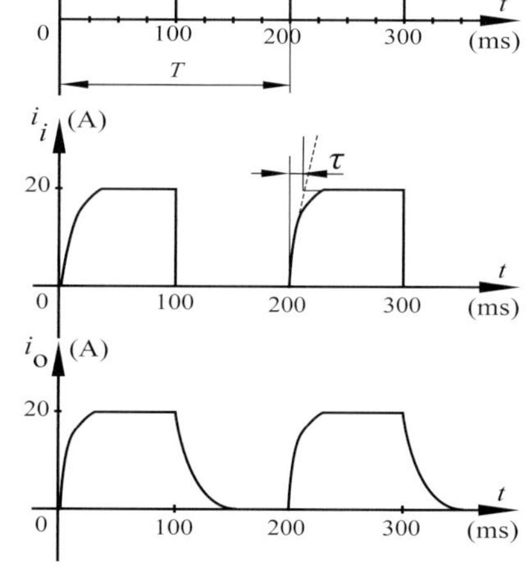

Fig 12-5: Waveforms associated with fig. 12-4

3.3 Resistive inductive load with $\tau \gg T$

In fig. 12-65 the time constant $\tau = \dfrac{L_b}{R_b} = \dfrac{100 \cdot 10^{-3}}{10} = 10\text{ms}$. The period time which operates on the switch is: $T = 0.1\text{ms}$, so that $\tau = 100 \cdot T$ or : $\tau \gg T$.

As a result of this large time constant discontinuous current does not occur in the output.

If the MOSFET is closed then the current i_o ($= i_i$) increases exponentially. If we neglect the voltage drop across the MOSFET then: $v_o = V_i$.

When the MOSFET is open then the current i_o decays exponentially via the free wheel diode D. Then $i_o = i_D$ and $v_o = - v_D \approx 0$.

We can now draw v_o , i_i , i_o and i_D in fig. 12-6. For simplicity sake in our drawing we assume that there is nominal service immediately from $t = 0$.

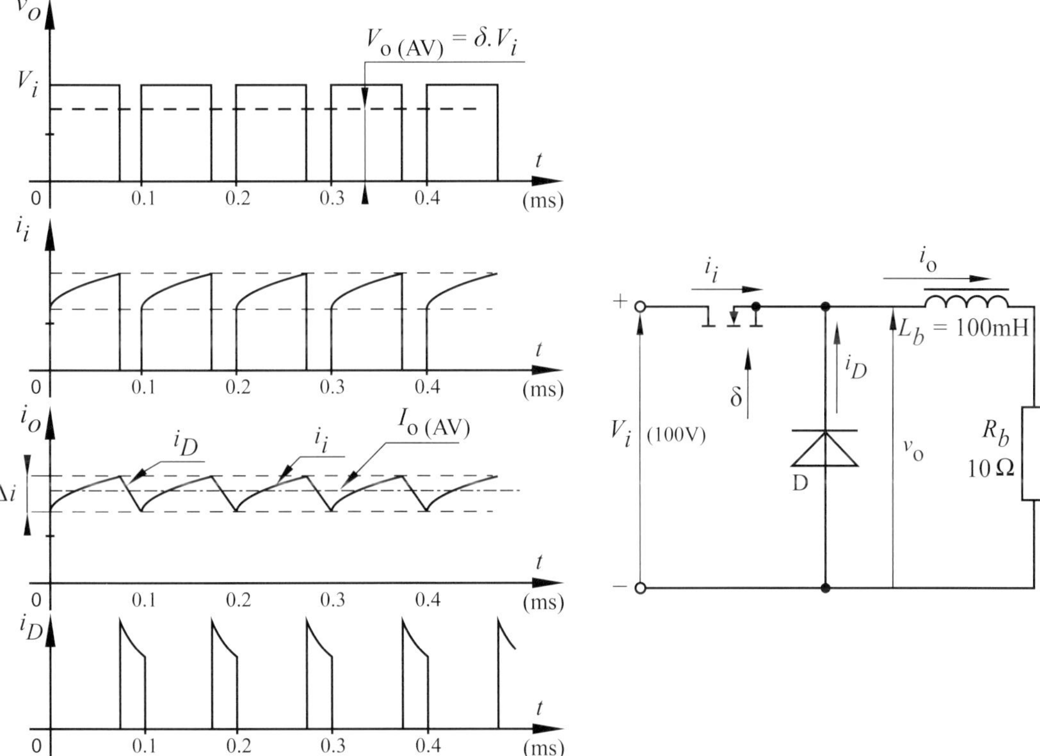

Fig. 12-6: Waveforms associated with highly inductive load

Calculation shows the ripple in the current to be:
$$\Delta i \approx \frac{V}{L_b} \cdot T \cdot (\delta - \delta^2) \tag{12-6}$$

The average output voltage remains: $V_{oAV} = \delta \cdot V_i$

From the principle of equal surface areas it follows that $v_{L_b AV} = 0$. The average output voltage V_{oAV} is completely across the resistor R_b , therefore:
$$I_{oAV} = \frac{V_i}{R_b} \cdot \delta \tag{12-7}$$

Numeric example 12-2:

Given:

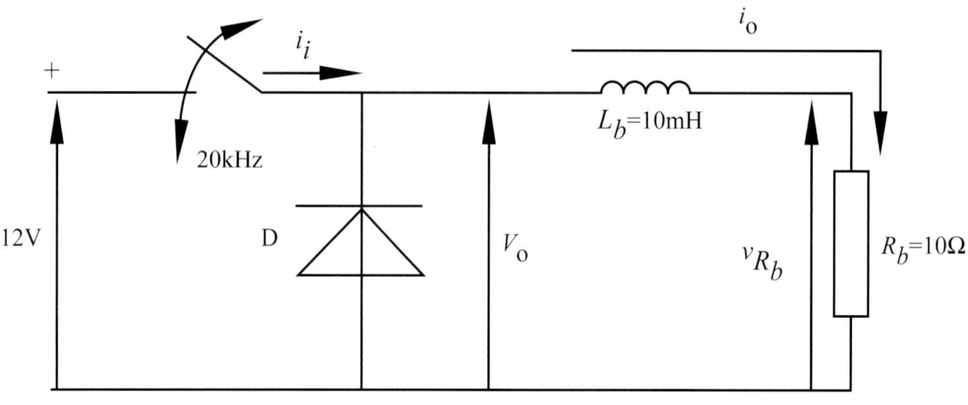

Fig. 12-7: Inductively loaded chopper

Required:

a.   with duty cycles of $\delta = 0$ and 1 , determine: I_{oAV} , Δi and v_o

b.   with $\delta = 0.1$ and $\delta = 0.7$, draw : v_o , i_o and v_{R_b} as a function of time

Solution:

a.   1. $\tau = \dfrac{L_b}{R_b} = \dfrac{10 \cdot 10^{-3}}{10} = 1\text{ms}$; $T = \dfrac{1}{f} = \dfrac{1}{20 \cdot 10^3} = 50\mu\text{s}$; $>> T$!

2. $\delta = 0$: switch is open most of the time: $I_{oAV} = 0$; $\Delta i = 0$; $v_o = 0$

3. $\delta = 1$: switch is closed most of the time:

$$\Delta i \approx \frac{V}{L_b} \cdot T \cdot (\delta - \delta^2) = 0 \;;\; I_{oAV} = \delta \cdot \frac{V_i}{R_b} = 1 \cdot \frac{12}{10} = 1.2\text{A}$$

b.   4. $\delta = 0.1$:

$$\Delta i \approx \frac{V}{L_b} \cdot T \cdot (\delta - \delta^2) = \frac{12}{10^{-2}} \cdot 50 \cdot 10^{-6} \cdot (0.1 - 0.01) = 5.4\text{mA}$$

$$I_{oAV} = \frac{V_i}{R_b} \cdot \delta = \frac{12}{10} \cdot 0.1 = 0.12\text{A} = 120\text{mA}$$

5. $\delta = 0.7$:

$$\Delta i \approx \frac{12}{10^{-2}} \cdot 50 \cdot 10^{-6} \cdot (0.7 - 0.49) = 12.6\text{mA}$$

$$I_{oAV} = \delta \cdot \frac{V_i}{R_b} = 0.7 \cdot \frac{12}{10} = 0.84\text{A}$$

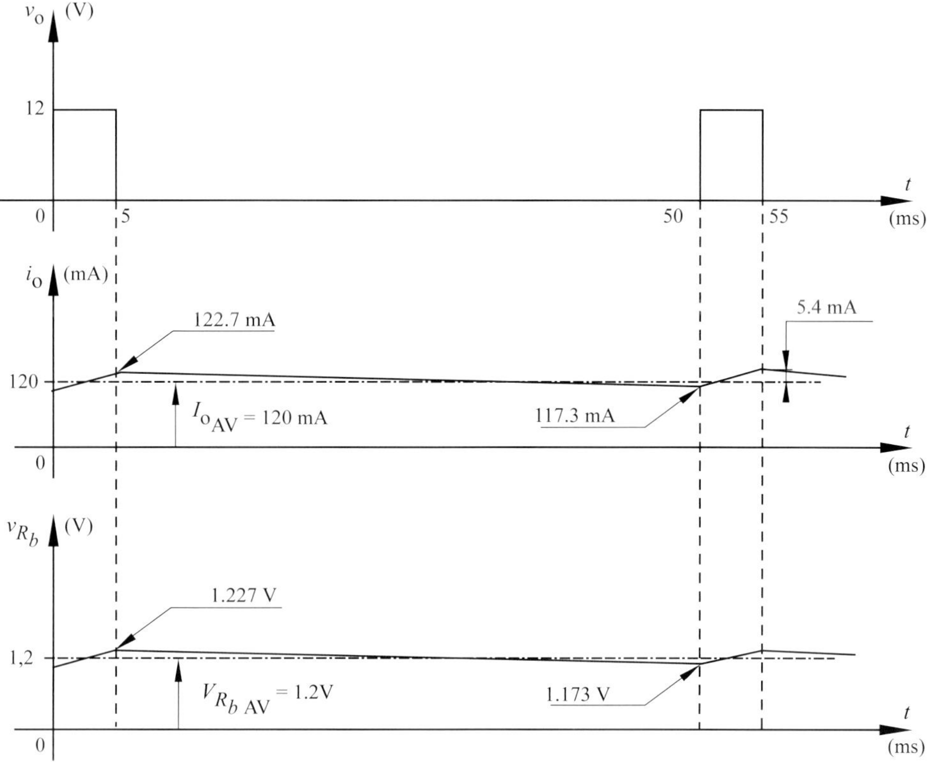

Fig. 12-8: Voltage and current waveforms in fig. 12-7 when $\delta = 0.1$

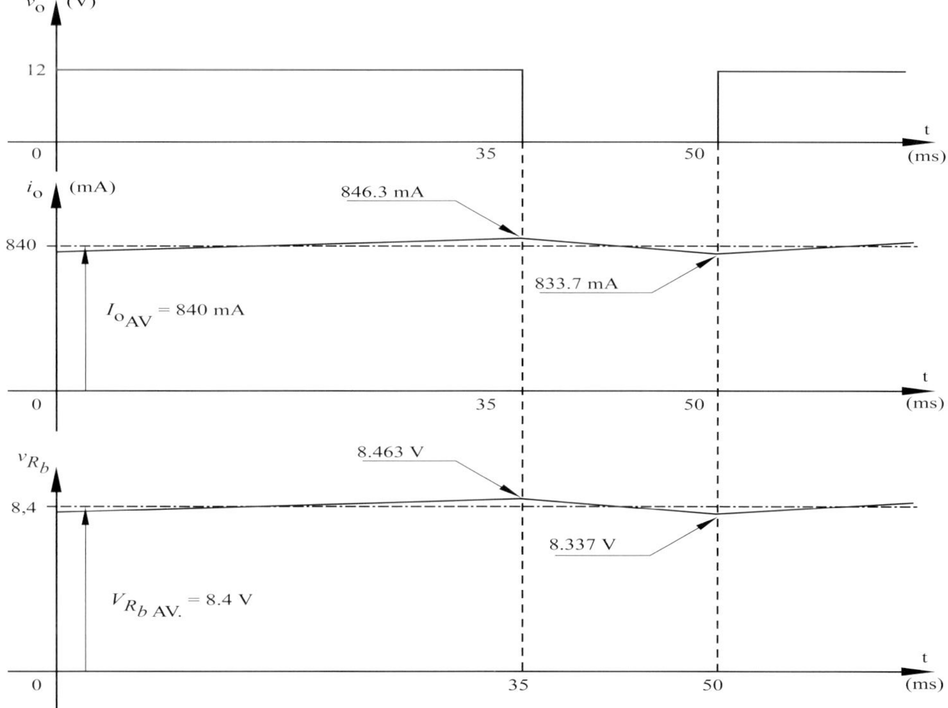

Fig. 12-9: Voltage and current waveforms in fig. 12-7 when $\delta = 0.7$

3.4 Calculating the current waveform through a resistive-inductive load when $\tau \gg T$

To mathematically describe the current flow we review the theory of a step response of a resistive inductive circuit.

3.4.1. Step response resistive-inductive circuit

If a current I_o is flowing at the instant that a step V is superimposed (fig. 12-10), then we can write:

$$I(s) = \frac{(V - I_o \cdot R_b)}{s} \cdot \frac{1}{s \cdot L_b + R_b} + I_o/s = \frac{V}{R_b} \cdot \frac{R_b/L_b}{s \cdot (s + R_b/L_b)} - I_o \cdot \frac{R_b/L_b}{s \cdot (s + R_b/L_b)} + I_o/s$$

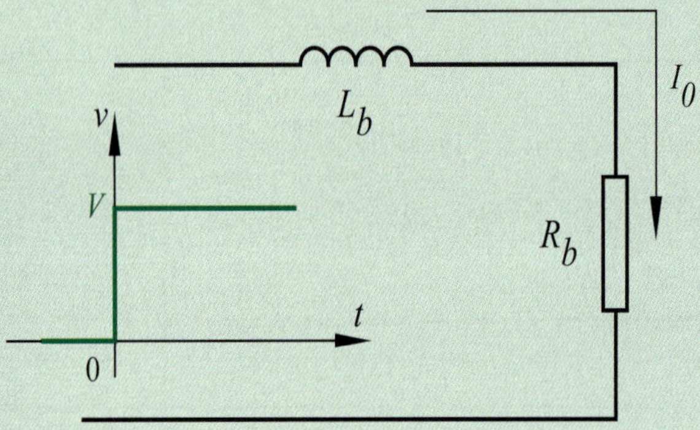

Fig. 12-10: Step response R-L - circuit

Reverse Laplace transform: $\mathscr{L}^{-1}\left[\dfrac{\alpha}{s \cdot (s + \alpha)}\right] = 1 - e^{-\alpha t}$ and with $\tau = \dfrac{L_b}{R_b} \longrightarrow\longrightarrow$

$$i(t) = \frac{V}{R_b} \cdot (1 - e^{-t/\tau}) - I_o \cdot (1 - e^{-t/\tau}) + I_o \longrightarrow\longrightarrow \qquad \boxed{i(t) = \frac{V}{R_b} + (I_o - \frac{V}{R_b}) \cdot e^{-t/\tau}} \qquad (12\text{-}4)$$

3.4.2. Current waveform

Due to the large time constant of the load discontinuous current does not occur. With a closed MOSFET (- switch) i_o increases exponentially and with an open switch it decreases exponentially through the freewheel diode. This current is drawn in fig. 12-11 and for simplicity sake we assume nominal values from the instant $t = 0$. The chopper duty ratio is δ .

Interval $0 < t < t_1$: Current increases: $(12\text{-}4) \rightarrow i(t) = \dfrac{V}{R_b} + (I_1 - \dfrac{V}{R_b}) \cdot e^{-t/\tau}$

Interval $t_1 < t < t_2$: Current decreases: $(12\text{-}4) \rightarrow i(t) = 0 + (I_2 - 0) \cdot e^{-(t - \delta \cdot T)/\tau}$

Increasing current: on $t = t_1 = \delta \cdot T \rightarrow i(t) = I_2 = \dfrac{V}{R_b} + (I_1 - \dfrac{V}{R_b}) \cdot e^{-\delta \cdot T/t}$ $\qquad (1)$

Decreasing current: on $t = t_2 = T \rightarrow i(t) = I_1 = I_2 \cdot e^{-(T - \delta \cdot T)/\tau} = I_2 \cdot e^{-(1 - \delta) \cdot T/\tau}$ $\qquad (2)$

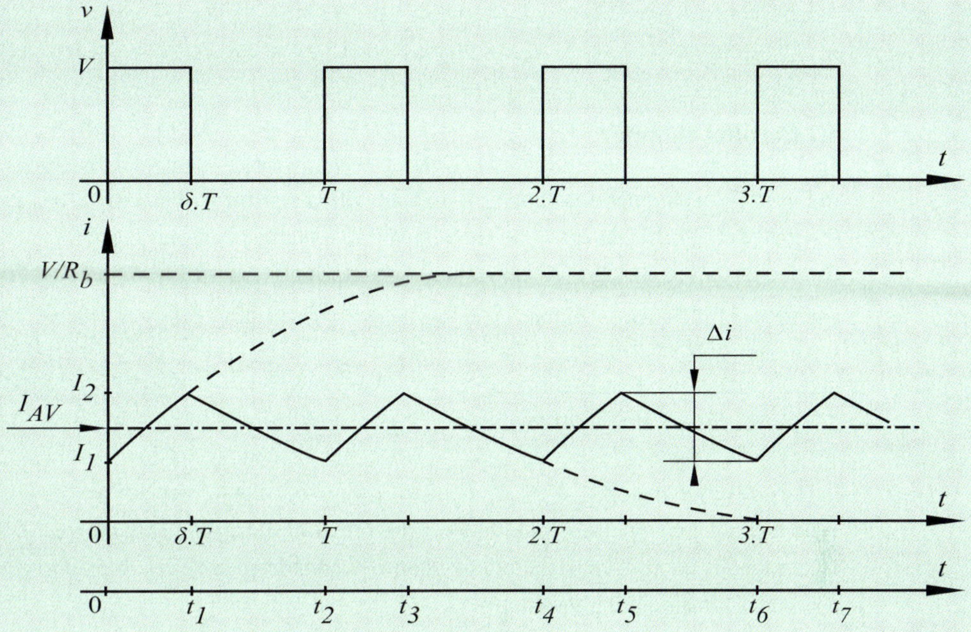

Fig. 12-11: Current flow in a highly inductive circuit

(2) in (1) $\rightarrow\rightarrow$ $I_2 = \dfrac{V}{R_b} + (I_2 \cdot e^{-(1-\delta).T/\tau} - \dfrac{V}{R_b}) \cdot e^{-\delta.T/\tau}$; $I_2 = \dfrac{V}{R_b} \cdot \dfrac{(1 - e^{-\delta.T/\tau})}{(1 - e^{-T/\tau})}$

a) Current ripple: $\Delta i = I_2 - I_1 = I_2 - I_2 \cdot e^{-(1-\delta).T/\tau}$

$$\Delta i = \frac{V}{R_b} \cdot \frac{(1 - e^{-\delta.T/\tau})}{(1 - e^{-T/\tau})} \cdot (1 - e^{-(1-\delta).T/\tau}) \tag{12-5}$$

From the Mac Laurin series it follows:

$e^{-x} \approx 1 - x$ (for small values of x) , so that: $1 - e^{-x} \approx x$

Since $0 < \delta < 1$ we find with $\tau >> T$ that : $\delta.T/\tau << 1$ and $(1-\delta).T/\tau << 1$

The current ripple becomes then: $\Delta i \approx \dfrac{V}{R_b} \cdot \dfrac{[(\delta.T/\tau) \cdot (1-\delta).T/\tau]}{T/\tau}$

$$\Delta i \approx \frac{V}{L_b} \cdot T \cdot (\delta - \delta^2) \tag{12-6}$$

b) Average current:

Using the equal surface area criterion (p. 7-8) it follows that $V_{L_b AV} = 0$.

The average output voltage $V_{oAV} = \delta \cdot V$ is fully across the resistor R_b, so that :

$$I_{AV} = \frac{V}{R_b} \cdot \delta \tag{12-7}$$

4. CHOPPED RESISTOR

A resistor can be varied by placing a chopper either in series or in parallel with the resistor.

4.1 Parallel chopper

In fig. 12-12 a chopper S is shown in parallel with a resistor R. The circuit has an equivalent resistance R_v which can easily be calculated. We assume a constant current I_a and consider that the chopper is closed during $t_{ON} = \delta \cdot T$ and open during $t_{OFF} = (1 - \delta) \cdot T$. We can now write the equations of the energy balance:

$$\underbrace{I_a^2 \cdot \delta \cdot T \cdot 0}_{(\text{ S closed })} + \underbrace{I_a^2 \cdot (1 - \delta) \cdot T \cdot R}_{(\text{ S open })} = I_a^2 \cdot R_v \cdot T \quad \text{so that:} \quad \boxed{R_v = (1 - \delta) \cdot R} \qquad (12\text{-}8)$$

In (12-8) we see that: $0 < R_v < R$.

With a parallel chopper it is possible to vary a resistor of say $R = 10\Omega$ between 0 and 10Ω. To keep the current as good as constant a choke coil L is connected in series. This prevents also the chopper from short-circuiting the supply.

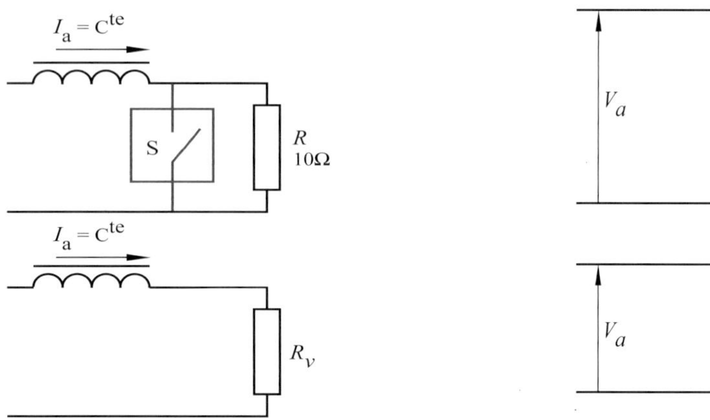

<table>
<tr><td>Fig. 12-12: Parallel chopper</td><td>Fig. 12-13: Series chopper</td></tr>
</table>

4.2 Series chopper

In fig. 12-13 chopper S is connected in series with resistor R. If we assume a constant V_a, then we can calculate the equivalent resistance R_v from the energy balance :

$$\frac{V_a^2}{R} \cdot \delta \cdot T + \frac{V_a^2}{\infty} \cdot (1 - \delta) \cdot T = \frac{V_a^2}{R_v} \cdot T \quad \text{so that:} \quad \boxed{R_v = \frac{R}{\delta}} \qquad (12\text{-}9)$$

From (12-9) it follows that $R < R_v < \infty$.

With a chopper in series we can for example vary a resistance $R = 100\Omega$ between 100Ω and ∞.

4.3 Applications

Chopped resistors occur for example as brake resistors in variable speed drives.

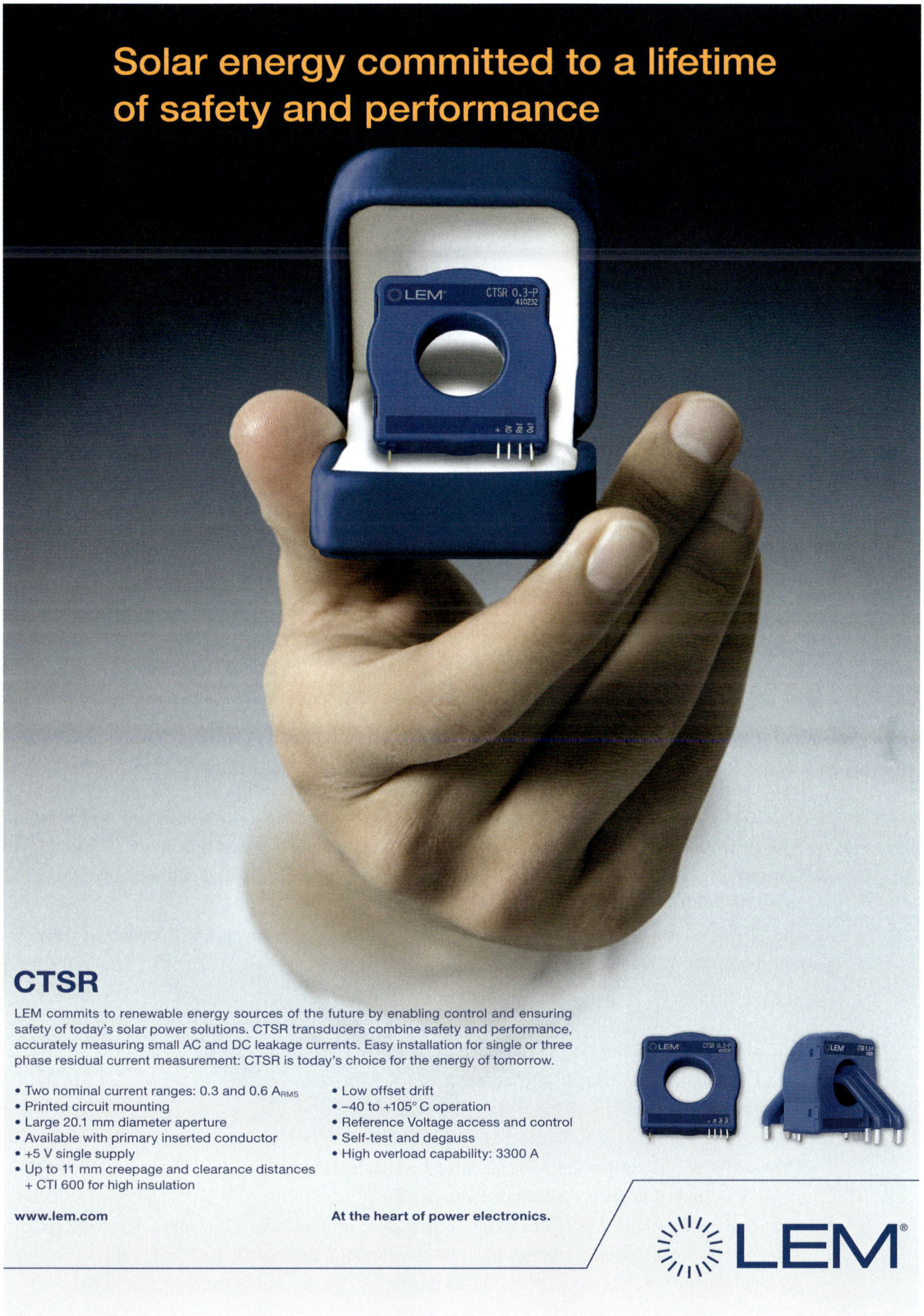

5. CHOPPER CONTROL IC'S

5.1 Description of block diagram

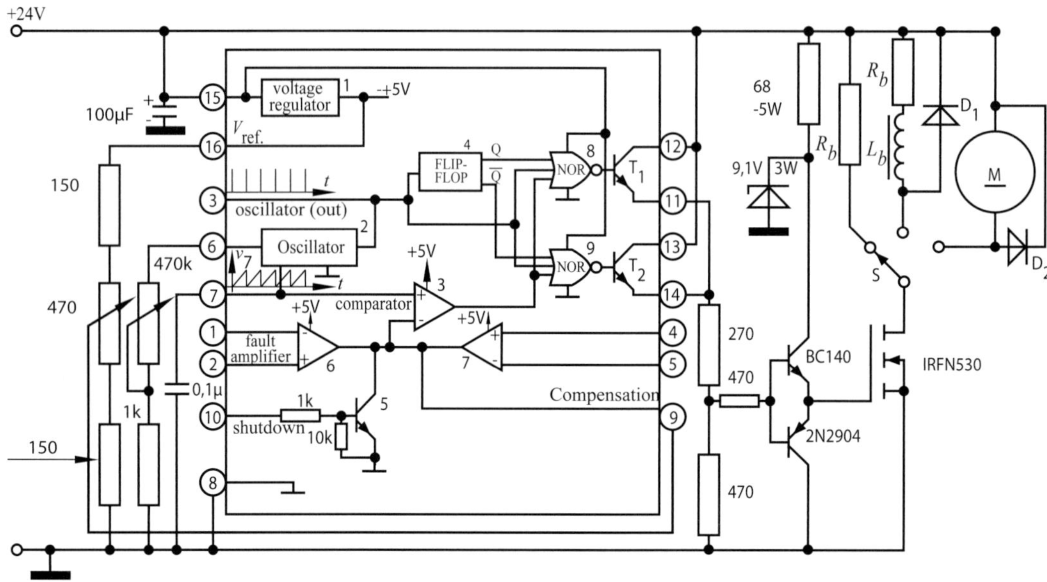

Fig. 12-14: A didactic model of a chopper with a "1524" control IC. This IC works with a supply between 8 and 40V(max), has an oscillator frequency between 100Hz and 500kHz and has two output transistors (100mA) that can handle a maximum collector voltage of 60V

The "1524" and its derivatives is a much used IC-controller. It first appeared in 1976 as the "SG1524" (SG = Silicon General) and quickly became the industry standard and to this day is offered by many chip producers (CA1524, LM1524, UC1524, XR1524,…).

Fig. 12-14 shows a didactic model which allows us to study a chopper circuit when PWM is used. With switch S we can choose between a resistive load , an inductive load or a motor as load. The interface between IC and switch S we recognise as a control circuit from the study of the powerMOSFET.

The "1524"is a pulse width modulated controller for voltage controllers. The frequency is determined from the following expression:

$$f = \frac{1.18}{R_6 \cdot C_7}$$

(12-10)

Here in is f in kHz with R_6 in kΩ and C_7 in μF.

With R_6 and C_7 we mean the discrete components that are connected between pin 6 and 7 of the IC. Practical values are: $1.8 \text{ kΩ} < R_6 < 500 \text{ kΩ}$

$$0.001\,μF < C_7 < 0.1\,μF$$

The value of R_6 determines the constant current which linearly charges C_7. The sawtooth of +1V to +3.5V is applied to the (+) terminal of comparator 3.

A flip flop makes sure that only one of the NOR gates 8 or 9 receives a low input. The NOR gate which has all inputs low will saturate its respective output transistors T_1 or T_2. For this to occur the following must be the case: $V_3 = 0V$ and the output of comparator 3 should also be 0V.

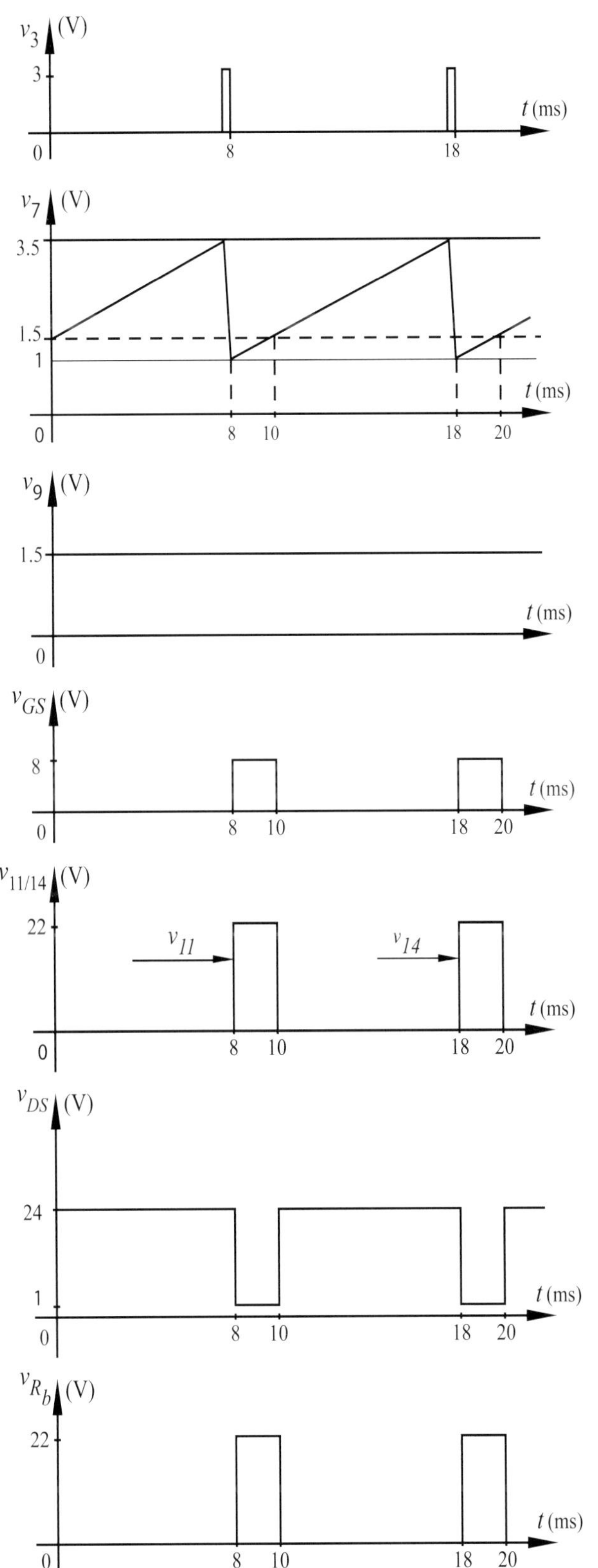

Fig. 12-15: Waveform associated with fig. 12-14

An internal voltage controller 1 regulates the 5V($\pm$ 1%) supply of the IC and which can be externally loaded to 20mA.

The output pulse V_3 of oscillator 2 is a "blanking" pulse who's function is to prevent T_1 and T_2 conducting at the same time. This is necessary if the transistors are not in parallel as shown in fig. 12-14, but where for example they operate separately to control a balance circuit. In this case the frequency of the output voltage is half of the sawtooth frequency.

5.2 Operation of comparator 3

C_7 provides a sawtooth voltage of +1 to 3.5V to the non inverting terminal (+) of comparator 3. On the inverting input ($-$) of the comparator a voltage is applied with four possibilities:

A. The inverting input is smaller than +1V so that the output of the comparator is high (+5V) and both output transistors are blocking.
   This can be achieved by:
   1) saturating transistor 5 via pin 10 so that V_{CE} < 1V and the IC blocks within 200ns (shut down protection).
   2) using opamp 7 as a current limiter.

B. A voltage of between +1V and +3.5V is added to the inverting input of the comparator. As a result the pulse width at the IC output is controlled.
   We can achieve this by:
   3) linearly controlling the duty cycle via the fault amplifier (6).
   4) adding a voltage V_9 as shown in fig. 12-14 which also regulates the duty cycle.
   In this manner we see in fig. 12-15 that with $V_9 = +1.5$V there is conduction between 8 and 10ms (V_{11}) and between 18 and 20 ms (V_{14})

6. EVALUATION

12.1   What does PWM and PFM mean ?

12.2   a.   What is the purpose of a free wheel diode in fig. 12-4 ?
b.   Sketch the current flow through this diode.

12.3   In the configuration shown in fig. 12-6 the chopper frequency is 10kHz and $\delta = 0.05$. Determine V_{R_b} and the current ripple Δi . Sketch v_o and v_{R_b} as a function of time.

12.4   Determine the amplitude of the ripple current and the average current for the configuration shown in fig. 12-6 if the chopper frequency is 10kHz and the duty cycle is respectively 0, 0.2, 0.5, 0.8 and 1.

12.5   If both potmeters in fig. 12-14 are set exactly to the half, determine:
a.   the frequency at which the MOSFET is operated.
b.   the duty cycle of the chopper.

12.6   Assume the same details as in question 12.5. The switch S is set to the inductive load. If $R_b = 3\Omega$ and $L_b = 1.5H$, determine the average current through the load and also the amplitude of the ripple current.

12.7   What is the purpose of a chopper ?

12.8   Where are chopped resistors used ?

12.9   How do we prevent a (parallel) chopper switch from short-circuiting the supply ?

12.10   What is a comparator? A NOR gate ?

Photo Elipse: A 19 inch rack supply from the brand Premium for use in trains. It is developed to withstand high levels of vibration and mechanical shocks and is insensitive to high input peaks.
Operates across a wide temperature range.

13 SWITCH-MODE POWER SUPPLIES

CONTENTS

1. Basic principles of switch-mode power supplies
2. Basis converter configurations
   . Buck converter
   . Boost converter
   . Buck-Boost converter
3. Isolated switch-mode power supplies
4. Flyback converter
5. Forward converter
6. Converter control strategies
7. Two-transistor SMPS of the forward type
8. Forward with multiple outputs
9. Full bridge of the buck type
10. Synchronous SMPS
11. SMPS components
12. Overview of SMPS up to 2500W
13. Non ideal waveform
14. Digital control of an SMPS
15. Evaluation
16. The design of switch-mode power supplies

To stabilise a DC voltage, linear series controllers are often used. The efficiency of linear voltage regulators varies between 30 and 50%. With a switch-mode power supply under the same conditions the efficiency varies between 70 and 90%.

The use of linear voltage regulators is therefore limited in practice to power levels under 30W. Linear voltage regulators have a larger volume and are heavier than their switch-mode power supply brothers, but they are cheaper and less complicated and do have better EMI performance. Applications for switch-mode power supplies include motor control, battery chargers, low voltage installations (e.g. halogen lamps), TV-receivers, film projectors, calculators, computer power supplies, voltage and current stabilisers for industrial and laboratory applications, galvanically isolated power supplies with respect to the network supply, power supplies in satellites (converting 150V from the solar panels to the 5V-rail) etc.

1. BASIC PRINCIPLES OF SWITCH-MODE POWER SUPPLIES

The main component of a switched mode controller is the chopper. Normally the chopper operates with a fixed frequency. In order to obtain a variable output use is made of pulse width modulation. Filtering the chopper output results in a stabilised DC voltage.

Fig. 13-1 shows the basic operating principle of a switch-mode power supply.

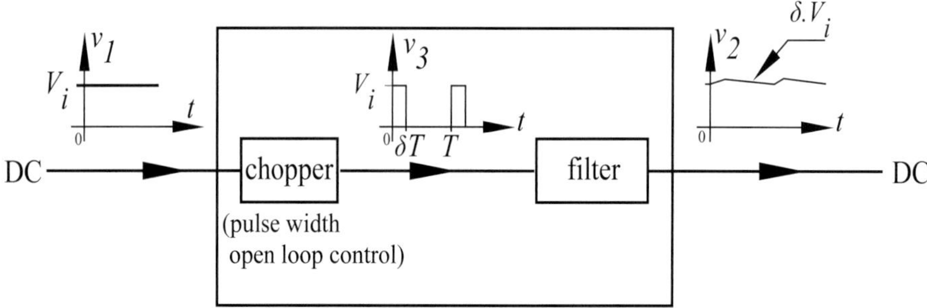

Fig. 13-1: Principle of a switch-mode power supply

2. BASIC CONVERTER CONFIGURATIONS

There are two main operating principles of switched mode power supplies:
- "buck" converter.
- "flyback" converter. The flyback type is split into boost and buck-boost types. In the literature the term flyback is mostly reserved for isolated buck-boost configurations which will be studied later.

2.1 Buck converter

2.1.1 Configuration and operation
Fig. 13-2 shows the basic principle and configuration.

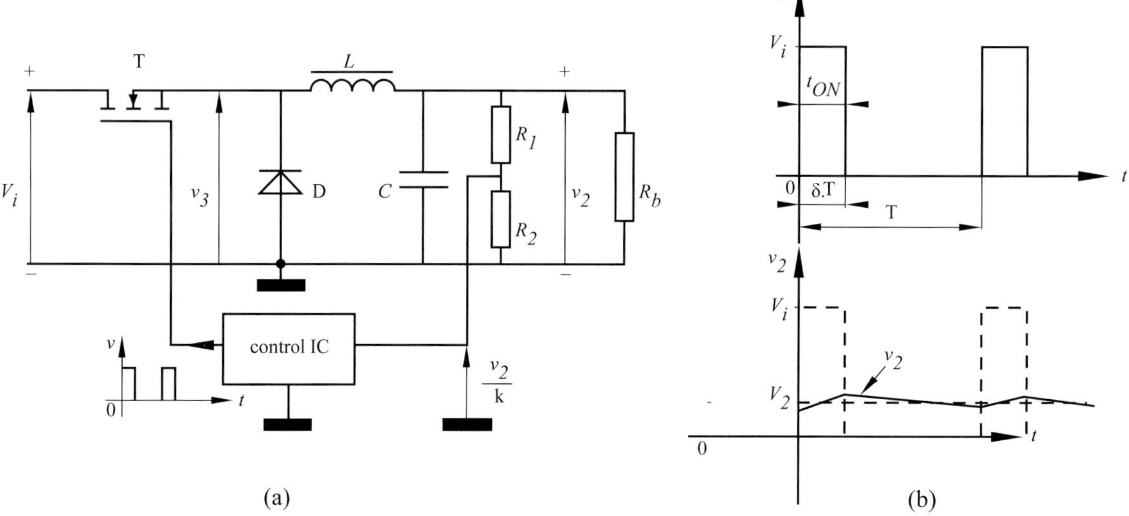

(a)   (b)

Fig. 13-2: Drawing of the basic operation principle of a buck converter

A series transistor T functions as a switch. A control circuit, usually in the form of an IC
(TDA4601 Infineon, SL442 Plessey, "3524" from RCA, Silicon General, Unitrode,…) allows the
duty cycle $\delta = t_{ON}/T$ to be varied, so that a voltage v_3 is obtained as shown in fig. 13-2b.
Consider a conducting MOSFET. The voltage drop V_{DS} is small with respect to the practical values
of V_i so that a good approximation is found with

$$V_{3AV} = V_i \cdot \frac{t_{ON}}{T} = V_i \cdot \delta.$$

To obtain a constant output DC voltage V_2 a filter is required. The simplest type of filter is an
LC-configuration. Usually the switching frequency is between 20 and 100kHz resulting in filter
elements with small values and small dimensions. Diode D is a freewheel diode and serves to
discharge the magnetic energy of the inductor. This is explained later.
If the filtering is sufficient $V_2 \approx V_{3AV} = V_i \cdot \delta$. Here in V_2 is the average output voltage of the
converter.

From $V_2 \approx V_i \cdot \delta$ it follows that:
1. V_2 is variable by varying δ,
2. fluctuations in V_2 can be compensated for by varying δ. This occurs as a result of the feedback
    of v_2/k.

Before considering the basic operating principle with the associated formulas we review the
protocols for indicating the voltages and currents associated with the passive elements R, L, C.
This is shown in fig. 13-3.

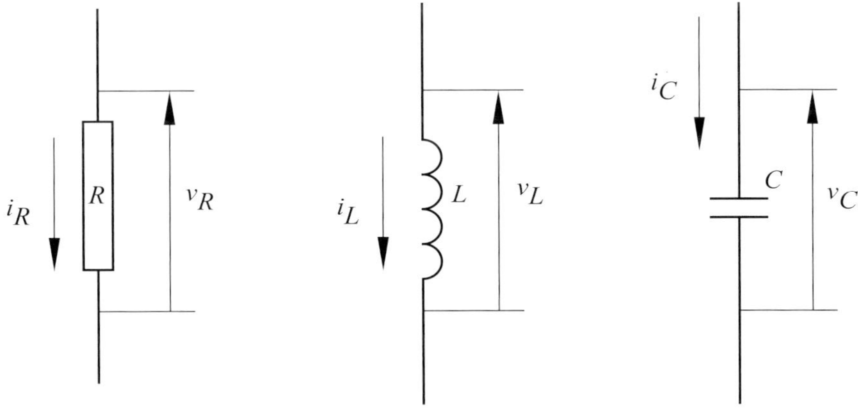

Fig. 13-3: Notation for current and voltage drop with R, L and C

By fig. 13-3 we can write:

$$v_R = i_R \cdot R \ ; \ \ v_L = L \cdot \frac{di}{dt} \ ; \ \ v_C = \frac{1}{C}\int_0^t i \cdot dt$$

The polarity of v_L is indicated in a similar fashion to v_R, even though v_L is not dependent on the
current direction of i_L but rather on the current rate of change di/dt ! In fact the polarity of
v_L is determined by the polarity of di/dt.

We redraw fig. 13-2 now in fig. 13-4. In this drawing we consider the transistor as a switch. This transistor is controlled by a pulse width modulated signal. The duty cycle δ will determine the average value of output voltage. This average value is referred to as V_o. When the switch S closes a current i_L flows and a V_o is produced with the polarity as indicated. Diode D is reverse biased. When S opens the magnetic energy, stored in the core of the coil, will discharge via $C//R_b$ and D. D is therefore referred to as a "free-wheel" or "flywheel" diode.

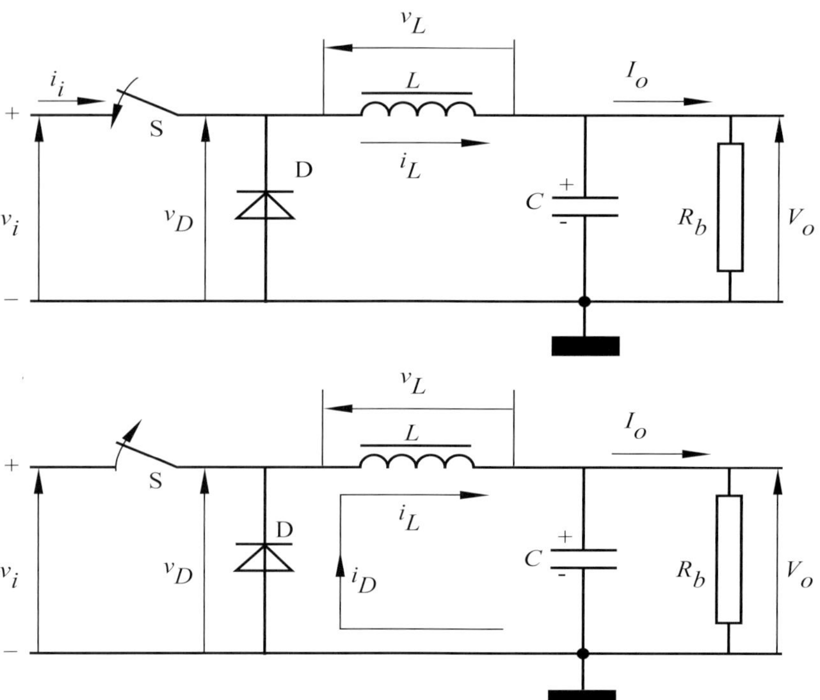

Fig. 13-4: Operating principle of a buck converter

The input current i_i flows uninterrupted while the current through the inductor may be continuous or discontinuous. If all the energy in the coil is discharged before the transistor switch closes we get discontinuous current in the inductor. This is referred to as discontinuous current. If S is closed before i_L becomes zero we get continuous current (rising and falling) through the inductor. Fig. 13-5 shows a number of waveforms for these operating modes (continuous and discontinuous current mode).

2.1.2 Continuous current through the inductor

We consider the case when steady state has been reached.
When the switch is closed a voltage $V_i - V_o$ is across the inductor and may be written as:

$$v_L = V_i - V_o = L \cdot \frac{\Delta i}{\Delta t} = L \cdot \frac{(I_1 - I_2)}{\delta \cdot T} \quad \text{or:} \quad L \cdot (I_1 - I_2) = \delta \cdot T \cdot (V_i - V_o) \tag{13-1}$$

When the switch is open $v_D \approx 0$ and $- V_o$ is across the inductor so that :

$$V_L = - V_o = L \cdot \frac{\Delta i}{\Delta t} = L \cdot \frac{(I_2 - I_1)}{(1 - \delta) \cdot T} \quad \text{or:} \quad L \cdot (I_1 - I_2) = V_o \cdot (1 - \delta) \cdot T \tag{13-2}$$

From (13-1) and (13-2) it follows:

$$(V_i - V_o) \cdot \delta \cdot T = V_o \cdot (1 - \delta) \cdot T \quad \text{or:} \quad \boxed{V_o = \delta \cdot V_i} \tag{13-3}$$

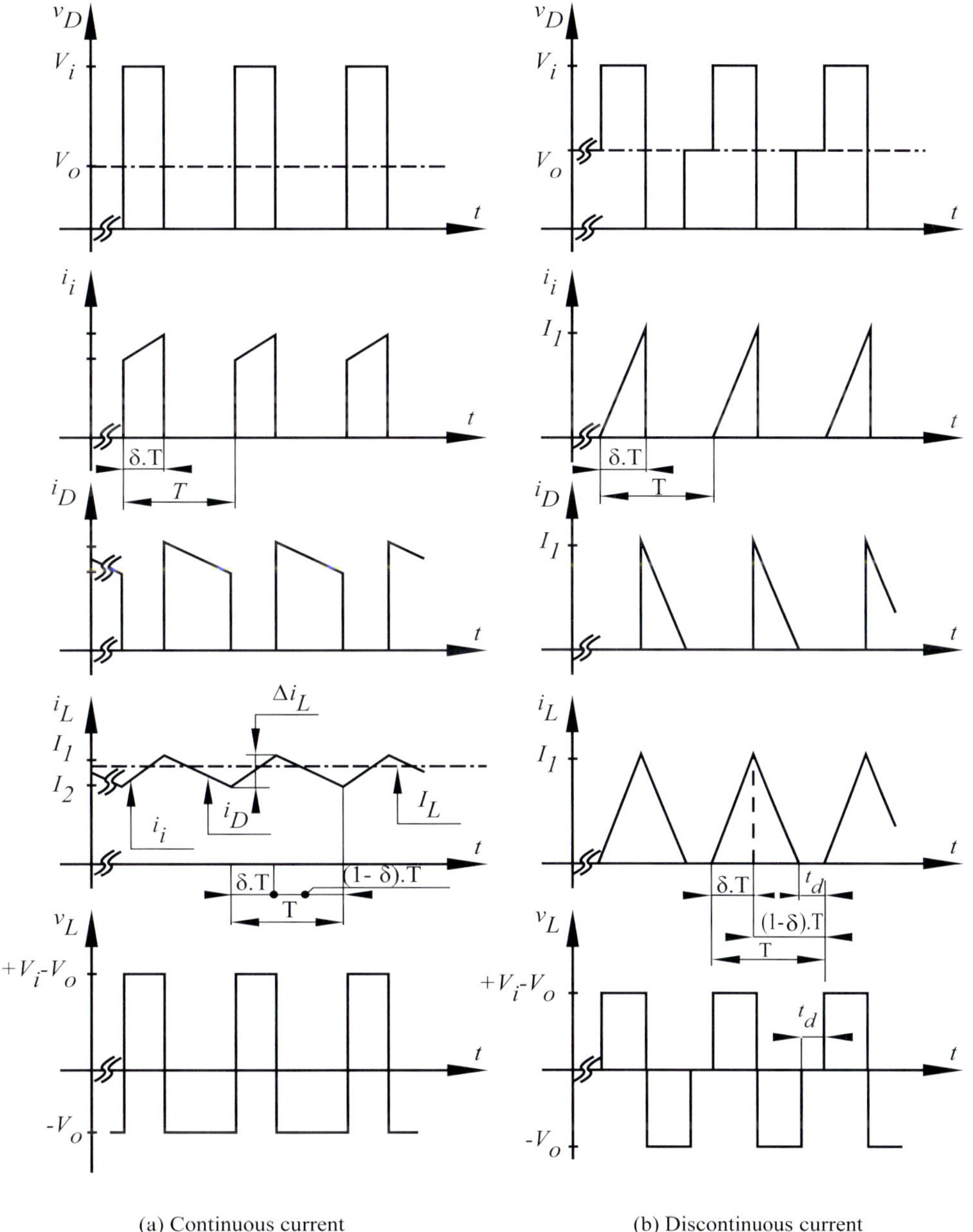

(a) Continuous current   (b) Discontinuous current

Fig. 13-5: Waveforms of fig. 13-4

2.1.3 Discontinuous current in the inductor

We now consider the case of discontinuous current. The current in the inductor or coil is interrupted. The switch is closed for a time $\delta.T$ and i_L will rise from zero to maximum in this time $\delta.T$. When S opens I_1 will fall during the time $(1-\delta).T-t_d$.

Here in is t_d the dead time in which no current flows in the inductor.

In the time interval t_d the output voltage V_o is also across the diode D.

In accord with fig. 13-4 and fig. 13-5b we find for the rising current (closed switch):

$$V_L = V_i - V_o = L.\frac{\Delta i}{\Delta t} = L.\frac{(I_1-0)}{\delta.T} \quad \text{or:} \quad L.I_1 = (V_i - V_o).\delta.T \tag{13-4}$$

When the switch opens and the current I_L falls to zero we can write:

$$V_L = -V_o = L.\frac{\Delta i}{\Delta t} = L.\frac{(0-I_1)}{(1-\delta).T-t_d} \quad \text{or:} \quad L.I_1 = V_o.\left[\,(1-\delta).T-t_d\,\right] \tag{13-5}$$

From (13-4) and (13-5) it follows that: $(V_i - V_o).\delta.T = V_o.(T-\delta.T-t_d)$

$$V_i.\delta.T = V_o.(\delta.T+T-\delta.T-t_d) \rightarrow\rightarrow \quad \boxed{V_o = V_i.\frac{\delta}{1-t_d/T}} \tag{13-6}$$

2.1.4 Remarks

1. With continuous current in the inductor is $V_o = \delta.V_i < V_i$
2. With discontinuous current it is always the case that $t_d < (1-\delta).T$ or $t_d/T < 1-\delta$; $-t_d/T > \delta - 1 \rightarrow\rightarrow (1-t_d/T) > \delta$ so that from (13-6) it follows that $V_o < V_i$
3. From the previous remarks it is obvious that a buck converter is a voltage reducing converter. That explains the term "buck". Sometimes referred to as a "step-down" converter.
4. The maximum voltage across the transistor is V_i since $v_D \approx 0$ with an open switch.
5. The maximum reverse voltage across the diode is V_i , this with a closed switch!.
6. Filter capacitor.

In fig. 13-6 we redraw the output circuit with associated waveforms. With sufficient capacitance the ripple in the output current (fig. 13-6) is negligible since $i_C = \Delta i_L$.

During $\dfrac{t_{ON}+t_{OFF}}{2} = \dfrac{T}{2}$ the average capacitor current flows $I_{C(AV)} = \dfrac{\Delta i_L}{2}/2 = \dfrac{\Delta i_L}{4}$.

This corresponds to a quantity of electricity : $\Delta Q = \dfrac{T}{2}.\Delta i_L/4 = \dfrac{\Delta i_L}{8.f}$. The voltage fluctuation

across the filter capacitor is then: $\Delta u_C = \dfrac{\Delta Q}{C} = \dfrac{\Delta i_L}{8.f.C}$,

so that: $\boxed{C_{min} = \dfrac{\Delta i_L}{8.f.\Delta u_{Cmax}}} \tag{13-7}$

The smoothing capacitor also has an ESR (Equivalent Series Resistance), so that the ripple current $i_C = (\Delta i_L)$ results in a voltage drop across this resistance: $\Delta v_{ESR} = R_{ESR}.\Delta i_L$.

The output voltage ripple across R_b is the vector sum of Δv_C and Δv_{ESR} .

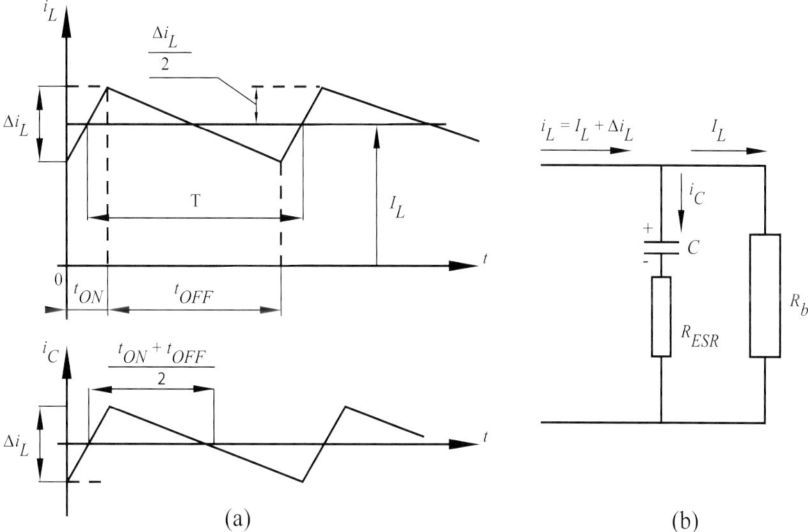

Fig. 13-6: Currents in the output circuit

Often $\Delta v_{ESR} >> \Delta v_C$ so that in the case of peak to peak output ripple a good a approximation is provided by: $\Delta v_o \approx \Delta i_L \cdot R_{ESR}$, it follows: $R_{ESRmax} = \dfrac{\Delta v_{o\,max}}{\Delta i_L}$.

2.1.5 Numeric example 13-1:

The following data are given concerning a buck converter:

- input voltage: $V_i = 8$ to 15V
- output voltage: $V_o = 5V \pm 0.1\%$
- maximum load current: $I_o = 2A$
- chopper frequency: 100kHz

Required:

- calculate the value of the choke and the filter capacitor.

Solution:

From (13-2) it follows: $L = \dfrac{V_o \cdot (1 - \delta) \cdot T}{\Delta i_L}$. The largest value of L is clearly the case where $\delta = \delta_{min}$ (for $V_i = V_{i\,max}$), so that: $L_{max} = \dfrac{V_o \cdot T \cdot (1 - \delta_{min})}{\Delta i_L}$ or: $L = \dfrac{V_o \cdot t_{off\,max}}{\Delta i_L}$.

A rule of thumb is $\Delta i_L \leq 25\% \cdot I_o$.

With $\Delta i_L = 20\% \cdot I_o = 0.4A$ and $t_{off\,max} = (1 - \delta_{min}) \cdot T = (1 - \frac{5}{15}) \cdot 10^{-5} = 6.67\mu s$ we then find for the self inductance : $L = \dfrac{5 \times 6.67 \times 10^{-6}}{0.4} = 83.4\mu H$.

With $\Delta v_{Cmax} = \Delta v_{o\,max} = 0.1\% \cdot 5 = 5mV$, we find with (13-7) $\rightarrow\rightarrow$

$$C_{min} = \dfrac{0.4}{8 \cdot 10^5 \cdot 5 \cdot 10^{-3}} = 100\mu F \ .$$

Finally: $ESR \leq \dfrac{\Delta v_{o\,max}}{\Delta i_L} = \dfrac{5 \cdot 10^{-3}}{0.4} = 0.0125\Omega$.

2.2 Boost converter

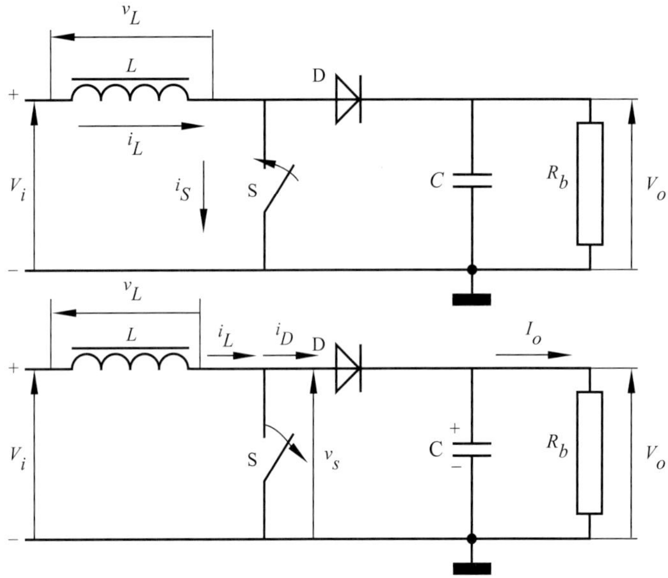

Fig. 13-7: Operating principle of a boost converter

Rearranging the elements (coil, switch and diode) of fig. 13-4 results in a boost converter (fig. 13-7) and a buck-boost converter (fig. 13-9).
Both types are flyback converters. The explanation of the term flyback can be found on p. 13.11.

2.2.1 Operation of a boost converter

During the time that the switch S is closed (fig. 13-7) a current i_L flows and magnetic energy $\dfrac{L \cdot i_L^2}{2}$ is stored in the coil. When the switch S opens the inductor energy together with V_i of the supply is responsible for the output voltage with average value V_o . Here again we distinguish two operating modes, namely continuous and discontinuous current modes.

2.2.2 Continuous current in the coil

Fig. 13-8a shows a number of waveforms for continuous current mode.

When the switch S is closed (during a time $\delta.T$) we see in fig. 13-7 that: $v_L = V_i = L \cdot \dfrac{\Delta i}{\Delta t}$.

From fig. 13-8a it follows that $\Delta i_L = I_1 - I_2$ and $\Delta t = \delta.T$, so that:

$$\Delta i = I_1 - I_2 = V_i \cdot \frac{\Delta t}{L} = V_i \cdot \frac{\delta.T}{L}. \qquad \text{From which: } L \cdot (I_1 - I_2) = V_i \cdot \delta \cdot T \qquad (13\text{-}8)$$

During a time $(1 - \delta).T$ the switch is open and the voltage across the coil is $(V_i - V_o)$ so that:

$$v_L = V_i - V_o = L\frac{\Delta i}{\Delta t} = \frac{L \cdot (I_2 - I_1)}{(1 - \delta).T} \quad \longrightarrow\longrightarrow \quad L \cdot (I_1 - I_2) = (V_o - V_i) \cdot (1 - \delta) \cdot T \quad (13\text{-}9)$$

From (13-8) and (13-9) it follows that: $V_i \cdot \delta \cdot T = (V_o - V_i) \cdot (1 - \delta) \cdot T$

from which:

$$V_o = \frac{V_i}{1 - \delta}$$

(13-10)

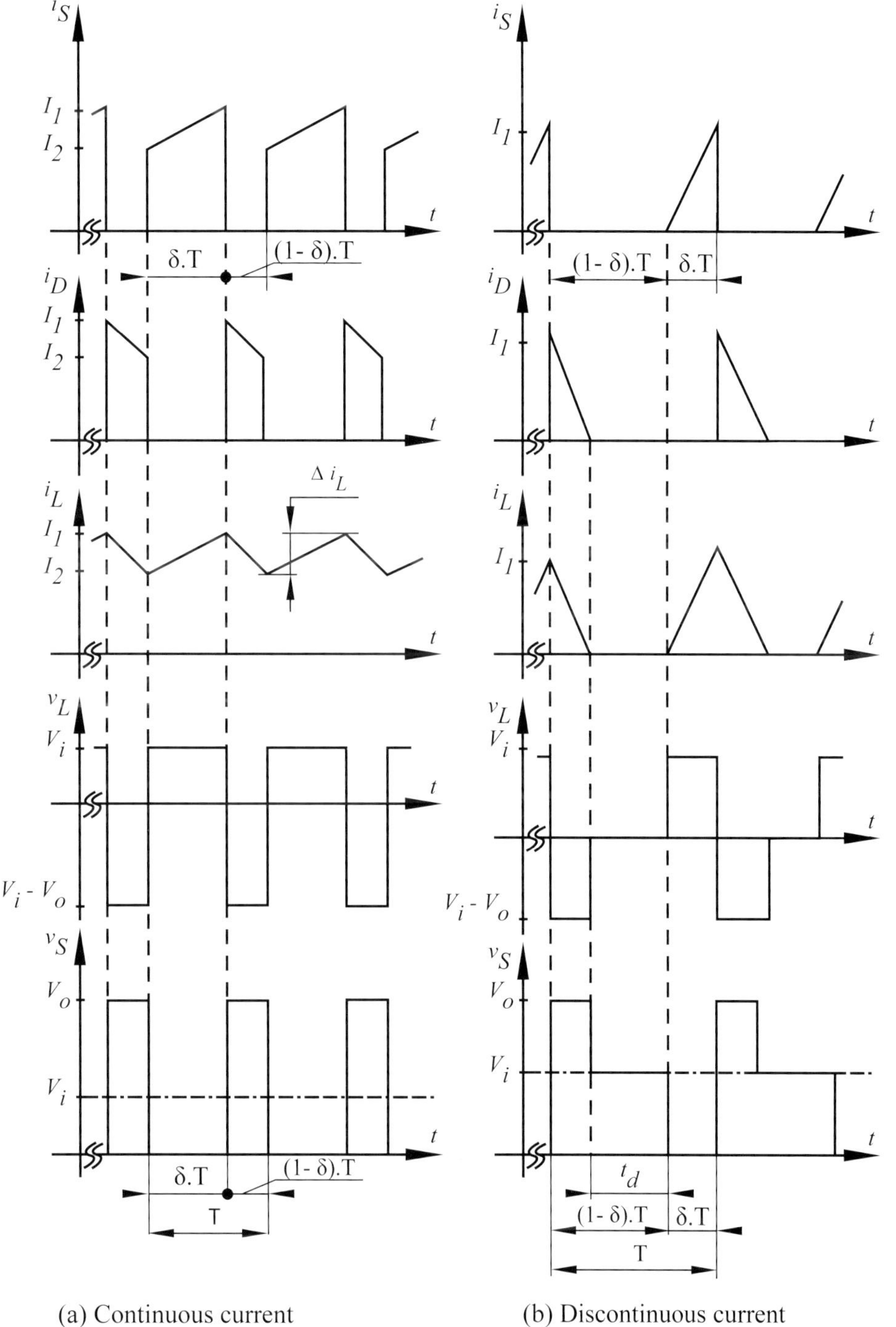

(a) Continuous current   (b) Discontinuous current

Fig. 13-8: Waveforms of fig. 13-7

2.2.3 Discontinuous current in the coil

This is shown in fig. 13-8b. In this case all the energy in the coil is discharged before the switch recloses. Now there is a dead time t_d.

When the switch is closed: $v_L = L \cdot \dfrac{\Delta i}{\Delta t} = V_i = L \cdot \dfrac{I_1}{\delta.T}$ or: $L \cdot I_1 = V_i \cdot \delta \cdot T$ (13-11)

When the switch is opened v_L becomes: $v_L = L \cdot \dfrac{\Delta i}{\Delta t}$ or: $(V_i - V_o) = L \cdot \dfrac{(0 - I_1)}{(1 - \delta).T - t_d}$ from

which: $L \cdot I_1 = (V_o - V_i) \cdot \left[(1 - \delta).T - t_d \right]$.   (13-12).

From (13-11) and (13-12) it follows that: $V_i \cdot \delta \cdot T = (V_o - V_i) \cdot \left[(1 - \delta).T - t_d \right]$

or: $V_i \cdot \delta \cdot T = V_o \cdot T - V_i \cdot T - V_o \cdot \delta \cdot T + V_i \cdot \delta \cdot T - V_o \cdot t_d + V_i \cdot t_d$

$V_i \cdot (T - t_d) = V_o \cdot (T - \delta.T - t_d) \longrightarrow$

$$V_o = V_i \cdot \frac{\left(1 - \dfrac{t_d}{T}\right)}{\left(1 - \delta - \dfrac{t_d}{T}\right)} \qquad (13\text{-}13)$$

2.2.4 Remarks

1. In (13-10) and (13-13) it appears that $V_o > V_i$. The boost converter increases the output voltage which explains the term "boost". The boost converter is sometimes referred to as a "step-up" converter.
2. The maximum voltage across the (transistor-) switch is V_o .
3. The maximum reverse voltage across the diode is V_o .

2.2.5 Numeric example 13-2:

For a boost converter with continuous current the following is known:
- input voltage: $V_i = 3$ to $5V$
- output voltage: $V_o = 9V \pm 0.1\%$
- maximum output current: $I_o = 1A$
- chopper frequency: $f = 50kHz$

Required: Determine the values of L and C.

Solution:

From (13-8) $\longrightarrow$ $L = \dfrac{\delta.V_i.T}{\Delta i_L}$. With $V_o = \dfrac{V_i}{1 - \delta}$ $\longrightarrow$ $L = \dfrac{V_o.(\delta - \delta^2).T}{\Delta i_L}$.

The maximum of $(\delta - \delta^2)$ is found when $\delta = 0.5$, so that $L_{max} = \dfrac{0.25 \times V_o \times T}{\Delta i_L}$.

If $\Delta i_L = 20\% \cdot I_o = 0.2 \times 1 = 200mA$, then $L = \dfrac{0.25 \times V_o \times T}{\Delta i_L} = \dfrac{0.25 \times 9 \times 2 \times 10^{-5}}{0.2} = 225\mu H$.

With $\Delta v_{C\,max} = \Delta v_{o\,max} = 0.1\% \cdot 9 = 9mV$ from (13-7) gives :

$C_{min} = \dfrac{\Delta i_L}{8.f.\Delta v_{C\,max}} = \dfrac{0.2}{8.50.10^3.9.10^{-3}} = 55.6\mu F$ and $ESR_{max} = \dfrac{9.10^{-3}}{0.2} = 0.045\Omega$.

2.3 Buck-boost converter

2.3.1 Operation of buck-boost converter

Fig. 13-9 shows the basic configuration. Closing S causes a current i_L to flow and magnetic energy is stored in the coil ($\frac{1}{2} . L . i_L^2$).

When the switch opens, the magnetic energy can only be discharged in the load and this results in a reverse polarity of the output voltage V_o with respect to V_i. The capacitor once again functions as a filter capacitor.

As with a boost converter we first store energy in the coil and **pump this energy to the load when the switch is open**. This explains why both converters are classified as **flyback** converters.

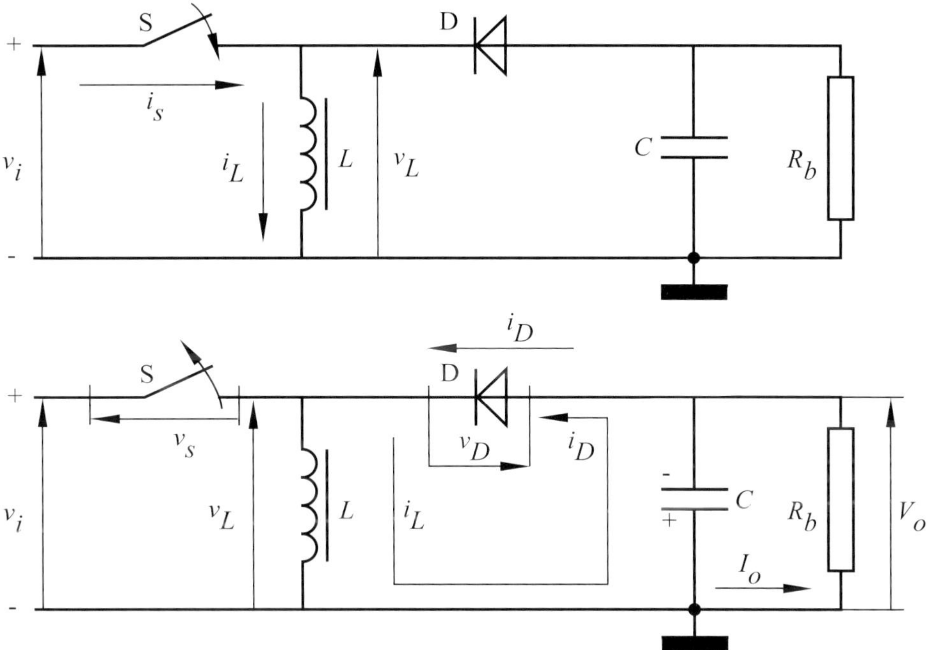

Fig. 13-9: Basic principle of buck-boost converter

Once again we distinguish between two operating modes, namely continuous and discontinuous current mode in the coil.

2.3.2 Continuous current in the coil

Fig. 13-10a shows the associated waveforms. When the switch closes a voltage results across the

coil: $\quad v_L = V_i = L . \dfrac{\Delta i}{\Delta t} = L . \dfrac{(I_1 - I_2)}{\delta . T}\quad$ or: $\quad L . (I_1 - I_2) = \delta . T . V_i \qquad$ (13-14).

When the switch opens after neglecting v_D : $v_L = V_o = L\dfrac{\Delta i}{\Delta t} = L .\dfrac{(I_2 - I_1)}{(1 - \delta) . T}\quad$ or:

$$L . (I_2 - I_1) = V_o . (1 - \delta) . T \qquad (13\text{-}15)$$

From (13-14) and (13-15) it follows: $\qquad V_o = - V_i . \dfrac{\delta}{1 - \delta} \qquad$ (13-16)

2.3.3 Discontinuous current in the coil

In this case the energy in the coil is completely discharged before the transistor switch recloses. Once again there is a dead time t_d (discontinuous current). Fig. 13-10b shows the associated waveforms. Closing the switch S results in a voltage across the coil, whereby

$$V_i = v_L = L \cdot \frac{\Delta i}{\Delta t} = L \cdot \frac{I_1}{\delta . T} \quad \text{or:} \quad L \cdot I_1 = \delta . T . V_i \tag{13-17}$$

After the switch opens during the time interval $(1 - \delta) . T - t_d$ almost all the energy in the coil is gone. The voltage across the coil is now V_i and the current changes from I_1 to zero, so that:

$$V_i = v_L = L \cdot \frac{(0 - I_1)}{[(1 - \delta) . T - t_d]} \quad \longrightarrow \longrightarrow \quad L . I_1 = - V_o . \left[(1 - \delta) . T - t_d \right] \tag{13-18}$$

From (13-17) and (13-18) it follows that: $\delta . T . V_i = - V_o . \left[(1 - \delta) . T - t_d \right]$

from which:

$$V_o = - V_i \cdot \frac{\delta}{1 - \delta - t_d /T} \tag{13-19}$$

2.3.4 Remarks

1. From (13-16) and (13-19) it follows that the buck-boost converter results in voltage inversion.
2. Depending on the value of δ we have $| V_o | \leq$ or $\geq | V_i |$. This is therefore a step-up /step-down converter.
3. The maximum voltage across the (transistor) switch is $V_i + V_o$ as can be seen in fig. 13-10 at the bottom .
4. The maximum reverse voltage across the diode is $V_i + V_o$.

2.3.5 Numeric example 13-3:

For a buck boost converter with continuous current mode we are given:
- input voltage: $V_i = 3$ to 15V
- output voltage: $V_o = 9$V $\pm$ 0.1%
- max. output current: $I_o = 3$A
- chopper frequency: $f = 100$kHz

Required: determine the values of L and C

Solution:

From (13-16) $\longrightarrow \longrightarrow$ $V_o = V_{i\,max} \cdot \dfrac{\delta_{min}}{(1 - \delta_{min})} \longrightarrow \longrightarrow \delta_{min} = \dfrac{V_o}{V_{i\,max} + V_o} = \dfrac{9}{15 + 9} = 0.375$.

Assumption $\Delta i_{L\,max} = 0.2 \times I_o = 0.6$A

From (13-15) $\longrightarrow \longrightarrow L_{max} = V_o \cdot \dfrac{(1 - \delta_{min}) . T}{\Delta i_L} = \dfrac{9 . (1 - 0.375) . 10^{-5}}{0.6} = 93.75$µH

From (13-7) $\longrightarrow \longrightarrow C_{min} = \dfrac{\Delta i_L}{8 . f . \Delta v_{C\,max}} = \dfrac{0.6}{8 . 10^5 . 9 . 10^{-3}} = 83.4$µF

and: $\quad ESR_{max} = \dfrac{\Delta v_{o\,max}}{\Delta i_L} = \dfrac{0.1\% . 9}{0.6} = 0.015 \,\Omega$

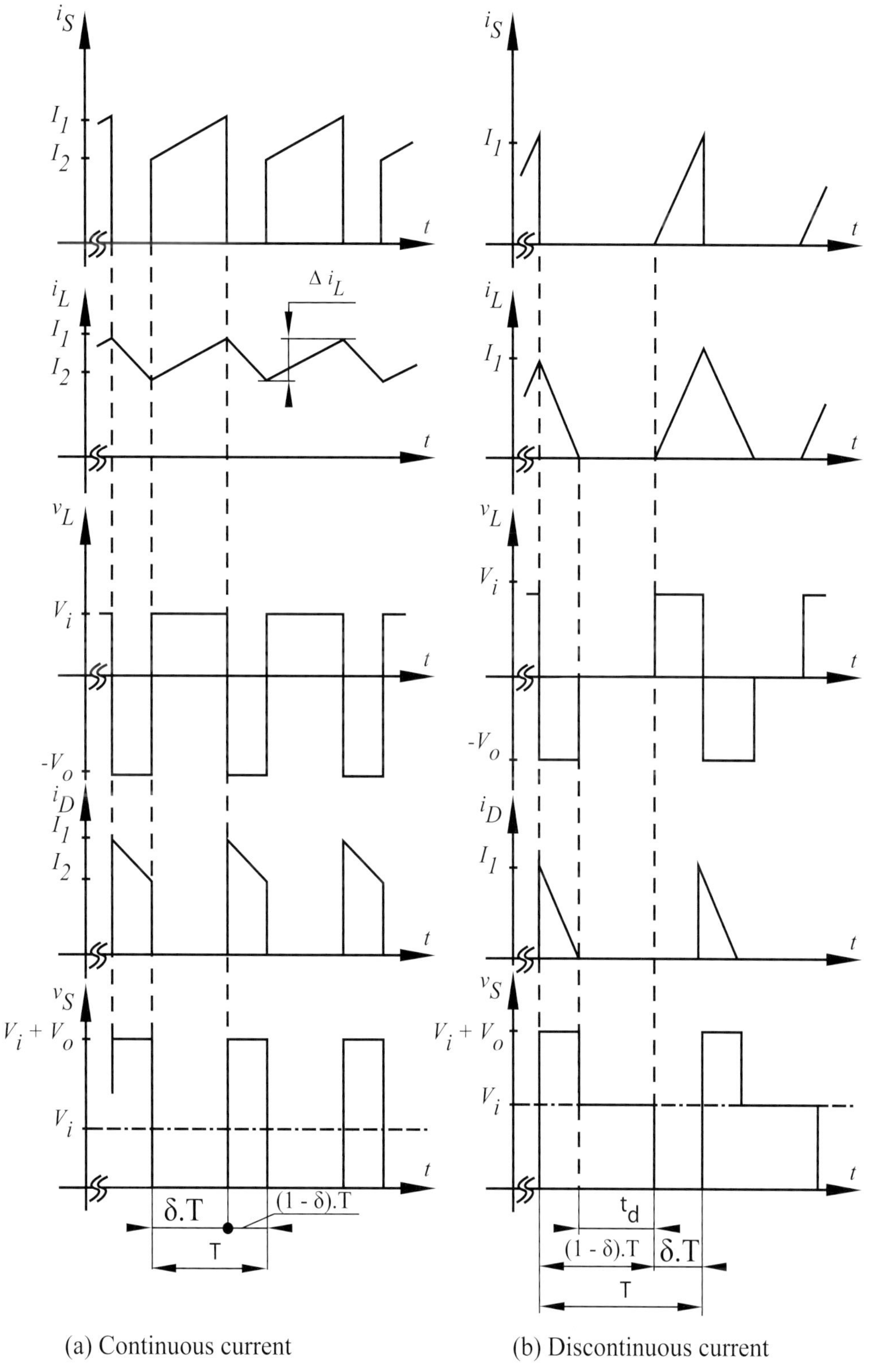

(a) Continuous current   (b) Discontinuous current

Fig. 13-10: Waveforms of fig. 13-9

3. ISOLATED SWITCH-MODE POWER SUPPLIES

The DC voltage V_i at the input of the SMPS is provided as a result of two methods :

1. Via a rectifier that is galvanically separated from the supply network or via a battery. In this case one of the three previously mentioned SMPS is used:
   - *voltage step down:*   buck
   - *voltage step up:*   boost
   - *voltage inversion:*   buck-boost.

2. By directly rectifying from the supply net (off-the-line!). It is desirable that the stabilized output voltage be galvanically separated from the supply net. The practical circuits used in this case are the flyback and the forward converter.

The block diagram of an off-the-line switching power supply is shown in fig. 13-11.

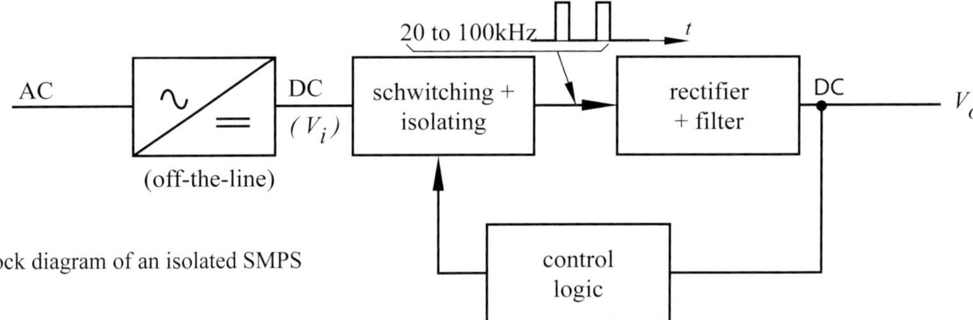

Fig. 13-11: Block diagram of an isolated SMPS

4. FLYBACK CONVERTER

4.1 Basis configuration

With an off-the-line rectifier there is no separation between the stabilised output voltage V_o and the supply network. To avoid this situation an isolation transformer is used after the switching elements. This is shown in fig. 13-12b and c.

From fig. 13-12a/b it appears that especially the buck-boost converter is suitable for creating an isolated output. The coil in fig. 13-12a is replaced by a transformer in fig. 13-12b. This transformer is now not only operating as a choke but also as an isolating transformer. That is why a transformer with air gap is used.

Rearranging the elements in fig. 13-12b and we arrive at the basic configuration of an isolated flyback converter which is shown in fig. 13-12c. This is simply called a "flyback". The term flyback is usually used for an isolated buck-boost converter.

During conduction of the transistor the current i_p increases linearly and magnetic energy is built up in the core of the transformer. Due to the polarity of the secondary voltage the diode D cannot conduct (fig. 13-12c). Switching off the transistor T results in an emf across the primary winding (with the positive terminal connected to the collector of the transistor). The positive side of the secondary emf is now connected to the anode of the diode D. The secondary emf E_S rises to a value that allows diode D to conduct, in fact at the instant that $E_S > V_o$. The secondary start current is $I_S = n.I_1$. Here in n is the winding ratio of the transformer. The stored magnetic energy in the transformer is now electrically discharged in the capacitor C and load R_b .

We recognize in fig. 13-12c a number of waveforms from fig. 13-10b.

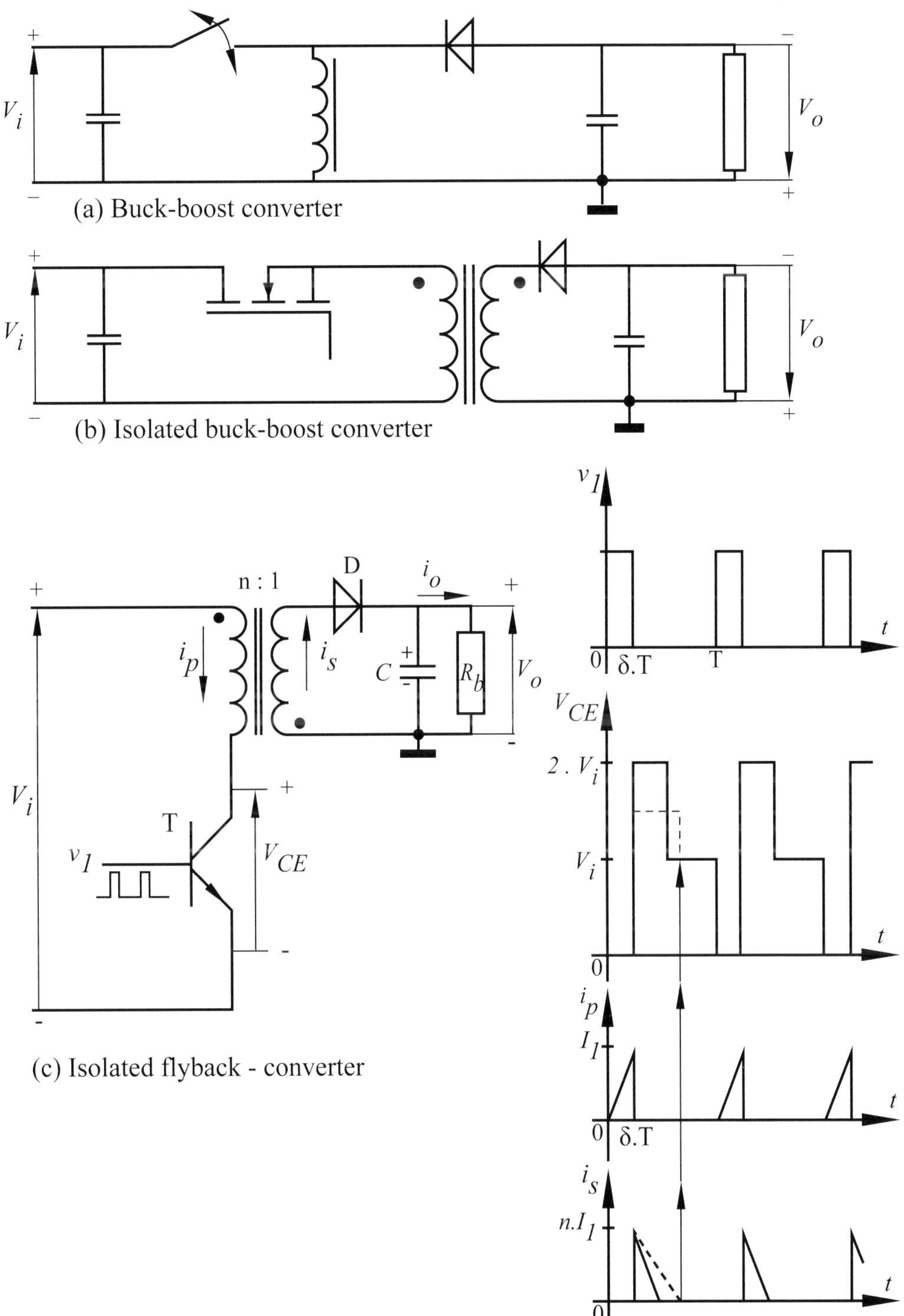

Fig. 13-12: Flyback converter with a sawtooth current waveform

Since this transformer also operates as a choke it is referred to as a "transformer-choke".
Just as with the previous converters we distinguish two operating methods for the flyback
converter. We are now going to take a look at these methods.

4.2 Operating modes

4.2.1 Sawtooth current through the transformer windings

The sawtooth waveform of the transformer currents in fig. 13-12c corresponds to the discontinuous current mode of the buck-boost.

We now determine the output voltage. We know that for a coil $v_L = L \cdot \dfrac{di}{dt}$

With a constant voltage V_L the current is $\dfrac{di}{dt} = \dfrac{V_L}{L} = $ constant. The current i increases linearly in time: $di = \dfrac{V_L}{L} \cdot dt$. The current waveform in the primary winding as shown in fig. 13-12c is given by $i_p(t) = \dfrac{V_i}{L_1} \cdot t$. Here L_1 is the self inductance of the primary winding.

After the time $\delta \cdot T$ is reached the primary currents value $I_1 = \dfrac{V_i}{L_1} \cdot \delta \cdot T$ $\qquad$ (13-20)

The stored energy is then: $W_i = \dfrac{L_1 \cdot I_1^2}{2} = \dfrac{V_i^2 \cdot (\delta.T)^2}{2 \cdot L_1}$. Assuming an average output voltage V_i during the entire period T, then the stored energy is: $W_o = \dfrac{V_o^2}{R_b} \cdot T$.

With an efficiency η this becomes $W_o = \eta \cdot W_i$ and: $\dfrac{V_o^2}{R_b} \cdot T = \eta \cdot \left[\dfrac{V_i^2 \cdot (\delta . T)^2}{2 \cdot L_1} \right]$.

After rearranging: $\qquad$ (13-21)

$$V_o = \delta \cdot V_i \cdot \sqrt{\dfrac{\eta \cdot R_b \cdot T}{2 \cdot L_1}}$$

4.2.2 Remarks

1. Load resistor of flyback converters

With a constant duty cycle δ the output voltage varies with the load resistor R_b.

Flyback converters may ordinarily not operate without a load resistor ($R_b = \infty$!).

An exception is if a closed control loop is used to control the output voltage (via δ !).

2. Turn-on losses of the transistor

The converter with a sawtooth current form has the advantage that the current is zero at the instant that the transistor conducts so that turn-on losses are minimal.

3. Blocking voltage across the transistor

During the time that the secondary current flows in the transformer a voltage is induced in the primary so that the blocking voltage across the transistor is larger than the supply voltage V_i.

We are going to calculate the primary transformer voltage. Fig. 13-13 shows the current and voltage waveforms of the transformer in fig. 13-12, with R_b as a parameter.

During the time $\delta \cdot T$ energy is stored in the transformer coil: $W_i = V_i \cdot \dfrac{I_1}{2} \cdot \delta \cdot T$.

With a load R_{b1} it takes t_1 to discharge this energy:

$$W_o = V_{o1} \cdot \dfrac{n \cdot I_1}{2} \cdot t_1 = \eta \cdot W_i = \eta \cdot V_i \cdot \dfrac{I_1}{2} \cdot \delta \cdot T \quad \text{so that:} \quad V_{o1} = \eta \cdot \dfrac{V_i \cdot \delta \cdot T}{n \cdot t_1} \qquad (1)$$

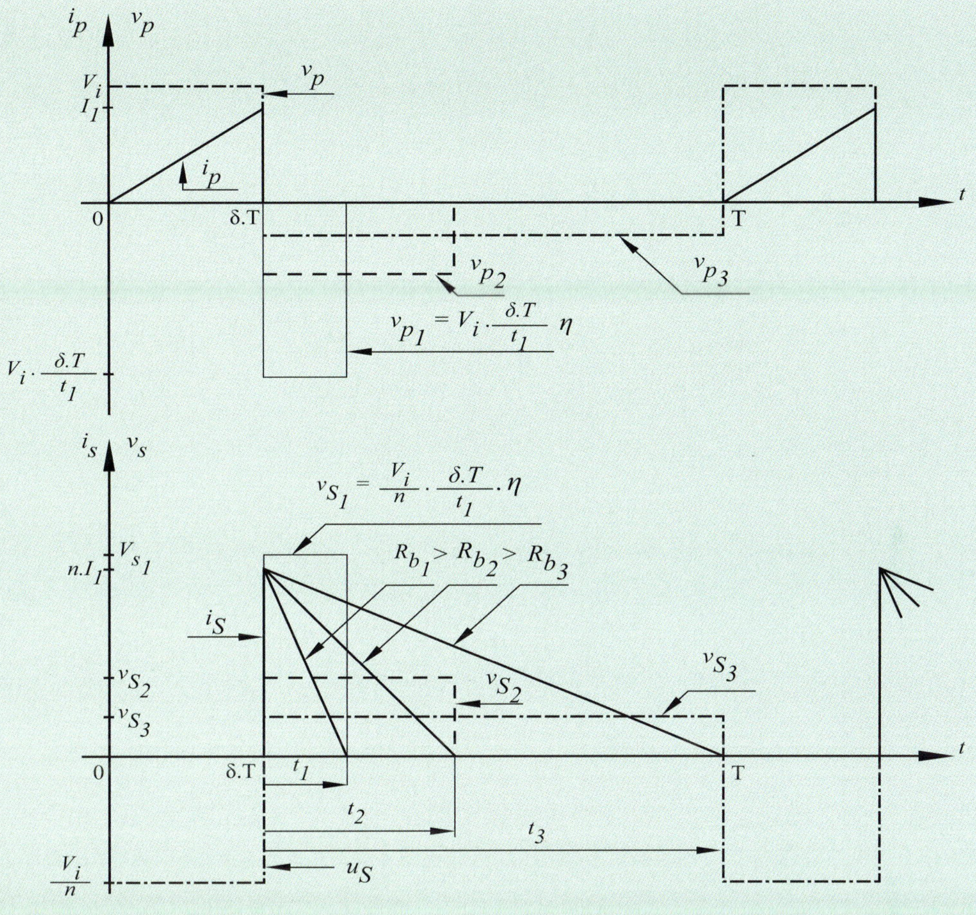

Fig. 13-13: Transformer currents and voltages in fig. 13-12 with R_b as parameter

From $W_o = \dfrac{V_{o1}^2}{R_{b1}} \cdot T$ and $W_o = V_{o1} \cdot \dfrac{n \cdot I_1}{2} \cdot t_1$ it follows: $t_1 = \dfrac{V_{o1} \cdot 2 \cdot T}{n \cdot I_1 \cdot R_{b1}}$

With $I_1 = \dfrac{V_i}{L_1} \cdot \delta \cdot T$ becomes $t_1 = \dfrac{V_{o1} \cdot 2 \cdot T \cdot L_1}{n \cdot V_i \cdot \delta \cdot T \cdot R_{b1}}$ (2)

(1) in (2) gives after calculation: $t_1 = \dfrac{1}{n} \cdot \sqrt{\dfrac{2 \cdot L_1 \cdot T \cdot \eta}{R_{b1}}}$ (3)

Further : $V_{o1} = V_{S1} - V_D$. Neglecting the diode voltage drop V_D gives:

$V_{S1} \approx V_{o1} = \dfrac{\eta \cdot V_i \cdot \delta \cdot T}{n \cdot t_1}$ (4)   and: $V_{p1} = n \cdot V_{S1} = \dfrac{\eta \cdot V_i \cdot \delta \cdot T}{t_1}$ (5)

The blocking voltage on the transistor is: $V_{CE} = V_i + V_{P1} = V_i + \eta \cdot \dfrac{V_i \cdot \delta \cdot T}{t_1}$

With $t_1 = \delta \cdot T$ and $\eta = 1$ results in $V_{p1} = V_i$ and $V_{CE} = 2 \cdot V_i$. This case is shown in fig. 13-12. If $R_{b2} < R_{b1}$ then $t_2 > t_1$; $V_{p2} < V_{p1}$ and $V_{CE} < 2 \cdot V_i$.

4. Maximum transistor current

From $W_o = \eta \cdot W_i = \eta \cdot \left[\dfrac{L_1 \cdot I_1^2}{2} \right]$ it follows that: $P_o = \dfrac{W_o}{T} = W_o \cdot f = \dfrac{f \cdot \eta \cdot L_1 \cdot I_1^2}{2}$.

Here I_1 is the maximum primary current and also the maximum collector current I_C of the transistor, so that: $I_{C\,max}^2 = \dfrac{2 \cdot P_o}{f \cdot \eta \cdot L_1}$.

From (13-20) it follows that: $I_1 = \dfrac{V_i}{L_1} \cdot \delta \cdot T$ .From this we can determine the maximum

collector current: $I_{C\,max} = \dfrac{V_i}{L_1} \cdot \delta_{max} \cdot T = \dfrac{V_i \cdot \delta_{max} \cdot}{L_1 \cdot f} \longrightarrow f \cdot L_1 = \dfrac{V_i \cdot \delta_{max}}{I_{C\,max}}$

If we substitute this in the above expression for I_C^2 , then we find: $I_{C\,max}^2 = \dfrac{2 \cdot P_o \cdot I_{C\,max}}{V_i \cdot \delta_{max} \cdot \eta}$,

from which :

$$I_C = \frac{2 \cdot P_o}{\eta \cdot V_i \cdot \delta_{max}}.$$

(13-22)

with :

P_u	=	output power of SMPS (W)
η	=	efficiency of converter (e.g. $\eta = 80\%$)
V_i	=	input DC voltage (V)
δ_{max}	=	maximum duty cycle (e.g. 0.4 or 0.5)

4.2.3 Trapezium shaped current in the transformer windings

If the transistor is switched on before the secondary current is zero then the primary and secondary currents have a trapezium shape, see fig. 13-14. This is continuous current mode as was the case with the buck-boost.

With $t_3 = (1 - \delta) \cdot T$ in fig. 13-13 it follows from (1) at the top of p. 13-17:

$$V_{o3} = \eta \cdot \frac{V_i \cdot \delta \cdot T}{n \cdot t_3} \quad \text{so that:} \quad V_o = \eta \cdot \frac{V_i}{n} \cdot \frac{\delta}{(1 - \delta)}$$

(13-23)

With:

V_o	=	average output voltage
V_i	=	supply voltage at the input
n	=	winding ratio of the transformer (primary/secondary)
δ	=	duty cycle of chopper
η	=	efficiency of converter

If δ approaches 1 the output voltage would become infinitely large, as well as the blocking voltage across the transistor.

To limit the blocking voltage δ is limited to a maximum of $\delta_{max} = 0.5$.

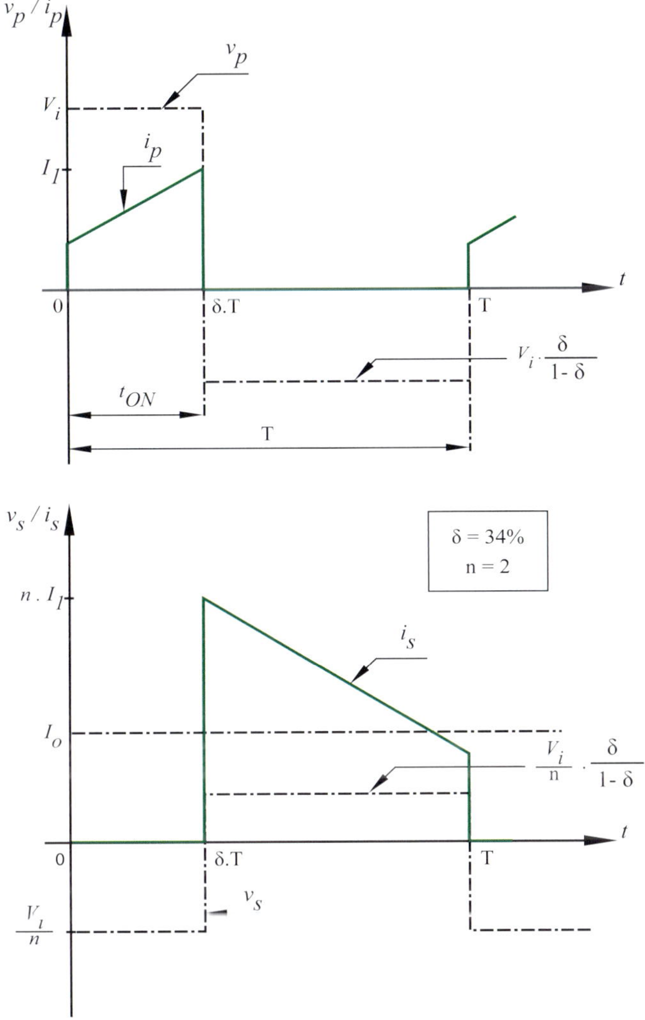

Fig. 13-14: Voltage and current waveforms of the flyback transformer with trapezium shaped current

The waveforms in fig. 13-14 are comparable to those in fig. 13-10a. Expression (13-23) is equivalent to expression (13-16)!

4.2.4 Remarks

1. *Maximum blocking voltage across the transistor*

    The maximum voltage across the secondary of the transformer when the transistor is switched off is given by (13-23). Setting $\eta = 1$ and $\delta = \delta_{max}$: $\rightarrow \rightarrow \rightarrow \rightarrow$

    $$V_{p\,max} = (n . V_o)_{max} = V_i \cdot \frac{\delta_{max}}{1 - \delta_{max}} \; ; \; V_{CEmax} = V_i + (n . V_o)_{max} = \frac{V_i}{1 - \delta_{max}}$$

    so that : $$V_{CEmax} = \frac{V_i}{(1 - \delta_{max})} \tag{13-24}$$

We maintain this maximum blocking voltage within acceptable limits by for example assuming $\delta_{max} = 0.5$, then $V_{CEmax} = 2 \cdot V_i$.
Direct rectification from the 230V supply voltage results in $V_i \approx 310V$. In this case we need to select transistors with a working voltage of 800V.

2.   The advantage of the trapezium current waveform is that the peak values of the current are much lower than that of the sawtooth waveform.

3.   *Transformer with air gap*
     The *BH* - curve of the transformer choke is followed in only one direction. The volume and the air gap need to be sufficiently large to prevent saturation.

4.   *Ferrite core transformer*
     In addition to the transformer losses the apparent power also plays a role as far as the volume of the transformer is concerned. When we compare an SMPS and a linear voltage supply with the same output power we find that due to the low efficiency of the linear supply the (50Hz-) transformer will have a larger apparent power than the (20 to 100kHz-) transformer of the SMPS. With its minimal volume the ferrite core transformer plays an important role in the compact nature of the SMPS.

5.   *Flyback converters with multiple outputs*
     An advantage of flyback converters is the simplicity of having multiple outputs. Since the transformer operates as a choke we only need a diode and a capacitor per extra output (fig. 13-15).

6.   *Two-transistor flyback converter*
     To limit the working voltage across the transistor we can use a configuration with two transistors (fig. 13-16). The transistors T_1 and T_2 are switched on and off simultaneously. Diodes D_1 and D_2 serve to limit the maximum voltage across a single transistor to V_i . Indeed, at switch off of T_1 and T_2 an emf is generated in the transformer primary with the positive terminal connected to the anode of D_1 and the negative terminal to the cathode of D_2 .
     If the emf is larger than V_i then diodes D_1 and D_2 conduct so that the input voltage V_i is across each transistor .
     Diodes D_1 and D_2 clamp the transistor voltage at V_i , hence the name clamping diodes.

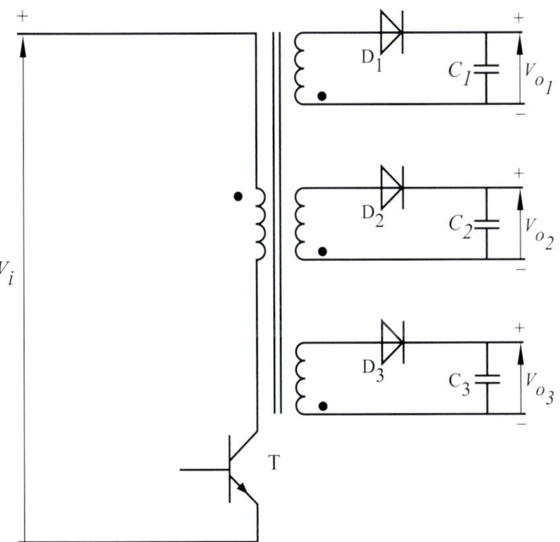

Fig. 13-15: Flyback converter with multiple outputs

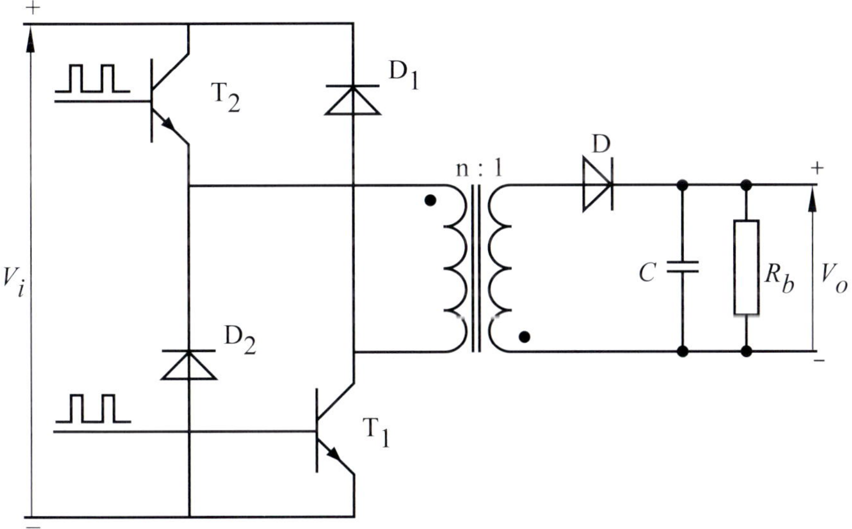

Fig. 13-16: Two-transistor flyback converter

Photo Schaffner: Two stage EMC filter. Maximum voltage 250V-50/60Hz. Current level 1 to 30A. MTBF up to 1.3 million hours. Up to 90dB damping. Applications: consumer appliance, data equipment, machines etc.

5. FORWARD CONVERTER

5.1 Configuration and operation

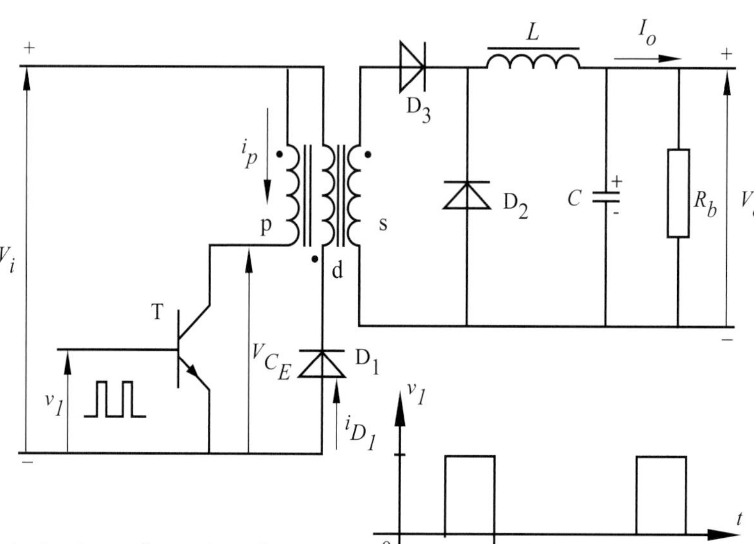

Fig. 13-17: Forward converter

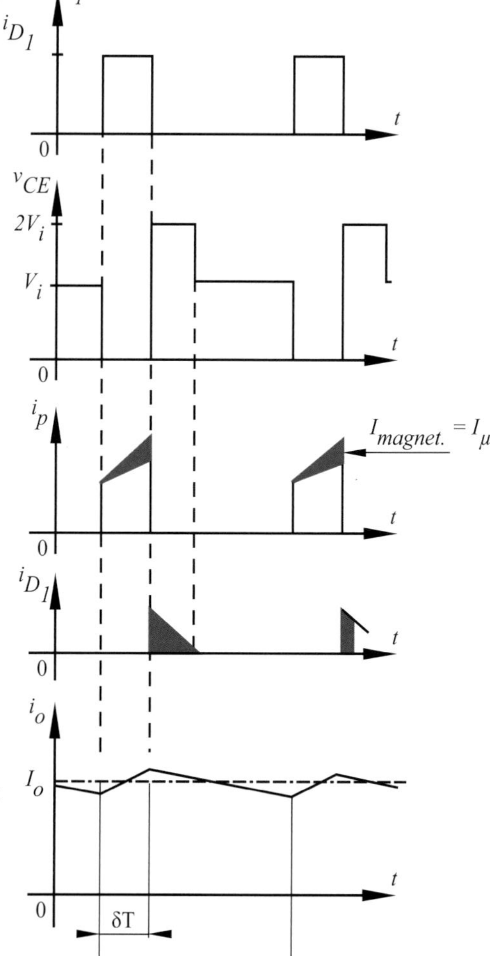

Fig. 13-17 shows the basic configuration of an isolated "forward" converter. Here the elements D_2, L, C and R_b correspond with D, L, C and R_b of fig. 13-4. A forward is therefore an isolated version of a buck converter. The winding ratio of the transformer plays a role in determining the V_o as was the case with the flyback. Once again the transistor operates as a switch. Galvanic separation from the supply is provided by the transformer with primary p and secondary s. When the transistor T conducts, magnetic energy is stored in the primary. Due to the polarity of the secondary voltage D_3 can conduct and the energy flows via D_3 to R_b where some of it is stored in coil L. When the transistor switches off the polarity of the transformer voltage reverses and D_3 blocks. As a result the coil L will discharge its energy via R_b and freewheel diode D_2. When T and D_3 block the remaining energy in the transformer core can not go any where. That is the reason that a demagnetizing winding d is included. This winding d has the same number of turns as the primary p.

The purpose of this third transformer winding d together with D_1 is that when T blocks, the magnetic energy can be drawn out of the transformer core and back to the source V_i . The transformer has no air gap since the transformer has to have a large inductance. Also here ferrite cores are used.

5.2 Remarks

1. *Maximum transistor voltage:*

   Since the third transformer winding has the same number of turns as the primary, the voltage across the transistor T when blocking is limited to

$$V_{CEmax} = 2 . V_i \qquad (13\text{-}25)$$

2. Since d has the same number of turns as p, when the transistor conducts a voltage V_i is across p and an inverse voltage of $2.V_i$ is across D_1 .

3. *Magnetisation current:*

   Magnetising current (green part of i_p in fig. 13-17) is brought to zero via de demagnetization winding d (the same number of turns as p but thinner wire!) while the transistor blocks. Since the volt-second product during demagnetization of the transformer has to be at least as large as during magnetization the δ_{max} is limited to 50%.

$$I_\mu = \frac{\delta_{max} . T . V_i}{L_0}$$

L_0 = magnetizing inductance of the transformer, seen from the primary side.

$$L_0 = N_p^2 . \frac{4 . \pi . 10^{-7} . \mu_r . A_e}{l_e}$$

With:
$\quad N_p \quad = \quad$ number of primary turns
$\quad A_e \quad = \quad$ effective cross sectional area of core
$\quad l_e \quad = \quad$ average length of field lines
$\quad \mu_r \quad = \quad$ relative permeability of core.

4. *Maximum collector current:*

   The maximum collector current is: $I_C = I_p + I_\mu$ or:

$$I_C \approx \frac{I_o}{n} + \frac{\delta_{max} . T . V_i}{L_0} \qquad (13\text{-}26)$$

   With:
   $\quad I_\mu \quad = \quad$ output current of the supply (A)
   $\quad n \quad = \quad$ transformer ratio primary/secondary

5. *Output voltage:*

   Since the circuit on the secondary side is comparable to a buck converter is: $V_o = \delta . V_{sec}$

   The output voltage can be described by:

$$V_o = \frac{V_i}{n} . \delta \qquad (13\text{-}27)$$

6. CONVERTER CONTROL STRATEGIES

In every SMPS the output voltage is compared with a constant reference voltage. The resulting error voltage is amplified and serves as a control voltage V_C for the closed loop controller.

We distinguish three control methods:

6.1 Direct duty cycle control

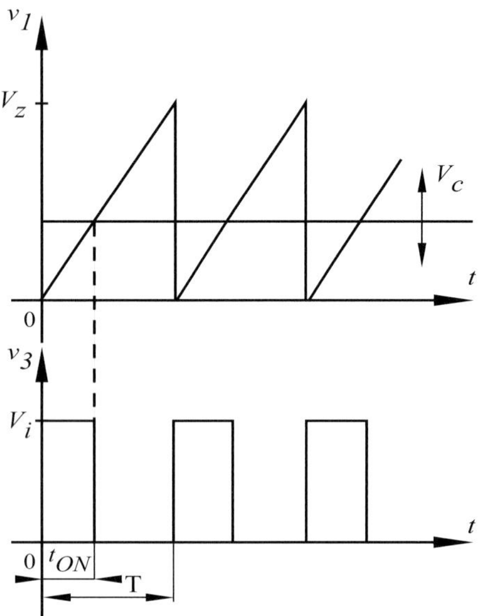

Fig. 13-18: Basic direct duty cycle control of an SMPS

The transistor duty cycle $\delta = \dfrac{t_{ON}}{T}$ is altered proportionally to the control voltage V_C (fig. 13-18). This is the oldest and most common method. The amplitude of the sawtooth is constant (see for example operation of the control IC "3524" on p. 12.12).

With this method there is a slow reaction to input voltage variations. This means this control method has poor dynamic behaviour.

6.2 Feed forward (forward voltage feedback)

In this case the amplitude of the sawtooth is varied proportionally with the input voltage V_i so that
$V_Z = k \cdot V_i$ (fig. 13-19) .

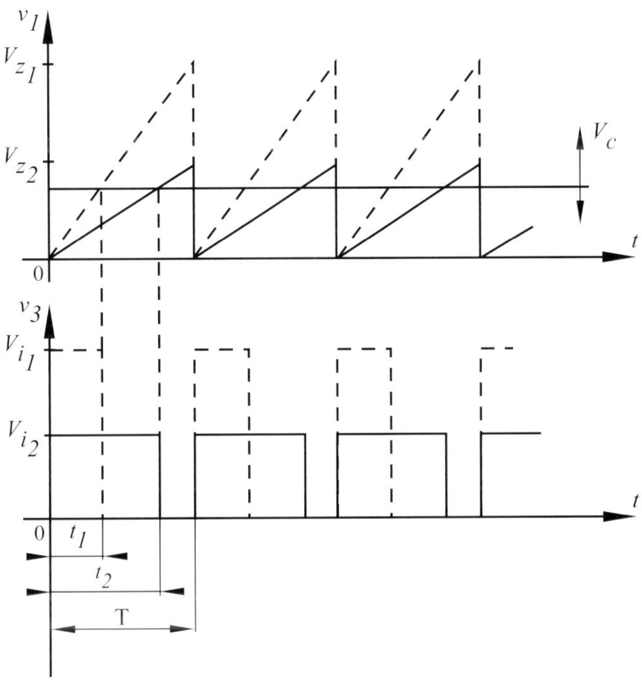

Fig. 13-19: Feed forward control strategy for an SMPS

From fig. 13-19 it follows that: $\delta_1 = \dfrac{t_1}{T}$; $\delta_2 = \dfrac{t_2}{T}$; $\dfrac{t_1}{t_2} = \dfrac{\delta_1 \cdot T}{\delta_2 \cdot T} = \dfrac{V_{Z2}}{V_{Z1}}$

$\delta_1 \cdot V_{Z1} = \delta_2 \cdot V_{Z2}$; $\delta_1 \cdot k \cdot V_{i_1} = \delta_2 \cdot k \cdot V_{i_2}$

or $V_{i_1} \cdot \delta_1 = V_{i_2} \cdot \delta_2 = V_i \cdot \delta$ = constant.

The supply voltage stability is much better than with the direct duty cycle control strategy .
A UC1840 (Unitrode) and a TDA4718 (Infineon) are examples of control IC's suitable for feed forward control.

6.3. Current mode control

Here use is made of two feedback circuits:

1. Internal loop with current feedback
2. External loop with voltage feedback

Fig. 13-20 shows an example. For the DC-DC converters of the flyback family the current through the transistor is measured using a sensing resistor. If the voltage V_S reaches V_C the transistor T blocks. With the next clock pulse T conducts etc.

The voltage feedback loop determines the level (V_C) at which the internal loop controls the peak current through the coil and transistor. As a result we see that pulse by pulse the current is limited and optimal dynamic behaviour is ensured. The opto-coupler provides galvanic separation between the output and the supply network.

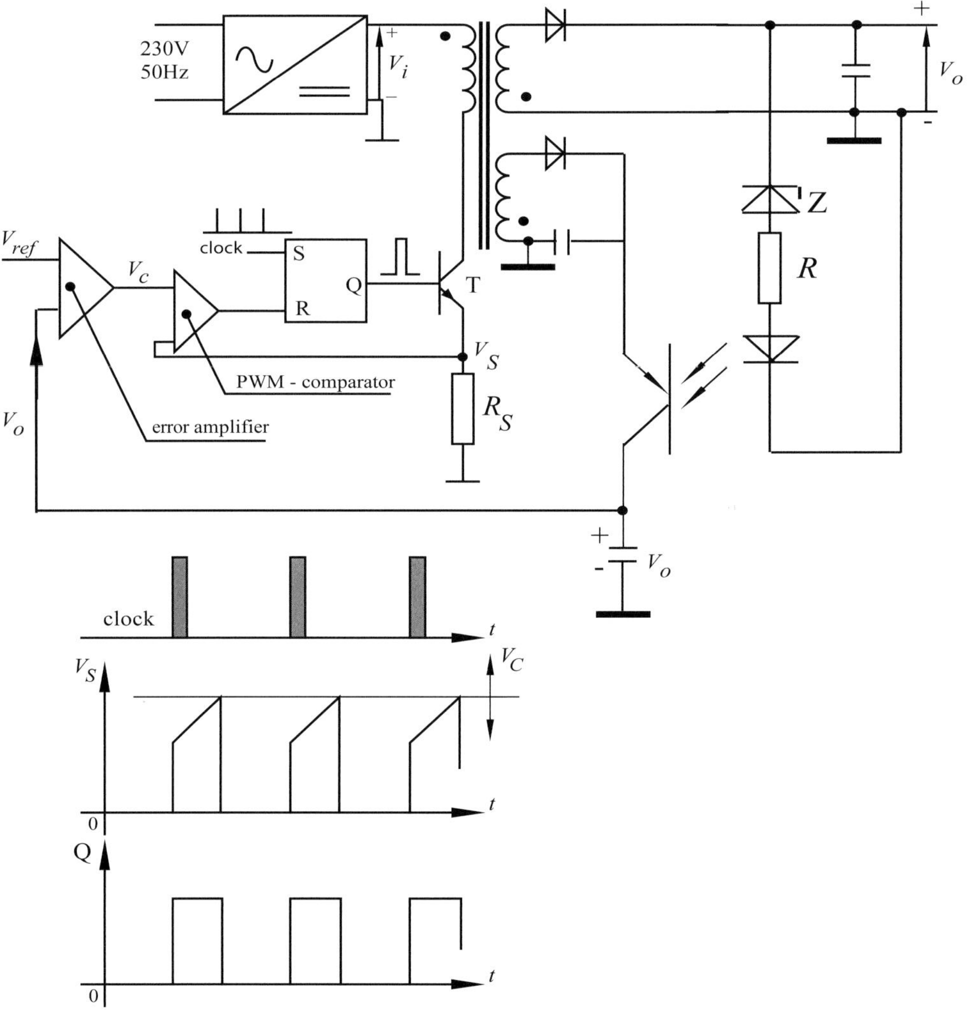

Fig. 13-20: Current mode control of an SMPS

7. TWO-TRANSISTOR SMPS OF THE FORWARD TYPE

When the input voltage is high we can use a configuration with two transistors (fig. 13-21) and limit the voltage V_{CE} to the value V_i.

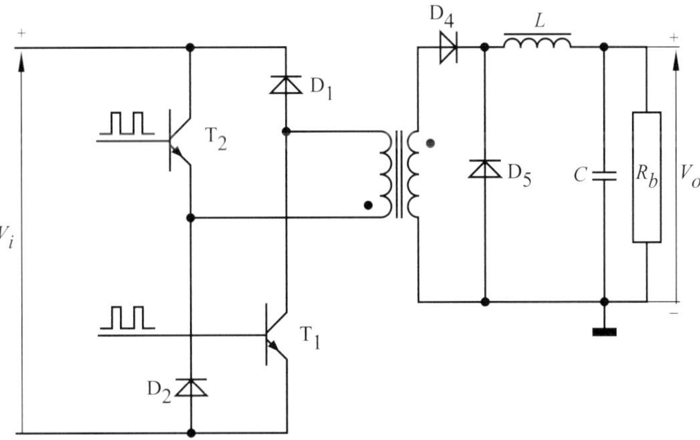

Fig. 13-21: Forward converter with two transistors

- *Freewheel diode:*
  The freewheel diode D_5 must be able to handle the full output current since this flows through the diode when the transistor blocks.
- *Clamp winding:*
  The requirement of a demagnetizing winding is a disadvantage of a forward converter. For the converter shown in fig. 13-21 the third winding (= clamp winding) is not required , diodes D_1 and D_2 clamp the transistor voltage V_{CE} to V_i.

8. FORWARD WITH MULTIPLE OUTPUTS

This type of DC-DC converter can also have multiple outputs.
Every output needs its own choke L_1 , L_2 , ..., as shown in fig. 13-22 .

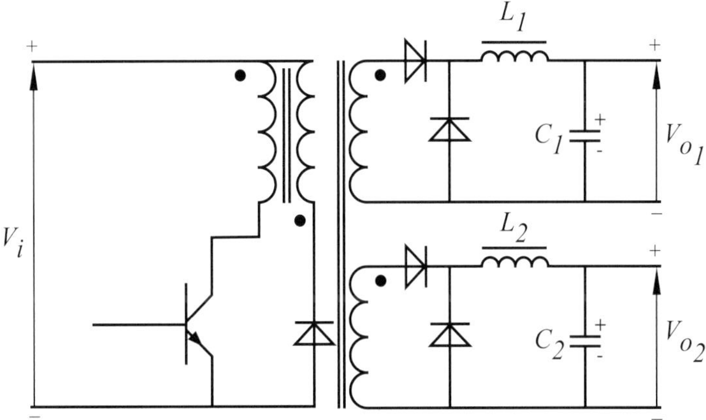

Fig. 13-22: Forward converter with multiple outputs

9. FULL BRIDGE OF THE BUCK TYPE

A full bridge can be used in order to limit the current through the transistors of a large power SMPS. With this configuration the current through the transistors is halved (with respect to the half bridge). Two extra transistors are required though with commutation diodes. The transistors T_1 and T_4 conduct simultaneously and flip flop with the transistor pair T_2 / T_3 (fig. 13-23) .

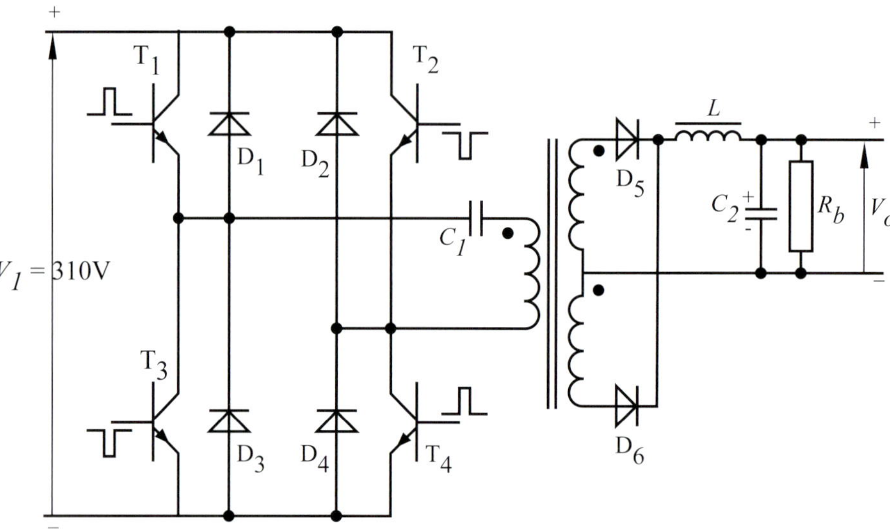

Fig. 13-23: Full bridge converter

Remarks

1. *Ferrite core:*
    For the isolated SMPS normally a transformer with ferrite core is used. This material with its low losses is suited for the typical switching frequencies which vary between 20 and 100kHz.

2. *Short circuiting the transistor bridge:*
    With a full bridge (fig. 13-23) galvanic separation of the control circuits of the transistors is essential. Another problem for every transistor bridge is simultaneous conduction of the transistors in one half (vertical) leg of the bridge. This has to be avoided since it results in a direct short circuit of the supply. The necessary protection logic needs to be included in the control circuitry.

10. SYNCHRONOUS SMPS

More and more SMPS have an extremely low output voltage which means the power loss in the output diodes is an important factor in the efficiency of the circuit. Even Schottky diodes with their low losses come under pressure to maintain circuit efficiency. A solution can be found by replacing the diodes with MOSFETS with a very low $R_{DS(ON)}$. These MOSFETS need to conduct and block synchronised with the SMPS. This is referred to as a synchronous SMPS or synchronous rectifier.

Consider an SMPS with output values 3.3V / 25A, therefore a P_o = 82.5W. We are using the forward converter from fig. 13-17. When D_3 is a Schottky diode with for example a voltage drop of 0.5V, then we find for the power losses in the diode at full load: P_1 = 25 x 0.5 = 12.5W.

If we replace D_3 with a MOSFET T_1 (fig. 13-24) with an $R_{DS(ON)}$ = 0.011Ω then the power losses become: P_2 = 25 x 25 x 0.011 = 6.875W .

The percentage power loss in the diode is: $\frac{12.5}{82.5}$ = 15.5% and in the MOSFET $\frac{6.875}{82.5}$ = 8.3%. Another important factor is: $\frac{6.875}{12.5}$ = 55% less heat losses with the MOSFET.

As a result the dimensions of the SMPS can be further reduced.

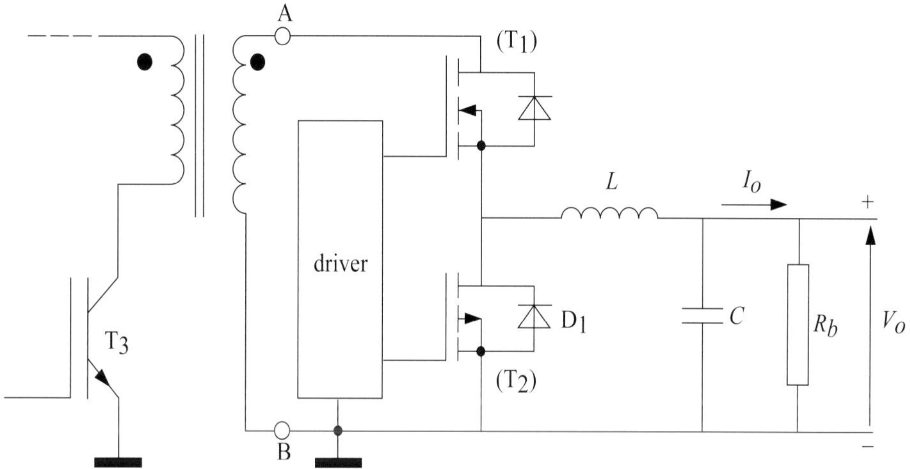

Fig.13-24: Synchronous forward DC-DC converter

The MOSFET T_2 is not conducting. When T_1 (together with T_3) is turned on, then a current begins to rise through the coil. When T_1 turns off D_1 conducts, but after a dead time of a number of ns T_2 is brought into conduction. As a result of the low $R_{DS(ON)}$ the voltage drop across T_2 is too small to keep D_1 conducting: T_2 conducts the current. When T_2 switches off D_1 starts conducting the current and after a small dead time T_1 starts to conduct again.

We can even remove the transformer in fig. 13-24 and apply a DC voltage V_i between A and B which gives in effect the buck converter from fig. 13-4. Now T_1 is the switch S and T_2 replaces the diode D. The MOSFETS can be dual MOSFETS, two in one casing. Additionally there may be Schottky diodes included in the casing.

The dead time is needed since T_1 and T_2 can not conduct simultaneously to avoid a short circuit. To solve the problem of producing a voltage lower than 3.3V a special control IC can be used. An example of this is the TPS54x1y SWIFT-family from Texas Instruments. This enables output voltages respectively of 0.9 V/1.2V/1.5V/1.8V/2.5V and 3.3V to be produced with a supply voltage between 3 and 6V.

SWIFT = switcher with integrated FET-technology.

In the type code the symbols have the following meaning:

- x = output current (up to 6A)
- y = 1 to 6 in accord with the above mentioned output voltages

y = 0 = variable output voltage in accord with an external reference.

Example: TSP54312: DC-DC converter with output values 3A and 1.2V.

11. SMPS COMPONENTS

Diodes

On the supply side (input from net) a classic rectifier bridge can be used.

As far as the output rectifier is concerned attention should be paid to the following:

- the reverse recovery time should be very small since a voltage with a very high frequency is being rectified (up to 100kHz!). FRD's and SiC Schottky diodes are used when the output voltage is relatively high.
- with a low output voltage (e.g. 5V) the rectifier diodes should have a low forward voltage drop, if not 20 to 30% of the output power is lost in the diodes. The synchronous SMPS offers a solution for this problem.

Switching transistors

In most cases this will be power MOSFETS or IGBTs.

Control circuit

A control IC can, in addition to controlling the duty cycle, provide over voltage protection, under voltage protection, soft start , etc…

Optocouplers

With isolated converters the output variable (current, voltage) can easily be fed to a control IC via an optocoupler. In this way the galvanic separation of input and output is maintained (e.g. see fig. 13-20).

Ferrite core transformers and coils

Ferrites have a DC resistance which is about 10^9 times larger than normal metal transformer cores. As a result there eddy current losses are practically negligible and these ferrites prove useful as cores of transformers and coils operating at high frequencies (10 to 1000 kHz). Saturation occurs with an inductance of 0.1 to 0.4 T with respect to the 1 to 1.5 T of regular transformer cores. For chokes in addition to ferrite as a core material iron powder and MPP (molypermalloy) is also used. Iron powder and MPP are mostly used for ring shaped cores. As far as transformers are concerned you are referred again to remark no. 4 on page 13.20.

Net filters

Switch mode power supplies are a source of unwanted high frequency harmonics which need to be suppressed. This undesired radiation can be combated with good screening of the SMPS. To reduce the EMI effects in the supply cables an LC-filter is used.

Filter capacitors

An electrolytic capacitor may be considered as a series combination of a capacitance, an ESR and self inductance. For series resonance the impedance is minimal. Above that the impedance rises. To prevent this, another capacitor with good R.F. properties is connected in parallel with the electrolytic capacitor. Often a multilayer ceramic capacitor is used for this purpose. This type is less sensitive to frequency effects so that the ESR of this parallel circuit remains low at high frequencies.

12.   OVERVIEW OF SMPS UP TO 2500W

Fig. 13-25 provides an overview of the common SMPS

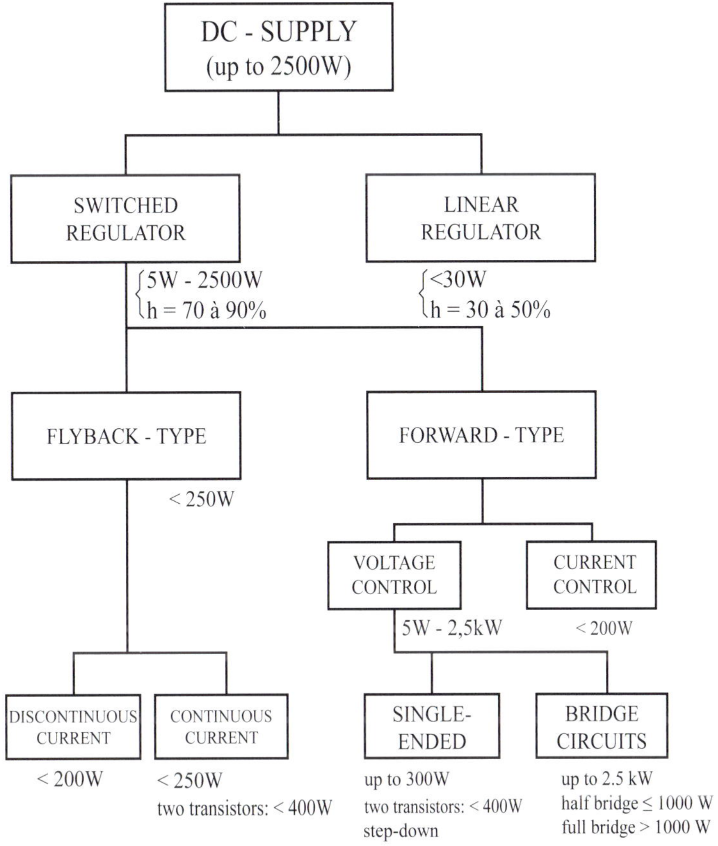

Fig. 13-25: Overview of SMPS technology (up to 2500W)

Table 13-1

Output power	Circuit	Operating principle	Control method
$P_b < 200W$	flyback	discontinuous current	• forward voltage feedback • current control
$200W < P_b < 1kW$	forward	continous current	• current control
$P_b > 1kW$	forward (half of full bridge)	continuous current	• current control

Remarks

The flyback topology is much used as a DC-DC converter for low power levels. Advantages include the simple design and low price of the components. The limited number of components and the availability of sophisticated IC's and high voltage MOSFETS make the flyback converter useful in the most demanding applications.

13. NON IDEAL WAVEFORM

In fig. 13-26 we redraw the flyback converter, but now a number of parasitic elements are included. These have been drawn in green.

Where:

L_{cb} :   parasitic self induction of the printed circuit board;
L_0 :   magnetizing inductance of the transformer;
L_{lk} :   leakage inductance of transformer;
C_{prim} :   capacitance of the primary windings;
C_{sec} :   capacitance of the secondary windings;
C_{tr} :   feedback capacitance between secondary and primary transformer;
C_D :   diode capacitance;
C_o :   output capacitance of transistor.

Fig. 13-26 is an approximated drawing since these parasitic elements do not have a constant value and are often voltage and frequency dependent.

Fig. 13-27a shows for example the voltage across the MOSFET. It is clear that the oscillations cause the voltage across the MOSFET to increase. A snubber across the MOSFET can prevent it being destroyed.

If we take a look at the first two periods of oscillation (fig. 13-27b), then we see in this example a period of 100ns. This corresponds to 10MHz.

A rule of thumb indicates that the oscillation frequency is approximately 100 times the switching frequency.

Fig. 13-27b is the result of resonance between the leakage inductance of the transformer on the one hand and the self inductance of the current loop on the printed circuit and on the other hand the output capacitance of the FET and the transformer capacitances. The distortion of the sinusoids is the result of the voltage dependent output capacitance of the MOSFET.

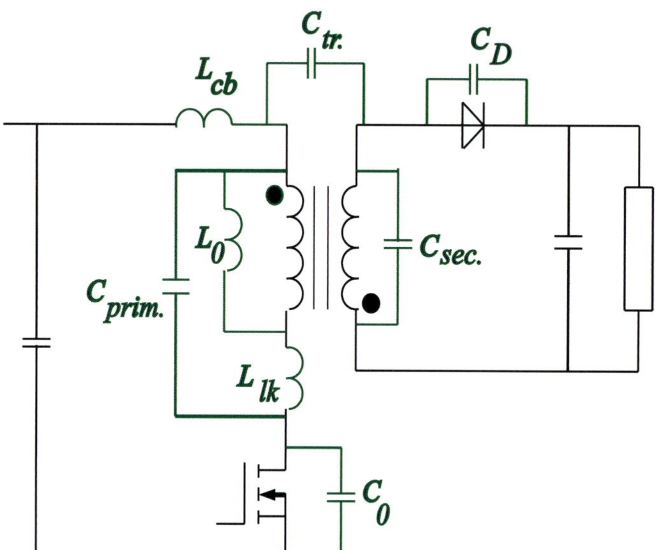

Fig. 13-26: Flyback converter with parasitic elements

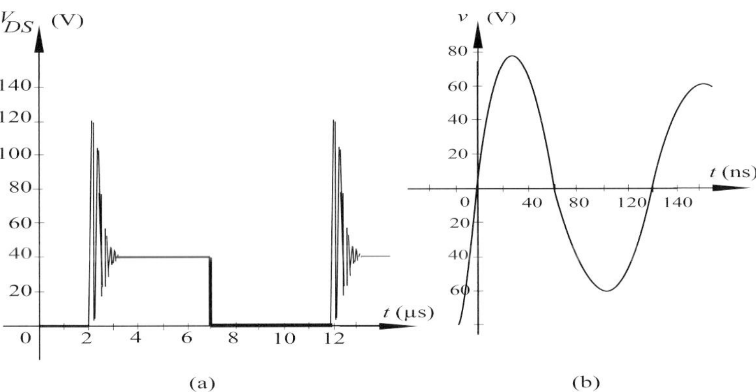

(a)   (b)

Fig.13-27a: Waveform on the FET in fig. 13.26   Fig. 13-27b: First period of the oscillations in fig. 13-27a

14. DIGITAL CONTROL OF AN SMPS

With a new type of digital signal controller (DSC) there is now enough calculation power on board to execute the software quickly enough to handle closed loop control. This DSC has in addition built-in intelligent modules for PWM, and an analogue to digital converter (ADC).

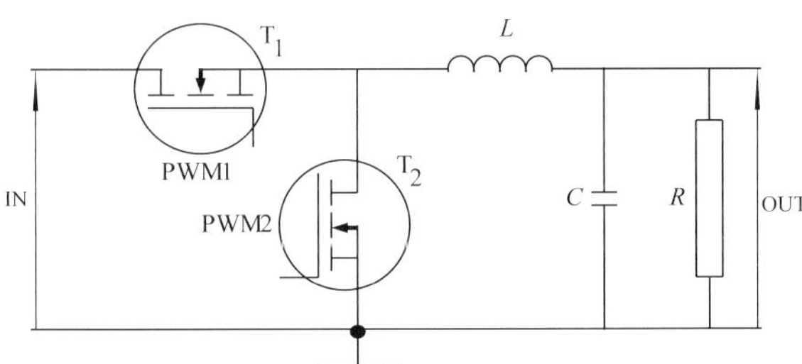

We look at the example of the synchronous buck converter from fig. 13-24 and apply a voltage between the points A and B. This brings us to fig. 13-28. The analogue control circuit from fig. 13-24 has now been replaced with a digital control system with a DSC (fig. 13-29).

Fig. 13-28: Basic configuration of a synchronous buck converter

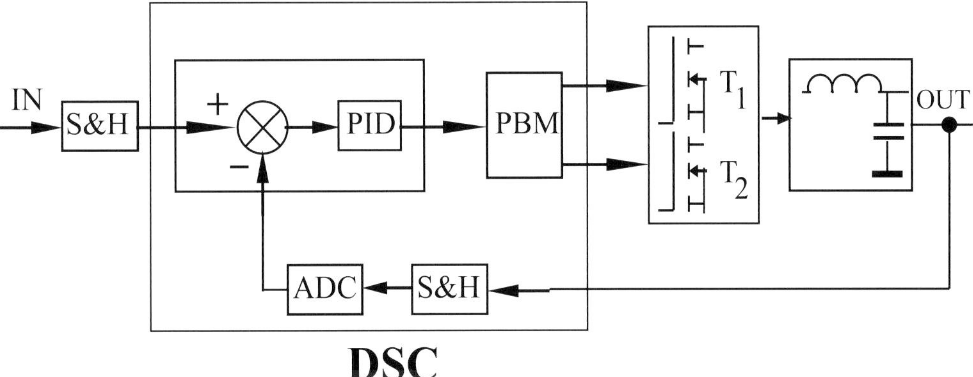

Fig. 13-29: Digitally controlled buck converter

The processing in the digital control loop results in delays. A PID controller that runs in the DSC has a typical calculation time of $1\mu s$. The PID control itself takes about $2\mu s$. The analogue-digital converter (ADC) takes samples and converts in about 500ns. Consider in addition a transistor switching time of 100ns, then we arrive at a total delay of $3.6\mu s$. From this we can determine an optimum sampling rate of 277kHz. In order not to loose any information, the Shannon theorem tells us the sampling frequency has to be at least twice the highest frequency that Fourier analysis reveals in the analogue signal. With this " twice" sampling rate the phase displacement is $180°$. Due to the associated delay in the system we exceed the maximum allowed $180°$ required for a stable system. For this reason in digital control systems the analogue signal is over sampled with a factor ten.

15. EVALUATION

13.1   With which isolated converter is the supply voltage across the transistor and for which converter is the double supply voltage across the transistor.

13.2   What does discontinuous current mean for a converter?

13.3   Which types of flyback converter are you familiar with?

13.4   What is the difference between a flyback and a forward type?

13.5   What does "feed forward" control of an SMPS involve?

13.6   What type of output diode is used for an SMPS with a high output voltage ?

13.7   What is the big advantage of ferrite core transformers compared to standard iron core transformers?

13.8   What does off-the-line mean?

13.9   Why should flyback converters never have an open output, or be without a control loop?

13.10   Up to which power level are linear power regulators used? What is the efficiency of such a regulator?

13.11   a) What does EMI stands for?

b) What for the most part determines the weight and volume of a linear stabilised power supply?

13.12   Prove that expression (13-21) is equivalent to expression (13-19).

16. THE DESIGN OF SWITCH-MODE POWER SUPPLIES

This can be found on the website of CASPOC. www.caspoc.com/education under the heading books "Pollefliet".

14 INVERTERS

CONTENTS

A. THREE-PHASE INVERTERS

B. SINGLE PHASE INVERTER

An inverter is a DC-AC converter. It converts a constant DC voltage into a block shaped voltage, trapezium shaped voltage or a sinusoidal shaped voltage.

Single and three-phase inverters exist.

Single phase types are usually constructed for a fixed frequency while three-phase inverters usually are part of a frequency converter intended to produce a variable frequency voltage useful for drive technology.

As is the case with a chopper an inverter uses artificial or forced commutation.

The standard switch for small to large applications is the IGBT. For very large power levels, thyristors can be used (GTO, IGCT, SCR).

Inverters are used for:
- frequency converters in the control of AC motors.
- UPS supplies or "no-break" sets to supply computers and measurement and control installations in case of supply interruption. Also used for emergency supplies in hospitals, etc...
- renewable energy sources (solar panels, windmills, ...) in which the energy generated is converted to AC energy that is then mostly transferred to the supply grid.
- power supplies in caravans, boats, etc...

A. THREE-PHASE INVERTERS

1. VOLTAGE AND CURRENT SOURCE INVERTER

Three-phase inverters converter a DC voltage or current into a three-phase AC voltage. In most cases the phase voltage has to be as sinusoidal as possible.

Depending on the supply source at the converter input we distinguish between a voltage source inverter and a current source inverter (fig. 14-1).

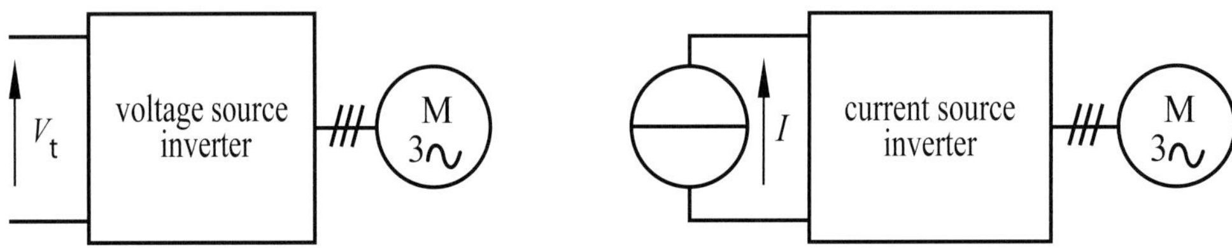

Fig. 14-1: Voltage source and current source inverter

The current source inverter has a limited number of applications, such as controlling extremely large AC motors. The current source inverter is studied on p. 20.90. In the current chapter we limit the study to the voltage source inverter.

2. SWITCHING MATRIX OF A VOLTAGE SOURCE INVERTER

The input voltage V_t can be obtained via:
- or from a battery
- or from a rail and feed wire from a traction system
- or from a rectifier from an AC supply net.

Fig. 14-2 shows the switching matrix of a three-phase bridge inverter.

There are two fundamental methods to complete a switching cycle: the 120°- and the 180°- conduction time per switch. In the 120°-type two switches are always closed: one in the top half of the bridge and one in the bottom half. In the case of the 180°-type inverter three switches are always closed which obviously means two switches closed in one half of the bridge. We will study the voltage waveforms for both types in the case of a symmetric load.

3. 180°-TYPE INVERTER

3.1 Operating principle

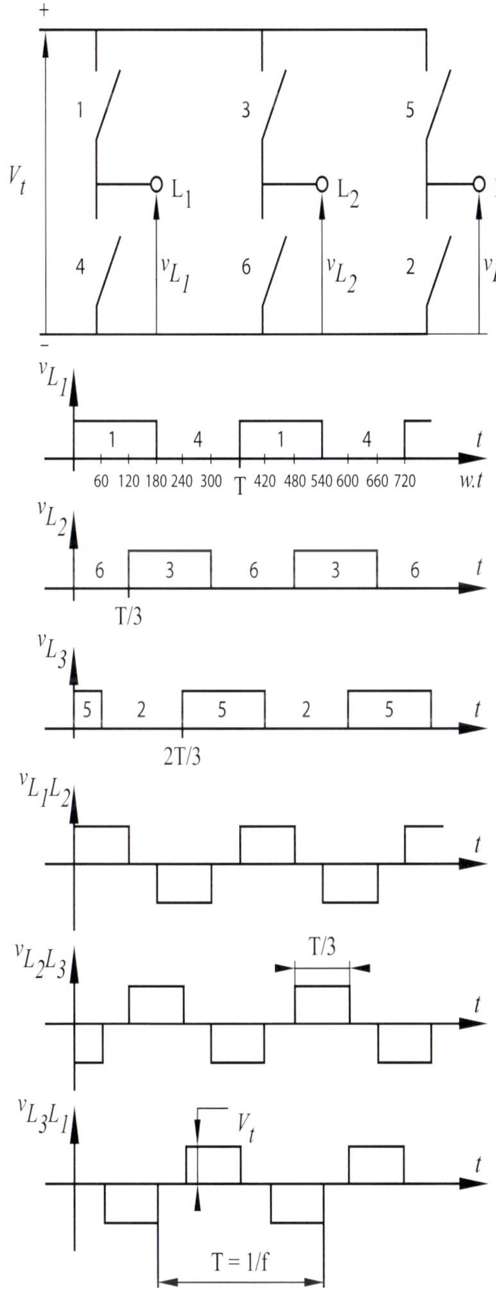

Fig. 14-2: Switching matrix three-phase inverter

The switches 1 and 4 are alternately opened and closed during a half period of the desired AC voltage. We draw the potential difference of L_1 with respect to the negative terminal of the DC voltage source. If the switches 3 and 6 operate in a similar manner to 1 and 4 (of course T/3 later), then we can also draw the potential of L_2. A third of a period later the pattern repeats, but now with switches 5 and 2 so that v_{L_3} is obtained. The line voltage $v_{L_1 L_2}$ is the difference between v_{L_1} and v_{L_2} or:

$$v_{L_1 L_2} = v_{L_1} - v_{L_2}.$$

If we also draw the other potential differences $v_{L_2 L_3}$ and $v_{L_3 L_1}$, then we notice that the inverter shown in fig. 14-2 produces a square wave three-phase waveform. In table 14-1 we note the switches which conduct simultaneously per 60° of the angle ωt.

Table 14-1

ωt	closed switches
0 - 60°	5 - 6 - 1
60 - 120	6 - 1 - 2
120 - 180	1 - 2 - 3
180 - 240	2 - 3 - 4
240 - 300	3 - 4 - 5
300 - 360°	4 - 5 - 6

Note that three switches with consecutive numbers are closed simultaneously.

This type of inverter is used to control three-phase induction motors

3.2 Voltage waveforms

If we connect a load configured in delta to the three-phase inverter of fig. 14-2 then the phase voltages of the load are identical to the line voltages. A load configured in star (fig. 14-3a and fig. 14-5) produces a step shaped phase voltage as shown in fig. 14-5. The line voltages in fig. 14-5 have been reproduced from fig 14-2.

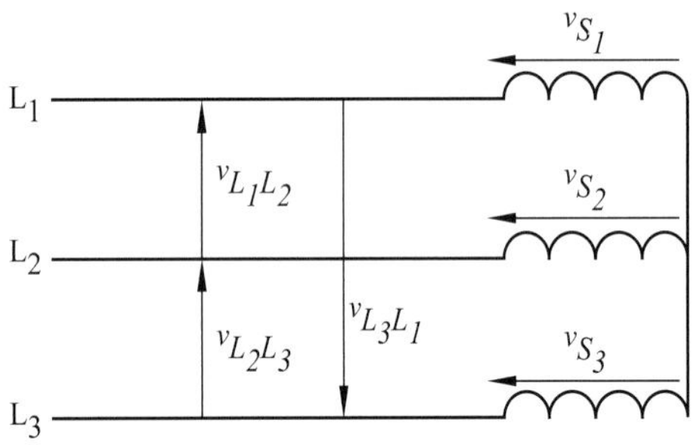

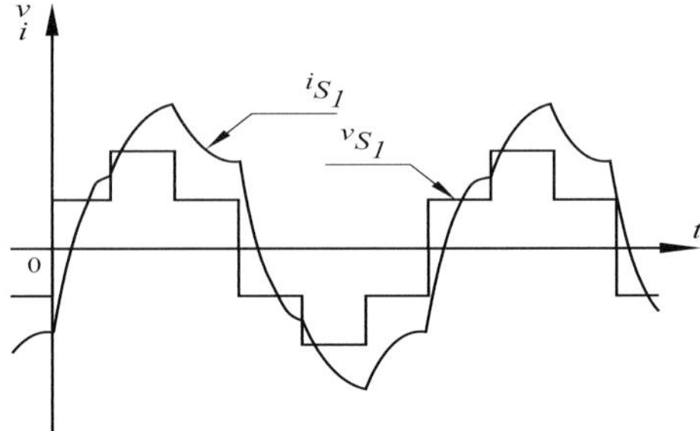

a: Symmetrical three-phase system   b: Stator current and voltage of a single motor phase of fig. 14-5

Fig. 14-3: Inverter load connected in star

The phase voltages are easily determined graphically. In fig. 14-4 we have drawn the situation for the circuit shown in fig. 14-5 between 0 and T/6.

We note that: $v_{S_1} = \frac{1}{3} . V_t$, $v_{S_2} = -\frac{2}{3} . V_t$ and $v_{S_3} = \frac{1}{3} . V_t$. It is clear that with three equal motor impedances the voltage across a single winding is always $\pm \frac{V_t}{3}$ or $\pm \frac{2}{3} . V_t$. The polarity and magnitude of the phase voltage are dependent upon the switching combination that is used at that instant by the switching matrix.

The staircase pattern in fig. 14-5 is the translation of these $\pm \frac{V_t}{3}$ and $\pm 2 . \frac{V_t}{3}$.

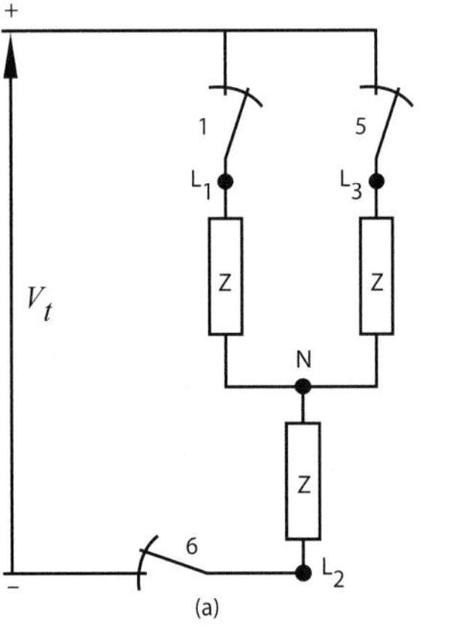
(a)

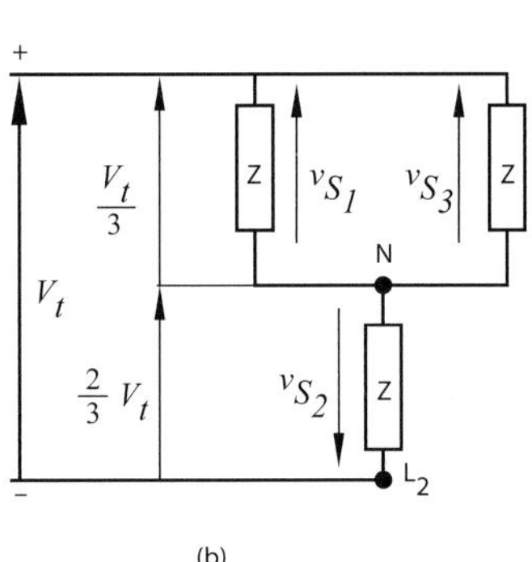
(b)

Fig. 14-4: Motor phases of fig. 14-5 between 0 and T/6

Applying a step shaped voltage to an inductive circuit results in an exponential current form. This explains the form of the phase current as shown in fig. 14-3b. In this case the line current is the same as the phase current since it is connected in star. Calculating the surface area of surface A of the resulting voltage waveform in fig. 14-5 provides a measure for the flux in the AC motor. This surface area represents a certain number of voltseconds (= Weber!). We return to this subject extensively in chapter 20.

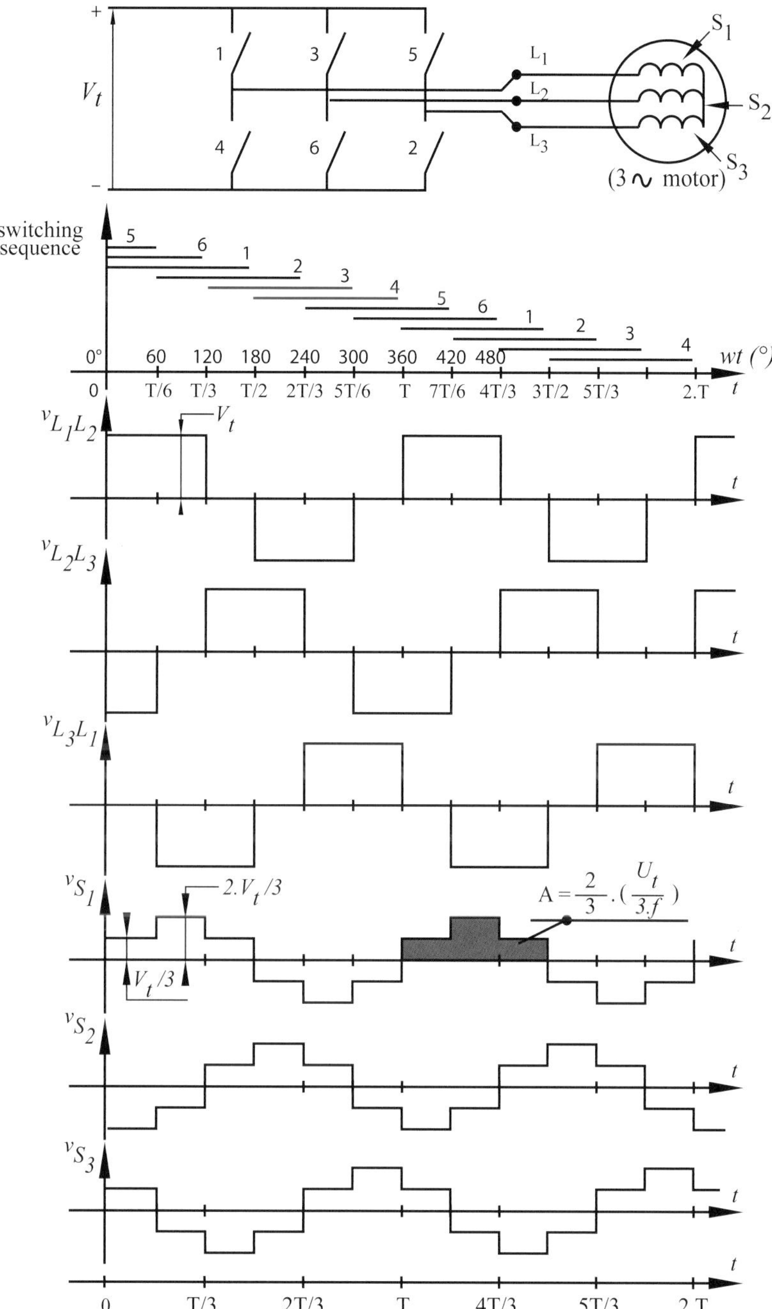

Fig. 14-5: Voltage waveforms of a 180°-type three-phase inverter, with a star connected motor as load

3.3 Dead time

Per bridge leg no two switches may conduct simultaneously otherwise we create a short circuit for V_t. In fig. 14-5 note that switch 4 closes at the instant that switch 1 should open. This opening of the switch requires a certain amount of time. To prevent a short circuit a dead time t_d is used between the opening of the one switch and the closing of the other switch per bridge leg. For transistor switches between 1 and 3μs is implemented for t_d.

3.4 Harmonics of a 180°-type inverter

3.4.1 Consumer connected in delta

Fig. 14-6a has the same form as fig. 8-35. The Fourier series is identical:

$$v_{L_1 L_2} = \frac{2 \cdot \sqrt{3} \cdot V}{\pi} \left(cos\omega t - \frac{cos5\omega t}{5} + \frac{cos7\omega t}{7} - \frac{cos11\omega t}{11} + \dots - \dots \right) \qquad (8\text{-}28)$$

The amplitude of the kth harmonic of the line voltage is:

$$\hat{v}_{L_1 L_2}(k) = \pm \frac{2 \cdot \sqrt{3} \cdot V}{k \cdot \pi} \qquad \begin{array}{l} + \ \text{ for } \ k = 1/7/13/\dots \\[2mm] - \ \text{ for } \ k = 5/7/11/\dots \end{array}$$

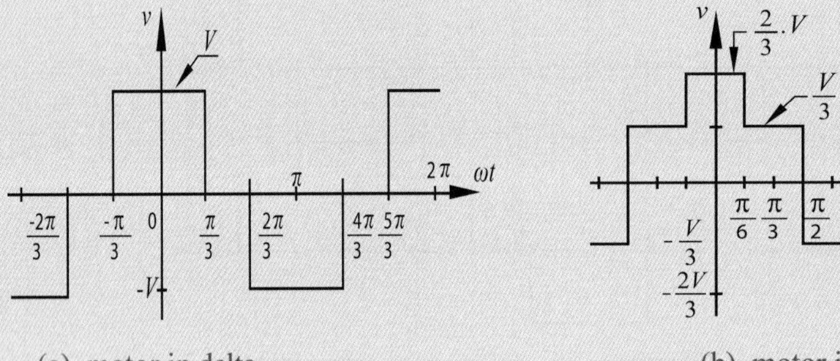

(a) motor in delta   (b) motor in star

Fig. 14-6: Motor voltages

3.4.2 Consumer connected in star

In a similar fashion to fig. 14-6a we find as solution for fig. 14-6b only odd numbered cosine terms.

1. Amplitude:

$$B_k = 2 \cdot \frac{1}{\pi} \int_0^{\pi/6} \left(\frac{2}{3} \cdot V\right) cos\,k\omega t \cdot d\omega t \ + \ 2 \cdot \frac{1}{\pi} \int_{\pi/6}^{\pi/2} \left(\frac{V}{3}\right) \cdot cos k\omega t \cdot d\omega t$$

$$+ \ 2 \cdot \frac{1}{\pi} \int_{\pi/2}^{5.\pi/6} \left(-\frac{V}{3}\right) \cdot cos k\omega t \cdot d\omega t \ + \ 2 \cdot \frac{1}{\pi} \int_{5\pi/6}^{\pi} \left(-\frac{2}{3} \cdot V\right) \cdot cos k\omega t \cdot d\omega t$$

$$B_k = \frac{2}{\pi.k} \cdot \frac{2}{3} \cdot V . sin k\omega t \Big|_0^{\pi/6} + \frac{2 \cdot V}{3 \cdot \pi \cdot k} . sin k\omega t \Big|_{\pi/6}^{\pi/2} - \frac{2 \cdot V}{3.\pi.k} . sin k\omega t \Big|_{\pi/2}^{5\pi/6} - \frac{4 \cdot V}{3.\pi.k} . sin k\omega t \Big|_{5.\pi/6}^{\pi}$$

$$B_k = \frac{4 \cdot V}{3.\pi.k} \cdot \left(sin\,k.\frac{\pi}{6} + sin\,k.\frac{5\pi}{6} \right) + \frac{2 \cdot V}{3.\pi.k} \cdot \left(sin\,k.\frac{\pi}{2} - sin\,k.\frac{\pi}{6} - sin\,k.\frac{5\pi}{6} + sin\,k.\frac{\pi}{2} \right)$$

$$B_k = \pm \frac{2 \cdot V}{\pi \cdot k} \qquad \begin{array}{l} + \ \text{ for } \ k = 1/5/13/17 \dots \\[2mm] - \ \text{ for } \ k = 7/11/19/23 \dots \end{array}$$

2. Fourier series:

$$u_R(t) = \frac{2 \cdot V}{\pi} \left(cos\,\omega t + \frac{cos\,5\omega t}{5} - \frac{cos\,7\omega t}{7} - \frac{cos\,11\omega t}{11} + \dots + \dots - \dots - \dots \right) \qquad (14\text{-}1)$$

4. PULSE FREQUENCY CONVERTER WITH A CONSTANT DC VOLTAGE

In fig. 14-5 a constant DC voltage V_t is formed with the aid of a three-phase diode bridge. The block diagram is shown in fig. 14-7.

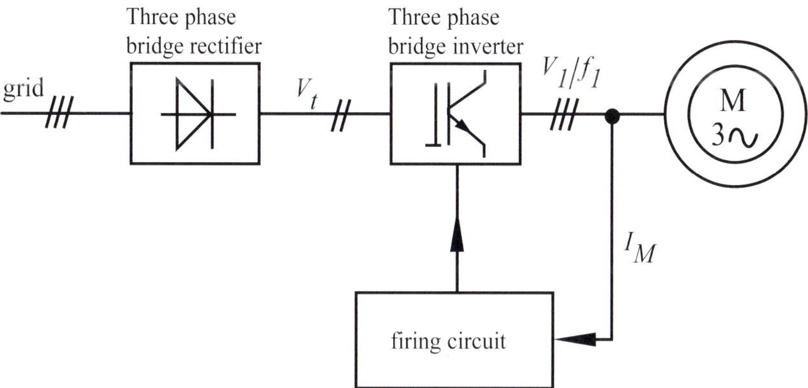

Fig. 14-7: Impulse control (quasi square wave) of an AC motor

Fig. 14-7 is redrawn in schematic form in fig. 14-8.

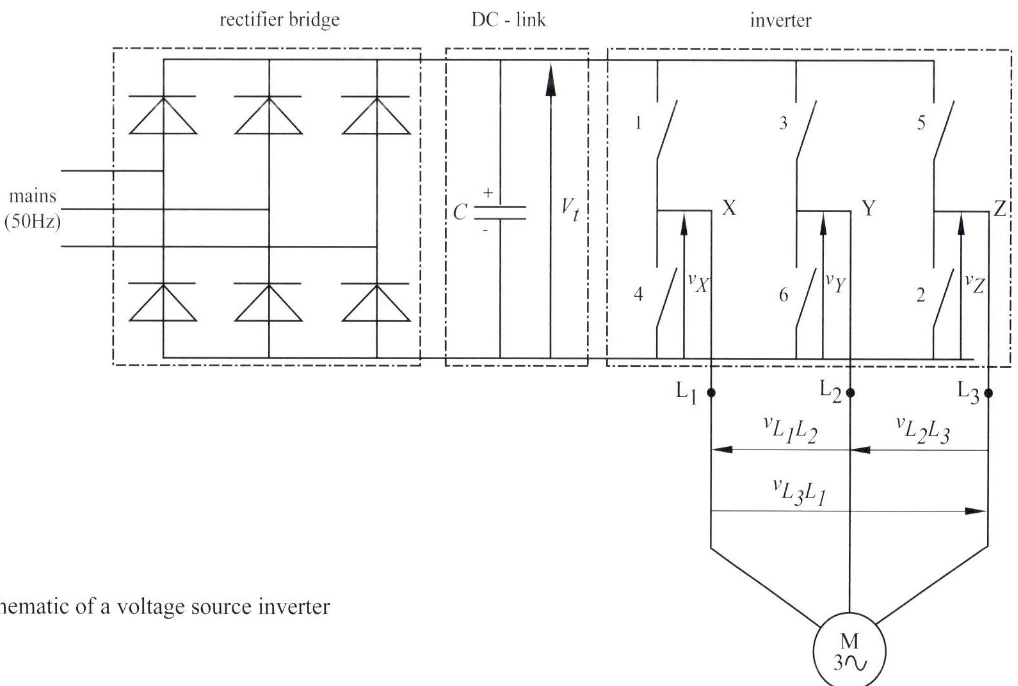

Fig. 14-8: Schematic of a voltage source inverter

In fig. 14-5 switch 1 is closed for a half period alternately with switch 4. These switches are open and closed n times per half period instead of closing each switch once. Closing time d is a constant. The voltages v_x, v_y, and v_z then appear as shown in fig. 14-9.

The line voltages are formed due to the difference between v_x, v_y and v_z.

$$v_{L_1L_2} = v_x - v_y, \quad v_{L_2L_3} = v_y - v_z, \quad v_{L_3L_1} = v_z - v_x.$$

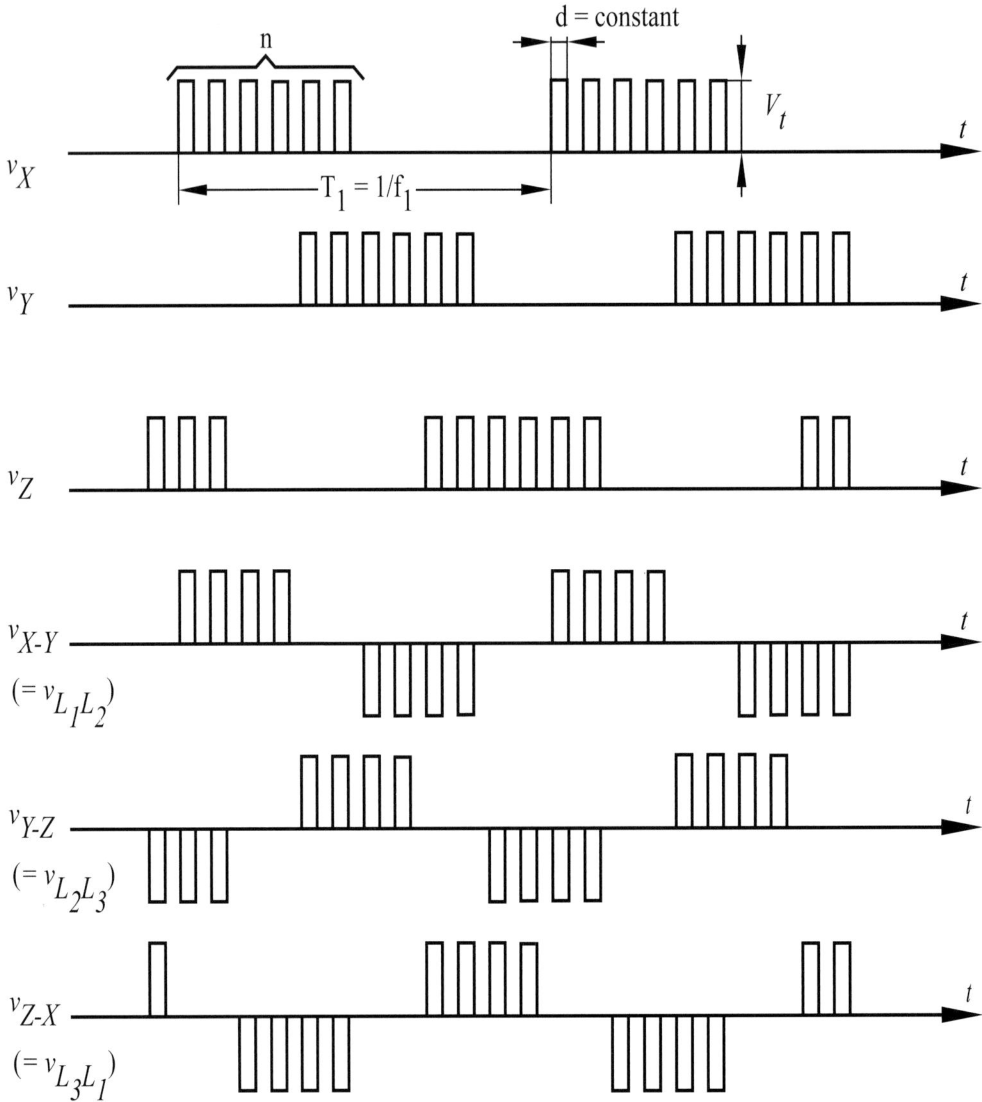

Fig. 14-9: Constant pulse train for the control of induction motors

The surface area of the line voltage per half period is: $\text{Area} = V_t \cdot n \cdot \dfrac{2}{3} \cdot d$.

Notice that just $\dfrac{2}{3} \cdot n$ pulses remain in the line voltage.

$$\left(v_{L_1L_2}\right)_{\text{average}} = \frac{\dfrac{2}{3} \cdot n \cdot d \cdot V_t}{\dfrac{T_1}{2}} = \frac{4}{3} \cdot n \cdot d \cdot V_t \cdot f_1$$

In paragraph no. 5.2 we will see that the instantaneous average of $v_{L_1L_2}$ is the fundamental harmonic, $(V_{L_1L_2})_1$, so that:

$$\left[\frac{(V_{L_1L_2})_1}{f_1}\right] = \frac{4}{3} \cdot n \cdot d \cdot V_t = C^{te}$$

(14-2)

Since $(\frac{V_1}{f_1})$ is an indication for the stator flux Φ, this will be constant.

In fig. 14-10 we see that when $T_1 = T_{1\,min}$, the pulses are next to each other and form full square wave blocks as the output voltage. This is called "quasi square wave".

The smallest period $T_{1\,min}$ which can be achieved is given by $\frac{2}{3} \cdot n \cdot d = \frac{T_{1\,min}}{3}$, from which: $T_{1\,min} = 2 \cdot n \cdot d$. The largest frequency is : $f_{1\,max} = \frac{1}{2 \cdot n \cdot d}$.

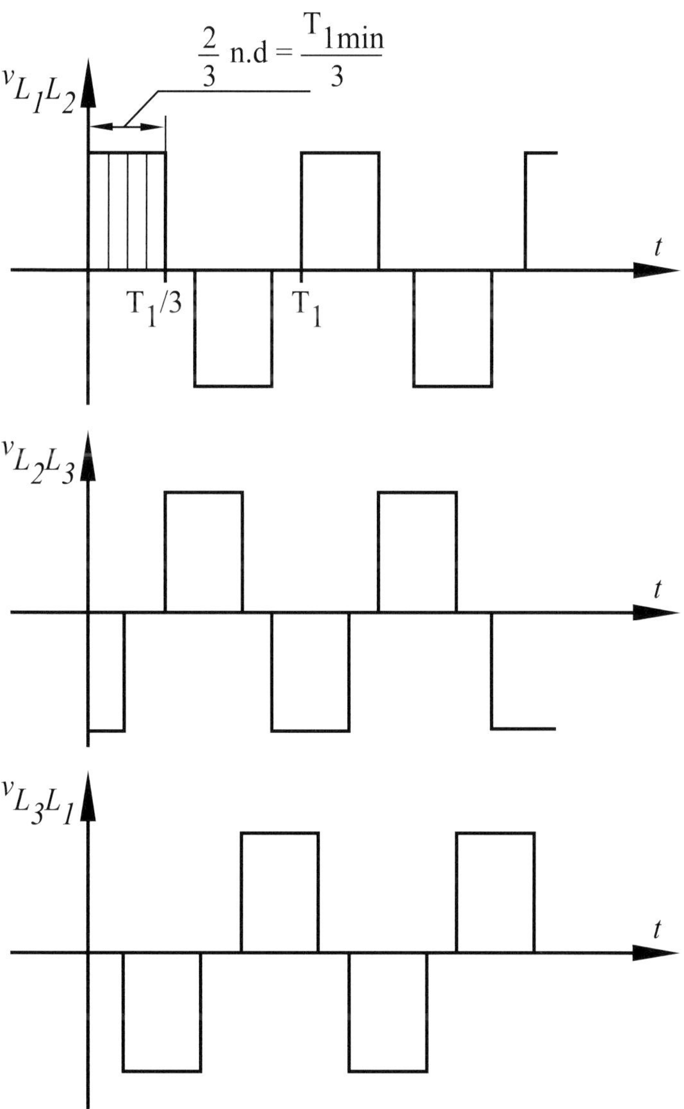

Fig. 14-10: Quasi square wave used to control a three-phase motor. The square waves represent the line voltages

5. PULSE WIDTH MODULATION (PWM)

5.1 Waveforms

The waveforms shown in fig. 14-9 generate a lot of harmonics. This results in high losses occuring in the (motor) load. By modulating the width of the pulses it is possible to achieve a better motor performance. This modulation technique is known as pulse width modulation or PWM for short.

A **didactic example** of PWM is shown in fig. 14-11a. The intersection of a triangular wave and a sinusoidal wave determine the times that the inverter switches are operated. We investigate the intersection points with the sine wave X:

Points A / C / E / G / I: closes switch 1 in fig. 14-8, switch 4 is opened.

Points B / D / F / H: opens switch 1 and after a dead time t_d switch 4 is closed.

For the sine wave Y which is displaced by 120° the operation is identical.

The voltages v_x and v_y in fig. 14-11b and -c now appear to be composed of pulses for which the pulse width varies according to a sinusoidal law. This is referred to as sinusoidal PWM.

The sine waves X and Y are modulated using frequency f_1.

The ratio $f_d / f_1 = N$ ($= 9$ in fig. 14-11) = pulse number.

The ratio m $= \hat{v}_M / V_d$ = modulation depth.

In fig. 14-11 the modulation depth m $= 0.6 = 60\%$.

We have only determined $v_{L_1 L_2}$. Using a third modulating wave Z (lags Y by 120°) we obtain a three-phase PWM wave with sinusoidal modulation.

The frequency f_d' is the switching frequency of the inverter. This is the frequency with which the switches are controlled. The frequency f_d is called the carrier frequency.

We refer to f_1 as the modulating frequency of the inverter. This is the frequency of the fundamental harmonic for the load which most often is a three-phase motor. In such a case f_1 determines the speed of the rotating (three-phase) field.

Since for this type of inverter the output voltage switches between zero and $+ V_t$, or between zero and $- V_t$, it is referred to as unipolar PWM as shown in fig. 14-11d.

In fig. 14-11 we also see that the inverter output is not a perfect sinusoid and therefore will also contain harmonics.

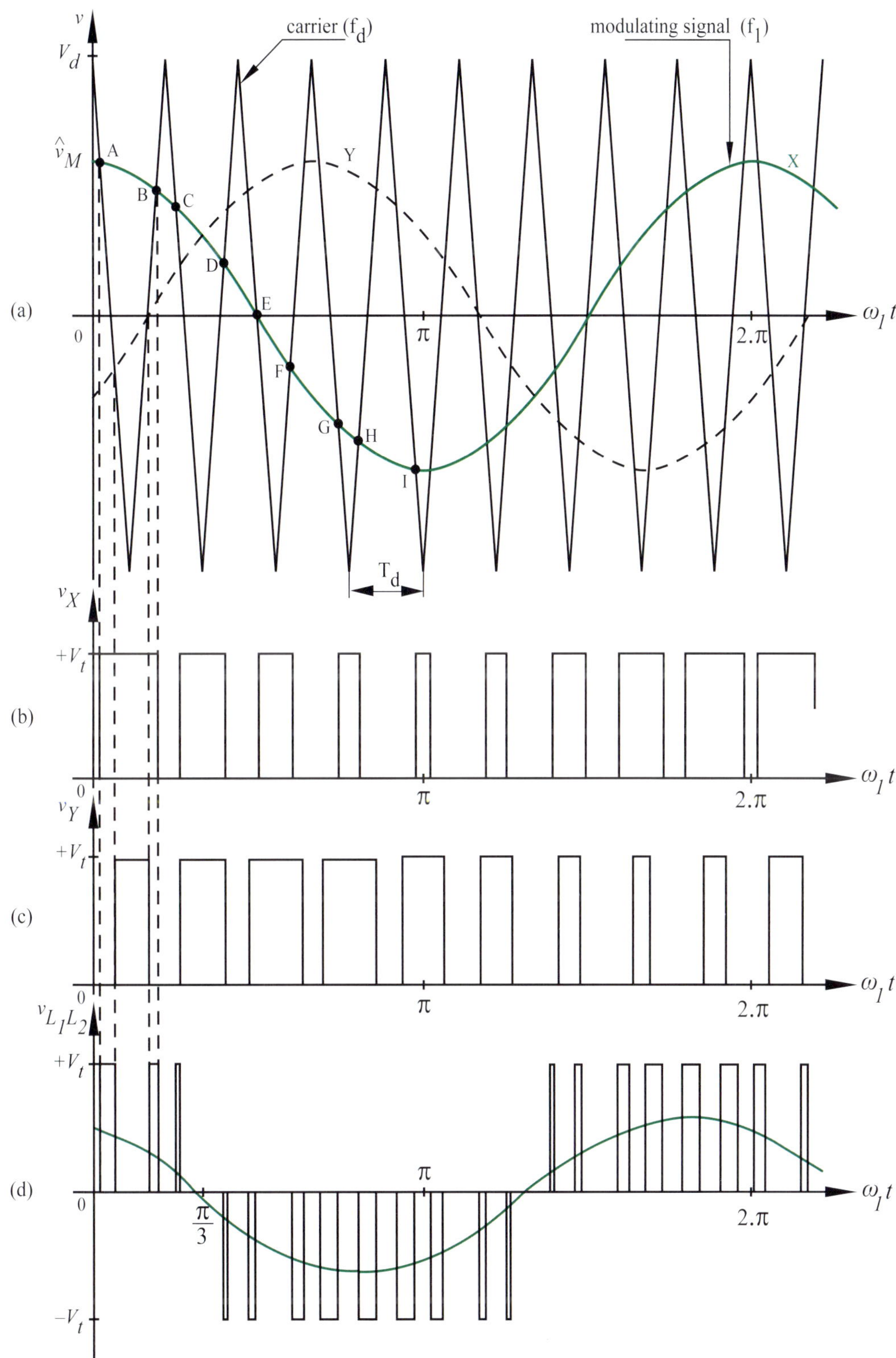

Fig.14-11: Sinusoidal PWM-waveform with nine pulses created using double sided modulation using a triangular wave

5.2 Calculating the output voltage of a voltage source inverter

Consider the triangular waveform shown in fig. 14-11a which is intersected by a constant control voltage instead of the modulating sinusoid X. This is drawn in fig. 14-12a.

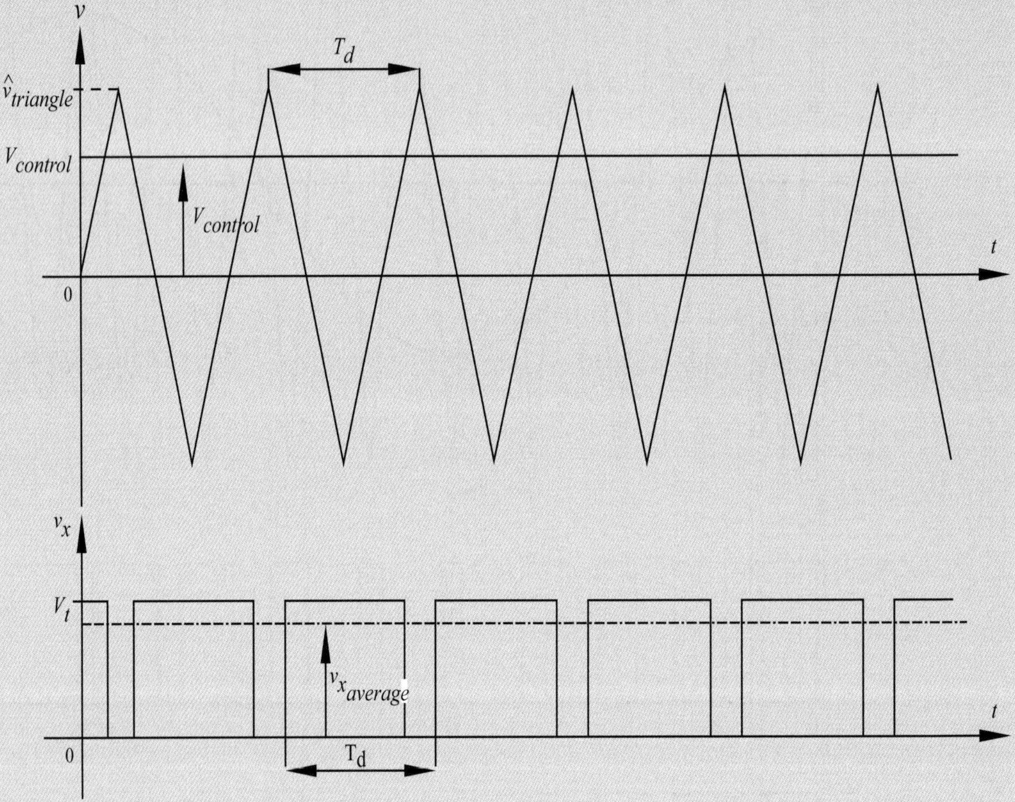

Figures 14-12a: PWM wave created using a constant control voltage

We now calculate the average value of v_x per period T_d.

From the detail drawing of fig. 14-12b we first determine t_x / T_d:

$$\frac{\hat{v}_{triang.} - v_{contr.}}{t_x/2} = \frac{\hat{v}_{triang.}}{T_d/4} \quad \text{from which:} \quad t_x = \frac{T_d}{2} \cdot \left[1 - \frac{v_{contr.}}{\hat{v}_{triang.}} \right]$$

$$\text{with } m_1 = \frac{v_{contr.}}{\hat{v}_{triang.}} \quad \longrightarrow\longrightarrow \quad \frac{t_x}{T_d} = \frac{1}{2} \cdot (1 - m_1)$$

In the lower part of fig. 14-12b we clearly see the average , per period T_d, of the voltage v_x:

$$(v_x)_{average} = V_t \cdot \frac{(T_d - t_x)}{T_d} = V_t \cdot (1 - \frac{t_x}{T_d}) = V_t \cdot \left[1 - \frac{1}{2} \cdot (1 - m_1) \right] = \frac{V_t}{2}(1 + m_1) \quad (14\text{-}3)$$

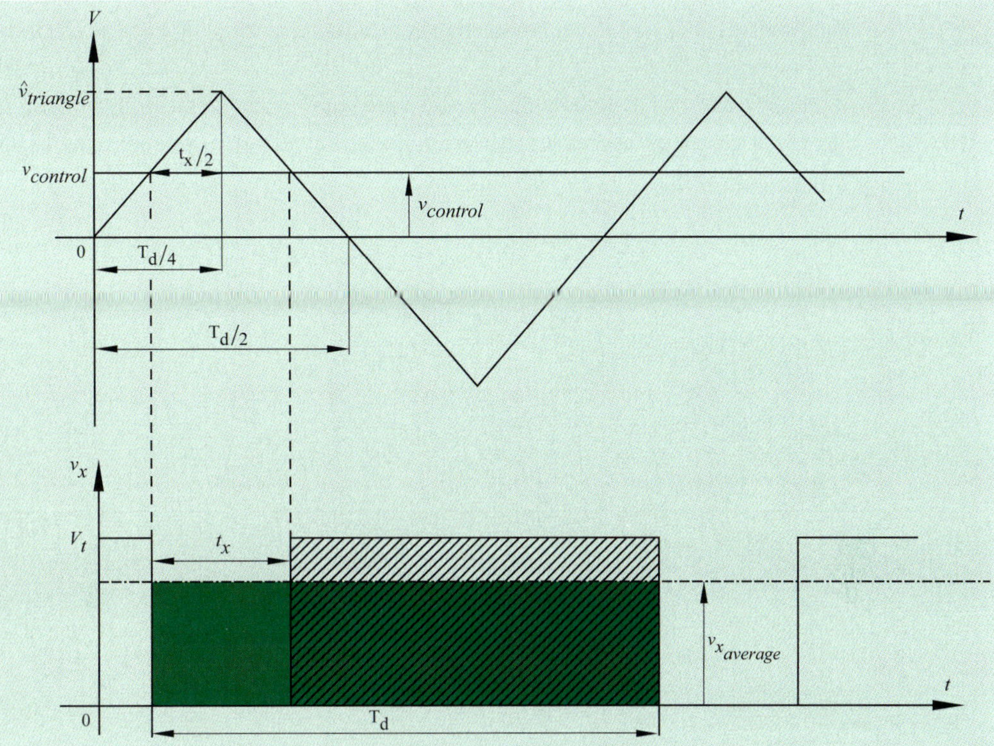

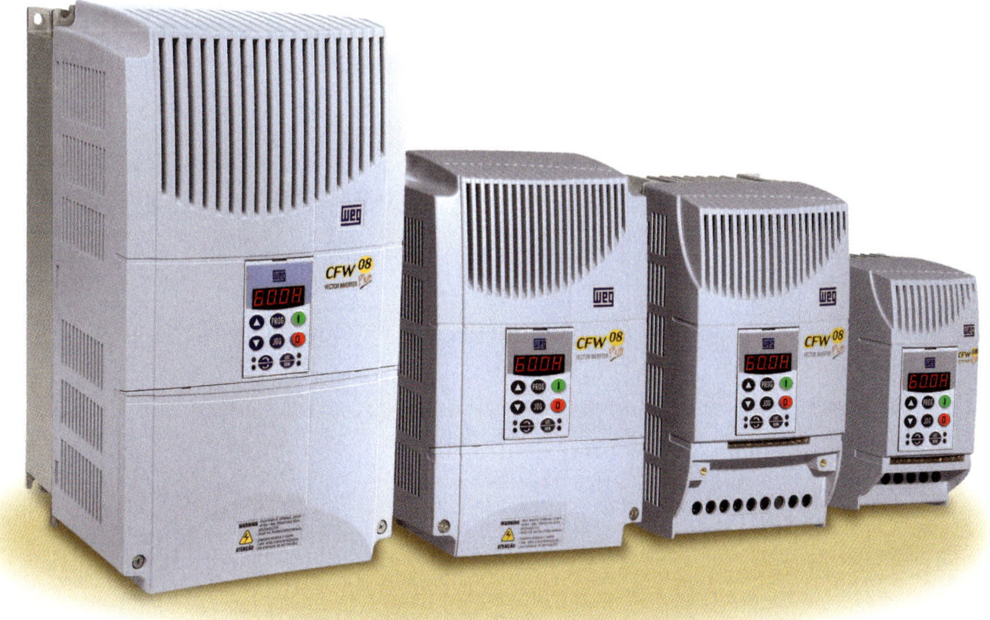

Fig 14-12b: Detail drawing of fig. 14-12a

Photo WEG Europe: The CFW08 is a standard frequency controller

Now consider the case that a modulating control voltage $v_{contr.} = v_M = \hat{v}_M \cdot \sin\omega_1 t$ is used in fig.14-11a.

If the period of this modulating voltage is large with respect to the period time T_d then we notice in fig. 14-13 that the modulating signal may be considered practically constant during one period T_d.

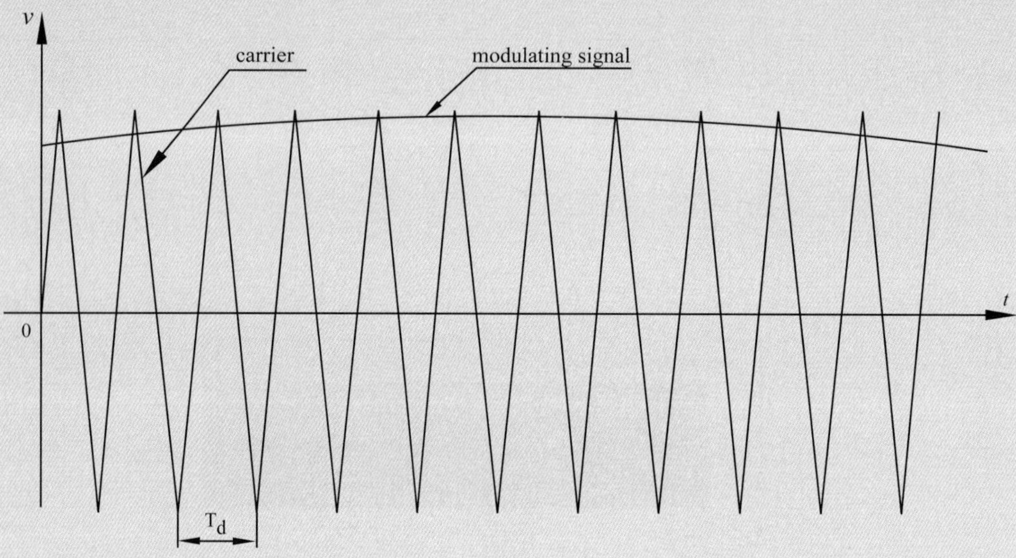

Fig.14-13

The expression (14-3) describes the average waveform v_x during one period. This expression tells us how the instantaneous average of v_x changes from one switching period T_d to the next. This "instantaneous average" is nothing more than the fundamental harmonic of the voltage v_x .

With $m_1 = \dfrac{v_M}{\hat{v}_{triang.}} = \dfrac{\hat{v}_M \cdot \sin\omega_1 t}{\hat{v}_{triang.}}$, (14-3) becomes: $v_x = \dfrac{V_t}{2} \cdot \left[1 + \dfrac{\hat{v}_M}{\hat{v}_{triang.}} \cdot \sin\omega_1 t \right]$

v_x is composed of :

- an average value $\dfrac{V_t}{2}$
- a sinusoidal voltage pulsating at ω_1 and with an instantaneous value: $v_{x_1} = \dfrac{V_t}{2} \cdot \dfrac{\hat{v}_M}{\hat{v}_{triang.}} \cdot \sin\omega_1 t$

The fundamental harmonic of the voltage is v_x and the modulation depth is $m = \dfrac{\hat{v}_M}{\hat{v}_{triang.}}$.

The amplitude of the fundamental harmonic is then: $\hat{v}_{x_1} = m \cdot \dfrac{V_t}{2}$

(14-4)

The RMS value of the fundamental harmonic is : $V_{x_1} = \dfrac{m \cdot V_t}{2 \cdot \sqrt{2}}$

(14-5)

The index 1 refers to the fundamental harmonic (= first harmonic!)

If we now consider the modulating Y in fig. 14-11a, then following an identical line of reasoning for the RMS value (of the fundamental harmonic) of the voltage v_y in fig. 14-11c gives :

$$V_{y_1} = \frac{m \cdot V_t}{2 \cdot \sqrt{2}} \cdot$$

The voltage v_{y_1} lags v_{x_1} by 120° so that the vector difference $(\vec{V}_{L_1 L_2})_1 = \vec{V}_{x_1} - \vec{V}_{y_1}$ introduces a factor $\sqrt{3}$ for the line voltage:

$$(\vec{V}_{L_1 L_2}) = m \cdot \frac{\sqrt{3} \cdot V_t}{2 \cdot \sqrt{2}} \tag{14-6}.$$

This fundamental harmonic is coloured green in fig. 14-11d and has an $m = 0.6$.

5.3 Remarks

1. Three-phase output

In fig. 14-11 only one line voltage $V_{L_1 L_2}$ is included at the output. In reality there are three wave forms ($V_{L_1 L_2}$, $V_{L_2 L_3}$, $V_{L_3 L_1}$), each displaced 120° with respect to each other. Due to the sinusoidal character of the modulation the output current will closely approximate a sinusoid.

2. Modulation depth m

Expression (14-6) shows that the RMS value of the fundamental harmonic in the output line voltage of the inverter is proportional to the modulation depth and has as a maximum value:

$$V_{L_1 L_2} = \frac{\sqrt{3}}{2 \cdot \sqrt{2}} \cdot V_t = 0.612 . V_t. \quad \text{This is valid as long as } m \leq 1 \,.$$

If the modulation depth is made larger than 1, then it is over modulated, with as a limit a square wave in the output as shown in fig. 14-6a.

Expression (8-28) provides the amplitude of the fundamental harmonic: $\hat{v}_{L_1 L_2} = \frac{2 \cdot \sqrt{3}}{\pi} \cdot V_t$.

The RMS value of the fundamental harmonic is then $V_{L_1 L_2} = \frac{2 \cdot \sqrt{3}}{\sqrt{2} \cdot \pi} \cdot V_t = \frac{\sqrt{6}}{\pi} \cdot V_t = 0.78 \cdot V_t$.

This square wave is obtained if in the didactic example of fig. 14-11 the reference cosine is made so large that the pulse train (v_x, v_y) becomes a square wave.

It is clear that the modulation depth at which this occurs is dependent upon the pulse number N. For example we find that with N = 15 that a square wave is produced in the inverter output for $m \approx 3.2$. This permits us to draw fig. 14-14.

3. Switching frequency f_d

Standard switching frequencies f_d are between 2 to 4 kHz with 8, 16 and 32 kHz also possible. Practical pulse numbers are 9/15/21/... It are usually odd multiples of 3 (see "remarks" under no. 7: harmonics of a PWM-wave on p. 14.23).

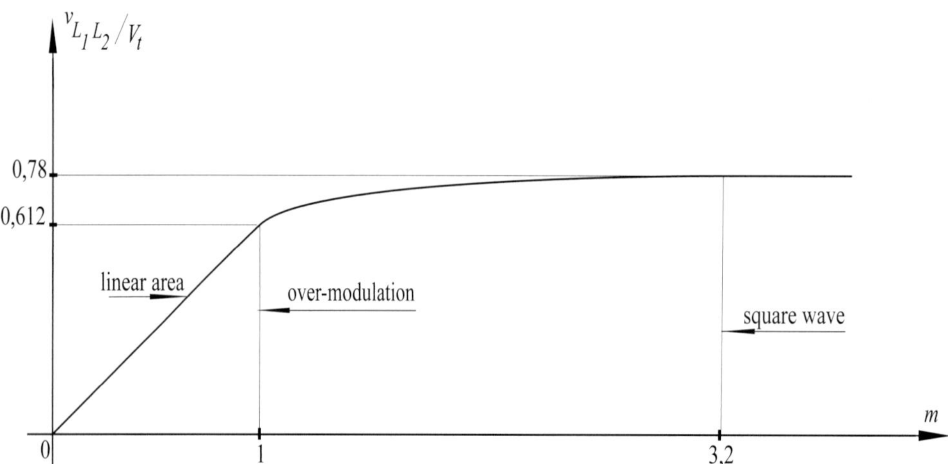

Fig. 14-14: Output line voltage as a function of modulation depth m (for N = 15)

6. PWM STRATEGIES

Up until 1980 the pulse patterns for PWM were produced using analog techniques. Between 1980 and 1990 digital technology started to be used to produce the PWM waveforms in three-phase inverters for motor control. In the early years 8 bit microprocessors were used. After that DSP's (digital signal processors) appeared on the scene.

6.1 Analog technology (system with modulated output voltage)

This system with a modulated output voltage was common in early analog technology. A reference voltage was compared with a carrier wave of higher frequency. This is comparable to the didactic example of fig. 14-11.

6.2. Digital technology (system with calculated voltage pattern)

The desired condition of the inverter switches is calculated using a DSP. For the three voltages there is a varying single step shaped reference, this can for example be derived from a "look-up" table. This (hypothetical) sinusoidal variable reference is corrected during every pulse with a "zero sequence" value. This zero sequence value is determined per pulse by dividing by two the sum of the most positive value and the most negative value and adding this zero sequence value to the three basic references.
In this way the zero pauses at the start and end of every pulse are synchronised.

6.3 Value of the output voltage of a PWM inverter

Fig. 14-5 shows the output voltage of an inverter with a quasi square wave in the output.

From (8-28) the amplitude of such a waveform can be found: $\hat{v}_1 = \dfrac{2 . \sqrt{3} . V_t}{\pi}$,

so that the RMS value is: $V_{1_{eff.}} = \dfrac{\sqrt{6}}{\pi} . V_t$. Here in : $V_t \approx \hat{v} = \sqrt{2} . V_{net}$.

The theoretical maximum of the fundamental harmonic in the output line voltage is

$$\left[\, V_{1\,eff.}\,\right]_{max} = \frac{\sqrt{6}}{\pi}\,.\,\sqrt{2}\,.\,V_{net} = 1.1\,.\,V_{net}$$

PWM is implemented to obtain a more sinusoidal motor current. In the case of a sinusoidal modulated PWM with a modulation index of 100% we find from (14-6): $V_{L_1L_2} = \dfrac{\sqrt{3}\,.\,V_t}{2\,.\,\sqrt{2}}$

For a single phase supply with $V_t = \sqrt{2}\,.\,V_{net}$ this becomes $V_{L_2L_1} = \dfrac{\sqrt{3}}{2}.\,V_{net}$.

A supply voltage of 230V is thus converted to a fundamental harmonic of 3x199V.

This is the case for (the theoretical maximum that) $V_t = \sqrt{2}\,.\,V_{net}$.

From $\dfrac{230}{199} = 1.155$ it follows that we need to over modulate with a single phase supply of 230V to obtain 3x230V as output.

With analog technology used a third harmonic (16% of the fundamental harmonic!) is added to create the reference modulating signal.

In the digital solution a number of the PWM pulses can be made somewhat wider.

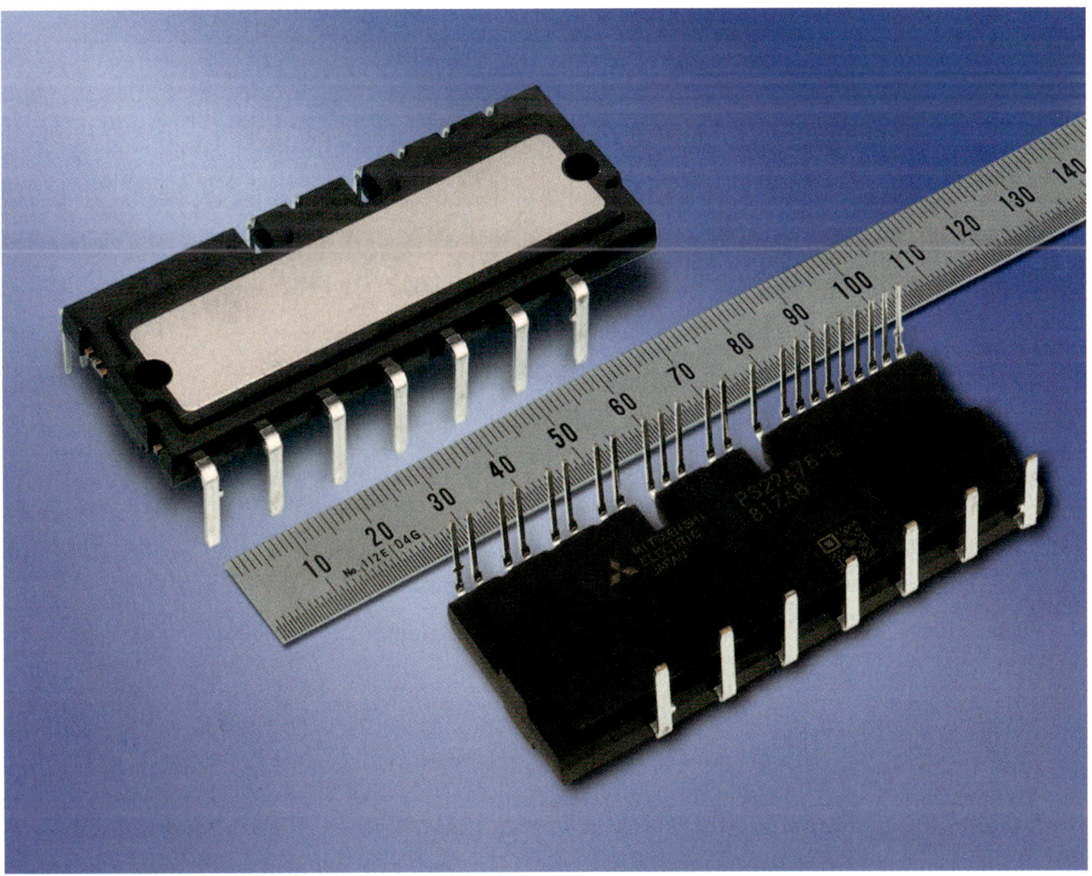

Photo Mitsubishi: The PS22A78-E is a DIP-IPM (dual in line package intelligent power module). This is a complete three-phase inverter with six IGBT's and the control incorporated in a single casing. Suitable for 1200V-35A. Dimension of casing: 79x31mm

6.4 The VCC principle of Danfoss

Drive manufacturer Danfoss uses a special procedure called VCC (Voltage Vector Control). In this system every switch in the inverter is closed sequentially for 60°, while the switching duration of the other two operating switches that switch at that instant is determined by a sine law and its width is modulated.

Fig. 14-15a shows the waveforms at the point X with respect to the negative voltage rail of V_t.

We also draw v_y and v_z (fig. 14-15b) and the voltage differences $v_{L_1L_2} = v_x - v_y$; $v_{L_2L_3} = v_y - v_z$ and $v_{L_2L_3} = v_z - v_x$. A three-phase voltage results.

In this way the full value of the supply voltage can be obtained at the output of the converter, while the average output line voltage remains sinusoidal. This results in the motor having a much more sinusoidal current. Since each switch is off for 60° of the 180° the switching losses are reduced.

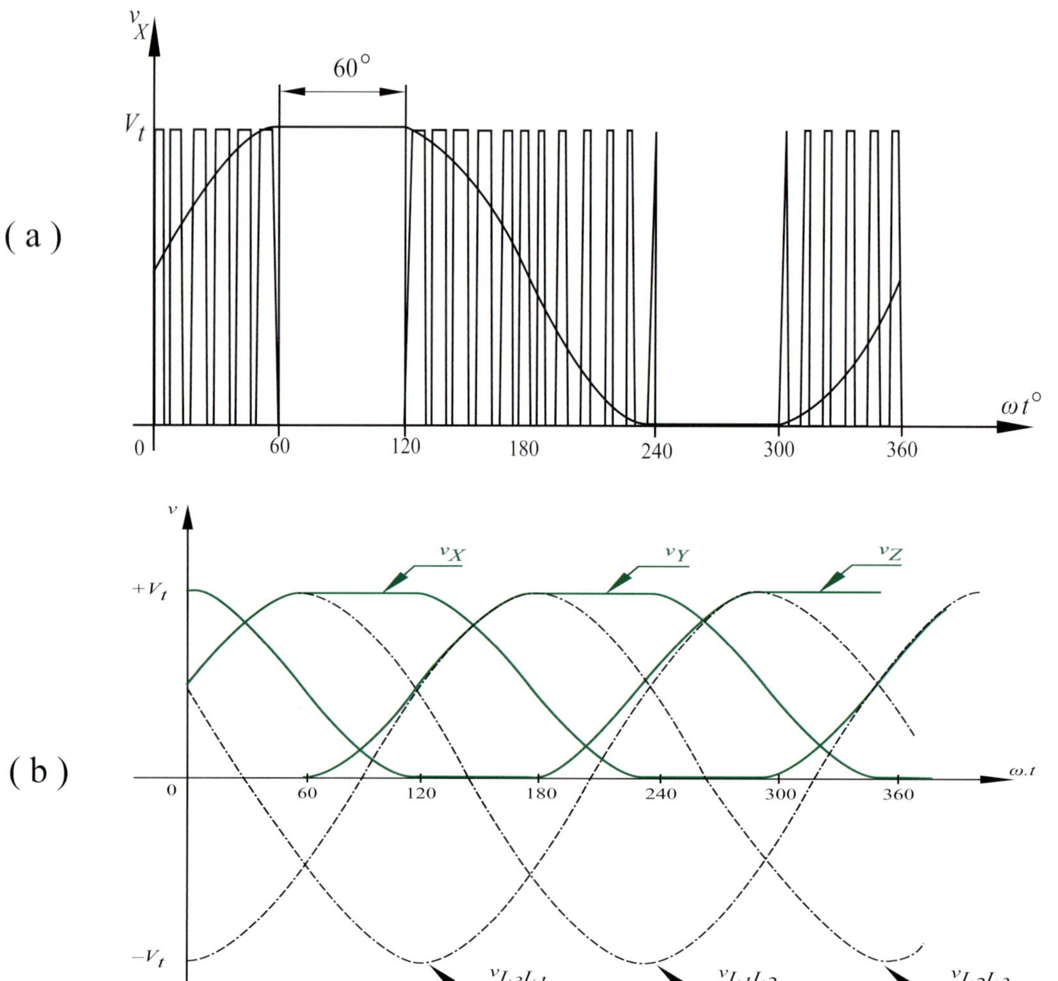

Fig. 14-15: Ground wave PWM converter using the VVC principle

7. HARMONICS IN A PWM-WAVE

If the sampling method is known that is used to generate the PWM wave then via numerical methods the harmonics can be determined.

For didactic reasons (to develop the thoughts involved) we consider the case of an analog generated PWM-wave as shown in fig. 14-11. This is a sinusoidal PWM-wave with nine pulses obtained via so called double modulation with a triangular wave.

For clarity this is redrawn here in fig. 14-11.

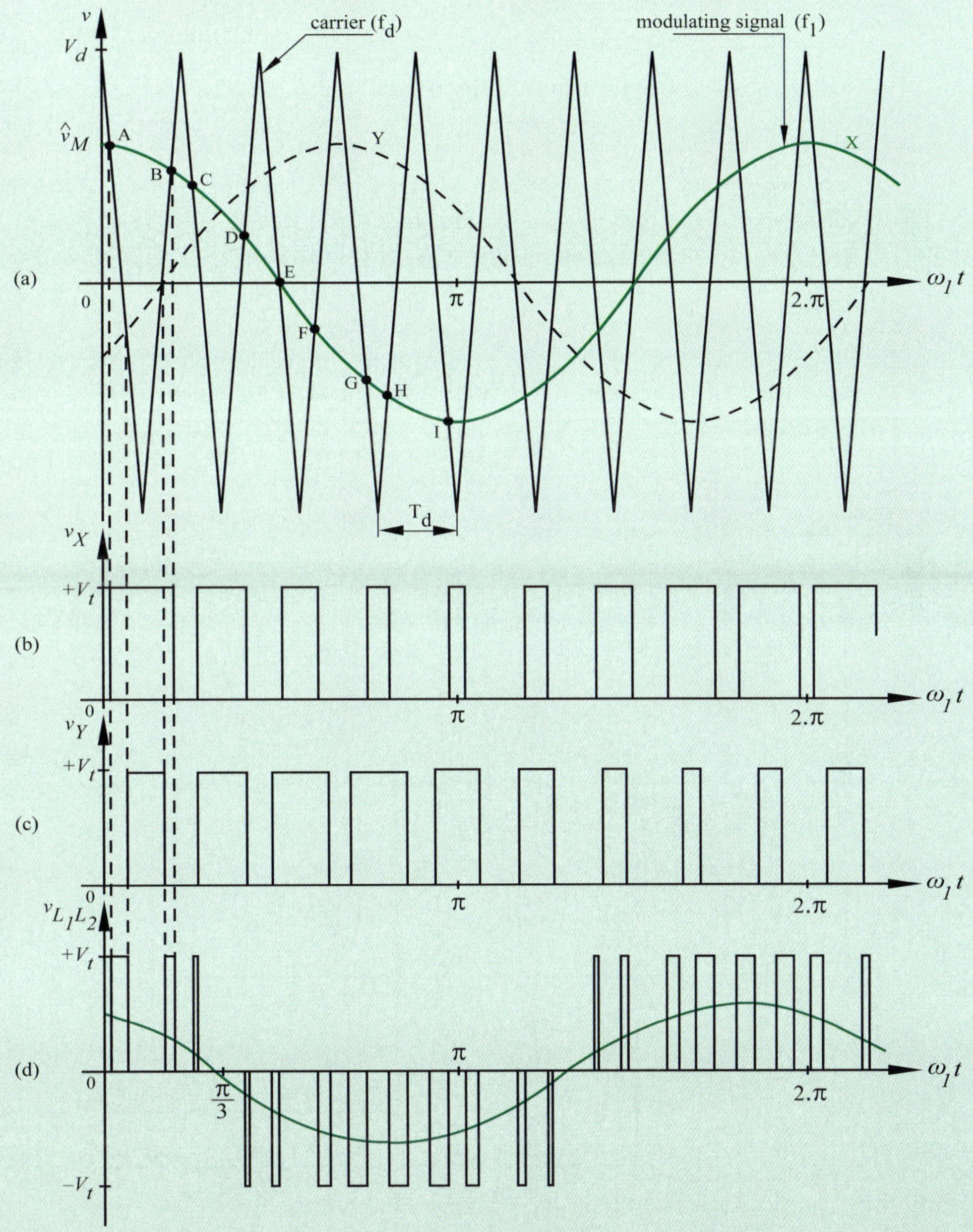

Fig.14-11

The relationship between the motor frequency f_1 of the sinusoidal modulator and f_d of the triangular shaped carrier is: $N = f_d/f_1$. The ratio between the amplitudes of the modulator and carrier is the modulation index $m = V_M/V_d$.

In fig. 14-11 the maximum of the triangular wave and the maximum of the carrier are synchronized. This is an example of synchronous modulation. The intersection points A, B, C… determine the times at which the switches in a half leg of the bridge are controlled. A second reference sinusoid Y which lags X by 120° will control the other bridge half of the three-phase inverter based on its intersection with the triangular wave. A third sinusoid Z (not shown) lags a further 120° behind and controls the last bridge leg.

The voltages v_x and v_y are the voltages with respect to the negative supply terminal of the DC link voltage V_t (fig. 14-16) so that for every instant: $v_x - v_y = v_{L_1L_2}$.

In fig. 14-11 the line voltage $v_{L_1L_2}$ is graphically derived. The average value of the line voltage follows a sine law, as a result of the modulating reference voltages X and Y. The intersection of the modulator and the triangular waves are easy to calculate.

From fig. 14-17 it follows for point A: $m \cdot \cos\omega_1 t = 1 - \dfrac{2 \cdot N}{\pi} \cdot \omega_1 t$.

With m = 60% and N = 9 we find: $0.6 \cdot \cos\omega_1 t = 1 - \dfrac{18}{\pi} \cdot \omega_1 t$.

The easiest way to solve this is by iteration since $\omega_1 t$ appears in both left and right terms of the equation. We find: $\omega_1 t = 4.015°$. In an identical manner we find the location of the other intersection points.

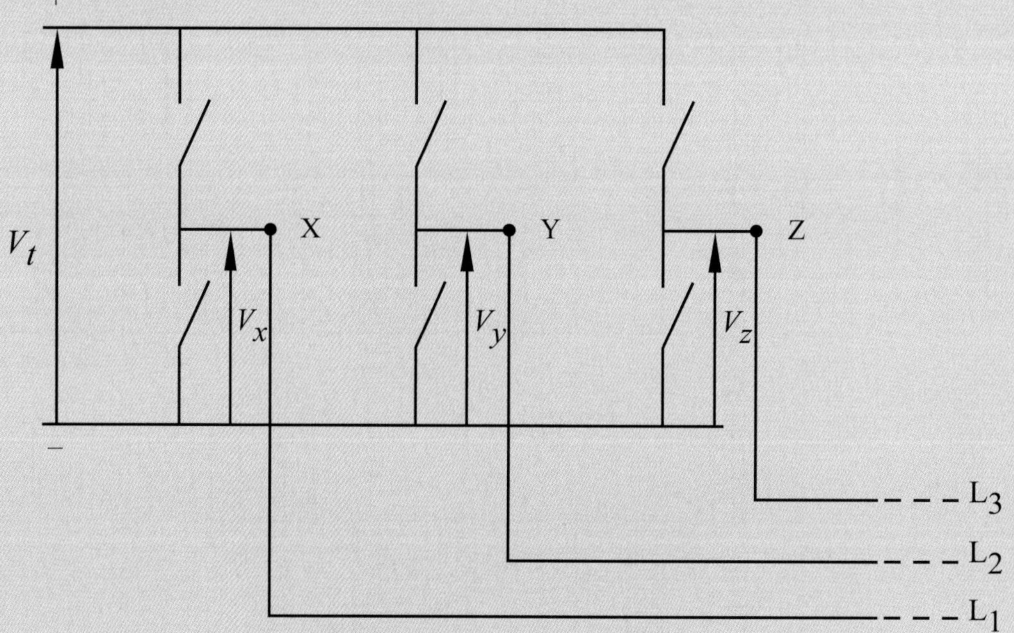

Fig. 14-16: Switching matrix of a three-phase bridge inverter

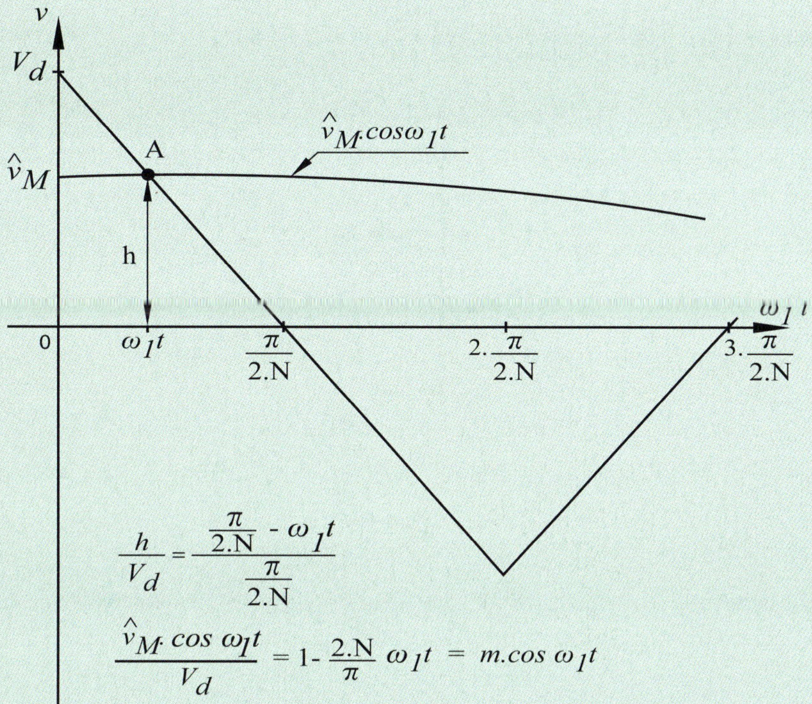

Fig. 14-17: Determining the intersection point of the triangular wave and the sinusoidal voltage X from fig. 14-11

The intersection points between the triangular wave and sinusoid X in fig. 14-11 are found from:

Point A: $\quad m \cdot \cos\omega_1 t = 1 - \dfrac{2 \cdot N}{\pi} \cdot \omega_1 t \quad \rightarrow \quad \omega_1 t = 4.015°$

Point B: $\quad m \cdot \cos\omega_1 t = \dfrac{2 \cdot N}{\pi} \cdot \omega_1 t - 3 \quad \rightarrow \quad \omega_1 t = 34.92°$

Point C: $\quad m \cdot \cos\omega_1 t = 5 - \dfrac{2 \cdot N}{\pi} \cdot \omega_1 t \quad \rightarrow \quad \omega_1 t = 45.82°$

Point D: $\quad m \cdot \cos\omega_1 t = \dfrac{2 \cdot N}{\pi} \cdot \omega_1 t - 7 \quad \rightarrow \quad \omega_1 t = 71.87°$

Point E: $\quad m \cdot \cos\omega_1 t = 9 - \dfrac{2 \cdot N}{\pi} \cdot \omega_1 t \quad \rightarrow \quad \omega_1 t = 90°$

Point F: $\quad m \cdot \cos\omega_1 t = \dfrac{2 \cdot N}{\pi} \cdot \omega_1 t - 11 \quad \rightarrow \quad \omega_1 t = 108.13°$

Point G: $\quad m \cdot \cos\omega_1 t = 13 - \dfrac{2 \cdot N}{\pi} \cdot \omega_1 t \quad \rightarrow \quad \omega_1 t = 134.18°$

Point H: $\quad m \cdot \cos\omega_1 t = \dfrac{2 \cdot N}{\pi} \cdot \omega_1 t - 15 \quad \rightarrow \quad \omega_1 t = 145.08°$

Point I: $\quad m \cdot \cos\omega_1 t = 17 - \dfrac{2 \cdot N}{\pi} \cdot \omega_1 t \quad \rightarrow \quad \omega_1 t = 175.98°$

Using these results the intersection points in fig. 14-11 can easily be determined for the Y-modulator and the triangular wave. This enables us to determine the borders of the pulses which create the line voltage.

Solution: (border of the pulses which create $v_{L_1L_2}$ in fig. 14-11).

Pulse 1:	4.015°	to	11.87°	Pulse 7:	124.015°	to	134.18°
Pulse 2:	30.0°	to	34.92°	Pulse 8:	145.08°	to	154.92°
Pulse 3:	45.82°	to	48.13°	Pulse 9:	165.82°	to	175.98°
Pulse 4:	71.87°	to	74.18°	Pulse 10:	184.015°	to	191.87°
Pulse 5:	85.08°	to	90.0°	Pulse 11:	210.0°	to	214.92°
Pulse 6:	108.13°	to	115.985°	Pulse 12:	225.82°	to	228.13°

To simplify the process of Fourier analysis we determine a number of pulses to the left of the zero of the time axis. We then find a pulse pattern as shown in the upper half of fig. 14-18.

By shifting the abscis with 30° we arrive at the situation shown in the lower half of fig. 14-18.

If all the pulses of $v_{L_1L_2}$ had to be drawn in fig. 14-18 then we should note that:

1) the line voltage is a symmetrical function with respect to the ordinate.
2) this function is reversible per half period with respect to the time axis. From both of these properties it follows that the Fourier series will only contain odd cosine terms. The goal of rearranging the pulses as shown in the bottom of fig. 14-18 was to facilitate this simplified calculation.

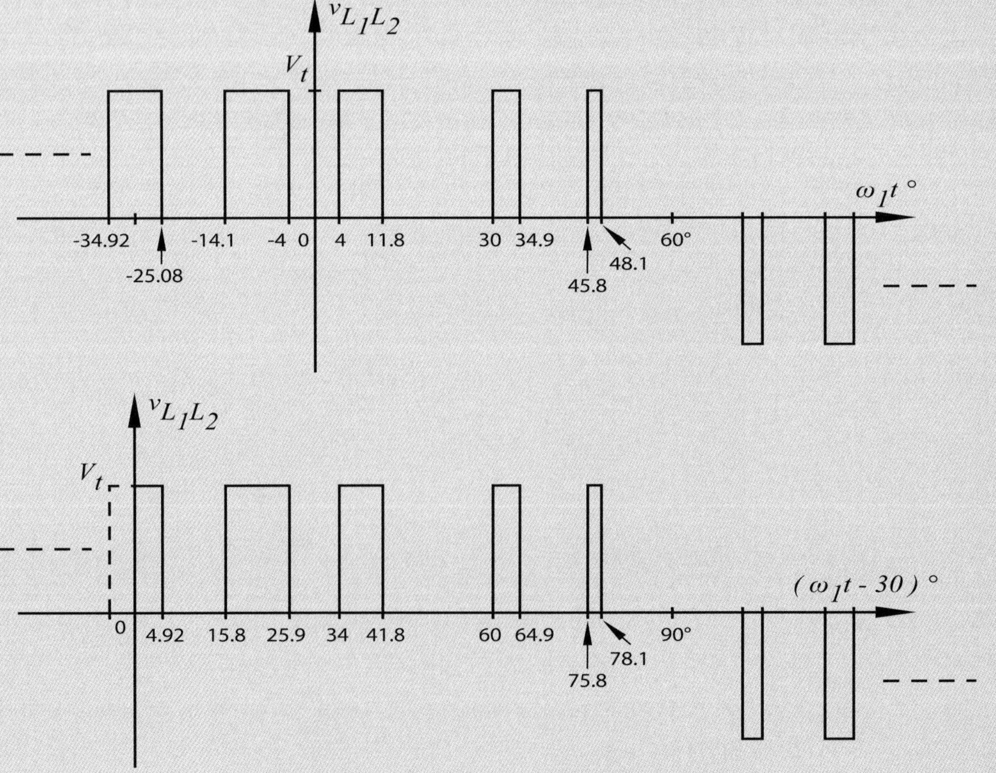

Fig. 14-18: Line voltage of the PWM-wave of fig. 14-11, rearranged for a simplified Fourier analysis

The amplitudes of the odd cosine terms in the Fourier series are given by:

$$B_k = \frac{1}{\pi} \int_{-\pi}^{+\pi} V_t \cdot \cos k\omega t \cdot d\omega t = \frac{4 \cdot V_t}{k \cdot \pi} \int_0^{\pi/2} \cos k\omega t \cdot d\omega t$$

$$B_k = \frac{4 \cdot V_t}{k \cdot \pi} [\ \sin k\omega t\ ] \quad \begin{vmatrix} 4.92° \\ 0° \end{vmatrix} \quad \begin{vmatrix} 25.985° \\ 15.82° \end{vmatrix} \begin{vmatrix} 41.87° \\ 34.015° \end{vmatrix} \begin{vmatrix} 64.92° \\ 60° \end{vmatrix} \begin{vmatrix} 78.13° \\ 75.82° \end{vmatrix}$$

$$B_k = \frac{4 \cdot V_t}{k \cdot \pi} (\ \sin k \cdot 4.92° + \sin k \cdot 25.985° + \sin k \cdot 41.87° + \sin k \cdot 64.92° + \sin k \cdot 78.13°$$
$$- \sin k \cdot 0° - \sin k \cdot 15.82° - \sin k \cdot 34.015° - \sin k \cdot 60° - \sin k \cdot 75.82°\ )$$

Replacing k with 1, 3, 5, 7, ... provides the amplitudes of the harmonic components.
The results are indicated here.

$B_1 = 0.5196 . V_t$	$B_{17} = 0.320 . V_t$	$B_{31} = 0.069 . V_t$
$B_3 = 0$	$B_{19} = 0.320 . V_t$	$B_{33} = 0$
$B_5 = 0.002 . V_t$	$B_{21} = 0$	$B_{35} = 0.007 . V_t$
$B_7 = 0.113 . V_t$	$B_{23} = 0.037 . V_t$	$B_{37} = 0.007 . V_t$
$B_9 = 0$	$B_{25} = -0.176 . V_t$	...
$B_{11} = 0.113 . V_t$	$B_{27} = 0$	
$B_{13} = 0.005 . V_t$	$B_{29} = 0.179 . V_t$	
$B_{15} = 0$		

The RMS value of the fundamental harmonic in the line voltage is determined as:

$$V_{1\,eff} = \frac{0.5196}{\sqrt{2}} \cdot V_t = 0.3674 \cdot V_t$$

Remarks

1. To limit the harmonics we take an odd whole number for N ($= f_d / f_1$). As a result we have only odd cosine terms as shown in the numeric example above. If in addition, N is an odd numbered multiple of three, then all the harmonics that are multiples of three disappear (B_3 , B_9 , B_{21} , ...). The dominant harmonics in the line voltage disappear as a result. The most dominant harmonics are now $(2.N \pm 1). f_1$.
   In our numeric example with N = 9 that is B_{17} and B_{19} !
2. The harmonics do not contain multiples of f_d ($= N . f_1$).
   With N = 9: $B_9 = 0$, $B_{18} = 0$, $B_{27} = 0$,...
3. If $V_t = 300V$ then $V_{1\,eff} = 0.3674 \times 300 = 110.22V$. Note that these values are valid for a modulation index of 60%.
   With $m = 100\%$ this becomes : $V_{1\,eff} = 183.4V$, and with $V_t = 325V$ and $m = 100\%$ we find $V_{1\,eff} = 199V$.
4. For a square wave output the line voltage contains harmonics with a frequency of $(2.k \pm 1) . f_1$ and with an amplitude that decreases inversely proportional to the order k of the harmonic.

Numeric example 14-1:

Given:

A single phase supply 230V-50/60Hz. A three-phase inverter with pulse number N = 9.
DC link voltage V_t = 310V. The load is a three-phase induction motor.

Required

If the output of the inverter is required to deliver 100Hz, calculate the RMS value and frequency
of the fundamental harmonic and the seven lowest harmonics when the modulation depth is set to
100%.

Solution:

From the Fourier analysis on p. 14.23 it follows:

$B_1 = 0.5196.V_t$; $B_5 = 0.002.V_t$; $B_7 = 0.113.V_t$; $B_{11} = 0.113.V_t$; $B_{13} = 0.005.V_t$;
$B_{17} = 0.32.V_t$; $B_{19} = 0.32.V_t$; $B_{23} = 0.037.V_t$

These are the values when $m = 0.6$. For a modulation depth of 100% we find:

- *fundamental harmonic :*

$$V_{1\,eff} = \frac{0.5196}{0.6 \text{ x } \sqrt{2}} . 310 = 189V \text{ with } f_1 = 100Hz$$

- *harmonics:*

$$V_{5\,eff} = \frac{0.002}{0.6 \text{ x } \sqrt{2}} . 310 = 0.73V \text{ at } 500Hz \text{ ; } \qquad V_{7\,eff} = \frac{0.113}{0.6 \text{ x } \sqrt{2}} . 310 = 41.28V \text{ at } 700Hz \text{ ;}$$

$$V_{11\,eff} = \frac{0.113}{0.6 \text{ x } \sqrt{2}} . 310 = 41.28V \text{ at } 1100Hz \text{ ; } \qquad V_{13\,eff} = \frac{0.005}{0.6 \text{ x } \sqrt{2}} . 310 = 1.82V \text{ at } 1300Hz \text{ ;}$$

$$V_{17\,eff} = \frac{0.32}{0.6 \text{ x } \sqrt{2}} . 310 = 116.9V \text{ at } 1700Hz \text{ ; } \qquad V_{19\,eff} = \frac{0.32}{0.6 \text{ x } \sqrt{2}} . 310 = 116.9V \text{ at } 1900Hz \text{ ;}$$

$$V_{23\,eff} = \frac{0.037}{0.6 \text{ x } \sqrt{2}} . 310 = 13.5V \text{ at } 2300Hz \text{ .}$$

8. 120°-TYPE INVERTER

Every switch (fig. 14-19) conducts for 120° , of which 60° is simultaneous with a switch from the other bridge half. Notice that now two switches with consecutive numbers are conducting.

An advantage of this method is that per half bridge there is a dead time of 60° between the closing of the switches. This type of inverter is used in the control of three-phase brushless DC motors.

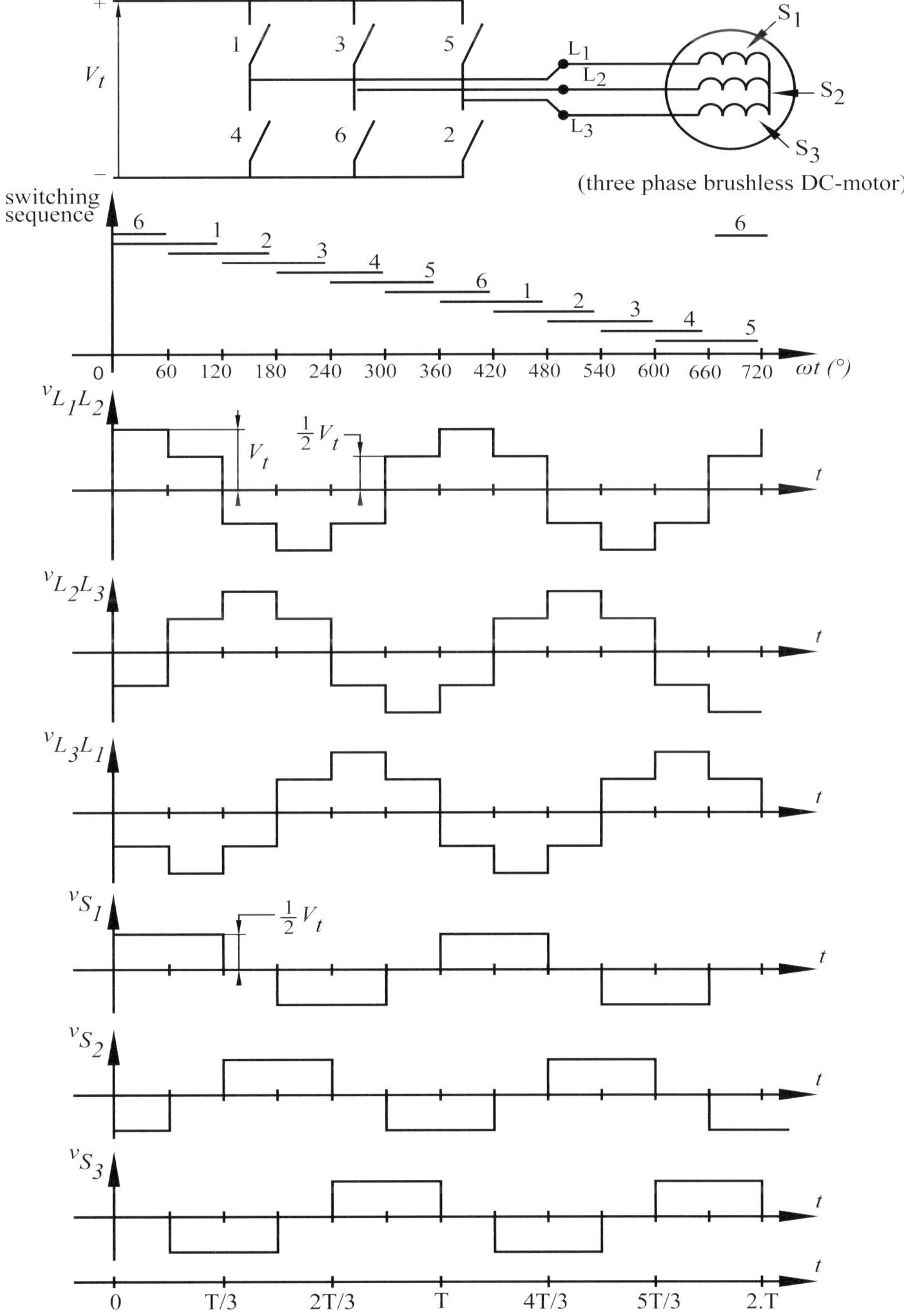

Fig. 14-19: Voltage waveforms of an 120°-type inverter

9. COMMON SWITCHES USED IN INVERTERS

Three-phase inverters are most frequently used in variable frequency drives for induction motors. The first frequency converters appeared on the market toward the end of the sixties. They were built using SCR switches. For the smaller power ranges in the first half of the seventies the switches were replaced with bipolar transistors. These BJT's were predominantly darlington types. They required sophisticated protection networks.

Towards the end of the eighties the robust IGBT and the controllable GTO began to be used instead of a BJT or SCR, and this for a power level up to 1MW. Above 1 MW the SCR may still be used.

On p. 14.27 the details of a modular IGBT are reproduced. It is a 23NAB126V1 from Semikron. A short analysis of the data tells us:

1.   The module is suitable for inverter use up to 16 kVA and for a motor of 7.5kW.
2.   The module contains a three-phase bridge rectifier, a brake chopper and a three-phase inverter.
3.   The inverter and brake chopper are constructed using fast trench IGBTs suitable for 1200V.
4.   Depending on the casing temperature (25°C or 70°C) the nominal current per transistor is 41 or 31 A.
5.   The brake chopper can conduct 30A (25°C).
6.   The input rectifier can rectify 46A ($T_S = 70°C$).
7.   Each IGBT has a typical voltage drop of 1.7V at 25°C up to 2.4V at 125°C and this with a collector current of 25A.
8.   The input capacitance is not much larger than with a MOSFET. The input capacitance is 1800pF compared to a MOSFET with 800 to 1000pF.
9.   On and off switching times are much larger than for a MOSFET.
      For this IGBT we find $t_{d(off)} = 400$ns and $t_f = 100$ns. For a MOSFET we find for example $t_{d(off)} = 40$ns and $t_f = 25$ns.
10.   The term SKiiP stands for Semikron integrated intelligent Power.
11.   CAL-technology with respect to the diodes means Controlled Axial Lifetime.

SKiiP 23NAB126V1

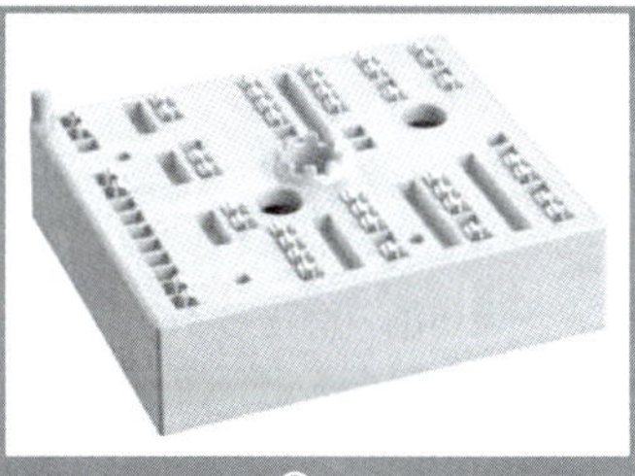

MiniSKiiP® 2

3-phase bridge rectifier + brake chopper + 3-phase bridge inverter
SKiiP 23NAB126V1

Features

- Fast Trench IGBTs
- Robust and soft freewheeling diodes in CAL technology
- Highly reliable spring contacts for electrical connections
- UL recognised file no. E63532

Typical Applications

- Inverter up to 16 kVA
- Typical motor power 7,5 kW

Remarks

- V_{CEsat} , V_F = chip level value

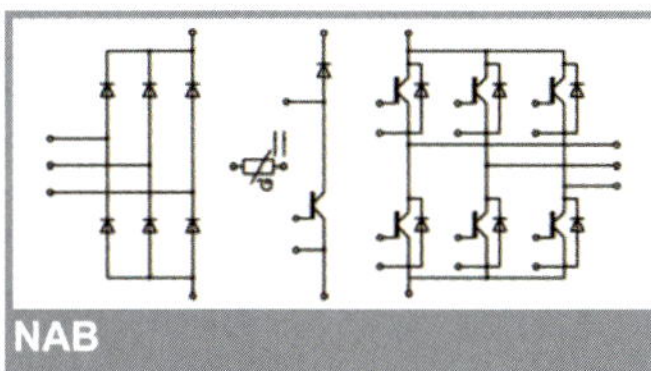

NAB

Absolute Maximum Ratings		T_s = 25 °C, unless otherwise specified	
Symbol	**Conditions**	**Values**	**Units**
IGBT - Inverter, Chopper			
V_{CES}		1200	V
I_C	T_s = 25 (70) °C	41 (31)	A
I_{CRM}		50	A
V_{GES}		± 20	V
T_j		- 40 ... + 150	°C
Diode - Inverter, Chopper			
I_F	T_s = 25 (70) °C	30 (22)	A
I_{FRM}		50	A
T_j		- 40 ... + 150	°C
Diode - Rectifier			
V_{RRM}		1600	V
I_F	T_s = 70 °C	46	A
I_{FSM}	t_p = 10 ms, sin 180 °, T_j = 25 °C	370	A
i^2t	t_p = 10 ms, sin 180 °, T_j = 25 °C	680	A²s
T_j		- 40 ... + 150	°C
Module			
I_{tRMS}	per power terminal (20 A / spring)	40	A
T_{stg}		- 40 ... + 125	°C
V_{isol}	AC, 1 min.	2500	V

Characteristics		T_s = 25 °C, unless otherwise specified			
Symbol	**Conditions**	**min.**	**typ.**	**max.**	**Units**
IGBT - Inverter, Chopper					
V_{CEsat}	I_{Cnom} = 25 A, T_j = 25 (125) °C		1,7 (2)	2,1 (2,4)	V
$V_{GE(th)}$	V_{GE} = V_{CE}, I_C = 1 mA	5	5,8	6,5	V
$V_{CE(TO)}$	T_j = 25 (125) °C		1 (0,9)	1,2 (1,1)	V
r_T	T_j = 25 (125) °C		28 (44)	36 (52)	mΩ
C_{ies}	V_{CE} = 25 V, V_{GE} = 0 V, f = 1 MHz		1,8		nF
C_{oes}	V_{CE} = 25 V, V_{GE} = 0 V, f = 1 MHz		0,3		nF
C_{res}	V_{CE} = 25 V, V_{GE} = 0 V, f = 1 MHz		0,2		nF
$R_{th(j-s)}$	per IGBT		0,9		K/W
$t_{d(on)}$	under following conditions		85		ns
t_r	V_{CC} = 600 V, V_{GE} = ± 15 V		30		ns
$t_{d(off)}$	I_{Cnom} = 25 A, T_j = 125°C		465		ns
t_f	R_{Gon} = R_{Goff} = 30 Ω		100		ns
E_{on}	inductive load		3,5		mJ
E_{off}			3		mJ
Diode - Inverter, Chopper					
V_F = V_{EC}	I_{Fnom} = 25 A, T_j = 25 (125) °C		1,8 (1,8)	2,1 (2,2)	V
$V_{(TO)}$	T_j = 25 (125) °C		1 (0,8)	1,1 (0,9)	V
r_T	T_j = 25 (125) °C		32 (40)	40 (52)	mΩ
$R_{th(j-s)}$	per diode		1,7		K/W
I_{RRM}	under following conditions		33		A
Q_{rr}	I_{Fnom} = 25 A, V_R = 600 V		5,7		µC
E_{rr}	V_{GE} = 0 V, T_j = 125 °C		2,5		mJ
	di_F/dt = 1140 A/µs				
Diode - Rectifier					
V_F	I_{Fnom} = 25 A, T_j = 25 °C		1,1		V
$V_{(TO)}$	T_j = 150 °C		0,8		V
r_T	T_j = 150 °C		13		mΩ
$R_{th(j-s)}$	per diode		1,25		K/W
Temperature Sensor					
R_{ts}	3 %, T_r = 25 (100) °C		1000(1670)		Ω
Mechanical Data					
w			65		g
M_s	Mounting torque	2		2,5	Nm

10.   CONTROL CIRCUIT FOR A THREE-PHASE INVERTER BRIDGE OF THE 180°-TYPE

10.1 Home built bridge with the help of a control IC

Inverter bridges are often built using IGBT modules. The control circuit for such a bridge is usually available in IC form. An example of this is the control-IC IR2130 from International Rectifier. Using the IR2130 a three-phase inverter bridge using MOS-switches (MOSFET, IGBT) can be controlled. Fig. 14-20 shows a functional block diagram. Extra information concerning the IR2130 can be found in the "Application Note AN-985" from International Rectifier or on the IR website.

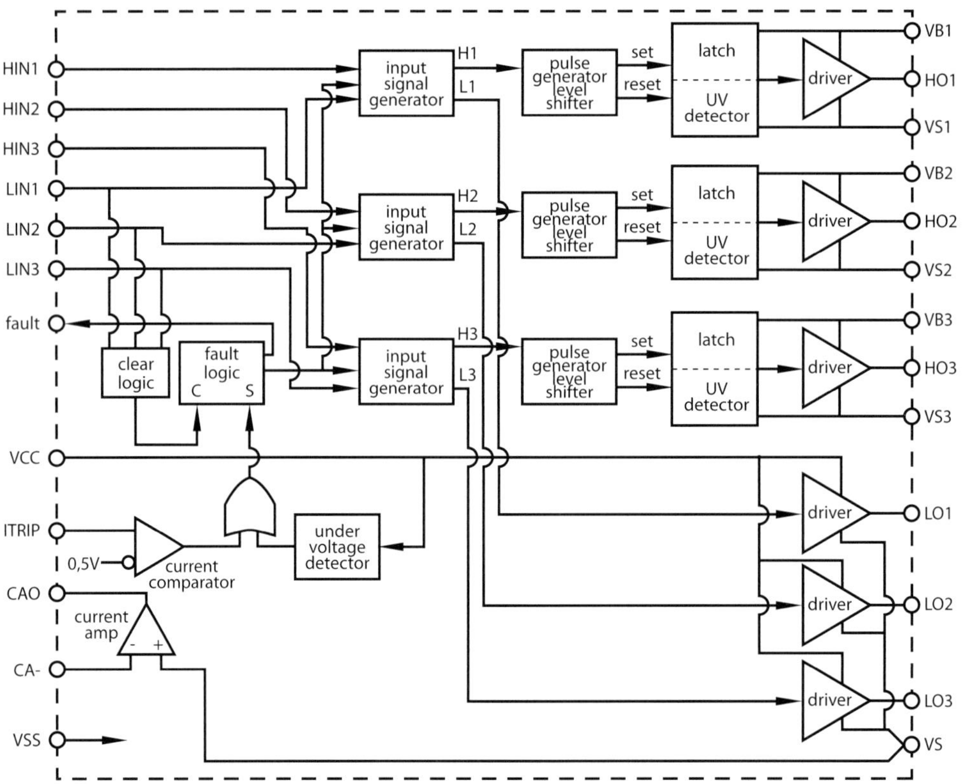

Fig. 14-20: Functional block diagram of the IR2130

Fig. 14-21 shows a configuration used for PWM control of a motor. The inverter bridge is built with IGBT's. The control IC is an IR2130.

The supply of the three upper control stages is obtained using "bootstrap" techniques.

When transistor 4 conducts for example C_1 is charged via D_1 to a potential of about 14V, in other words V_{B1}.

The voltage across C_1 provides the supply to the upper driver. The voltages V_{B2} and V_{B3} are obtained in similar fashion.

At switch on of the circuit C_4, has no charge and point A is low. As a result the interlock circuit comprising the NOR-gates B and C is reset.

As C_4 is charged the circuit $R_1 / D_4 / C_4 / D_5$ no longer plays a role. There only purpose is to reset the interlock by use of the supply voltage.

The module is in "normal" service when A is low, the NOR gates D to I cause the input control signals of the module to be present at the corresponding pins of the IR2130. If the module makes an error, then the fault pin goes low and the interlock sets A high. As a result the input pins 2 to 7 of the control IC are set low via de NOR gates. Additionally the fault-LED illuminates so that there is a visible indication of a control error. Point A can only go low by switching the circuit off and again on.

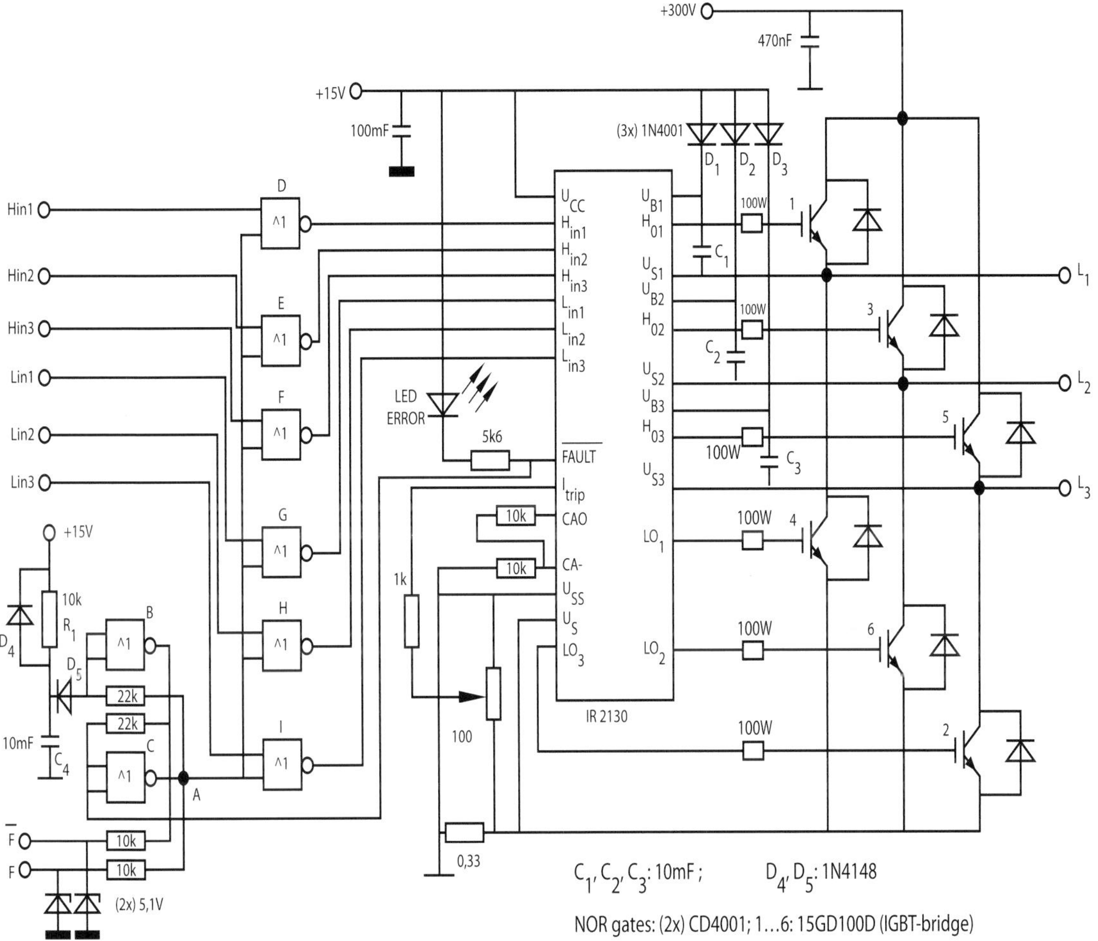

Fig. 14-21: Power module for PWM motor control

10.2 Professional version

Companies such as Semikron for example provide ready made control modules (SKHI-…) for IGBT- and MOSFET-bridges.
IXYS has for example the GDBD4410 control IC which provides galvanically isolated control signals (± 2A peak) for the six IGBTs in a three-phase bridge and in addition gate control of a "seventh" IGBT for the brake chopper.

With the SKYPER 52 control module Semikron choose the digital route (complete digital signal processing, less components , greater reliability, higher fault tolerance, adjustable depending on circuit properties, blocking levels,…).

B. SINGLE PHASE INVERTER

11. BASIC CONFIGURATION OF A FULL BRIDGE INVERTER

Figure 14-22 shows the basic configuration of a switched full bridge inverter. The switches are usually IGBTs, but for lower power levels MOSFETS may also be used. We will return to the choice of switches in the discussion of solar panels in chapter 15.

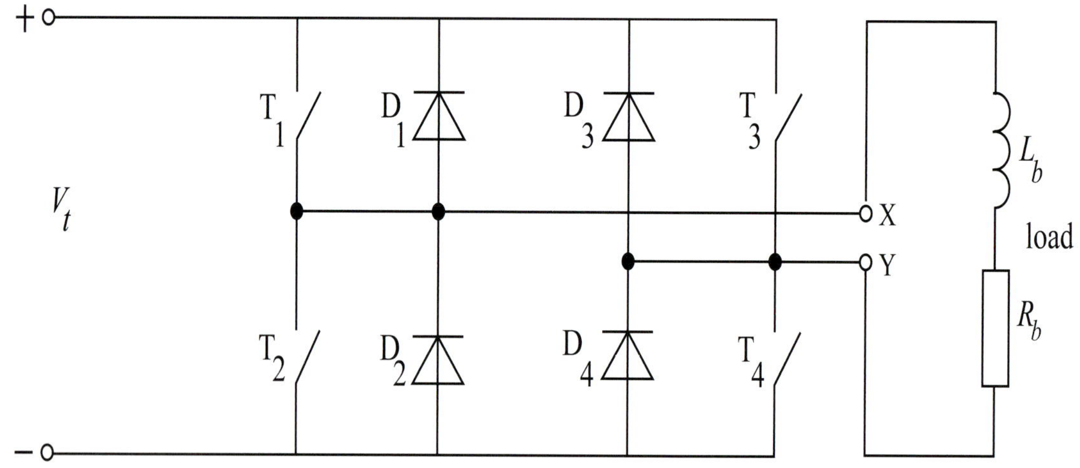

Fig. 14-22: Switched single phase inverter

12. UNIPOLAR AND BIPOLAR PWM

The bridge in fig. 14-22 has two fundamental methods of operation: unipolar and bipolar operation. In addition in order to have a sinusoidal output voltage we need to implement pulse width modulation PWM, as was the case with the three-phase inverter.

With **unipolar** PWM we work with the switches T_1 and T_4 **or** with $T_2 T_3$ When the pair $T_1 T_4$ are working the inverter output switches between 0 and $+ V_t$. With the pair $T_2 T_3$ the output switches between 0 and $- V_t$. This is drawn in fig. 14-23.

If the modulation signal is sinusoidal then we obtain a sinusoidal ground wave with frequency f_1 From fig. 14-23 we can conclude that during a half period of the ground wave (0 to $+ V_t$) the transistor pair $T_1 T_4$ operates and during the negative half period (0 to $- V_t$) it is the turn of pair $T_2 T_3$.

If per period of the switching frequency f_d we switch from $T_1 T_4$ to $T_2 T_3$ then the output voltage is as shown in fig. 14-24. We are switching from $+ V_t$ to $- V_t$ in synch with f_d, this is **bipolar** PWM.

It can be shown with bipolar PWM harmonics result according to $f_h = (j.N \pm k) . f_1$.
In this case j is every whole number from 1 and k is every whole number from zero.
The harmonics of the order h correspond with the kth sideband of j times the pulse number N.
For odd values of j only harmonics with an even value of k exist.
The following series of harmonics occur:

$$N ; \quad N \pm 2 ; \quad N \pm 4 ; \quad 2.N \pm 1 ; \quad 2.N \pm 3 ; \quad 3.N ; \quad 3.N \pm 2 ; ...$$

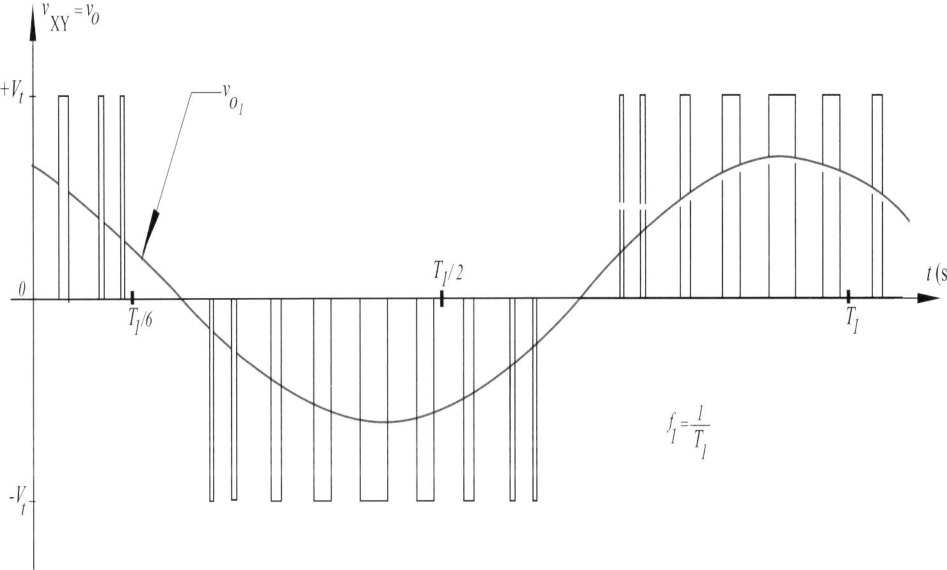

Fig. 14-23: Unipolar PWM

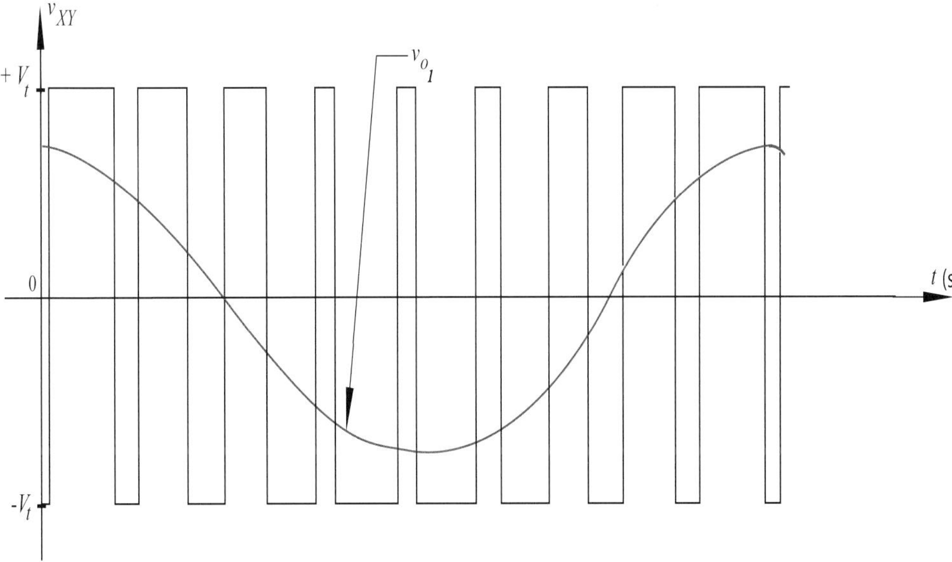

Fig. 14-24: Bipolar PWM.

13. FULL BRIDGE WITH UNIPOLAR PWM

For identical modulation depths unipolar and bipolar PWM produce the same amplitude of the fundamental harmonic. The unipolar PWM does result however in less harmonics. This is the reason that unipolar PWM is more interesting in practice. We limit ourselves therefore to the study of the unipolar bridge circuit.

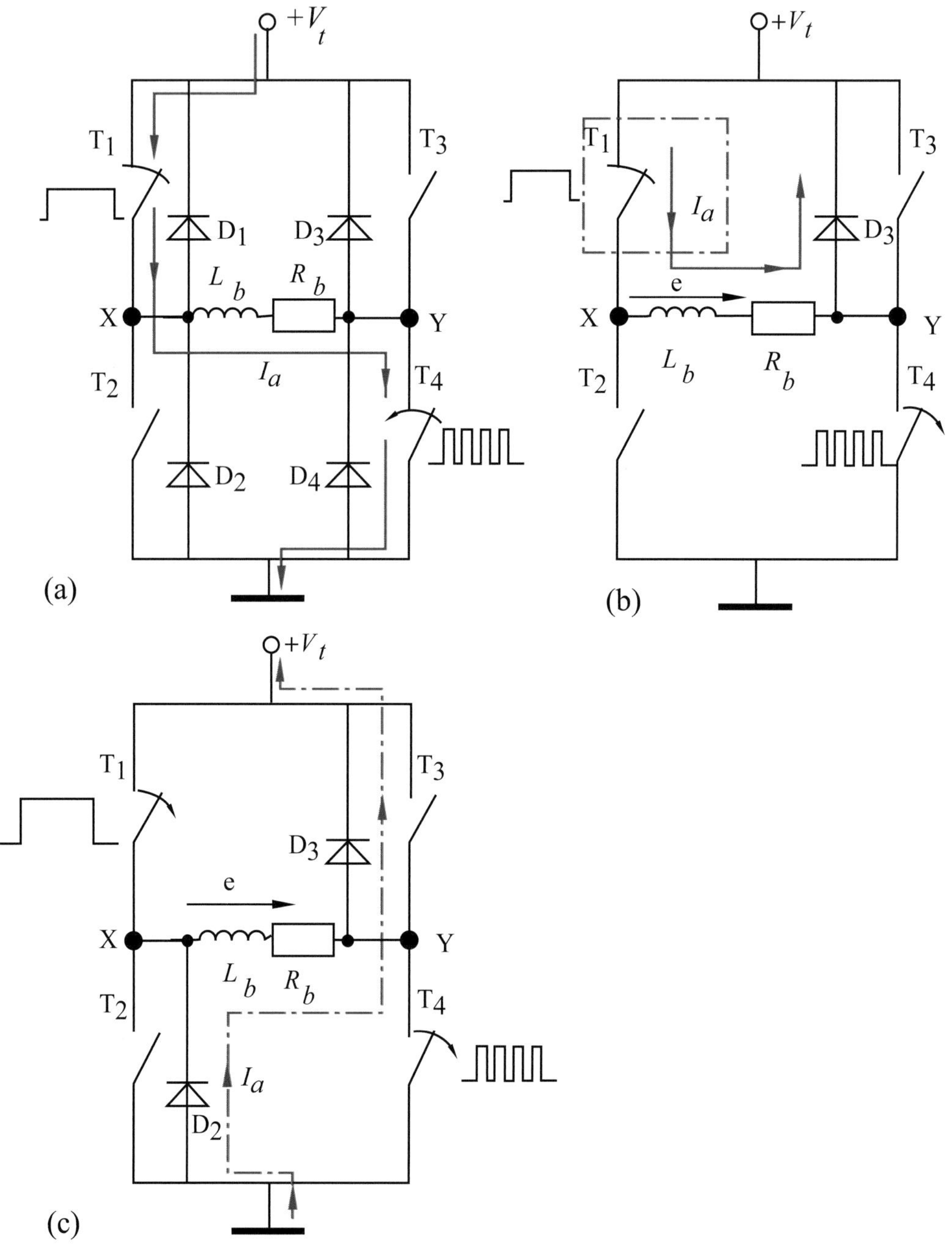

Fig. 14-25: Inverter bridge with unipolar PWM

We know that the transistors operate in pairs during a half period of the output signal. Consider when T_1 is closed and T_4 is PWM controlled (fig. 14-25a). After a half period T_4 is opened.

With an inductive load, which is frequently the case, the energy stored in the load will flow via D_3 and T_1 (fig. 14-25b). If now T_1 is also opened, then the possible remaining energy from the load will flow via D_2 and D_3 to the V_t - source (fig. 14-25c).

The positive half period of the voltage waveform across the load (fig. 14-26) is a PWM wave that switches between 0 and $+V_t$. The negative half period (0 to $-V_t$) is obtained with the transistor pair $T_2 T_3$ in an identical manner as described for $T_1 T_4$. Now the diodes D_1 and D_4 can possible play a role in allowing the (inductive) energy to flow from the load during the opening of T_2 and T_3

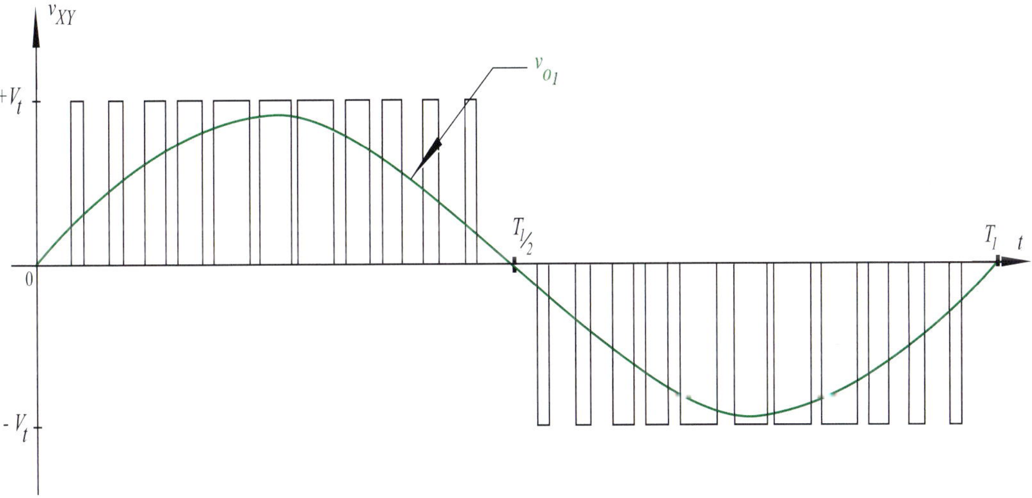

Fig. 14-26: Output signal unipolar inverter bridge

14. HARMONICS WITH UNIPOLAR PWM

A Fourier analysis of the unipolar PWM waveform results in:

- amplitude of fundamental harmonic: $\hat{v}_{o\,1} = m \cdot V_t$. (for $m \leq 1$)

- amplitude of fundamental harmonic: $V_t < \hat{v}_{o\,1} < \dfrac{4}{\pi} \cdot V_t$ (for $m > 1$)

- there are only odd harmonics of order h = j . (2.N) ± k .
  The harmonics consist of sidebands centred on $2.N.f_1$, $4.N.f_1$, $6.N.f_1$., etc...
  Since h is odd k will also be an odd number.

- the table (an extract) indicates the harmonics for larger values of N (≥ 9).

	modulation depth m	
	0.8	1
harmonic h	$(\hat{v}_o)_h / V_t$	
1 (fundamental)	0.8	1
2.N ± 1	0.314	0.181
2.N ± 3	0.139	0.212
2.N ± 5	0.013	0.033

Numeric example 14-2:

Given: The configuration in fig. 14-22. Supply voltage V_t = 300V. Modulation depth m = 1. Unipolar PWM with N = 15. Fundamental harmonic of output voltage f_1 = 60Hz.

Required: Determine the value of the fundamental harmonic and the first five consecutive harmonics.

Solution:

$$\hat{v}_{o(h)} = V_t \cdot \left[\hat{v}_{o(h)} / V_t \right]$$

The RMS value is then: $\quad V_{o(h)} = \dfrac{\hat{v}_{o(h)}}{\sqrt{2}} = \dfrac{V_t}{\sqrt{2}} \cdot \dfrac{\hat{v}_{o(h)}}{V_t} = \dfrac{300}{\sqrt{2}} \cdot \dfrac{\hat{v}_{o(h)}}{V_t} = 212.13 \cdot \dfrac{\hat{v}_{o(h)}}{V_t}$

With m = 1 we find:

- fundamental harmonic $V_{o(1)}$ = 212.13 x 1 = 212.13V with 60Hz
- 2.N - 5 : $\quad V_{o(25)}$ = 212.13 x 0.033 = 7V with 1500Hz
- 2.N - 3: $\quad V_{o(27)}$ = 212.13 x 0.212 = 44.97V with 1620Hz
- 2.N - 1: $\quad V_{o(29)}$ = 212.13 x 0.181 = 38.39V with 1740Hz
- 2.N + 1: $\quad V_{o(31)}$ = 212.13 x 0.181 = 38.39V with 1860Hz
- 2.N + 3: $\quad V_{o(33)}$ = 212.13 x 0.212 = 44.97V with 1980Hz

15. EVALUATION

A. THREE-PHASE INVERTERS

14.1   Where are "no-breaks" used?

14.2   What is the characteristic of a current source inverter?

14.3   What is a 180°-type inverter? A 120°-type inverter?

14.4   Which switch types are used for three-phase inverters?

14.5   What is the dead time of a three-phase inverter?

14.6   What does the abbreviation "PWM" stands for?

14.7   Give three examples of the source of the DC voltage used in a three-phase voltage source inverter?

14.8.   What types of motors are controlled by a 180°-type inverter and what types are controlled by a 120°-type?

14.9   Why do we replace a pulse frequency inverter using a fixed DC voltage with a PWM converter?

14.10   What forms the DC link of a PWM converter in the case of a DC voltage link?

14.11   What do you understand by the term quasi square wave of a frequency converter?

14.12   Does fig. 14-9 represent a current source or voltage source inverter?

14.13   a) Determine the amplitude of the fundamental harmonic of the line voltage in fig. 14-6a

     b) Calculate also the RMS value of the line voltage

     c) Determine the relationship between the RMS value of the fundamental harmonic and the RMS value of the complete line voltage?

14.14   Sketch the frequency spectrum of the phase voltages of a motor connected in delta. Make use of the ratio $\dfrac{\text{RMS value of the } k^{\text{th}} \text{ harmonic}}{\text{RMS value of the fundamental harmonic.}}$

14.15   a) Calculate the amplitude and RMS value of the fundamental harmonic (fig. 14-6b)

     b) What is the RMS value of the phase voltage?

     c) What is the ratio of the RMS value of the fundamental (harmonic) and the RMS value of the phase voltage.

     d) Sketch the phase voltage and line voltage and determine the phase displacement between the two voltages.

14.16   Calculate (fig. 14-6b) the ratio between line voltage and phase voltage for:
   a) the fundamental harmonic
   b) complete voltage waveform

14.17   Given a PWM waveform with N = 15 and m = 100%. DC link voltage V_t = 325V.
   Calculate: a) the amplitude of every harmonic up to the 31st harmonic
      b) the RMS value of the line voltage
      c) the RMS value of the fundamental harmonic.

B. SINGLE PHASE INVERTERS

14.18   Why is a unipolar single phase inverter preferred rather than a bipolar PWM inverter?

14.19   If we need a bipolar PWM with N = 15, what is then the frequency of the harmonics in the output of a single phase inverter?

14.20   What is the RMS value of the 35th harmonic in numeric example 14-2?
   What is the frequency of this harmonic?

Photo ABB (KWx): Semiconductor technology: BiMOS implanter.

15 APPLICATIONS OF POWER ELECTRONICS

CONTENTS

Up to now we have discussed the theory of power electronics and a number of applications. The present chapter's goal is the application of the circuits discussed in the previous chapters in seven specific situations. Note that two of the applications, electronic motor control (drive technology and motion control), are the subject of chapters 16 to 22 of the second book in this series. Motor control is the most widely used application of power electronics

1.   UNINTERRUPTABLE POWER SUPPLIES (UPS)

In applications such as industrial process (e.g. measurement and control circuits in refineries,...) telecommunications (tracking antennas in satellite communication...), operating theatres in hospitals, etc., a number of circuits need to continue operating even if the electrical supply fails. The equipment that facilitates this uninterruptable power is referred to as a "no-break" or UPS.
A UPS is also useful in improving the quality of the AC supply. This quality improvement consists of amongst other things filtering over and under voltage.

1.1 Rotating converter

An example is drawn in fig. 15-1. In this case M_2 is a three phase synchronous machine and M_1 a DC machine. In normal service M_2 operates as a synchronous motor and M_1 operates as a generator and keeps the batteries charged. When the supply is interrupted switch S_1 opens and M_1 starts to operate as a motor, feed by the batteries. Machine M_2 then operates as an alternator and provides current and voltage for the connected consumers.

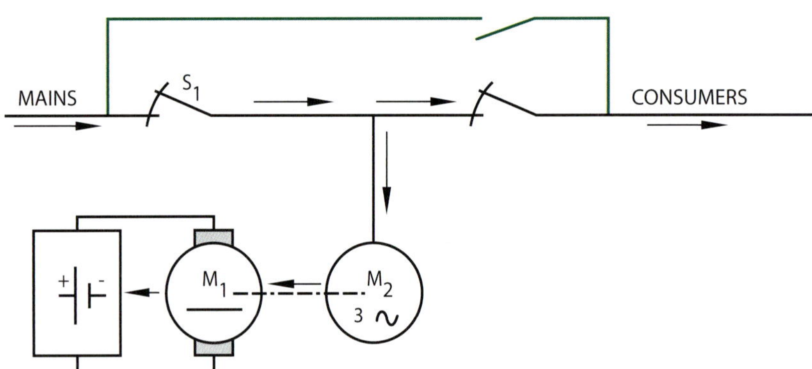

Fig. 15-1: Rotating converter in normal service

Units are available with power levels greater than 100kVA where the voltage has a maximum allowable tolerance of ± 1% and a maximum frequency tolerance of ± 0.5%. The autonomy can last between 15 to 30 minutes but can be extended to hours, for example by charging the batteries with a diesel generator.

1.2 Static UPS with auxiliary network - the double conversion topology

In comparison with a rotating converter the static UPS requires almost no maintenance, makes no noise and experiences little wear and tear. Another advantage of a static UPS is its high efficiency, even with low load. In addition the UPS operates almost instantaneously and has a high reliability. Fig. 15-2 shows the basic schematic of such a UPS, in this case there is an auxiliary grid.

The supply voltage is rectified, which enables a battery to be charged and to remain so. An inverter converts this DC voltage to the nominal AC voltage and frequency to supply the consumers. When there is a supply interruption in network 1 the supply continues via the batteries. In this manner there is autonomy for several minutes.

If the batteries are discharged or if the rectifier-inverter circuit is faulty then auxiliary network 2 can take over via the electronic bypass.

A double energy conversion takes place: AC-DC-AC, which explains the name "UPS with double conversion" topology.

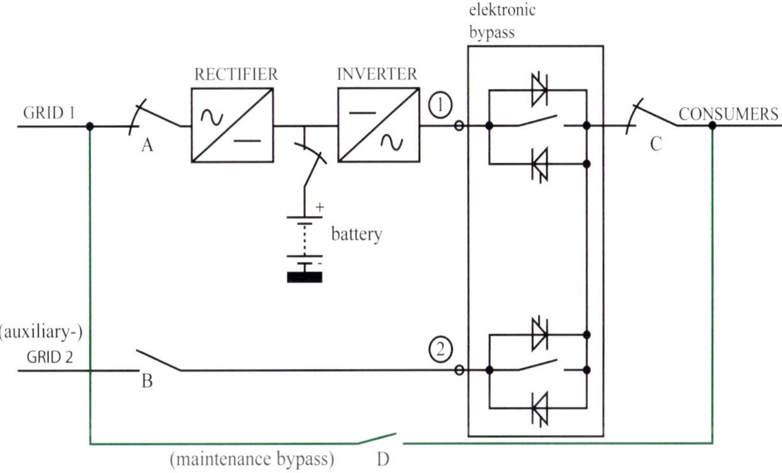

Fig. 15-2: UPS with double conversion topology

Due to the electronic bypass it is not necessary to over dimension the UPS. In the case of an extreme load it also switches over from 1 to 2. Since this needs to occur quickly and without interruption the bypass is implemented using thyristors. If the load returns to within the normal (inverter) power limits then it switches back from 2 to 1. Notice that the output voltage of the inverter needs to be in phase with that of the supply network.

By closing bypass D and opening A, B and C, the UPS can be made voltage free and maintenance or fault finding can be carried out.

Remarks

1. Battery

The usual battery types are open or closed lead batteries or NiCd batteries. Open lead batteries have a long lifetime (10 to 15 years) and are the cheapest solution for large powers. The disadvantage is the need for maintenance and the need for ventilation.

Closed lead batteries have a limited lifetime ($\pm$ 6 years) but are maintenance free and do not generate gas. NiCd batteries are more expensive, but have the advantage that they can be quickly charged.

Nowadays lithium ion batteries are also developed for the supply of UPS. Additional protection devices are needed (e.g. over-charge and over-discharge). This batteries are about half in volume and weight of a conventional one. Some manufacturers produce Li Ion battery drop-in replacements for Lead Acid batteries.

2. Inverter

Modern UPS-systems use microprocessor controlled PWM-inverters.

The advantages of this configuration are as follows:

- extremely good dynamic properties
- excellent reliability
- high efficiency

3. Electronic bypass

Usually an electromagnetic contactor is placed in parallel with the thyristors. The anti-parallel thyristors operate quickly (< 1ms) and the electromagnetic contactor which is energized at the same time takes between 50 and 100ms to close. When that occurs the SCR's are not required to operate any longer and the current will not flow through the contactor.

4. Parallel connection of UPS's and redundancy

When just one UPS is installed and an emergency occurs (in other words if the consumer draws more power than the UPS can supply) then the UPS changes over to the auxiliary network via the electronic bypass. In the case of a fault in the UPS it also changes over to the auxiliary network. Connecting n UPS's in parallel increases the reliability of the circuit. Every no-break is suitable for $(n - 1)^{th}$ part of the total power. Under normal circumstances every UPS supplies an nth part of the supply which is less than what it can supply. If one UPS fails, then this is very quickly disconnected and the remaining no-breaks together can take care of the complete load. Without a fault in one of the UPS's there is in fact a surplus of capacity of one UPS, or there is "redundancy".

Photo Elipse: An on-line UPS from Meta System for powers between 1kVA and 60kVA. When the supply fails the UPS delivers voltage from the energy stored in the batteries. The output is fully isolated from the input via the inverter so that this UPS neutralizes the effects of (input) transients and voltage fluctuations.

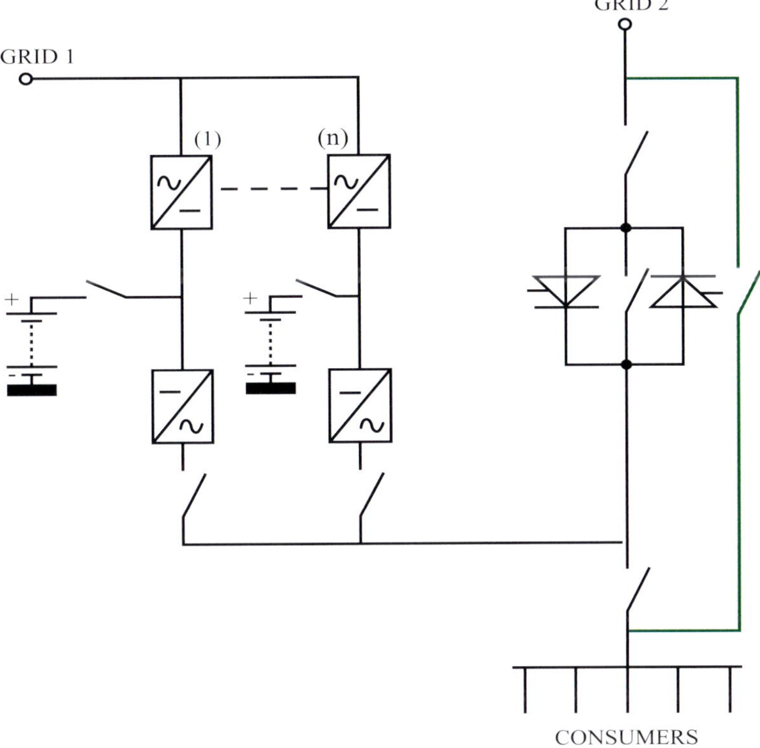

Fig. 15-3: Parallel operation of no-breaks. Redundancy

5. MTBF

The reliability R is the chance that a unit (UPS, DC-DC converter, compressor,...) will operate without fault for a specific time. We refer to the number of faults that occur in a specific time
period as the failure rate λ .

Fig. 15-4 shows the pattern of the number of faults. This is often referred to as the bath tub curve.
We distinguish three areas:

1.   the start-up period of the installation ("teething" problem period)
2.   the normal life cycle
3.   the aging period of the constituent components ("end of life" period).

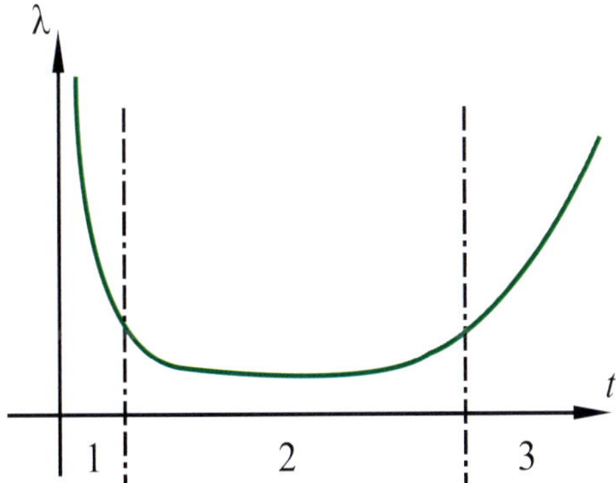

Fig. 15-4: Number of defects (in a UPS) as function of time

The average time between two consecutive failures is referred to as the mean time between failures (MTBF). For the reliability analysis it is usual to consider the "normal lifecycle " period. In this period the chance of failure is practically constant over a long period of time.

Considering such a constant failure rate λ (flat part of curve in fig. 15-4) then:

MTBF = 1/λ , or:
$$\lambda = \frac{1}{\text{MTBF}}$$
(15-1).

If the faulty component cannot be repaired then we talk of MTTF (mean time to failure).

The MTTF is the equivalent of the MTBF of the repaired unit.

Ignoring the repair time (MTTR) gives: MTBF = MTTF = 1/λ .

MTTR = mean time to repair.

The MTBF can be determined in two ways:

1. Theoretically calculated based on the reliability information of the constituent components of the no-break.
2. Trial and error based on the number of faults that have occurred in a large sample of installed UPS's of the same type.

The MTBF of the entire installation is not only dependent on the UPS type but also the configuration in which the no-break is installed (single UPS, multiple UPS's in parallel with or without redundancy) and can vary for example from 20,000 to 300,000 hours.

The MTBF is an important factor when weighing the technical and economic considerations in order to make a responsible choice.

6. Specifications

As an example we consider the specifications of the AEC-type T2003 of 50kVA (UPS using double conversion). Elipse is a supplier of these UPS's.

Three-phase input: variable 380/400/415 V AC 50/60 Hz.

Battery charger: 384 V battery voltage / max. charging current 10A / SLA (sealed lead acid) is a completely sealed and maintenance free lead battery.

Three-phase output:

- variable 380/400/415 V AC
- output frequency: variable 50 to 60 Hz
- nominal current 72 A , power factor 0.8
- waveform: perfect sinusoid
- total harmonic distortion: < 1% with linear loads; < 4% with non-linear loads
- load crest factor: 3:1
- frequency stability: 50Hz $\pm$ 0.01% (internal clock)
- maximum load: 0-110%
- overload capacity: 150% for maximum 45 seconds
- changeover time with supply failure: 0ms
- autonomy: up to 4 hours (with absence of supply voltage)

System details:

- efficiency: 92 % (with 75% load and 400V)
- MTBF: 110,000 (25°C ambient)
- 4 units can be connected in parallel

- software control: via RS232 and a local network. Via the built in web server the UPS can be remotely monitored and controlled.

1.3 Static UPS without auxiliary network

1.3.1 Passive stand-by topology

When there is no auxiliary network available when faults occur the batteries have to supply the energy. These batteries are referred to as the back-up or redundant current source. For computer power supplies extra requirements may be made.

In the supply network a number of disturbances may be present due to atmospheric influences (lightning,…) faults in the supply network (short circuits,…) industrial disturbance (welding machines, motors, lifts, fluorescent lamps,…). Over voltage and under voltage can occur as well as micro interruptions or long interruptions. These variety of disturbances are not compatible with good operation of (micro) computers.

The supply network of a computer should be free from the following:

- frequency and voltage variation
- voltage drop
- short power interruptions
- electromagnetic faults.

In addition there should be suitable autonomy to enable saving of data and the completion of calculations in the event of a power failure. An autonomy of between 10 to 15 minutes is usually the maximum. It takes several ms for the autonomy to kick in. Fig. 15-5 shows the basic schematic. A filter neutralizes the high frequency peaks. Sometimes an ultra isolator is included, This is a transformer with one or more screens so that short duration high frequency faults may be filtered.

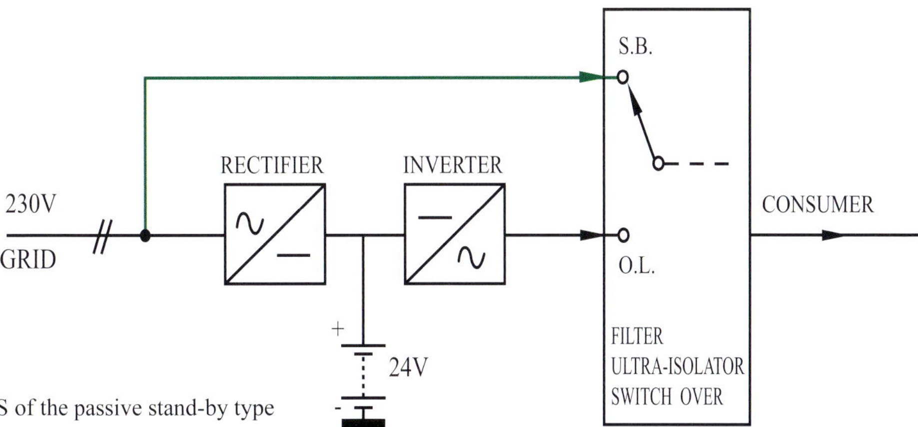

Fig. 15-5: UPS of the passive stand-by type

In normal service the consumer is supplied via a filter and ultra-isolator. When a voltage disturbance becomes too extreme the UPS switches to an inverter output. This inverter output is in continuous service together with the rectifier and battery.

The positions of the switch are indicated with S.B. (stand-by) and O.L. (on line).

Under 2kW UPS are used almost exclusively to supply computers.

1.3.2 Other topologies

A number of UPS's are difficult to classify with the previous types.
We distinguish:

- on line without bypass
- hybrid stand-by on line
- line interactive

A. On line without bypass

Fig. 15-6 shows this topology. The problem with this type is that with a power failure or mini power failure in the inverter the consumers experience a power failure.

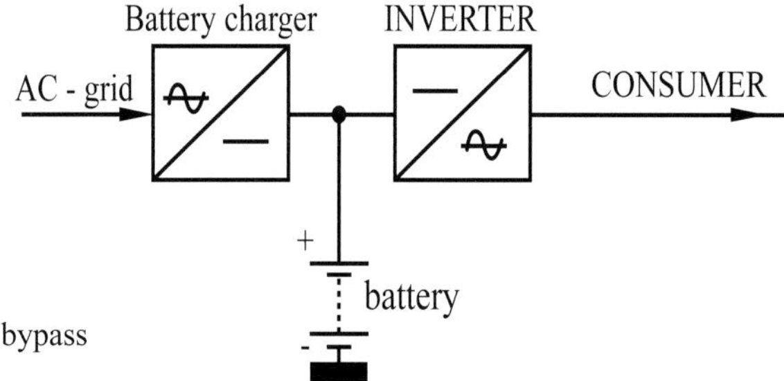

Fig. 15.6: On line without bypass

B. Hybrid on line with stand-by

The design as shown in fig. 15-7 contains an important change compared to the on line type of fig. 15-6.

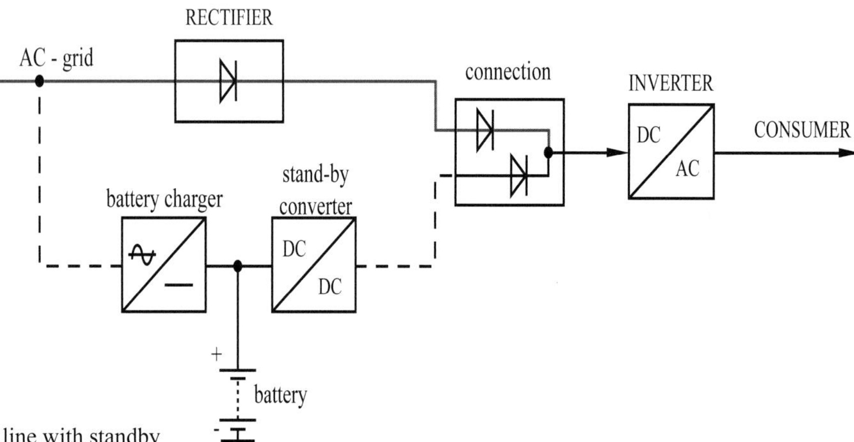

Fig. 15-7: Hybrid on line with standby

In addition to a by-pass (rectifier) there is a DC-DC converter in standby. When there is a supply interruption the stand by converter is brought on line. The transfer time is zero. In addition only a small battery charger is required as is the case in the stand-by UPS of fig. 15-5.
A weak point of this type of UPS is the inverter which operates all the time. When the inverter develops a fault the supply to the connected equipment is interrupted. In fact the most important supply route is on line and the connection with the battery is semi online (inverter is always switched in).

C. Line interactive UPS

This is shown in fig. 15-8. In this case the inverter is always connected to the output of the UPS. When a fault occurs in the AC net the transfer switch opens and the supply is via the batteries. The inverter is connected in such a way that if the inverter fails the connection between the AC-net and the consumer remains operative.

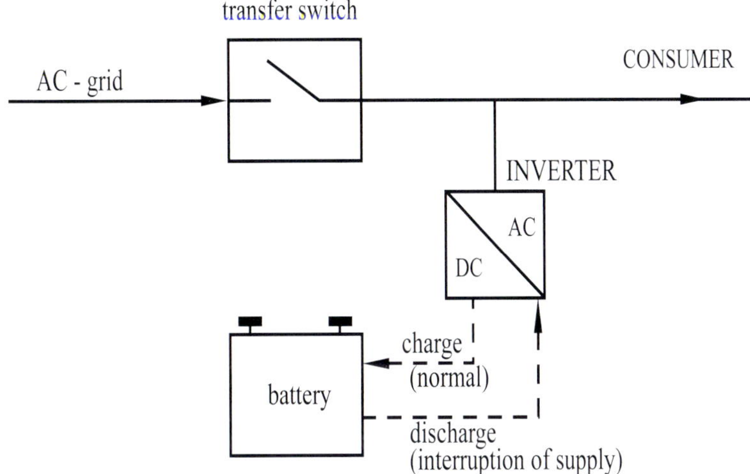

Fig. 15-8: Line interactive UPS

Remark

The IEC (International Electotechnical Commission) has recommended standards for testing a UPS.
The standard is IEC 62040-3. This standard defines three types of UPS:
- double conversion (fig. 15-2)
- passive stand-by (fig. 15-5)
- line-interactive (fig. 15-8).

For the sake of completeness we have included two other topologies , namely the on-line without bypass and the hybrid on-line with stand-by. These last two types are quite common.

2. HIGH FREQUENCY INDUCTIVE HEATING

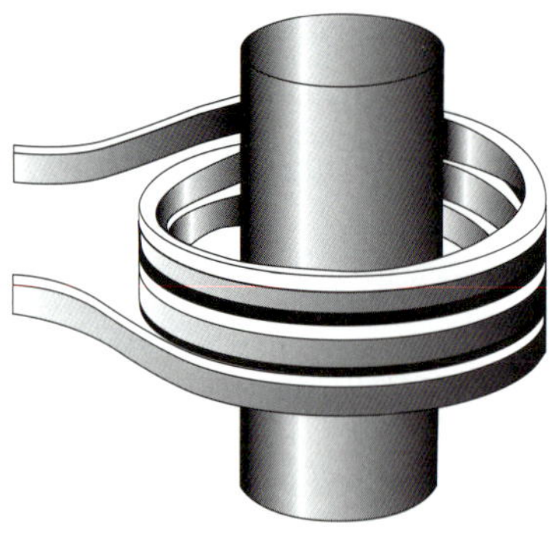

Fig. 15-9:   Object with an induction coil

Especially in the metal industry it is often necessary to heat objects, possibly for pouring, soldering, case hardening, melting or tempering of these parts.

Mostly the heat is generated outside the object and via radiation or convection the heat is transferred to the object or work piece. A much more elegant method is to generate the heat in the object (or part of the object) using eddy currents. These eddy currents are induced using high frequency magnetic fields produced in induction coils. This is called inductive H.F. heating. It is clear that only electrically conductive material can be heated in this way.

The induction coil that is placed around the object is called the work coil (fig. 15-9) This work coil is often a copper tube through which water flows to cool the coil.

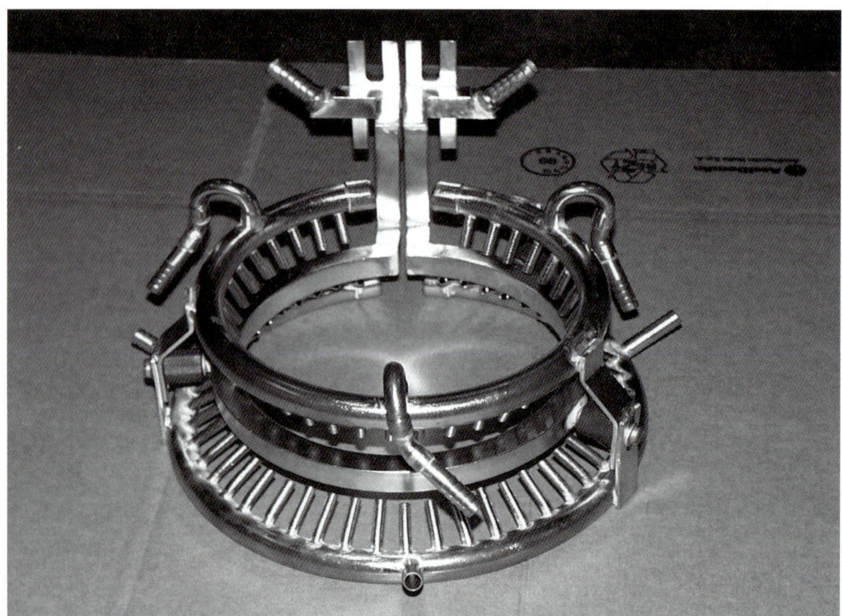

Photo Plustherm gmbh: Special induction coil for case hardening objects. Notice the integrated spray installation to shock the work piece. The (cool) water connections of the induction coil are visible at the top of the photo

The eddy currents in the work piece cause eddy current losses which result in heating. In ferromagnetic materials hysteresis losses also play a role. As we saw in chapter 5 depending on the frequency we have pronounced or non pronounced skin-effect. Induction ovens whose purpose is to melt metals operate with lower frequencies. In this case the skin effect plays virtually no role. Installations using higher frequencies (2 to 500 kHz) use the skin-effect to locally heat metals or to get them to glow. We limit our discussion to these installations. Due to the skin-effect the heating will be greatest on the outside of the work piece. This property is useful when we want to surface harden a work piece. The heat is concentrated in the outer layer (S_{di}) of the work piece.

The skin depth S_{di} follows from (5-32):

$$S_{di} = \sqrt{\frac{\rho}{\pi \cdot \mu_0 \cdot \mu_r \cdot f}} \qquad (5\text{-}32)$$

With $\mu_0 = 4.\pi.10^{-7}$ H/m this becomes:

$$S_{di} \approx \sqrt{\frac{\rho}{f \cdot \mu_r}} \quad \text{mm} \qquad (15\text{-}2)$$

Here f is in Hz and ρ in Ω mm²/m.

The magnetic field strength in the work coil is extremely high and with ferromagnetic work pieces magnetic saturation often occurs. This means amongst other things that the relative permeability μ_r is small. At the curie point (for steel at 760°C) the magnetic properties disappear ($\mu_r = 1$) and the penetration depth will increase for a specific frequency.

It is obviously also the case that if the high frequency field is maintained long enough that the work piece will quickly heat up.

Photo Plustherm gmbh: Special induction coil used to harden metal strips at 10m/min. In the middle of the photo six metal strips are visible passing through a special coil composed of 10 rectangular shaped (horizontal) windings

To work with a reasonable efficiency it can be shown that $\dfrac{S_{di}}{d} \approx \dfrac{1}{8}$. Here d is the diameter of the (round) work piece. The frequency is chosen such that the penetration depth is a maximum of $\dfrac{d}{8}$. From (15-2) we find then:

$$f_{min} = 16 \cdot 10^6 \cdot \frac{\rho}{\mu_r \cdot d^2} \qquad \text{(Hz)} \tag{15-3}$$

The efficiency decreases with increasing frequencies and in addition the efficiency of an R.F. generator also decreases with rising frequencies so that in most cases f_{min} is used.

Fig. 15-10 shows f_{min} as a function of the diameter of the work piece for different materials.

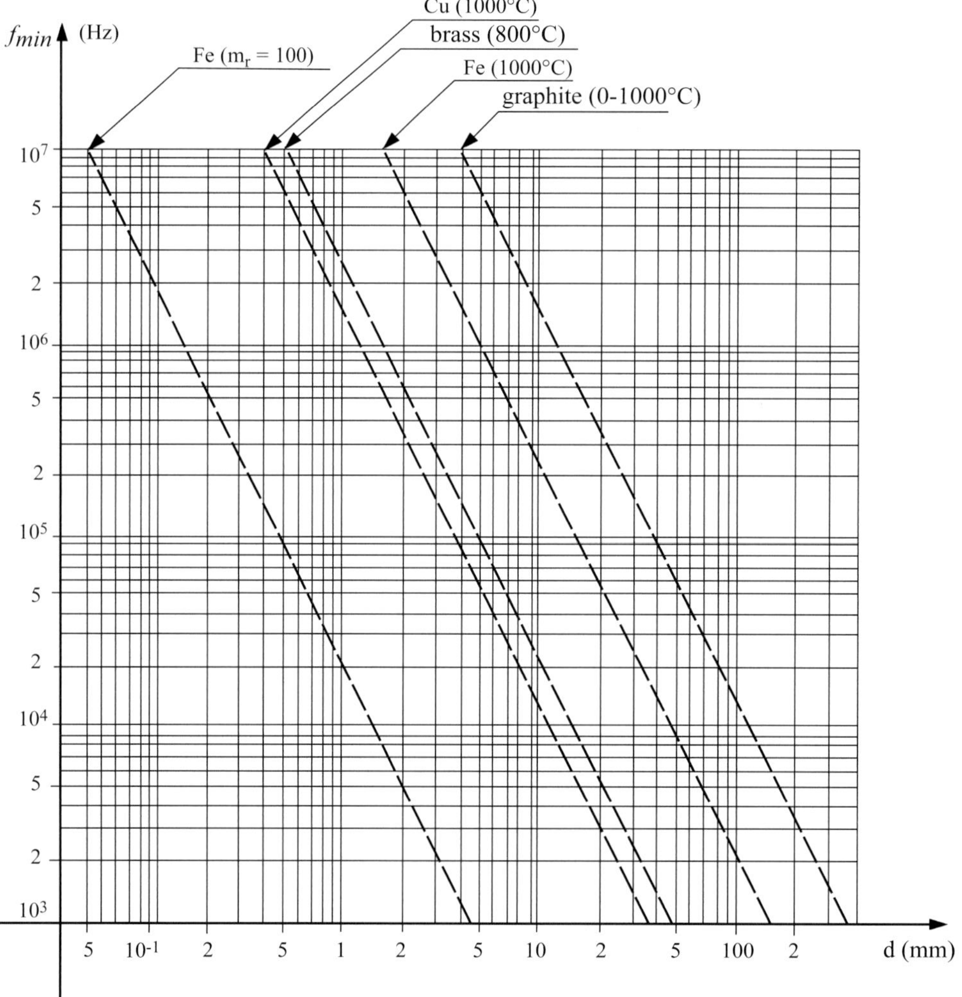

Fig. 15-10: Minimum frequency related to the diameter of the work piece

One of the most important applications of inductive high frequency heating is hardening steel. The heating duration is often a fraction of a second. Usually the power concentration is from 1 to 5 kW per cm^2 . Steel types with a carbon content above 0.3% can be hardened in this manner. Shocking the work piece with a liquid is easy with this system since it is possible to spray with the openings between the windings of the work coil (photo p. 15.10).

For the R.F. generator a so called resonant converter is used. Fig. 15-11 shows the basic schematic of a DC-AC current source inverter with a parallel resonant load.

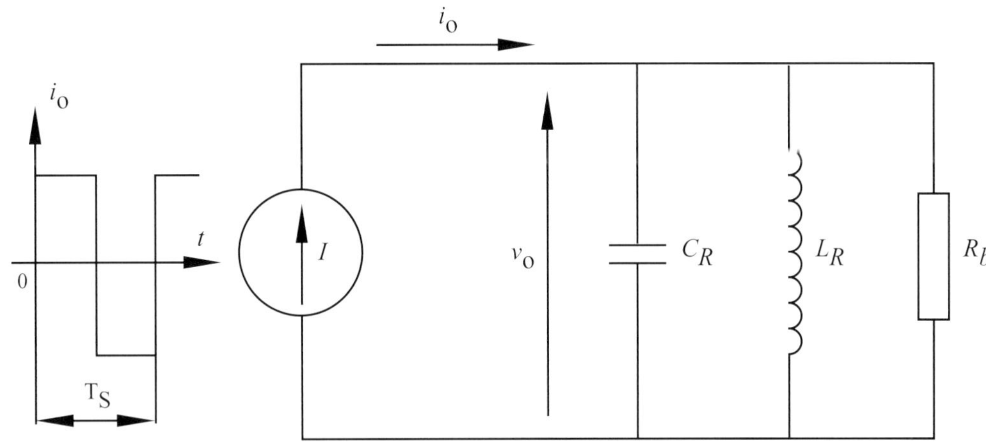

Fig. 15-11: Parallel resonant load, controlled by a current source

The induction coil and the load are replaced by an equivalent L_R and R_b. The capacitance C_R causes parallel resonance. With a sufficiently high Q-factor we obtain a practically sinusoidal voltage v_o across the parallel circuit. The current source inverter (three-phase bridge) and the square wave generator (single phase bridge) can be constructed using thyristors (fig. 15-12). To prevent high di/dt values through the thyristors a small coil L_2 is placed in series with the load.

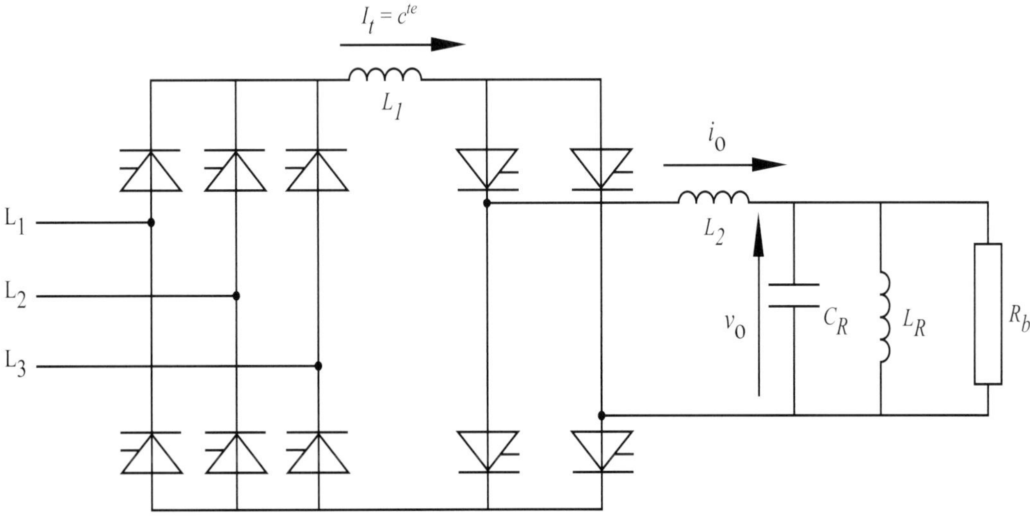

Fig. 15-12: Current source inverter with a parallel resonant load

3. POWER FACTOR CORRECTION (PFC)

In fig. 15-13 we see a classic bridge rectifier with smoothing capacitor. Whenever the capacitor voltage drops below the supply voltage the bridge draws current from the supply net. This pulsing current i_1 produces harmonics and a poor power factor:

$$\lambda = \frac{P}{S} \qquad (8\text{-}10)$$

The power factor is typically 0.6 to 0.65.

In fig. 15-14 a schematic is shown of an active harmonic filter with PFC (Power Factor Correction) recognisable as a boost converter.

An active PFC circuit is principally an AC-DC converter using an SMPS structure operating with PWM. The basis configuration is that of a boost converter. This converter can operate in continuous or discontinuous mode.

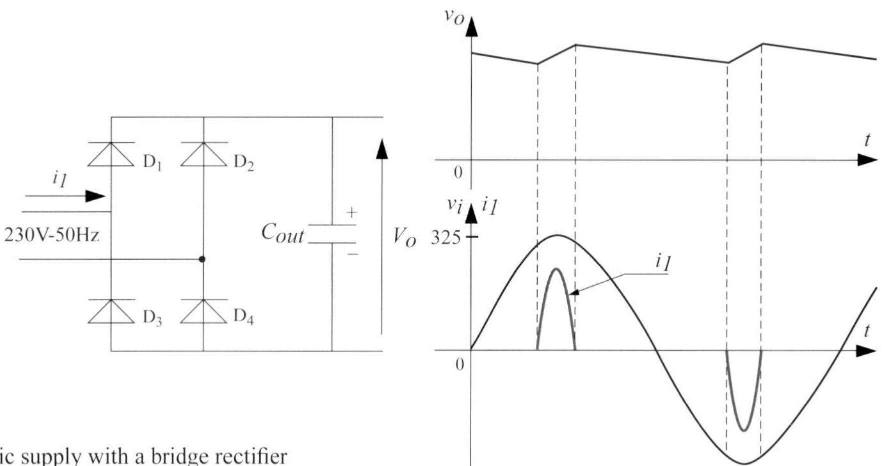

Fig. 15-13: Classic supply with a bridge rectifier

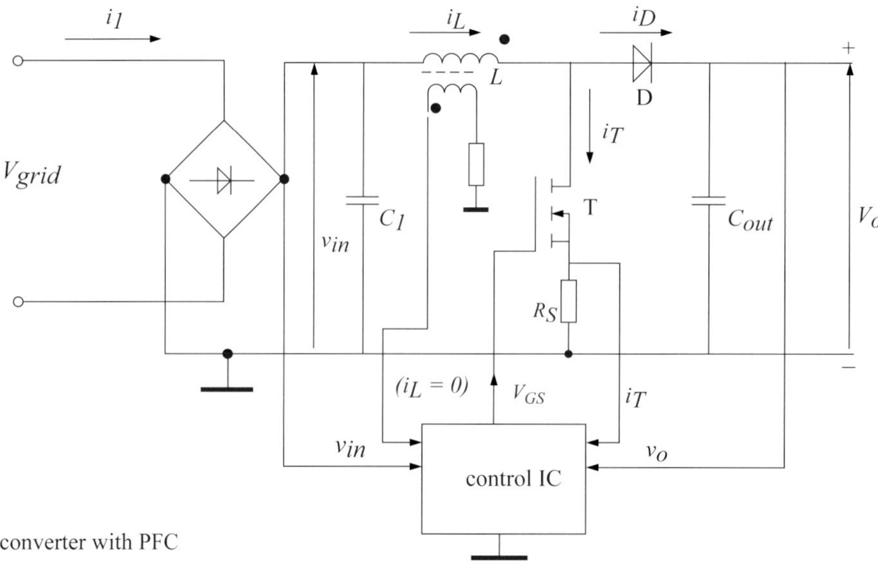

Fig. 15-14: Boost converter with PFC

3.1 PFC operation

We examine fig. 15-14. Assume a boost converter operating in discontinuous current mode. When MOSFET T is conducting the current i_L rises linearly. When the MOSFET is off , current falls linearly (via free wheel diode D). Since the switching frequency is high (60 to 100 kHz or more) these currents are practically linear.

The current is measured using a measuring resistance R_S and this is used for current feedback. A small coil is paired with the smoothing coil L of the converter. If the current through the coil L is zero the voltage across the coil is zero and also across the paired coil. This is zero current detection. If the current is zero the MOSFET is switched on. If after that the current i_T reaches the value $\hat{i}_L = v_{in} \cdot \dfrac{1}{L} \cdot t_{on}$, the MOSFET is turned off.

Since v_{in} is a half sinusoid , i_L will have the same shape (fig. 15-15). The average value of the input current i_I is half of the amplitude: $i_I = (i_L)_{average} = \dfrac{1}{2} \cdot \hat{i}_L$.

In addition i_I is in phase with v_{in} . The current i_I that is drawn from the net is sinusoidal and in phase with the supply voltage so that the power factor is very high.

In reality of course a ripple is present in i_L (fig. 15-16), but as a result of the high switching frequency a small capacitor C_I of about 1μF is sufficient to filter this.

Special control IC's have been developed (Infineon, Motorola, Tyco,…) to control converters with PFC.

An example of a single chip control IC is the UCC28060 from Texas Instruments (see power.ti.com). The TDA16888 from Infineon combines PFC and PWM-control and can work with a forward as with a flyback converter.

In continuous current mode the configuration is the same as in fig. 15-14, but now the current through the coil is continuous and varies around an average sinusoid (in fact the same result as in fig. 15-16!).

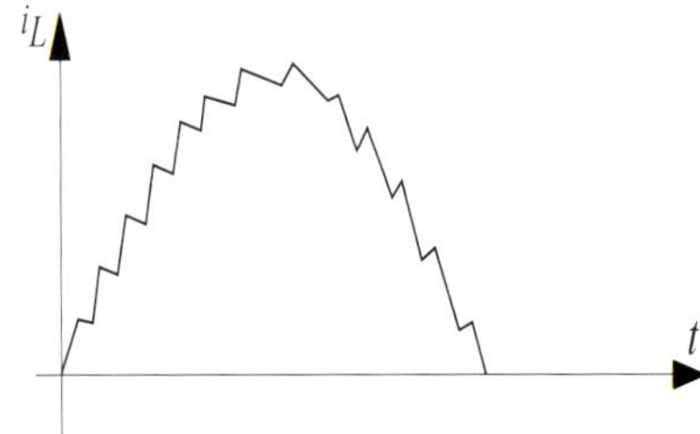

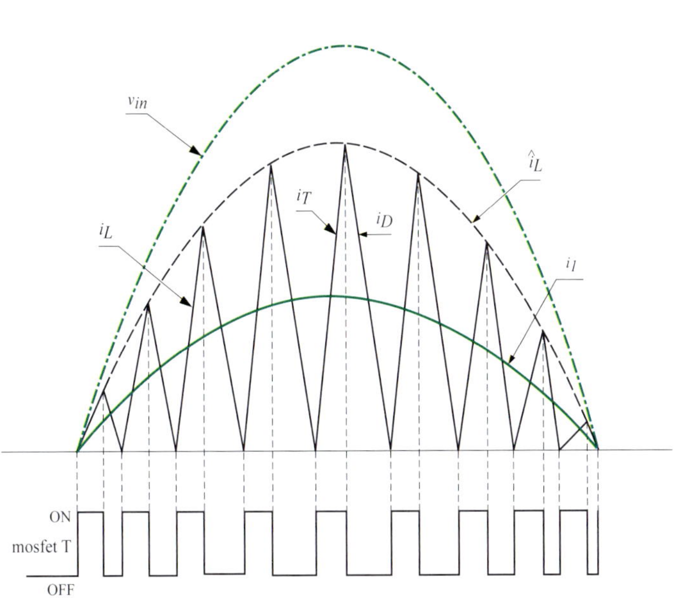

Fig. 15-16: Actual waveform of i_L

Fig. 15-15: Current and voltage waveforms of fig. 15-14

PFC advantages:

- even with large variations of the input voltage an almost constant DC voltage is maintained. That is why input voltages of from 90 to 270V may be connected without requiring alteration to the input.
- the power factor is 0.98 or higher.
- C_{out} is dramatically reduced in size. This capacitor may be 50% smaller than is the case with a classic rectifier.
- assume a supply of 230V-12A. Without PFC we can supply a DC load of about 650W since the power factor is typically 0.65. With PFC and drawing the same current from the supply we can connect a DC load of 1000W (power factor is now about 0.99!).
- due to the lower harmonics in the supply side there is less interference with radio, television and other communication systems.

3.2 Topologies

Fig. 15-14 and 15-17 show the most common topologies.

Topology 1 (fig. 15-14)
This is the conventional boost SMR (switch mode rectifier). Used for output powers up to 300W. It is current mode control with feedforward coupling of the input voltage.
For larger powers (300W to 1kW) a half controlled bridge is used. Two diodes are here replaced by MOSFETS.

Topology 2 (fig. 15-17)
Only used for large power levels with strict EMC-requirements. This quasi resonant circuit has the advantage of zero current switching. With ZCS (zero current switching) the losses in the MOSFET are minimal.

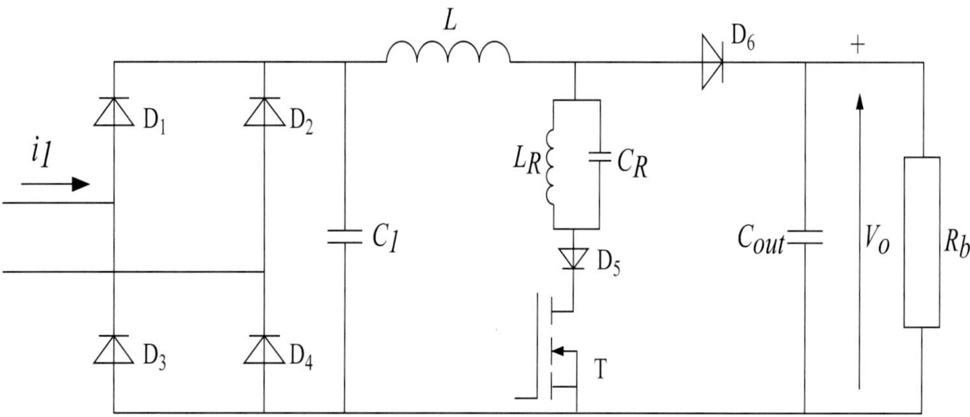

Fig. 15-17: Quasi-resonant SMR

4. LIGHTING

4.1 Electronic ballast of fluorescent lamps

Switching fluorescent lamps in series with the classic (choke) ballast has the advantage of simplicity and reliability but also has a number of disadvantages. These include amongst others flickering during turn on and towards the end of the lamps lifetime, sometimes there is a 100 Hz hum, stroboscopic effects, lighting level is dependent upon the value of the supply voltage.
If we increase the frequency the luminous flux of the lamp increases, as is indicated in fig. 15-18. Here the luminous flux is 100% at 50Hz.

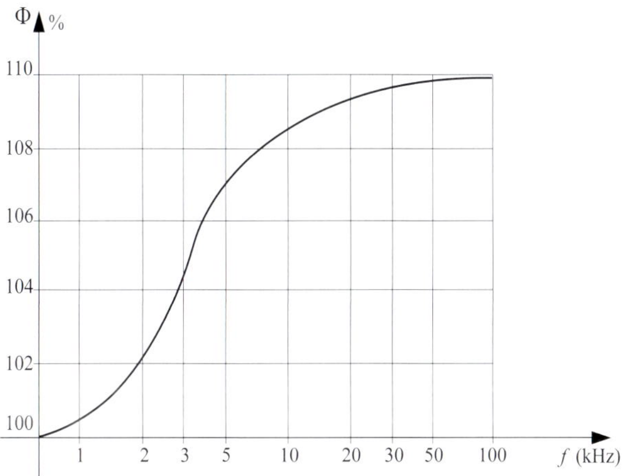

Fig. 15-18: Luminous flux as function of frequency

Replacing the ballast with an electronic starter allows us to use more than 50Hz with the lamp. Often 40kHz is used and the luminous flux increases by 10%.
A frequency higher than 20kHz is used since this is above the highest frequency which can be detected by the human ear.
The frequencies close to 36kHz are not used since this is the operating frequency of modern infrared remote controls

Photo Siemens: The rural population in developing countries predominantly use kerosene lamps for lighting since most dwellings are not connected to an electrical network. Using batteries charged via solar panels and an energy efficient light source is not only efficient for energy generation, but also the CO_2 emission is reduced

Fig. 15-19 shows a block diagram of an electronic ballast. This is sometimes referred to as a H.F.- supply for fluorescent lamps. H.F.= high frequency.

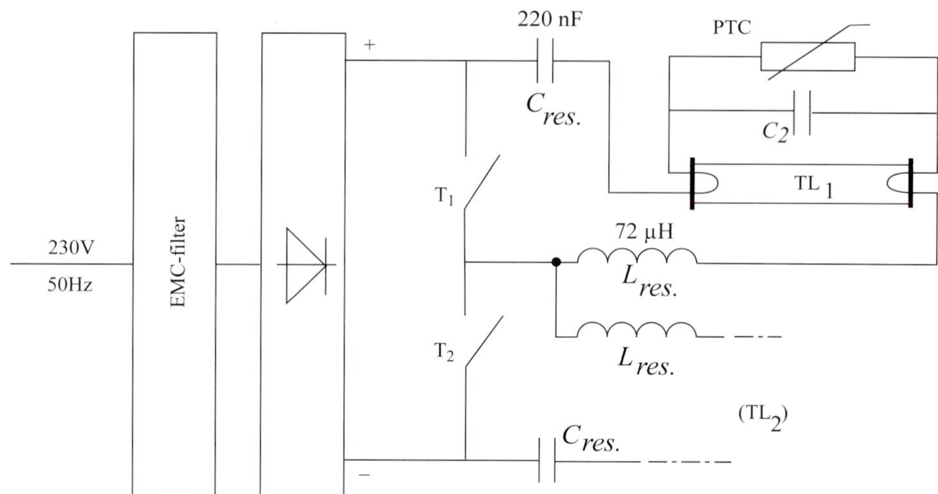

Fig. 15-19: Block diagram of electronic ballast

Mostly a resonant converter is used. The square wave voltage which is obtained from the switching of the MOSFET switches T_1 and T_2 is converted by a series resonant converter ($L_{res.}$ and $C_{res.}$) into a sinusoidal current. Often ZCS (zero current switching) is used by the transistors. By switching the transistors at the instant that the current is zero the losses are reduced and the efficiency is increased.

When T_2 is closed initially current flows through the PTC. This heats up and the resistance increases quickly. As a result of this the series resonant circuit comes on line and as a result of the high voltage rise $V_{C_{res.}}$ the lamp ignites.

Normally there is also a PFC in the circuit as discussed under point no.3.

The power factor of an electronic starter varies between 0.95 and 0.99.

Remarks

1. Dimming is not possible using the classic dimmer. Some electronic starters are dimmable. To find out if this is the case consult the manufacturers catalog.
    In such a case an analog 1-10V interface is used or a standard interface DALI (Digital Addressable Lighting Interface).
    DALI bridge the gap between the analog 0-10V systems and the complex bus systems.
2. In fig. 15-20 we see the relationship between luminous flux and consumed power during dimming.
3. The electronics in the electronic starter is not galvanically isolated from the voltage supply. Pay attention when exchanging lamps or fault finding.

4. Sometimes there are two fuses included. If there is only one this needs to be connected in the phase conductor (terminal "L").
5. The lamp housing needs to be connected to the safety earth.
6. There is an earth leakage current with this electronic starter (e.g. 0.25 mA) so that the number of these electronic starters that can be connected to an earth leakage circuit breaker is limited.
7. A lightmos is an IGBT specially designed for electronic starters. The lightmos is a cheap transistor that combines the good properties of an IGBT (low temperature dependence) and those of a MOSFET (low power loss).
8. In the case of a compact fluorescent lamp (CFL) shown in fig. 15-19 there is only one lamp (TL_1) and the electronic ballast is integrated into the socket (see photo energy saving lamp OSRAM).
9. If the lamp TL_1 is a CFL then it is sometimes manufactured in the form of a spiral.

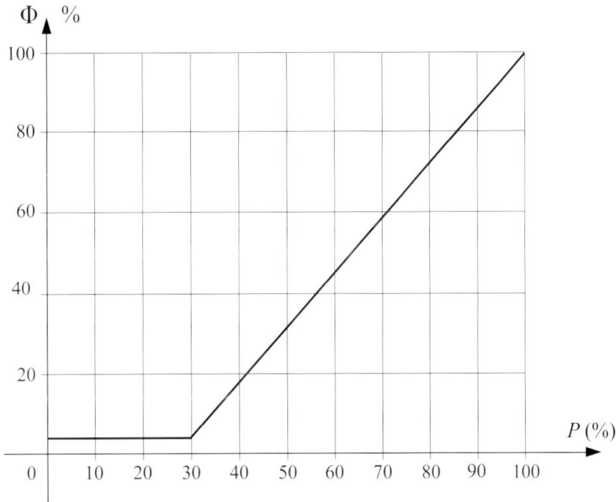

Fig. 15-20: Relationship between luminous flux and consumed power during dimming

Photo OSRAM: Exploded view of energy saving lamp

4.2 Electronic transformer

Halogen lamps can be supplied using an electronic transformer rather than an electromagnetic transformer. The advantages are a constant output voltage, better efficiency and a good power factor.

Fig. 15-21 shows a block diagram. An EMC filter is present in the input, followed by a rectifier bridge with smoothing capacitor, and then the DC-AC converter. This system classically operates at 35 kHz, so that a small RF-transformer is sufficient for the galvanic isolation between the lamp and the voltage supply

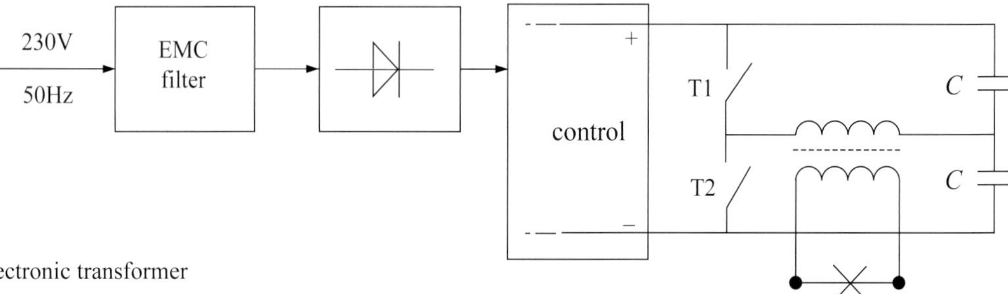

Fig. 15-21: Electronic transformer

The input of such an electronic transformer can behave as a capacitor, so that in that case a dimmer with phase chopping can be used. There are standardized tables of symbols that indicate which combination of dimmer and transformer are possible (table 15-1).

Table 15-1

Dimmer \ Transformer	L	C	L.C
R	—	—	—
R.L	yes	—	yes
R.C	—	yes	yes

Dimming is possible in the following circumstances:

1. Phase chopping:
- incandescent lamp 230 V
- halogen lamp with conventional transformer
- 12V halogen lamp with certain types of electronic transformer

2. Phase cut-off:
- incandescent lamp 230V
- 12 V halogen lamp with certain types of electronic transformer.

An electronic transformer is 80% lighter than a conventional transformer, is 40% smaller and has 60% less losses.

4.3 LED as light source

Light emitting diodes (LED) are currently used in displays, dashboards in aircraft and cars, in consumer electronics, in medical and industrial applications, for street lighting, etc.

As a light source, LEDs have a number of interesting properties:

1. In contrast to other light sources, LEDs do not suddenly fail or burnout. The luminous intensity of LEDs reduces gradually. They still emit 70% of their initial luminous intensity after 50,000 hours (on for 12 hours per day for 11 years!). The lifetime of LEDs is between 50,000 and 100,000 hours. By contrast, an incandescent lamp has a lifetime of about 3000 hours and a CFL (compact fluorescent lamp) lasts for up to 10,000 hours. There is therefore no necessity to replace a LED lamp, which is important for critical applications such as beacons, buoys, lighthouses, emergency exits, etc.

2. High efficiency. With current technology we have UHB-LEDs (ultrahigh brightness LED) which have 10 times the lumen/watt ratio of a halogen lamp. A LED headlight of 50 W in a car is therefore equivalent to a 500 W halogen lamp. A UHB-LED currently provides up to 100 lm/W (white light), and prototypes exist which provide 200lm/W. The original LED-indicator in the early 70s delivered 0.1 lm/W.

3. LEDs emit little heat.

4. They are flat elements with small dimensions.

5. Can withstand shocks and vibrations, which is important in cars.

6. They are semiconductors, which means there is no gas, mercury or filament present. The LED therefore has an environmental advantage.

7. Often multiple LEDs are connected in series, with a constant current. If one LED gets short-circuited the rest continue to operate since they are supplied with a constant current source.

8. By using a control IC, it is possible to control the luminous intensity using PWM. In practice, it is possible to control from 0 to the full luminous intensity. Using PWM it is possible to control the luminous intensity, while maintaining full colour integrity.

9. Operation is at low voltage levels and high voltage connections are therefore avoided. A LED requires about 3V (between 2.5 and 4.47V depending on the type) so that a series connection of tens of LEDs can easily be controlled from an IC that is supplied with an input voltage of 40 to 50V. The LT3595 (Linear Technology) is a 16 channel LED-driver capable of handling 160 LEDs from a DC source of 45V. This IC can have a duty cycle (dimming) of 5000:1.

Remarks

1. Another interesting application of LEDs is that an RGB-coloured light source can be built. To achieve this often two green LEDs with one blue and one red LED are used. This combination takes the sensitivity of the human eye to colour into account. A LED has a nominal wavelength and colour with a specific current. That's why we use a current source with the LEDs. Using colour mixing LEDs can create a continuous colour spectrum. To achieve this, the luminous intensity of each LED is controlled. This is not achieved by linearly controlling the current of each LED since this would change the colour. The solution is PWM of the current.

   Four channel LED-drivers exist in IC form to realize this RGB-luminous flux (for example the STP04CM596 from the firm ST, see www.st.com under "lighting").

2. Another example of lighting control using LEDs is the application of the control IC IRS2540 from International Rectifier. This IC can operate with a supply of 200V. A PWM-control signal uses burst mode control at 175kHz. The duty cycle of the PWM control is 100%.

3. An important niche for LEDs is the car industry. European cars were the first to use blue/green/white/amber coloured LEDs in the instrument panel. American and Japanese luxury car brands followed this trend. A brand such as Lexus uses 50W UHB-LEDS for the headlights. LEDs have started to replace halogen and Xenon lamps. Advantages include the streamlined designs and the longer lifetime of the lamp. LED technology is no longer just used in the luxury car market but is also used in the middle class segment. Applications vary from head lights to interior lighting. The LT3755 from Linear Technology is a control-IC intended for 50W headlights. This IC can boost the 12V battery voltage to 60 V to control a series circuit of 14 LEDS drawing 1A.

4. Another important terrain for LEDS is "backlighting" of HDTV LCD screens. For backlighting a 46" LCD TV requires about ten LT3595's per HDTV.

5. IC-manufacturers produce families of LED-drivers that comply with specific requirements for specific applications. Infineon for example has a low cost LED-driver (BCR450) for control of high power LEDs. In combination with an external transistor this BCR450 is capable of precise control of the current. In addition there is over current and over voltage protection, etc.

6. A LED cannot be dimmed using a triac dimmer. National has developed an IC (LM3445) that makes it possible to dim a LED in combination with a triac dimmer. (www.national.com/powerwise)

7. More and more public authorities are considering installing LED street lighting. A streetlight is routinely required to deliver 10,000 lumens. Current LED streetlights use between 50 and 200 LEDs with a current of 350mA. A solution for street lighting is to connect a number of series LED circuits in parallel. According to IEC specifications a transformer needs to be placed between the LEDs and the supply net and the secondary voltage may not be greater than 120V. Often 48V is used. In this manner 12 InGaN-LEDs can be connected in series (12 x 3.3V = 39.6V). A buck converter is recommended for the supply. It is the cheapest, simplest and most efficient of the classic converters. If a converter is used for every series circuit of LEDs then they can be dimmed independently. A disadvantage of this setup though is the cost price. When the buck converters are connected in parallel then EMC problems result. To combat this, filters containing at least a coil and a capacitor are required in the input of each converter. This increases the cost price even further. To preserve the quality of the supply, network power factor correction is also implemented. Street lighting is usually classified as HPWA (high power wide area lighting). In addition to energy efficient solutions for interior lighting, streets and tourist attractions at the World Expo 2010 in Shanghai, OSRAM together with its mother company Siemens, installed 150,000 LEDs in the pavilions and streets around the Expo.

5. RENEWABLE ENERGY

A. WIND TURBINES

5.1 History

Charles Brush (1849-1929) was one of the pioneers of the American electrical industry. In the winter of 1887-1888 Brush built the first automatic wind turbine to produce electricity. A rotor of 17m diameter with 144 rotor blades made of cedar wood drove a DC generator of 12kW. The turbine worked for 20 years.

Poul La Cour (1846-1908) was the pioneer of modern aerodynamics and built his own wind tunnel to use for experiments. In 1897 La Cour had experimental wind turbines in the Askov Folk High school (Denmark). In 1918 there were about 120 local wind turbines of between 20 and 35kW in Denmark. All the generators were DC machines. In 1951 a DC generator was replaced with a 35kW asynchronous machine. In 1980-81 a 55kW generator was developed that ultimately was responsible for the break through of the modern industrial wind turbine.

In special cases wind turbines up to 6MW are possible. On the other extreme of the power range there is also a market for small wind turbines (100kW) for private use on farms etc.

The majority of the wind turbines have a power capability between 500kW and 1MW.

The 6MW turbine from REpower installed in 2009 on the German Danish border has a rotor-diameter of 126m, rotor blades with a length of 61.5m and a hub height of 100m.

Currently 1% of all the worlds electricity is produced from wind energy, 16% from nuclear energy and 16% from hydro-electricity.

5.2 Construction of a wind turbine

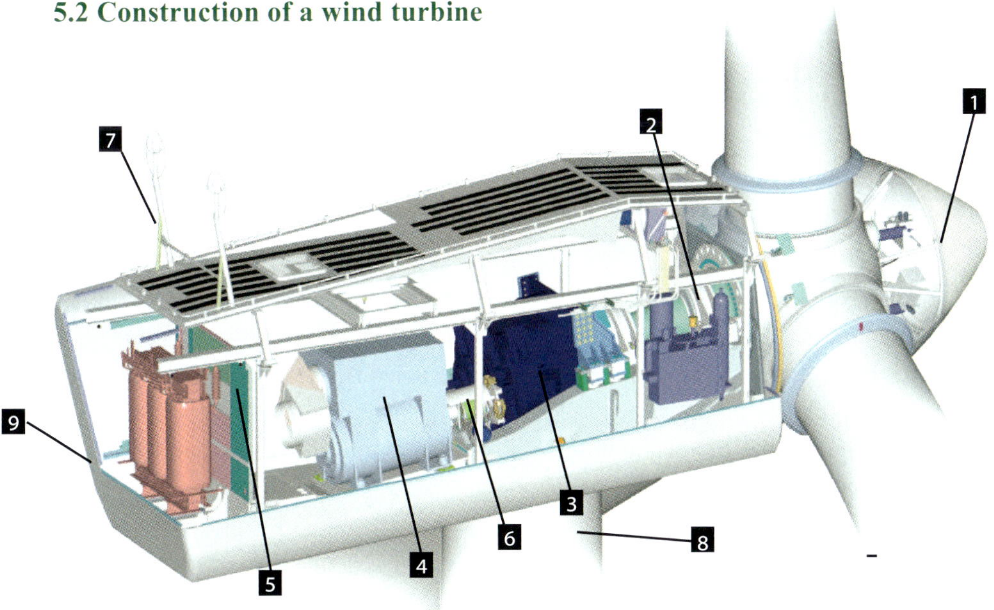

Fig. 15-22: Cross section of a 2MW wind turbine. Courtesy of Vestas Wind Technology A/S

The parts are :   1: rotor   4: alternator   7: measurement of wind speed and direction
                  2: main shaft   5: radiator   8: tower
                  3: gearbox   6: mechanical brake 9: gondola

5.3 Wind and Power: the Betz theorem

In 1926 Betz published a book "Wind-energy". He made calculations of the theoretical maximum power that could be obtained from the rotor of a wind turbine. The power absorbed from the wind that flows through the rotor is:

$$P_1 = \frac{1}{2} \cdot m \cdot (v_1^2 - v_2^2) \qquad \text{(W)} \qquad\qquad (15\text{-}4)$$

Here:
    m = mass of the air that flows through the rotor per second (kg/s)
    v_1 = entrance speed of the air in the rotor (m/s)
    v_2 = exit speed of the air from the rotor (m/s)

Herewith :
$$m = \rho \cdot F \cdot \frac{(v_1 + v_2)}{2} \qquad \text{(kg/s)} \qquad\qquad (15\text{-}5)$$

With:
- $\dfrac{v_1 + v_2}{2}$ average wind speed through the rotor (m/s)

- ρ = density of the air (kg/m^3)

- F = effective rotor surface area for the wind (m^2)

(15-5) in (15-4) $\longrightarrow$ $P_1 = \frac{\rho}{4} \cdot F \cdot (v_1^2 - v_2^2) \cdot (v_1 + v_2)$

Since m is here expressed in kg/s, $\frac{1}{2} \cdot m \cdot v^2$ represent power.

If we replace (15-5) in this expression then we find a power: $P = \frac{1}{2} \cdot \rho F \left[\cdot \frac{v_1 + v_2}{2} \right] \cdot v^2 \,.$

If the wind is not hindered by the rotor of our wind turbine then $v_1 = v_2 = v$.

The power for an equivalent surface area F is then determined as:

$$P_2 = \left(\frac{\rho}{2}\right) \cdot v_1^3 \cdot F$$

So that $\dfrac{P_1}{P_2} = \frac{1}{2} \cdot \left[1 - (v_2/v_1)^2 \right] \cdot \left[1 + v_2/v_1 \right] \,.$

This function is a maximum when $v_2/v_1 = \frac{1}{3}$, and then: $(P_1/P_2)_{\text{max}} = \frac{16}{27} = 0.59 \,.$

The theoretical maximum power that can be obtained from the wind is 59% of the total power in the wind.
The earth receives 174 petawatt ($=1.74 \times 10^{15}$W) of power from the sun. Of this about 1 to 2 % is converted to wind energy.

5.4 The rotor

5.4.1 Rotor blades

Modern rotor blades of large turbines are made from glass fibre reinforced plastics (GRP = Glass fibre Reinforced Plastics). These are polyester or epoxy blades reinforced with glass fibre. Carbon fibres or aramid (Kevlar) is also possible but not economic for larger machines.
Steel and aluminium bearings suffer problems with metal fatigue and weight. They are only used for small machines.
For turbine stability reasons a number of rotor types are avoided. A rotor with an odd number of blades (minimum of 3) may be considered as a disk as far as the dynamic properties of the machine are concerned.

5.4.2 Rotor surface area

The manufacturer optimizes the rotor diameter for the generator power and the wind strength. Consider a generator of 600kW. A typical rotor cross diameter is 44m.
For a location with limited wind strength the rotor may be 48m in diameter and in areas with high wind speed, 39m may be sufficient.

Courtesy of Vestas Windtechnology A/S

5.5 Generator

Double fed induction generators (DFIG) are most commonly used.

As shown in fig. 15-23a two inverters are present. Both inverters are implemented using 1700V IGBT modules. The base frequency of the inverter on the net side (INV 1) lies between 50 and 60Hz. The inverter on the machine side (INV 2) operates between 0 and 20Hz. This second inverter only needs to be dimensioned for a third of the output power of the alternator.

INV 2 is responsible for excitation of the machine and for the rotor field. The rotor field set point is determined by the difference between the synchronous stator rotating field (net) and the rotor speed (wind turbine W with gearbox T). The DFIG can therefore operate in both over and under synchronized mode.

The rotor currents are controlled in amplitude and phase, using vector control.

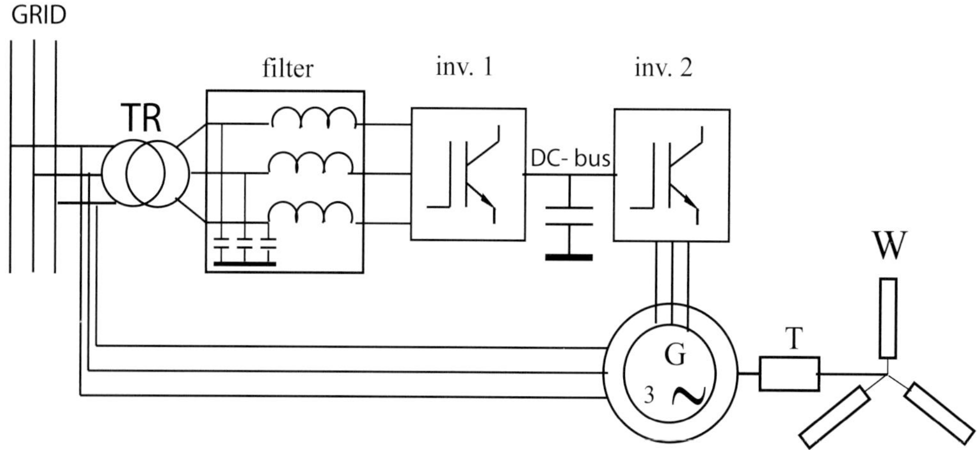

Fig. 15-23a: Wind turbine with DFIG (double fed induction generator)

Synchronous generators with excitation rotor are also used. They have a better efficiency than asynchronous alternators. The frequency generated (in relation to the driving speed) cannot be directly connected to the net.

Synchronous generators with permanent magnet excitation are being tested in many prototypes. With a synchronous generator no gearbox is needed, (fig. 15-23b).

Direct connection to the net is possible via an active front inverter, a DC bus and a second inverter as shown in fig. 15-23b.

It is not recommended to use optocouplers in the control circuitry since these age quicker than the standard 15 years for which the wind turbine has to operate.

A large ventilator in the gondola is responsible for cooling the alternator. Water cooled systems exist and are more compact than air cooled types but require a radiator to cool the water.

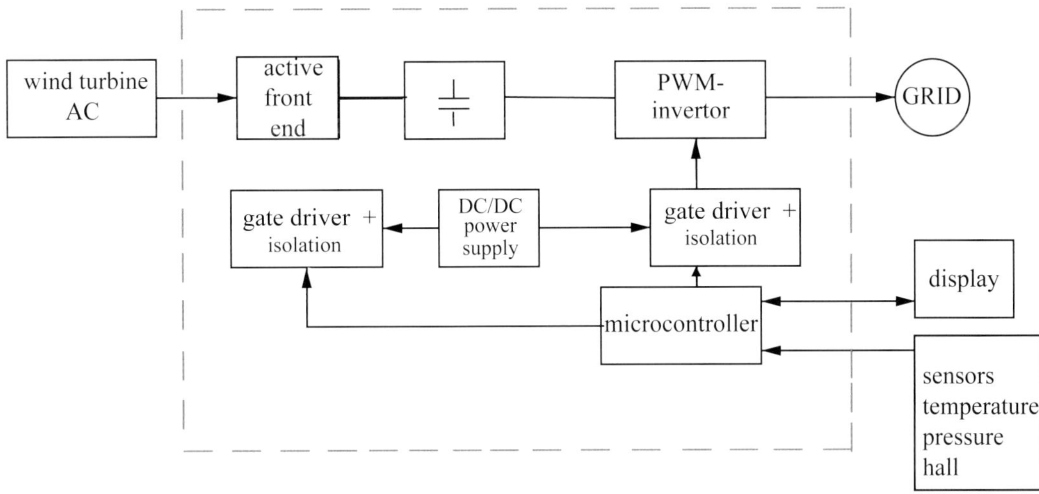

Fig. 15-23b: Wind turbine with synchronous generator

5.6 The yaw mechanism

This system is used to allow the turbine rotor to turn into the wind. A yaw error occurs when the rotor is not completely perpendicular to the wind direction. In this case less wind flows through the rotor blades.

Initially this seems like an ideal way to control the flow of wind energy. However forces are now generated on the rotor that change depending upon the location of the rotor blades. This results in every blade experiencing to and fro stresses for every revolution of the rotor. This in turn can result in fatigue of the rotor blades.

To prevent this every wind turbine is equipped with a yaw mechanism with electric motors and gearwheels to keep the turbine rotor "in the wind". To yaw = oscillate, to deviate from course. The yaw mechanism is supervised by an electronic controller that checks the position of the rotor several times per second.

5.7 Power Control

Wind turbines are generally designed to provide a maximum output power for a given wind speed of about 15m/s (= 33km/h).
In case of higher wind speeds it is necessary to dissipate the extra energy in order not to damage the turbine.
There are two basic methods to control the power.

5.7.1 Pitch control

The electronic turbine controller measures the output power of the turbine a number of times per second. If the output power is too high then a mechanism in the turbine will turn the rotor blades out of the wind along its longitudinal axis.
This turning is referred to as "pitching". The turn of the blade can be as little as a fraction of a degree. If the output power is too high then the rotor blade is turned into the wind !
This pitch mechanism can be hydraulic driven or with a stepper motor.

5.7.2 Active stall control

For turbines of 1MW or larger the manufactures tend to use active stall control. With low wind speeds the rotor blades are turned (pitch). When the machine approaches its nominal power then its operation changes dramatically from that of pitch control. When the generator is threatened by overload, the blades turn in the opposite direction to that of pitch control. The blades are in fact turned in a direction that induces a stall and in this way the excess energy is dissipated.

5.8 Electronic control

The electronic controllers make use of a number of computers that monitor and record the turbine data. The electronic controller checks the status of switches, hydraulic pumps, valves, and motors in the gondola. The controller communicates with the operator via a communication line. For a modern wind turbine between a 100 and 500 parameters are controlled: speed of the rotor, generator voltage and current, outside temperature, temperature of the electronic equipment, oil temperature in the gearbox, pitch angle of the rotor blades, yaw angle, wind direction and speed, tower door open or closed (alarm), etc…
The manufacturers secrets are the way in which the electronic controller interacts with the different components of the wind turbine. Voltage and current are typically measured 128 times per period. With this data the DSP calculates the stability of the frequency of the distribution network and the active and reactive power of the turbine.
Most wind turbine controllers are programmed for stationary rotation with low wind speed without connecting to the grid.

Modern wind turbines use a "soft start". The alternator is connected and disconnected from the distribution network by means of thyristors. A mechanical bypass switch is then operated after start-up is complete.

5.9 Aerodynamic brake

The primary braking system is aerodynamic. In the case of pitch control or active stall control the rotor blades are rotated 90° about its longitudinal axis. With stall control the tips of the rotor blades are rotated 90° around the longitudinal axis. These systems are automatically activated if the hydraulic system in the turbine looses pressure. In just a few rotations of the turbine this aerodynamic brake stops the rotor. A mechanical brake is used as backup for the aerodynamic brake and is also used as a parking brake when the turbine is stopped in the case of a stalled control turbine. Pitch controlled turbines only require this mechanical brake during maintenance work since the rotor can virtually not turn when the blades are "pitched" at 90°.

5.10 Noise pollution

5.10.1 Mechanical noises

These are caused by generator and gearbox. Research and development have resulted in wind turbines since 1995 which are virtually silent at a distance of 200m. The gearbox is no longer a standard product but is specially built for low noise operation. Part of the solution is to build gear wheels with a half soft, flexible core where only the tips are hardened.

5.10.2 Aerodynamic noise

Noise results whenever wind with a certain speed blows over an object. This is usually a mix of high frequencies, frequently referred to as "white noise". Rotor blades generate a quite hissing sound. Most of the noise comes from the back side of the blades. Also the rotor tips rotate much quicker than the part close to the shaft. The airflow around the tips is extremely complex compared to the rest of the rotor blade. The design of the rotor blade is becoming more sophisticated as research progresses. The result is on the one hand improved performance since the maximum torque is formed with the tips of the rotor blades and on the other hand a reduction in noise of the driven object.

5.11 Wind turbine tower

The tower bears the gondola and the rotor. For larger turbines it is common to use hollow steel tubes for the tower. The towers are constructed of 20 to 30m sections with flanges on each end. These flanges are fixed together on location. The towers are conical for strength and at the same time this saves on material.

The optimum tower height depends on:

- the cost per meter
- the change in wind speed at a specific height above ground
- the price the owner receives per kWh.

The components of the wind turbine are designed with a 20 year lifespan. This corresponds to 120,000 operating hours, usually in stormy weather conditions. A tower of 50m with a 600kW generator and a 44m rotor weights typically 44 ton. A 60m tower with a 72m rotor and a 2MW generator weights 80 ton.

5.12 Offshore

Megawatt turbines, cheaper foundations and better knowledge of the off-shore wind conditions have improved the economy of off-shore wind energy. By 2030 the Danish government plans to have installed 4,000 MW of off-shore generation with its "Energy 21" plan. The onshore capacity will be 1,500MW. Denmark will then be generating more than 50% of its electrical energy requirements from the wind. The energy is transported to land using submarine cables. These cables are buried in the sea bed to prevent damage (fishing equipment, anchors, etc…). The Horns Rev. Windfarm in Denmark has a 36-150kV transformer at its centre and a 150kV connection to transport the energy to land. Above a certain distance there is an alternative for this connection in the form of a HVDC (High Voltage Direct Current) connection between the windfarm and land.

5.13 Large or small turbines?

Large turbines are used especially:

- for offshore. The foundation costs, maintenance costs and electronic control systems do not rise in proportion to the power
- if there is not enough room to place multiple turbines. A large turbine on a high tower is more efficient for using the available wind energy.

Small turbines are used for the following reasons:

- if the electrical distribution network is too "weak" to cope with the energy of a heavy generator (e.g. in areas with low population density or low electrical usage)
- in a park with small generators. The output is more stable. In fact if the wind strength fluctuates, the fluctuations cancel each other
- to maintain a prettier landscape, even though larger machines normally rotate slower and are less disturbing than there smaller brothers
- the risk of failure is distributed if there are more generators.

B. PHOTOVOLTAIC SOLAR PANELS

The first solar panels were developed by Charles Fritts in 1880. In 1954 Gerald Pearson, Calvin Fuller and Daryl Chapin developed the first silicium solar panel. The first remarkable application for the solar panel was as an auxiliary power supply for the Vanguard 1-satellite in 1958. For this reason the satellite could continue to transmit for more than a year until the chemical batteries were exhausted. Photovoltaic cells were an essential part of the success of commercial satellites such as Telstar and they remain essential for the telecommunications industry of today. Germany leads the world in the use of photovoltaic systems as part of a renewable energy programme. Germany is followed by Italy, Spain, Japan and the USA as far as the photovoltaic market is concerned.

5.14 Solar panels

Photovoltaic cells are connected in series to raise the voltage. A number of series circuits are then connected in parallel to increase the output current. This in its entirety forms a solar battery or solar panel.

Solar panels have a number of applications:

1. To deliver energy to a distribution network (Grid-connected). We refer to "green" electricity.
2. To supply local users and to transfer the excess energy to the distribution network.
3. To charge a battery for local use (off-Grid system).

5.15 Supplying energy to the distribution grid

The present day solar panels produce a current of between 7 and 7.5A. The total voltage depends on the number of cells in the panel. Often used are 36, 54 or 72 cells which result in 22, 33 or 44V (open terminal emf).

These days the DC output voltage of the solar panel can be between 50 and 80V.

To produce an AC voltage of 230V a number of panels have to be connected in series. The subsequent DC voltage is then fed to an inverter which produces an AC voltage. Note that the voltage of a solar panel at any instant is dependent upon the sunlight. To have sufficient voltage at all times the number of panels connected in series is so high that with strong sunlight the input voltage to the inverter would be so high that (expensive) semiconductors with a high breakdown voltage rating would be required. The solution for this is to use less than the maximum required number of solar panels but to follow the voltage with a step up converter.

PWM control of this boost converter allows us to compensate for variations in sunlight.

The most efficient method of use is to load the PV-cell (PV = PhotoVoltaic) with an optimal resistor so that maximum power is transferred. This is called the MPP (maximum power point). The boost converter can operate as MPP converter. As a result the current drawn from the PV-cells is periodically altered.

To achieve a certain power level for the consumer a number of series connected cells can be connected in parallel. This is not desirable though since with unequal voltages from the different series circuits, the circuit with the lowest voltage would no longer participate in the power output. The trend is for every solar panel to have its own high performance low power converter. In this way we can strive to have maximum power transfer for every circuit (MPP!). If the individual panels have a different MPP, then the MPPs need to be adjusted before the power output from the individual panels are combined.

In fig. 15-24 a simple block diagram showing the energy transfer to a distribution network is shown.

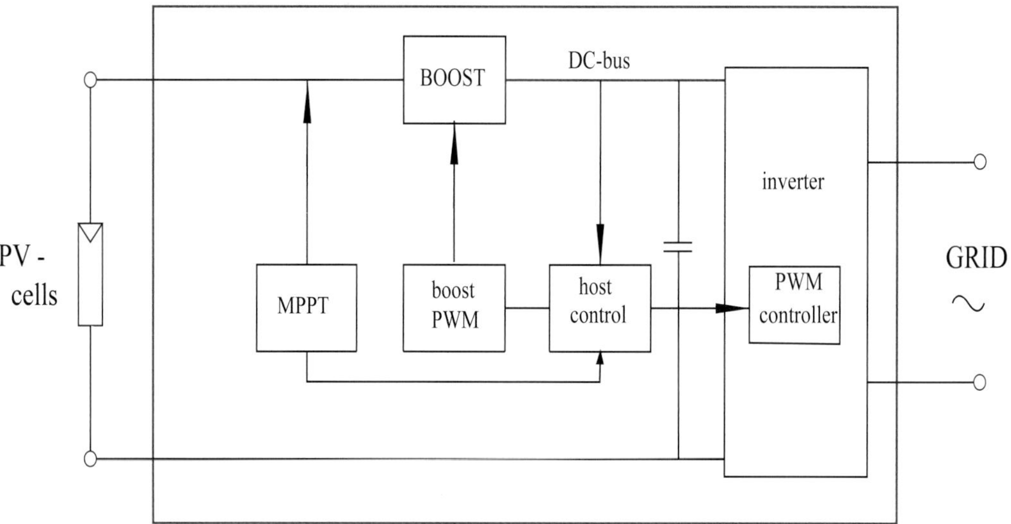

Fig. 15-24: Block diagram of energy transfer to a distribution network

To operate the PWM inverter with high efficiency it is best if the DC-bus voltage is much higher than the AC-supply voltage.

A microcontroller (or DSP) controls not only the output inverter but via a built-in algorithm the MPPT (maximum power point tracking) is arranged. This is achieved with the boost converter.

As far as the boost converter is concerned a large duty cycle is required with a low input voltage to have a high output voltage. In addition the conduction losses in the MOSFET can be high so that a low value of $R_{DS(ON)}$ is important. That is the reason that multiple MOSFETS are connected in parallel. A deep study also reveals that fast acting diodes are also important in connection with efficiency of the converter.

Summarising we can say that the input voltage is dynamically adjusted, usually with a boost converter. In the second stage the DC-voltage (from 400 to 500V) is converted to a sinusoidal voltage compatible with the network voltage.

5.16 Converter Topology

5.16.1 Single phase "solar" inverters without transformer

Up to about 4kW single phase inverters are used with a maximum DC-voltage of 500V.

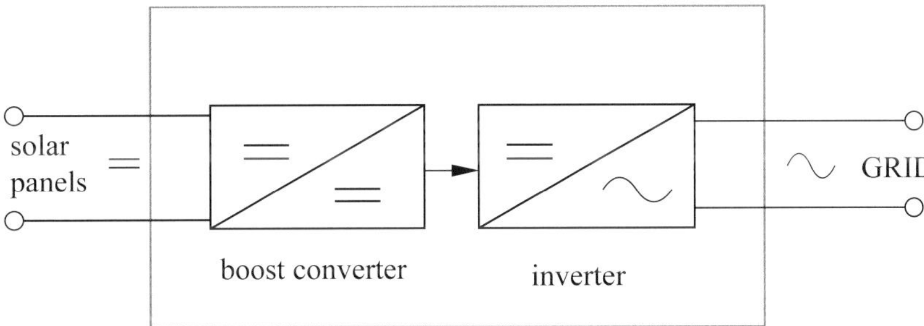

Fig. 15-25: Single phase solar inverter

The boost converter also operates as a MPPT as already mentioned. The output inverter is classically configured as a unipolar single phase bridge converter. Coils are included in the output to minimize EMI. In fig. 15-26 the basic schematic of a single phase solar inverter is shown.

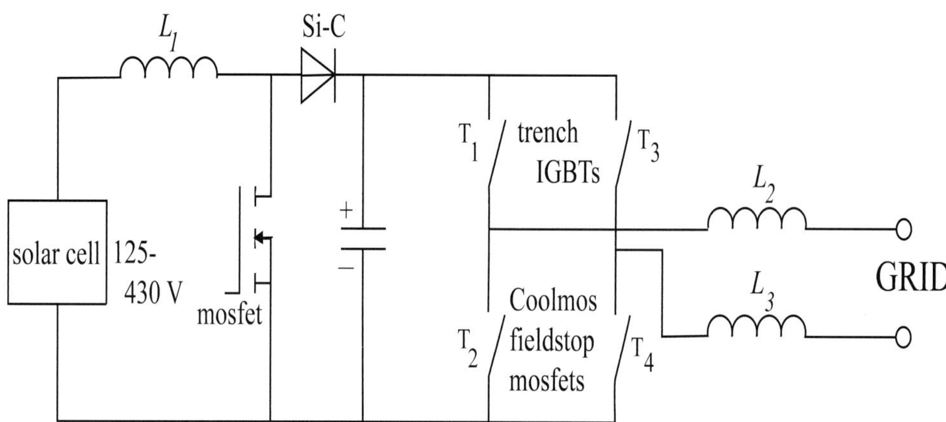

Fig. 15-26: Basic schematic of a single phase solar inverter

A high switching frequency (e.g. 16kHz which is above the audible range of humans) is used to limit the boost coil and the output filter coil. The output filter creates a pure sinusoidal voltage. An additional advantage of the coils in the output is that in the case of an external short circuit the current will rise linearly (di/dt) so that the protective devices have time to switch off the converter.

Standard modules (input voltage, output current) are for example 600V / 30 to 100A and 1200V / 25 to 50A.

5.16.2 Three-phase inverter without transformer

The maximum DC-voltage is now 1000V, which allows for a more extensive series circuit of PV-cells. Obviously a three-phase inverter is required (fig. 15-27).

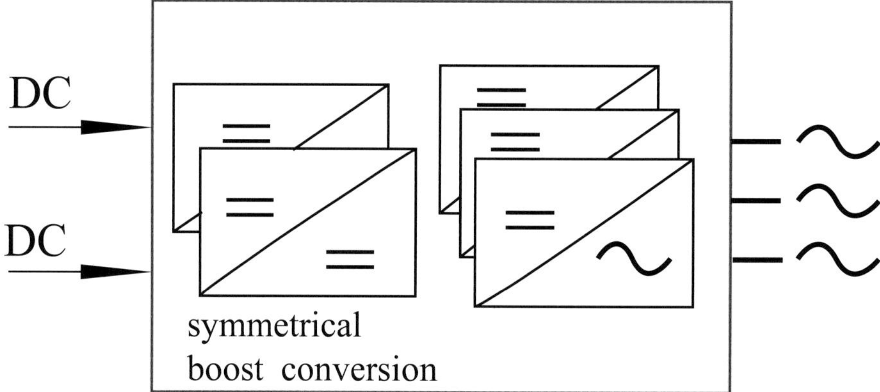

Fig. 15-27: Three-phase inverter

With 1000V DC voltage the IGBTs now need to be rated for 1200V. These are slower than the 600V type IGBTs that were used in the single phase inverter. To be able to use 600V semiconductors NPC technology is used (NPC = neutral point clamped).

5.17 Remarks

1. Net connected solar inverters have a typical output power of 1.5 to 6kW. Heavier inverter systems have an output power up to 100kW and sometimes higher. The unipolar bridge circuit provides the best efficiency for modern solar inverters.

2. For solar panel owners an exact current measurement is important as this forms the basis of the tariff for energy supply to the distribution grid. The maximum allowed DC current in the energy sent back varies from country to country. In Germany it is 1A and in most other European lands it is a few mA. This DC component of the network current can saturate distribution transformers or other transformers of consumers on the grid. Current meters in solar inverters therefore measure not only the AC current that is returned to the net but also the DC component of this current.

3. **Switching frequency and semiconductors**
    For the usual switching frequencies of between 16 and 50kHz the use of IGBTs is recommended. Normally the lower transistors ($T_2 T_4$) in the single phase bridge inverter (fig. 15-26) operate at this high frequency and the upper transistors operate at the network frequency. It is usually easier to design the control circuit for the high frequency circuit of the lower switches with ground as reference.

    The lower switches are usually fast NPT-IGBTs or MOSFETS. For the upper switches (at network frequency) IGBTs with a low saturation level are used such as Trench and Field-stop IGBTs. In the case of MOSFETS a Coolmos of 900V can be used whereby the converter can operate at the maximum 1000V which is allowed by IEC60364.

    The diodes in parallel with the upper IGBTs should have low recovery losses since they work in combination with the lower IGBTs. For this we use fast soft recovery types or Si-C Schottky diodes. These SiC diodes have practically no losses during forward or reverse recovery operation.

    The current 600V trench and fieldstop IGBTs are designed to operate up to 20kHz. A low V_{CEsat} together with extremely fast switching times make an efficiency of 96% possible. To further improve the efficiency (up to 98%) the frequency can be lowered to 16kHz (just above the audible range) without too much impact on the dimensions of the magnetic components (coils, EMI filters).

4. **Peak power**
    Since the power depends on the light levels the power is indicated in terms of the possible peak value of the solar panel, expressed as kWp (kW peak).

5. The efficiency of new solar panels is estimated at 77% (efficiency of converter, losses in cables,etc...). Further it is estimated that the solar panel degrades by 1% per year. After 20 years the efficiency has dropped to 77 x 0.8 = 61.60%.

6. One topology is for example that the various series circuits of PV-cells have an individual boost converter. The outputs from these boost converters are then connected in parallel to supply energy to the distribution network. Such configurations are suitable for up to 10kWp. **Alternatives are:**

   a. Every solar panel has its own inverter and produces a few hundred Wp.
   b. The series circuits of PV-cells are connected in parallel and supply a single inverter. This has as disadvantage that the PV cell with the lowest production does not contribute to the energy output.
   c. Separate DC-DC conversion for every solar panel and the outputs are then connected in parallel as an input to a single DC-AC converter. This is the most efficient topology and can deliver about 100kWp.

7. **Protection:** in the USA a transformer has to be used to guarantee galvanic isolation while in Germany, the most significant European market, no transformer is required. The converter does need to be protected against voltage spikes in the distribution network.
   A no current protection is required in all conductors including the neutral. This is especially important in systems without a transformer.

8. **DSP:** the control of a solar inverter is completely dependent on a number of real time challenges, on the one hand to maintain efficient DC-DC conversion, and on the other all the safety measures. Also included is the need for MPPT (maximum power point tracking) and additionally control of a possible battery charger. This requires processing a number of algorithms with substantial processing power.
   DSPs are therefore imminently suited for the control and protection of solar converters. They also can handle communication with the outside world.

5.18 Solar panels with isolated battery

Instead of supplying energy to a distribution network the energy can be used to charge a battery. Fig. 15-28 shows an example. The MPP converter arranges a constant DC-bus voltage. A battery charger charges the battery. In case the solar panels receive insufficient light or as a result of excessive demand by the load the DC-bus voltage will drop resulting in the battery supplying the DC-bus.

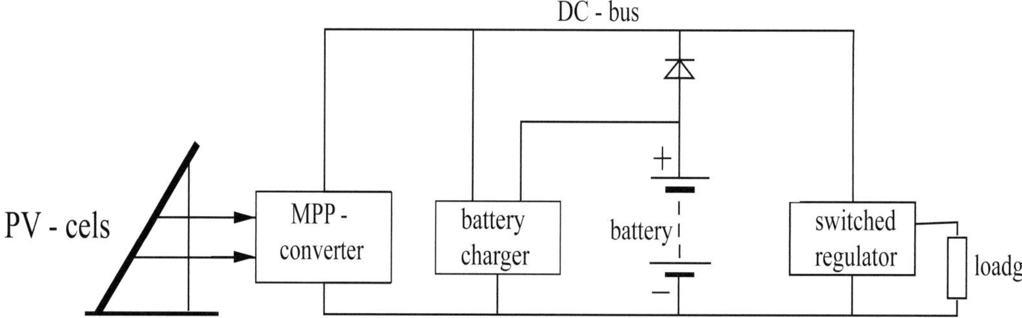

Fig. 15-28: Isolated battery charging via solar panels

Typical applications are those where solar panels are used to supply HBLEDs (High brightness LEDs). These LEDs are used for street lighting, emergency lighting, rural lighting where no electrical distribution network is available, etc...

Fig. 15-29 shows the principle configuration for such a LED lighting.

A state of the art version works with an 8-bit processor. This controls the battery voltage in the first case via an ADC (analog-to-digital-converter). If this voltage drops to 50% of full charge then via the duty cycle of a PWM-channel the luminous intensity of the lights is reduced to 50%. The goal is to operate the lighting for as long as possible even with less illumination. If the battery voltage drops to 10% then the LEDs are switched off to prevent the batteries totally discharging.

One of the projects of the IFC (International Finance Corporation), this is the private investment arm of the World Bank, is called " Lighting the bottom of the pyramid". With this project it is hoped to provide LED lighting via solar panels to 1.6 billion people across the entire world that are not connected to an electric distribution network. This can help reduce green house gas emissions that are produced from alternative lighting systems operating using kerosene.

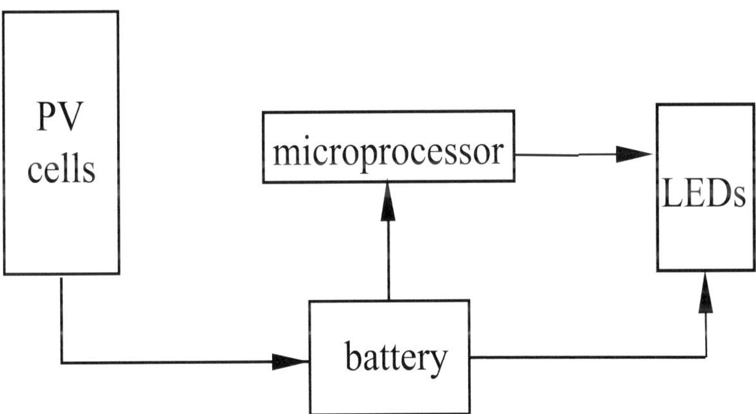

Fig. 15-29: LED lighting controlled via a solar panel

6. DRIVE TECHNOLOGY

The mechanical variables produced with the shaft of an electric motor are :

- speed
- torque
- shaft angle

The electrical control of speed and or torque is referred to as **drive technology**. Controlling the angle or shaft position is referred to as **motion control**. In this paragraph we look at speed control of a DC motor and an AC motor respectively.

6.1 Speed control of a DC motor

An ideal DC motor can be described by two equations:

$$E = k_1 . n . \Phi = K_G . \omega \qquad \text{(V)} \qquad\qquad (15\text{-}6)$$

$$M \approx M_{em} = k_2 . I_a . \Phi = K_M . I_a \qquad \text{(Nm)} \qquad\qquad (15\text{-}7)$$

In fig. 15-30 we have immediately made the armature voltage and the field voltage variable using the technique of the controlled rectifier discussed in chapter 8.

For simplicity these controllers are shown using SCRs. In reality the armature is usually supplied by a B_6 - controller while for the field circuit a single phase thyristor bridge is sufficient. The motor is driving a mechanical load and the mechanical properties of this load (moment of inertia J_m , reactive torque M_t) also need to be considered.

Fig. 15-30 shows the equivalent circuit for a DC commutator motor with load.

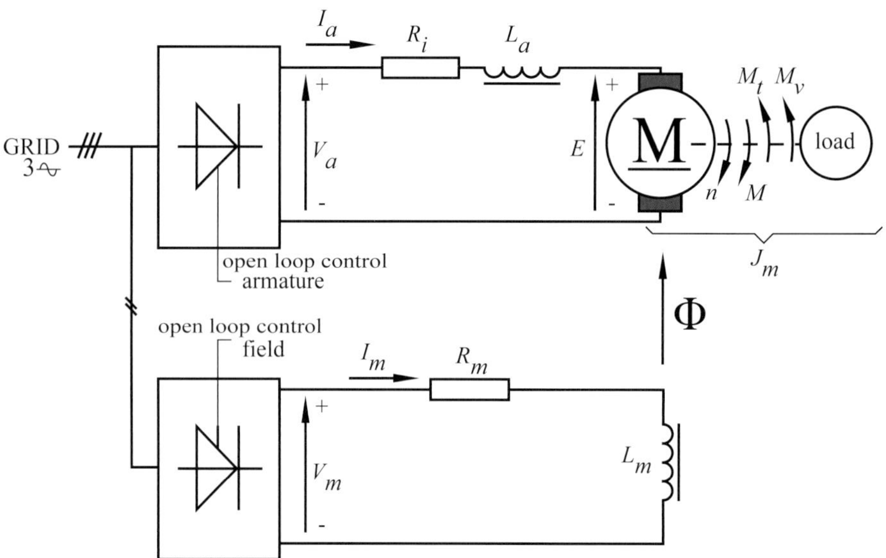

Fig. 15-30: Electronic control of an independently excited motor

In fig. 15-30 we have:

V_a	=	terminal voltage armature	Φ	=	machine flux
I_a	=	armature current	M_t	=	reactive moment
R_i	=	total resistance of armature circuit	M_V	=	moment (torque) of acceleration
L_a	=	total self induction of armature circuit	J_m	=	total moment (torque) of inertia
E	=	emf of armature	V_m	=	field excitation voltage
n	=	motor speed	I_m	=	excitation current
ω	=	angular speed of motor shaft	R_m	=	resistance of excitation circuit
M	=	moment (torque) on motor shaft	L_m	=	self inductance of excitation circuit
M_{em}	=	electromechanical moment (torque) motor			

Often Φ is kept constant so that expressions (15-6) and (15-7) can be rewritten as:

$$E = k_3 . n \qquad \text{(15-8)} \qquad\qquad M_{em} = k_4 . I_a \qquad \text{(15-9)}$$

With $E = V_a - I_a . R_a$ and neglecting $I_a . R_a$ we have as an approximation:

$$n = k_5 . V_a \qquad \text{(15-10)} \qquad\qquad M_{em} = k_4 . I_a \qquad \text{(15-9)}$$

Using the thyristor bridge enables regulation of V_a and as a result also the speed.
Regulating I_a allows the motor torque to be controlled. This is important for lifting machinery (cranes, elevators,etc ...).

6.3 Speed control of a three-phase asynchronous motor

Fig. 15-31 shows the basic configuration for speed control of an induction motor.

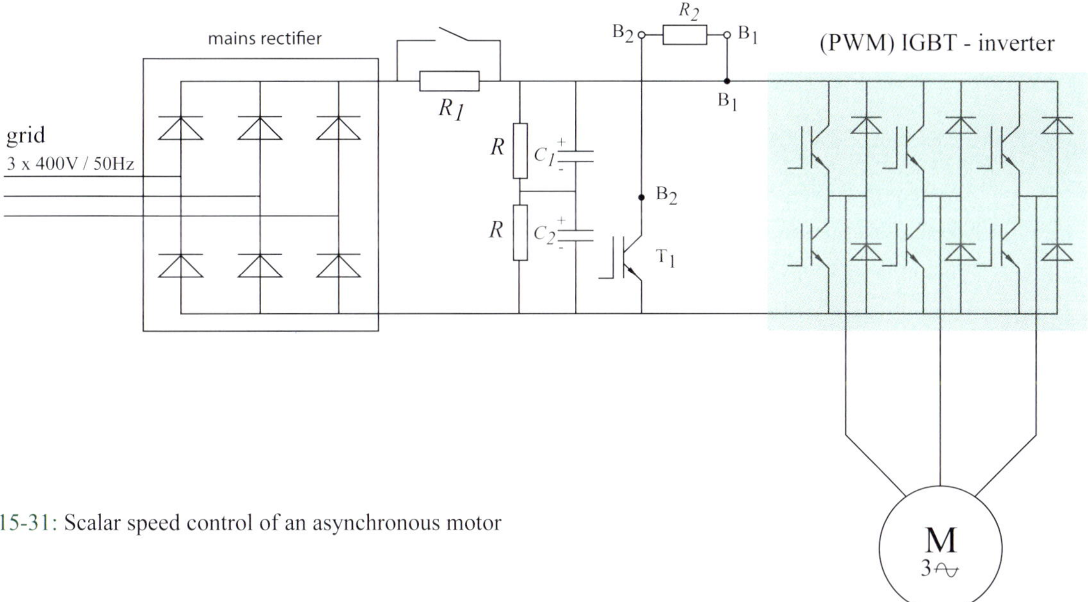

Fig. 15-31: Scalar speed control of an asynchronous motor

The speed of the asynchronous motor is:

$$n = \frac{60 . f_s}{p} . (1 - g) \qquad \text{(15-11)}$$

where : f_s = stator frequency (Hz)
  p = number of pole pairs
  g = rotor slip with respect to the rotating stator field.

With the PWM-inverter (see chapter 14) we can regulate the stator frequency f_s of the motor and consequently the speed.

Remark

For a detailed study of speed control of DC and AC motors you are referred to chapters 19 and 20 of volume 2 of this series.

7. MOTION CONTROL

Electric positioning systems are used in instrumentation and industrial applications.
In instrumentation we find for example plotters, printers, scanners, disc drives, etc... Industrial
applications include robotics, NC-machines, conveyer belts, flow control, actuators, etc...
Fig. 15-32 shows an example of an electric block diagram.

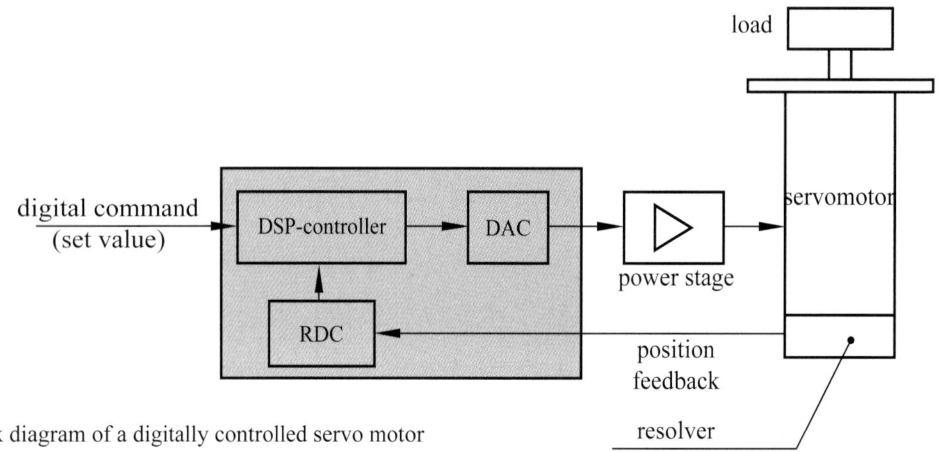

Fig. 15-32: Block diagram of a digitally controlled servo motor

The required angular position set point is determined. An angular position sensor is placed on
the shaft of the servomotor. In this case this is a resolver. This resolver generates an electric
signal that is proportional to the angle of the motor shaft. This actual angle is compared to the
set point and if both are not identical there is an error. As long as there is an error the power
stage is active, and it continues to rotate the motor. A power stage similar to this was introduced in chapter 14. In this case a 120°-inverter is used as shown in fig. 15-33.

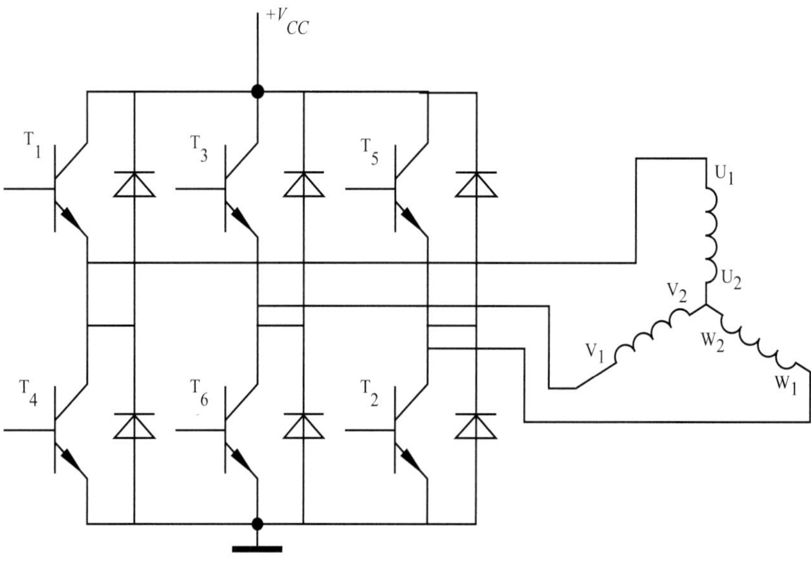

Fig. 15-33: Power bridge of a three-phase servomotor

In chapter 22 of volume 2 of this series electrical positioning systems are discussed in detail.

Photo Siemens: LED lighting for a "shopping experience" in the Trafford Centre in Manchester (UK).
High Power Flood LED fittings from OSRAM were used here

8. HIGH FREQUENCY INDUCTION COOKING PLATE

The European trend of cooking using HF-induction cook plates is starting to take off in the rest of the world. An important advantage of this type of appliance is the typical efficiency of 80% as compared to 40% for a gas stove. In addition the cooking time is shorter and the safety level is higher. This last point is especially interesting as far as children are concerned since all the heat is concentrated in the cooking pot itself.

8.1 Operating principle of an induction cook plate

Coils in an induction plate convert the electric energy into a magnetic field and this causes eddy currents in the metal pan which cause the pan to heat up. The pan plays the part of the secondary of a transformer, in which the coil functions as the primary. Typical frequencies of induction hot plates are between 25 and 40 kHz.

8.2 Topologies

A resonant converter is commonly used to generate the high frequency current in the coil which is in fact the induction element. We consider as an example two topologies, on the one hand an appliance with a series resonant circuit and on the other hand a quasi-resonant topology. In both cases a bridge rectifier is used to convert the AC supply into a DC-bus voltage.

8.3 Series resonant circuit topology

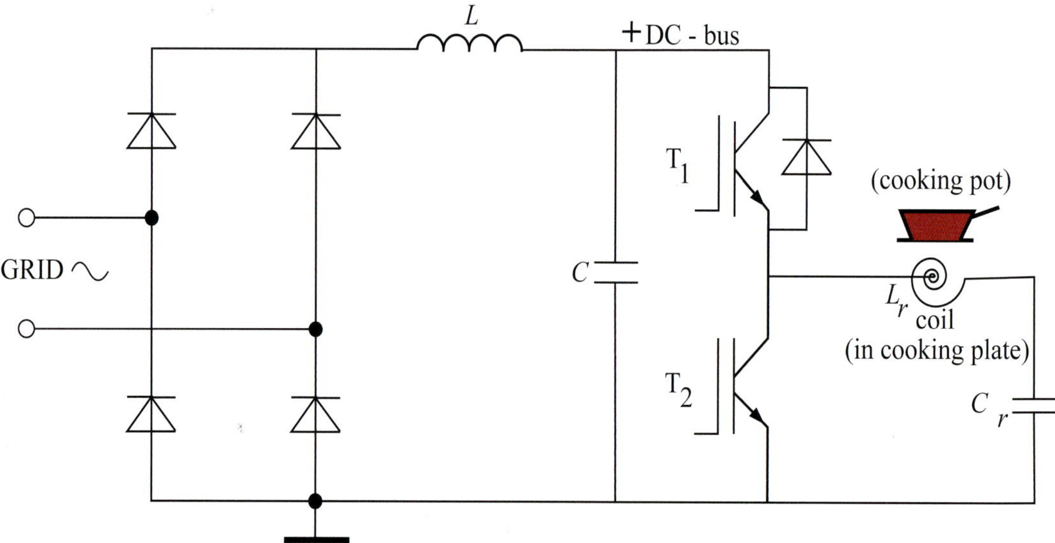

Fig. 15-34: Induction cooking plate with a series resonant circuit

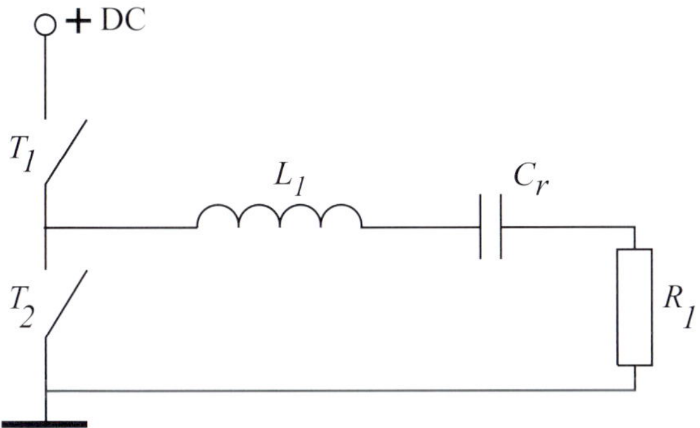

Consider the coil and the cooking pot. The coil represents the primary of a transformer and the cooking pot is a part of the secondary. The cooking pot can be viewed as a resistor.

If the secondary impedances are transferred to the primary, then the equivalent circuit is obtained as shown in fig. 15-35.

Fig. 15-35: Equivalent circuit of fig. 15-34

Closing T_1 causes current to flow through the series circuit of coil, resistor and capacitor. By opening T_1 and closing T_2 resonance can occur. That is why this type of converter is called a resonant converter. Switching of the IGBTs can occur at the zero crossover of the current (zero current switching = ZCS) or at the zero crossover of the voltage (zero voltage switching = ZVS). With ZCS the IGBT experiences no transient effects. An advantage of the half bridge with series resonant circuit is that the voltage in the circuit is never higher than the DC-bus voltage and the IGBT is therefore not exposed to high voltage. A disadvantage is that the control circuit of T_1 needs to be isolated from earth. This problem can be solved by equipping the control circuit with an optocoupler or a transformer.

8.4 Quasi resonant topology

Fig. 15-36 shows a quasi resonant circuit topology, the equivalent circuit is shown in fig. 15-37.

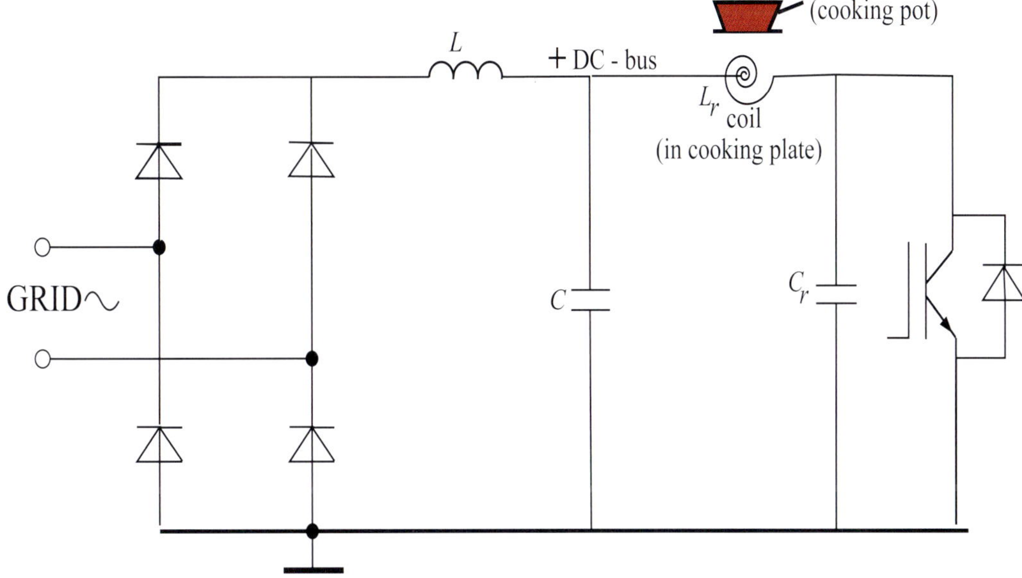

Fig. 15-36: Induction cooking plate using a quasi resonant circuit

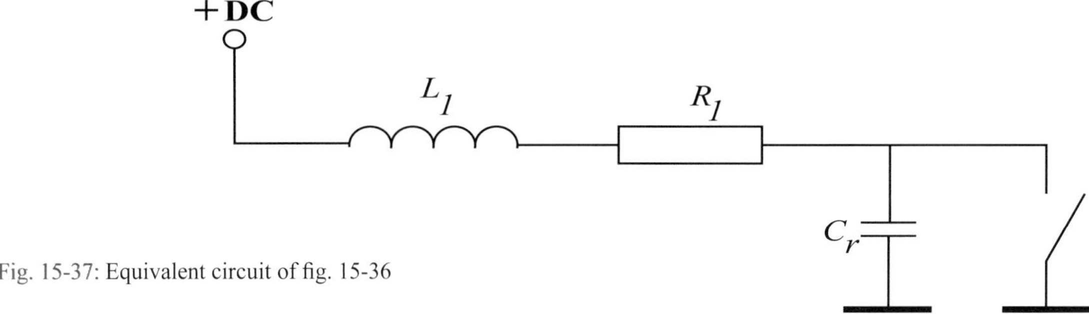

Fig. 15-37: Equivalent circuit of fig. 15-36

The advantage of this configuration is that only one IGBT is required. The disadvantage is that this circuit produces a voltage higher than the DC-bus voltage. This necessitates using an IGBT and capacitor that can handle a high blocking voltage.

Photo Mastervolt: In grid connected solar converters the solar energy is converted to an AC voltage and directly feed into the distribution network without an energy storage medium. The kWh-meter operates in reverse.

Grid connected converter series include: Mastervolt Sunmaster XS series to 5kW, Mastervolt Sunmaster XL series to 15kW and Master CP series to 200kW.

Technical data can be found on the website www.mastervolt.com, choose the product and go to "specifications" or "download".

EVALUATION : SOLUTIONS

CHAPTER 1

1.1   State of the art = the current level of knowledge and developments in technology

1.2   Chopper

1.3 $f = 1\text{kHz} \rightarrow\rightarrow T = 10^{-4} \rightarrow\rightarrow t_{OFF} = (1 - \delta) . T = (1 - 0.1) . 10^{-4} = 90\mu s$

1.4   During the 1940's during WWII

1.5   Power electronics can be defined as a power flux occurring in semiconductors

1.6   20kHz; IGBT = Insulated Gate Bipolar Transistor

1.7   Greater than 1000A

1.8   Natural or net commutation, artificial commutation and load commutation

1.9

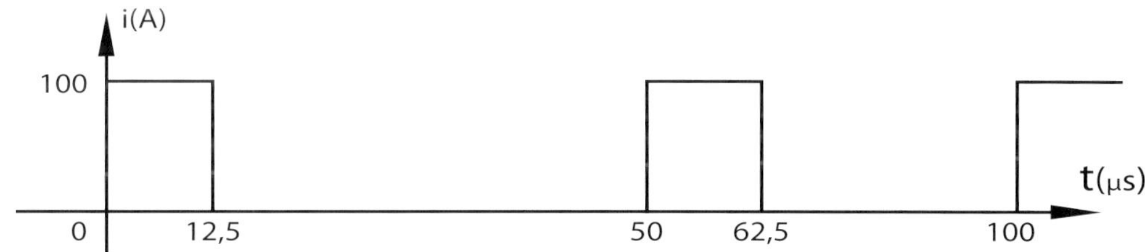

Fig. 1-12

1.10

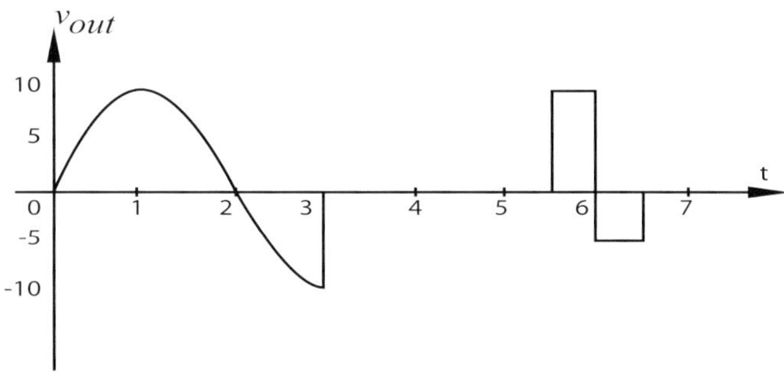

Fig. 1-13

1.11   1) $\delta = 0.25$ and $t_{on} = t_{off} = 1\mu s$

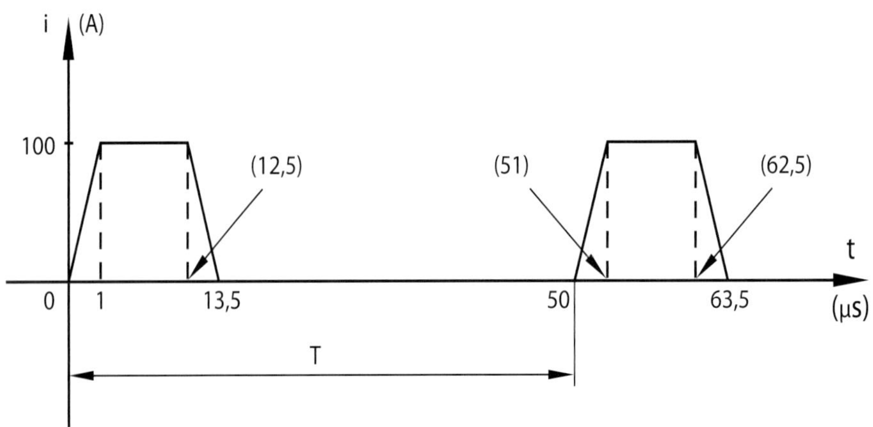

Fig. 1-14

2) $\delta = 0.2$ and $t_{on} = t_{off} = 12.5\mu s$

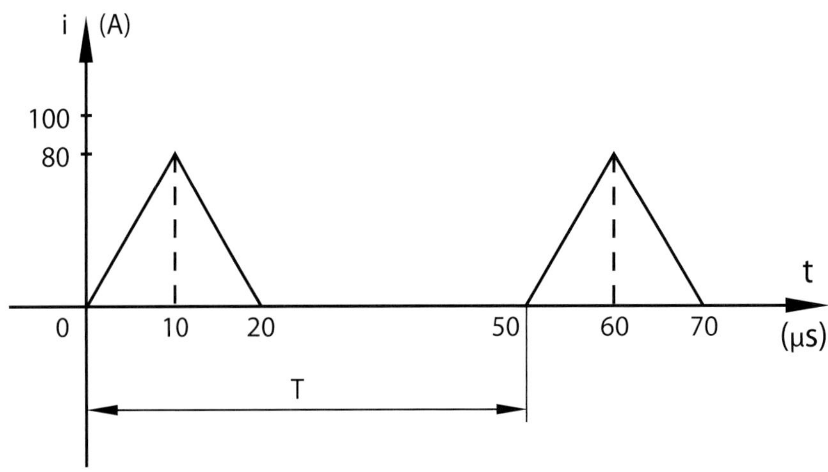

Fig. 1-15

1.12
AC   = alternating current
DC   = direct current
GTO   = gate turn off
SCR   = Silicon Controlled Rectifier
bipolar = BJT = bipolar junction transistor (see chapter 3!) = operating with two types of charge carriers
MOSFET = metal oxide silicon field effect transistor
IGCT = integrated gate commutated thyristor

CHAPTER 2

2.1 a) Has a dramatic change of conduction depending on direction
    b) At normal ambient temperatures the specific resistance can vary between that of metals and insulators

2.2 Conventional current flow

2.3 $I_{F(AV)} = I_0 = 1A$; $V_{RRM} = 1000V$

2.4 This is the time required for the space charge to dissipate so that the diode can go into blocking mode, and the nominal value of the reverse current is reached.

2.5 See p. 2.7

2.6 $q = 1.6 \times 10^{-19}C$; $m = 9.1 \times 10^{-31}kg$

2.7 Holes. Donor = a material that causes a majority of electrons to exist in the Si. Phosphor

2.8 $Q = I \cdot t = 10 \cdot 10^{-12} \cdot 10^{-3} = 10^{-14}C$; $n = \dfrac{Q}{q} = \dfrac{10^{-14}}{1.6 \times 10^{-19}} = 62500$ electrons

2.9 200V, 2700V.

2.10 1A, 420A ($T_c = 100°C$)

2.11 175°C

2.12 1.1V

2.13 1.05V; 0.98V; 1.15V

2.14 $r_T = \dfrac{\Delta V}{\Delta I}$

2.15 F = forward; S = surge; M = maximum.

CHAPTER 3

3.1   To prevent for the most part the switch off energy reaching the transistor. This enables a transistor with a lower power dissipation can be used

3.2 $V_{CB} = 0$

3.3 $t_{on} = t_d + t_r$ = delay time + risetime;
$t_{off} = t_s + t_f$ = storage time + fall time

3.4   1 to 2µs

3.5   To limit the collector emitter voltage at cut-off of the BJT. The energy is released from the coil and circulates through the diode, explaining the name "free wheel"

3.6   The time constant is smaller. Another explanation is that the energy is dissipated in two resistors

3.7   Between 10% of V_{CEoff} and 10% of I_c

3.8

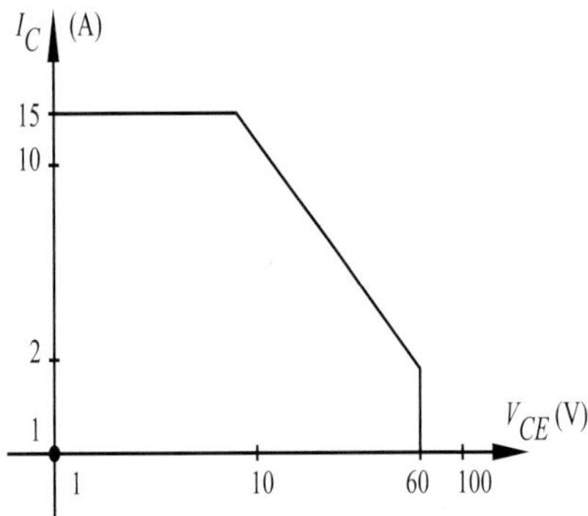

Fig. 3-52

3.9   turn-on / turn-off / delay- / storage- / fall- / rise- / TIME

3.10   Emitter / base / collector.   Source / drain/ gate

3.11   N- and P- channel MOSFET

3.12   MOS = metal oxide silicon (according to some metal oxide semiconductor);
FET = field effect transistor

3.13   See (3-6) on p. 3-15

3.14   Fig. 3-23

3.15   From the fact that the current flows vertically (under the gate and further) to the drain

3.16   It is the minimum value of the gate voltage for which the MOSFET can conduct. Since the threshold voltage is about 2 to 3V the gate voltage needs to be exceed a "safe" value to turn on the MOSFET. This renders the MOSFET immune for disturbances and in addition a logic zero of several tenths of a volt will not cause the MOSFET to conduct

3.17   International Rectifier (Hexfet); Infineon (Sipmos); Harris (Ultrafet); Trenchmos (Philips, IR, Infineon,…); trenchfieldstop (Infineon,…)

3.18 $C_{iss} = C_{GS} + C_{Mi}$

3.19   Delay time at switch-off

3.20   Maximum pulsed drain current

3.21	$R_{DS(ON)}$ and I_D
3.22	Under normal circumstance the charging of the C_{iss}-capacitance is determined by $C_{iss} \cdot R_G$ if the control source provides enough current, in other words $I_{source} \geq V_S / R_G$. If the given 10mA is too small then C_{iss} will charge linearly ($V_{GS} \approx I_G \cdot t / C_{iss}$) and the MOSFET will conduct from $V_{GS} > V_{threshold}$. With 100mA this will obviously happen a lot quicker
3.23	Electro static discharge. Internal zener diodes help protect the MOSFET from ESD.
3.24	Parallel switching of 100pF with 1.5Ω
3.25	They have a positive temperature coefficient which is responsible for an automatic uniform current distribution when switching in parallel
3.26	FREDFET = fast recovery epitaxial diode, field effect transistor
3.27	It is an PT-IGBT (PT = punch through) with maximum ratings of 6,500V and 750A
3.28	It is an area (with a low V_{DS}-voltage) where the ratio between I_D and V_{DS} is constant.
3.29	Only the polarity of the voltage and current is reversed. The *I-V* characteristic of a P-channel MOSFET lies in the third quadrant and that of an N-channel in the first quadrant.
3.30	A power electronic circuit that protects itself from faulty operation
3.31	$R_{DS(ON)} = 0.11Ω$
3.32	Miller capacitance: As a result of C_{GD} ($= C_{Mi}$) see fig. 3-27, a part of the output voltage is feedback to the input. This is the "Miller effect" and hence the name: Miller capacitance
3.33	

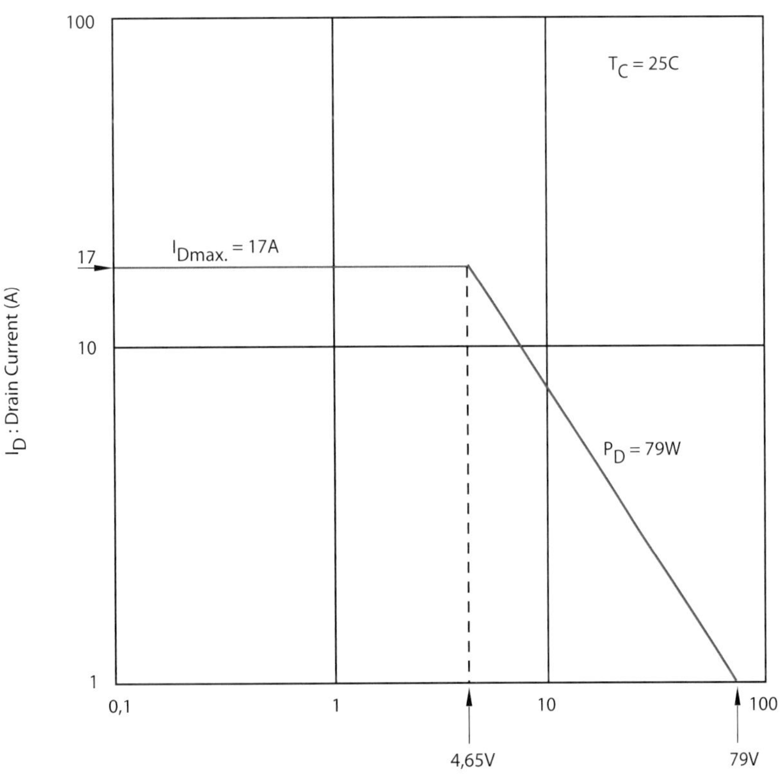

Fig. 3-53

3.34   With an unequal current density in the semiconductor crystal structure the place with the largest current will heat up the most. If the semiconductor has a negative temperature coefficient the resistance will locally decrease resulting in the current density further increasing. This results in "hot spots" and "second breakdown". This occurs in BJT's that operate close to the maximum allowable power

3.35   1.   With V_s = 5V the BC141 conducts and the input capacitance of the IRF530N will charge approximately to 5V, in other words the IRF530 will conduct.

2.   With V_s = 0V (and the control source is DC conducting, therefore base and collector of the PNP transistor are connected together!) the 2N2904 operates as a diode and the input capacitance of the MOSFET will discharge via this diode, in other words the IRF530N is blocking

3.36   With +5V on the input of the IRF530N it will conduct and with 0V it blocks.

3.37   An IGBT can withstand an over current of ten times its nominal collector current for 10µs.

3.38   The tail current is the result of the large number of minority charge carriers in the thick N^- -epilayer which need to recombine. This tail current lasts for 1 to 2µs

3.39   IGBT: Insulated gate bipolar transistor

3.40 V_{CEsat} = 2.1V

3.41   ± 20V

3.42   This is an inductive load with free wheel diode

3.43   A darlington configuration consists of a power transistor and its driver. Mostly both are in a single casing

3.44   10 to 30 Siemens.

3,45   Fall time: 90% to 10% I_c
Rise time: 10% to 90% I_c

3.46   DUT = device under test; a semiconductor under test in a test circuit.

3,47 E_{ON} expressed in mJ and stated between 5% of the test current and 5% of the test voltage

3.48 $W_{ON} = V_{CE(ON)} \cdot I_c \cdot \delta \cdot T$. Here $\delta \cdot T = t_{ON}$ so that the frequency plays no role in determining W_{ON}

3.49 t_f = 50 to 100ns

CHAPTER 4

4.1   A GTO requires a much larger gate power than an SCR

4.2   See p. 4-30

4.3   1.   An excessive anode voltage

    2.   An excessive junction temperature

    3.   An excessive voltage gradient (dv_a/dt)

4.4   Critical voltage gradient $(dv/dt)_{commutation}$

4.5   1.   Blocking SCR: subscript D

    2.   Conducting SCR: subscript T

    3.   Reverse polarized: subscript R

4.6   The creation of hot spots since the current initially flows through a limited area of the Si-tablet

4.7 $V_{GT}(I_{GT})$ = required trigger voltage (current) to trigger the specific component under consideration

    V_{GD} = minimum required gate voltage (T_J = 125°C) to trigger an SCR

4.8   Power dissipation in the thyristor combined with insufficient cooling

4.9   Triac

4.10   It is a thyristor control pulse which is extremely large but which reduces quickly after a few µs to a "normal" value

4.11   See p. 4-19

4.12   Because the $R_{DS(ON)}$ from the Off-MOSFET increases

4.13 $(dv/dt)_{commutation}$ = 60V/µs

4.14   Integration of the gate control circuit in the casing of the GCT

4.15   An MCT

4.16   DSC = Double Sided Cooling

4.17 D_1 = free wheel diode, D_2 = snubber diode

4.18   An IGBT does not require a snubber

4.19   For AC controllers of tens of kW's

4.20 $t_{on} = t_d + t_r$ = elapsed time between 10%.I_{GT} and 90%.I_T

    $t_{off} = t_s + t_f + t_t$ = elapsed time between 10%.I_{GR} and 2%.I_T

4.21   Symmetric GTO: $V_{RRM} = V_{DRM}$

    Asymmetric GTO: $V_{DRM} >> V_{RRM}$

4.22 V_T =1.2V max., see fig. 4-11

4.23   SCR   = Silicon Controlled Rectifier

    DIAC   = diode AC

    TRIAC   = triode AC

    GTO   = gate turn off

    GCT   = gate commutated thyristor

    MCT   = mos controlled thyristor

4.24   The current amplification factor α_2 of the transistor (cathode) is very large

4.25   As a result of the shorted emitter construction V_{RRM} and V_T of a GTO decrease

4.26 $t_{d(on)} = 3\mu s$; $t_{d(off)} = 6\mu s$; $t_r = 1\mu s$; $t_f = 1\mu s$; $(di/dt)_{max} = 425A/\mu s$;
$I_{T(RMS)max} = 655A$; I_{TSM} 15.7kA

4.27   a: I.E.C. = International Electrotechnical Committee
b: an NPN- and a PNP-transistor with the same properties (excluding the polarity)

4.28 I_a = anode current; I_L = latching current; I_H = holding current; I_G = gate current;
I_T = forward thyristor current; I_R = leakage current SCR blocking; V_a = anode voltage;
V_{BO} = break over voltage; V_{DRM} = maximum repetitive blocking voltage;
V_{RRM} = maximum repetitive reverse voltage; V_Z = zener voltage; V_T = voltage drop across
conducting thyristor; T_{vj} = junction temperature

4.29   a) A = ambient; AV = average; C = case ; G = gate; H = holding; J = junction; L = latching; M = maximum; R = reverse (first index); R = repetitive (second index); RMS =
root mean square; S = surge (one off peak in amplitude); T = thyristor
b)

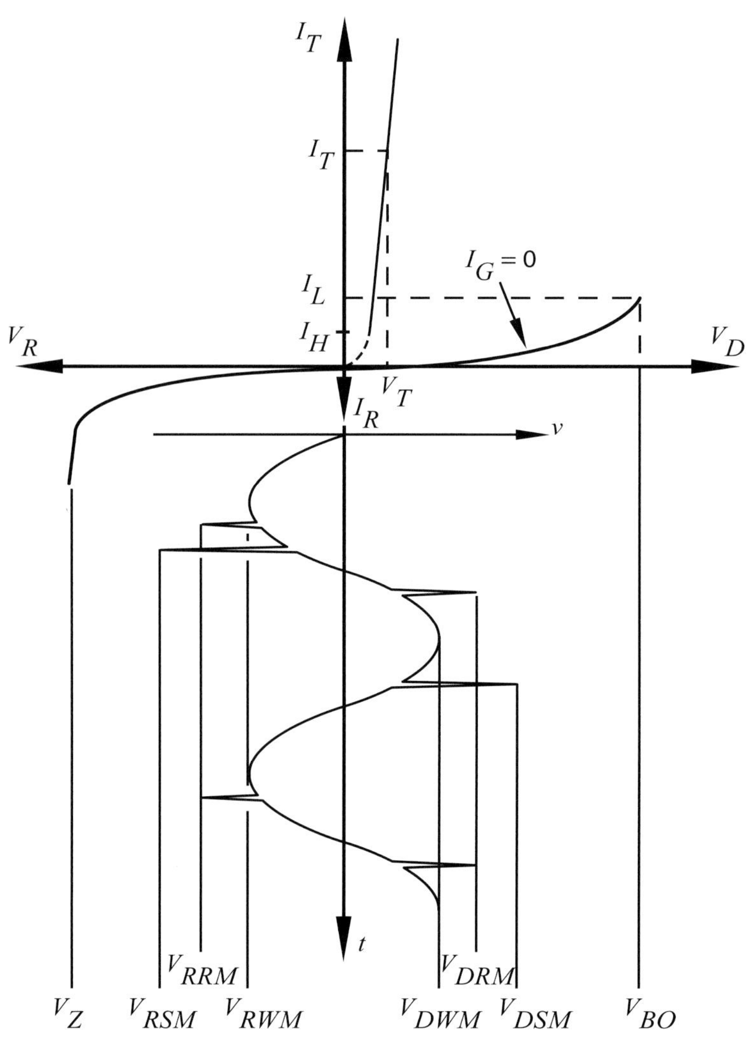

Fig. 4-60

4.30	At $T_{vj} = 25°C$
4.31	Triggering = synchronising or causing a certain signal to rise to a certain level

Examples:

a) triggering an oscilloscope means that we synchronise its time base with an applied voltage

b) triggering a thyristor means causing the thyristor to conduct at the instant that a control signal is applied to the gate

4.32   See p. 9-13

4.33   It is an AC switch: no $I_{T(AV)}$, but there is an $I_{T(RMS)}$ .Has a bidirectional operation, therefore no reverse effect on the current (no I_R)

4.34   Bidirectional switch; no reverse operation

4.35   Since the emitter of T_2 is 10V negative with respect to ground potential

4.36   Stud mounted = able to be fixed with a nut, see photo on p. 4-23 top right

Capsule = disk or hockey puck casing (see photo on p. 4-4)

4.37

1) By causing T_1 to conduct, the output voltage of the DC-DC converter is applied to the gate-cathode space of the GTO so that it is triggered. After that T_1 blocks

2) By switching T_2 on, the gate of the GTO is connected to a more negative voltage resulting in the GTO switching off

4.38   3 to 4µs (including the tail current!)

4.39 $V_{Tmax} \approx 2V$

CHAPTER 7

7.1

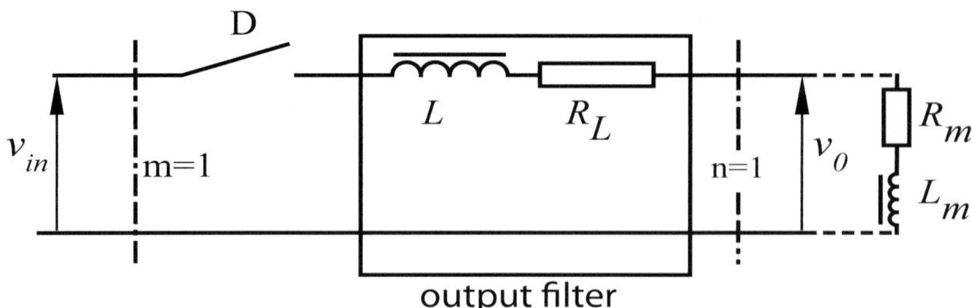

Fig. 7-28

We could also include L_m in the filter, but with the components shown the correct output voltage is obtained

7.2   a) $i_{p1} = i_{S1}$ and $i_{p2} = i_{s2}$.

From fig. 7-18 we have i_{S1} and i_{S2} and can construct i_{L2} $\longrightarrow$ fig. 7-29

b) from (5-6) $\rightarrow I_{L2} = \sqrt{2 . \left[(I_d/\sqrt{6})^2 + 2 . (2 . I_d/\sqrt{6})^2 + 2 . (I_d/\sqrt{6})^2 \right]} = \sqrt{2} . I_d$

c) apparent (primary) power: $S_{prim.} = \sqrt{3} . V_{line} . I_{line} = \sqrt{3} . V_f . \sqrt{2} . I_d = \sqrt{6} . V_f . I_d$

secondary apparent power: $S_{sec} = 3 . V_f . I_f = 3 . V_f . \sqrt{\frac{2}{3}} . I_d = \sqrt{6} . V_f . I_d$

therefore: $S_{prim} = S_{sec}$!!

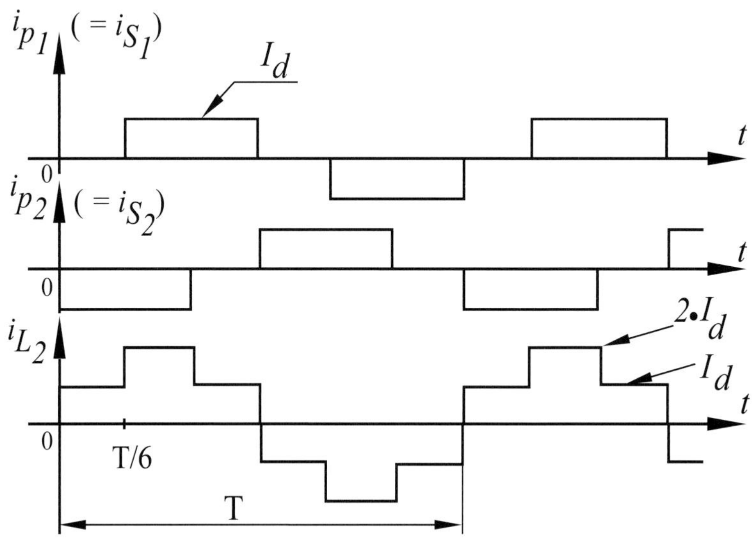

Fig. 7-29

7.3 $P = 0.9 \times 200 + 0.5 \times 10^{-3} \times 1.73^2 \times 200^2 = 239.86 \text{W}$

7.4   a x b = 3 x 2 = 6 switches. Agrees with the theory on p. 1.3.

7.5   This is the maximum apparent power the diode can switch

7.6 $$I_d = \frac{V_{di}}{R_b} = 100/5 = 20\text{A}$$

Table 7-4 (p. 7.22)

- diode specifications
  - $I_{FAV} = 0.333 \; x \; I_d = 6.66\text{A}$
  - $I_{F(RMS)} = 0.577 \; x \; I_d = 11.54\text{A}$
  - $V_{RRM} \geq 1.05 \; x \; V_{di} = 105\text{V}$
- transformer specifications
  - $V_{RMS} = 0.428 \; x \; V_{di} = 42.8\text{V}$
  - $I_{RMS} = 0.816 \; x \; I_d = 16.32\text{A}$
  - $S = 1.05 \; x \; P_{di} = 2100\text{VA}$

7.7 $$W = \frac{1}{2} . L . i_{max}^2 = \frac{0.1 \; x \; (13.75)^2}{2} = 9.453\text{J}$$

7.8   Bauleistung is the average apparent power of the primary and secondary of the transformer

7.9   The reference line voltage is, per diode, the line voltage which goes through zero at the instant the rectifier diode starts to conduct in the case of natural commutation

7.10   Since $e^{\frac{t_1 - 5.\tau}{\tau}} = e^{\frac{t_1}{\tau}} \approx 0.006 . e^{t_1/\tau}$ .Usual is $t_1 < \tau$ so that the complete exponential term of (7-1) can be practically ignored after $t = 5 . \tau$

7.11   The voltage drop of a Si rectifier diode with the nominal current is $V_D \approx 1.1\text{V}$

7.12   The average voltage across an ideal coil is zero

7.13   The conduction angle is the angle during which the diode is conducting

7.14   A natural commutation point for a rectifier diode is the instant that the diode naturally commutates due to the effect of the supply voltage

7.15

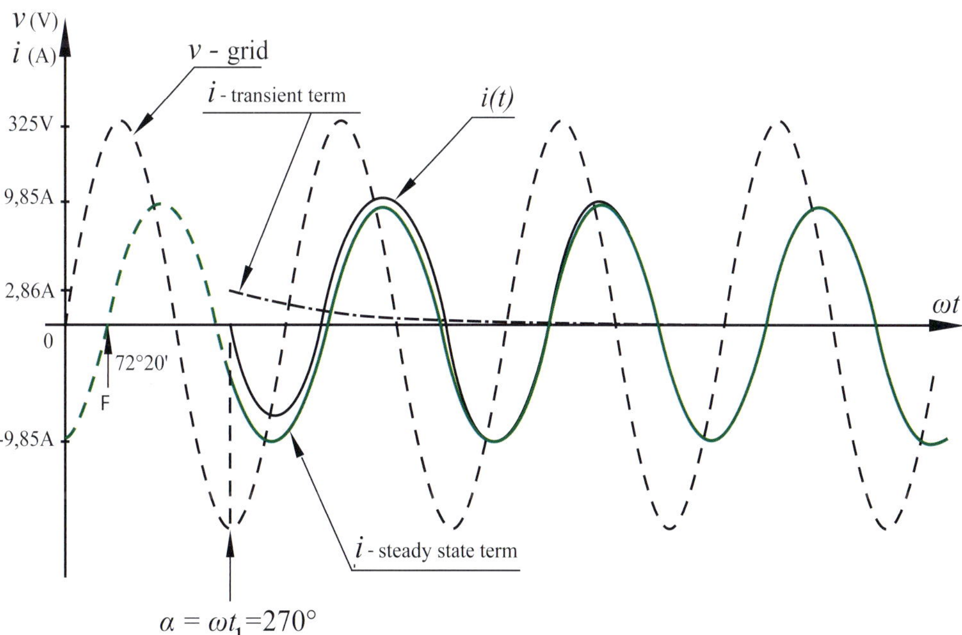

Fig. 7-30

7.16   1) $\tau = \infty$; $Z = \omega \cdot L = 31.41\,\Omega$; $\Phi = 90°$
              $\hat{v} = 325\text{V}$; $i(t) = 10.34 \cdot \left[\sin(\omega t - 90°) + \sin(90° - 0) \cdot e^{0} \right]$
              $i(t) = 10.34 \cdot (1 - \cos \omega t)$

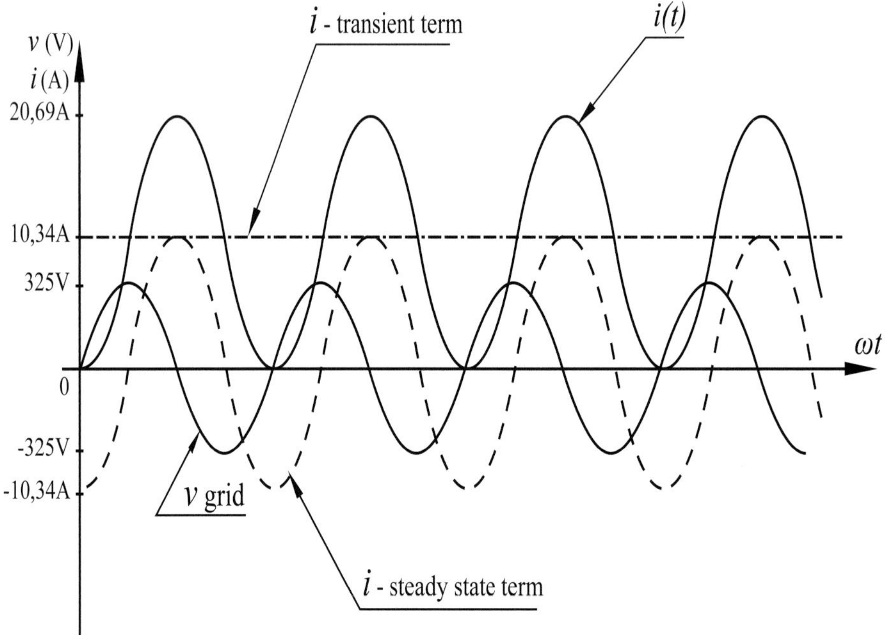

Fig. 7.31a

          2) $\tau = \infty$; $Z = \omega \cdot L = 31.41\,\Omega$; $\Phi = 90°$; $\hat{v} = 325\text{V}$;
              $i(t) = 10.34 \cdot \left[\sin(\omega t - 90°) + \sin(90° - 90°) \cdot e^{0} \right] = -10.34 \cdot \cos \omega t$

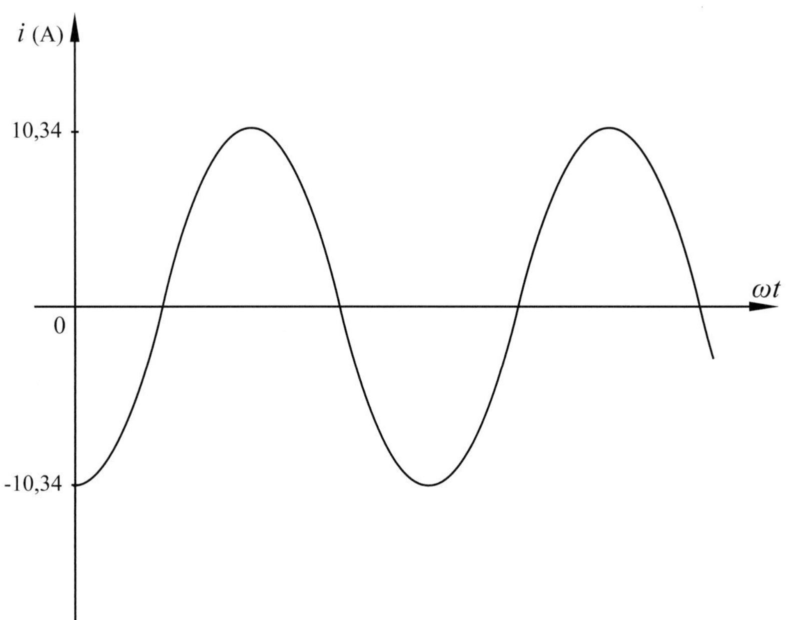

Fig. 7-31b

3) $\tau = \infty$; $Z = \omega . L = 31.41\,\Omega$; $\Phi = 180°$; $\hat{v} = 325V$;

$i(t) = 10.34 . \left[\sin(\omega t - 90°) + \sin(90° - 180°) . e^{0} \right]$

$i(t) = 10.34 . (-1 - \cos \omega t)$

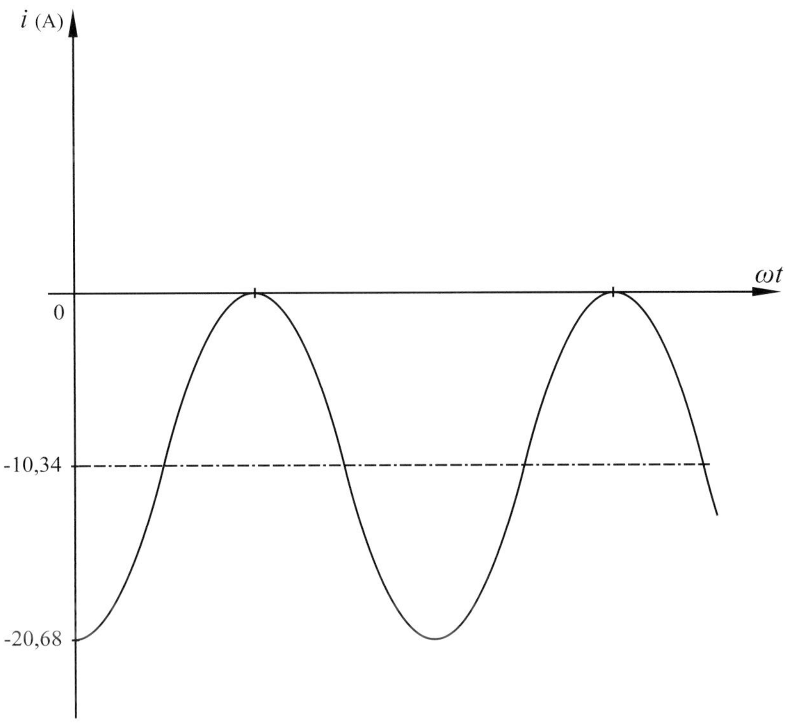

Fig. 7-31c

7.17   With this pattern of coding there will always be two diodes with consecutive numbers conducting: D_6-D_1 ; D_1-D_2 ; D_2-D_3 ; etc.

7.18

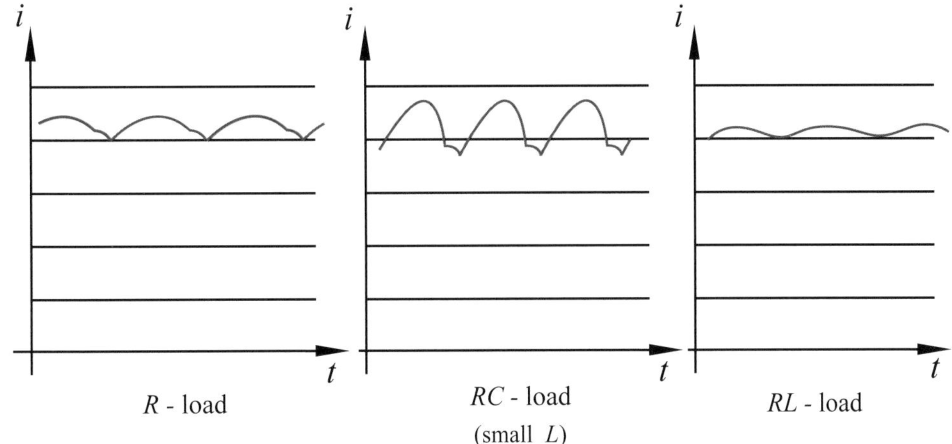

Fig. 7-32

7.19

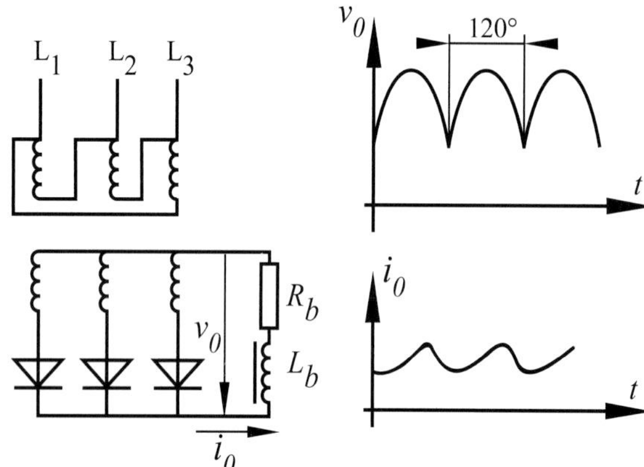

Fig. 7-33

7.20

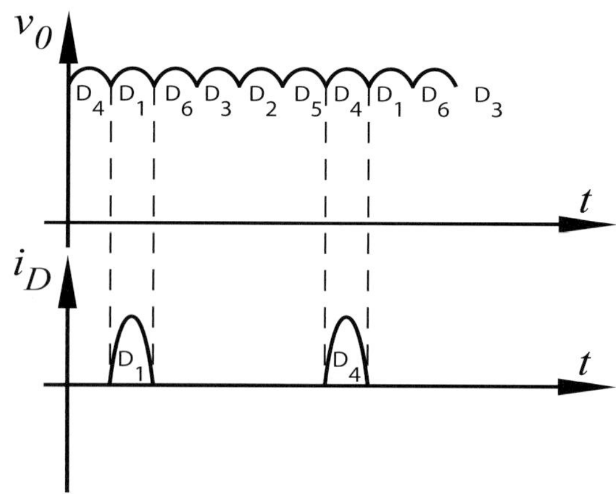

Fig. 7-34

7-21 $p = 300Hz/50Hz = 6$

7.22 $p = 3$

7.23   1) and 2): ideal diodes
3): ideal transformer
4): good filtering

CHAPTER 8

8.1   With a full controlled three-phase bridge since the circuit will not otherwise remain conducting, whether with the initial firing, or in the case of discontinuous current.

8.2   Disadvantages of phase control:

- low power factor

- high *di/dt* for the thyristor

- creation of harmonics causing radio and TV-interferences

8.3 $I_{T(AV)}$ = 560A (for continuous current); $P_{T(AV)} \approx$ 690W

8.4   The commutation of the switch is determined by the supply voltage

8.5   SKT10/08D $\rightarrow\rightarrow$ V_{RRM} = 800V; $V_{RRM} \geq$ g $. V_{net} . \sqrt{2}$

$$\text{With for example g} = 2 \rightarrow\rightarrow \ V_{net} \leq \frac{800}{2 . \sqrt{2}} = 282.8\text{V}$$

8.6   For $\alpha \geq 120°$ with a B_6-controller with resistive load: $V_{di\alpha} = 0$

$$\text{For } \alpha = 100° \rightarrow\rightarrow \ (8\text{-}18): \ V_{di\alpha} = V_{di} . \left[1 + \frac{\cos\alpha}{2} - \frac{\sqrt{3}}{2} \sin\alpha \right]$$

$$\text{With } V_{di} = \frac{3 . \hat{v}}{\pi} = \frac{3 . \sqrt{2} . 400}{\pi} \ \text{ and } \alpha = 100° \text{ we find: } V_{di\alpha} = 32.57\text{V}$$

8.7   See p. 8-38

8.8   See p. 8-13

8.9 $\alpha = 0° \ \rightarrow\rightarrow \ \lambda = \frac{3}{\pi} . \cos\alpha = 0.955$

8.10 $5 . \dfrac{L}{R} = 5 . \dfrac{150.10^{-3}}{10} = 75\text{ms} \gg 3.33\text{ms}$ (discontinuous service);

$$V_{di} = \frac{3 . \hat{v}}{\pi} = \frac{3 . 230 . \sqrt{2}}{\pi} = 310.6\text{V}$$

1) $\alpha = 0°$: V_{di} = 310.6V; $I_d = \dfrac{V_{di}}{R_b} = 31.06$A; $P_{di} = V_{di} . I_d = 9.75$kW;

$$\lambda = \frac{3}{\pi} . \cos\alpha = 0.955 \ ; \ \cos\varphi_1 = \cos\alpha = 1 \ ; \ \cos\delta = \frac{3}{\pi} = 0.955$$

$$Q_{1\alpha} = S_1 . \sin\varphi_1 = \sqrt{3} . V . I_1 . \sin\varphi_1 = 0 \ ; \ I_1 = \sqrt{6} . \frac{I_d}{\pi} = 24.21\text{A}$$

2) $\alpha = 60°$: $V_{di\alpha} = V_{di} . \cos\alpha = 155.3$V ; $I_{d\alpha} = \dfrac{V_{di\alpha}}{R_b} = 15.53$A ;

$$I_1 = \sqrt{6} . \frac{I_{d\alpha}}{\pi} = 12.1\text{A} \ \ ; \ \ P_{di\alpha} = V_{di\alpha} . I_{d\alpha} = 2.41\text{kW} \ ;$$

$$\lambda = \frac{3}{\pi} . \cos\alpha = 0.4775 \ ; \ \cos\varphi_1 = \cos\alpha = 0.5 \ ; \ \cos\delta = \frac{3}{\pi} = 0.955 \ ;$$

$$Q_{1\alpha} = S_1 . \sin\varphi_1 = \sqrt{3} . V . I_1 . \sin\varphi_1 = 4.177\text{kvar}$$

8.11 $V_{RRM} \geq$ g $. 230 . \sqrt{2} = 650$V ; $I_{F(AV)} = \dfrac{31.06}{3} = 10.35$A

8.12   Fig. 8-19: $\hat{v}$;   fig. 8-21: $\hat{v}$

8.13 $\Phi = 72°20'$; $\alpha = 60° \to\to$ (8-13): $V_{di\alpha} = 103.5$V

$\alpha = 105° \to\to$ (8-11): $V_{di\alpha} = 26.52$V

8.14 $\Phi = 60°$; fig. 8-4 $\to\to \delta = 240°$; (8-5) $\to\to V_{di\alpha} = 51.76$V

(8-5) $\to\to V_{di} = \dfrac{\hat{v}}{2\pi} \cdot (1 + 0.438) = 74.46$V; $I_{T(AV)max} = 10.26$A and $V_{RRM} \geq 325$V

8.15 $\Phi = 60° \to\to$ $\alpha = 30° \to\to$ (8-13) $\to\to V_{di\alpha} = 179.29$V

$\alpha = 60° \to\to$ (8-13) $\to\to V_{di\alpha} = 103.5$V

$\alpha = 90° \to\to \delta = 233° \to\to$ (8-11) $\to\to V_{di\alpha} = 62.30$V

8.16   a) $\hat{v}$

b) $\hat{v}$ with $90° \leq \alpha < 180°$

8.17   No. Consider for example Th$_1$ and Th$_6$ with there driving voltage v_{12} . At $\omega t = 120°$, v_{12} is negative so that Th$_6$ cannot fire. The same conclusion can be made for the other semi-conductor pairs

8.18   Start-up: $\alpha = \varphi_1 = 90° \to\to Q_{1\alpha} = V_{di} \cdot I_d$ (x 1.5) $= 150$kvar

8.19 $P = \dfrac{V_{RMS}^2}{R_b}$

HALF WAVE: $\alpha = 0°$: $P_{max} = 0.5/R_b$; $\alpha = 30°$: $P_1 = 0.4855/R_b$

$\alpha = 150°$: $P_2 = 0.0144/R_b$

for the ratios: $P_1/P_{max} = 97\%$ and $P_2/P_{max} = 3\%$

FULL WAVE: $\alpha = 0°$: $P_{max} = 1/R_b$; $\alpha = 30°$: $P_1 = 0.97/R_b$

$\alpha = 150°$: $P_2 = 0.03/R_b$

for the ratios: $P_1/P_{max} = 97\%$ and $P_2/P_{max} = 3\%$

8.20 $I_1 = \dfrac{\sqrt{6} \cdot I_{d\alpha}}{\pi} = 12.1$A; $I_k = \dfrac{I_1}{k} \to\to I_5 = 2.42$A; $I_7 = 1.73$A

Table 8-4 $\to\to$ 300Hz: $V_{k\alpha} = \dfrac{21}{100} \cdot V_{di} = 65.22$V;

600Hz: $V_{k\alpha} = \dfrac{10.29}{100} \cdot V_{di} = 31.96$V;

8.21   Fig. 8-6: $\alpha = 60° \to\to X_3 = 0.115$ x 230 x $\sqrt{2} \to\to \hat{i}_3 = 0.374$A

$\alpha = 90° \to\to X_3 = 0.153$ x 230 x $\sqrt{2} \to\to \hat{i}_3 = 0.497$A

8.22   (8-29) and (8-30) $\to\to I_7 = \dfrac{\sqrt{6} \cdot I_d \cdot \cos\alpha}{\pi \cdot 7}$ with $I_d = \dfrac{V_{di}}{R}$

(8-25) $\to\to V_{di} = 540$V $\to\to I_7 = 10.42$A

8.23 Fig. 8-41

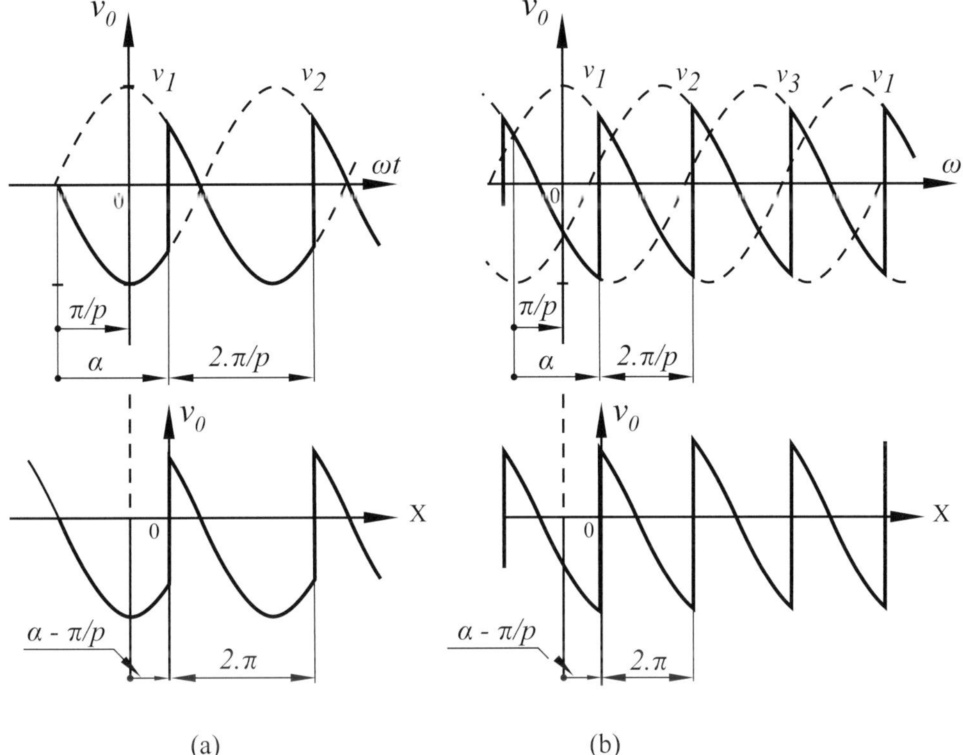

Fig. 8-41   (a)   (b)

The $v_o = f(x)$ axes is the same as previously found in fig. 8-34, so that the expression for $\omega t = f(x)$ in fig. 8-34 is also valid here

8.24 B_2-controller: $p = 2 \longrightarrow \hat{v} \cdot \dfrac{p}{\pi} \sin \dfrac{\pi}{p} = \hat{v} \cdot \dfrac{2}{\pi} \cdot \sin \dfrac{\pi}{2} = \dfrac{2 \cdot \hat{v}}{\pi}$

M_3-controller: $p = 3 \longrightarrow \hat{v} \cdot \dfrac{p}{\pi} \sin \dfrac{\pi}{p} = \hat{v} \cdot \dfrac{3}{\pi} \cdot \sin \dfrac{\pi}{3} = \dfrac{3\sqrt{3} \cdot \hat{v}}{2 \cdot \pi}$

B_6-controller: $p = 6 \longrightarrow \hat{v} \cdot \dfrac{p}{\pi} \sin \dfrac{\pi}{p} = \hat{v} \cdot \dfrac{6}{\pi} \cdot \sin \dfrac{\pi}{6} = \dfrac{3 \cdot \hat{v}}{\pi}$

8.25 $\alpha = 0° \longrightarrow \cos\alpha = 1 \longrightarrow \dfrac{V_{k0°}}{V_{di}} = \dfrac{\sqrt{2}}{(k^2 - 1)} \cdot \sqrt{k^2 - (k^2 - 1)} = \dfrac{\sqrt{2}}{(k^2 - 1)}$

$\alpha = 90° \longrightarrow \cos\alpha = 0 \longrightarrow \dfrac{V_{k90°}}{V_{di}} = \dfrac{\sqrt{2}}{(k^2 - 1)} \cdot \sqrt{k^2} = \dfrac{k \cdot \sqrt{2}}{(k^2 - 1)}$

$\dfrac{V_{k90°}}{V_{k0°}} = k.$ Table 8-4: example $k = 2 \longrightarrow$ $\left|\begin{array}{l} \alpha = 0° \longrightarrow \dfrac{V_{k\alpha}}{V_{di}} = 47.14\% \\[2mm] \alpha = 90° \longrightarrow \dfrac{V_{k\alpha}}{V_{di}} = 94.2\% \end{array}\right.$

CHAPTER 9

9.1 $\alpha \geq 90°$ (and 270°!) $\rightarrow\rightarrow$ $V_{triac(max)} = \hat{v} = 325V$

9.2 $\alpha = \Phi$ $\rightarrow\rightarrow$ nominal current: $\quad I = \dfrac{V}{Z} = \dfrac{230}{\sqrt{10^2 + (2 \times \pi \times 50 \times 0.1)^2}} = 6.97A$

9.3   There is an inductive load so that I_T lags v_{net} and also v_o. Since the thyristor only switches off at the instant that $I_T = I_H$, this happens quasi at the zero crossover of i

9.4 T_{min} = 13 seconds

9.5   EMI = electro magnetic interference
RFI = radio frequency interference

9.6   (9-1) $\rightarrow\rightarrow$ V_{eff} = 162.63V; I_{eff} = 1.626A

9.7   •   control of electric lighting

        •   control of motors in domestic appliances

        •   temperature control of convector heaters

        •   AC welders

9.8 $V_{RRM} \geq 230 \times \sqrt{2} \times 2.3 = 748V$; $I_{RMS} \geq 12A$

9.9   time ratio control / radio frequency interference / root mean square / zero crossover circuit / solid state relay.

9.10   DC- and AC-controllers. Net commutation

9.11 $\Phi = 85°27'$; $\alpha < \Phi$ $\rightarrow\rightarrow$ a) DC pulses in the supply net (see fig. 9-5)
                                           b) triac functions as a mechanical switch (fig. 9-6)

9.12   See p. 9-14

CHAPTER 11

11.1   They limit the voltage at pin 5 with respect to ground to 0.7V in both directions since both diodes are antiparallel

11.2   The BD675 functions as a switch. In the specifications of this transistor we find that

$$V_{CEsat} < 2.5\text{V when } I_C = 1.5\text{A and } h_{FE} = 750$$

In the configuration of fig. 11-8 we see that $I_{Cmax} \leq \dfrac{15 - 2.5}{12} \approx 1\text{A}.$

With $h_{CE} = 750$ the base current will be $I_B \leq \dfrac{1000}{750} = 1.33\text{mA}$. The TCA 785 is lightly

loaded at pin 14 and 15 while in the primary of the pulse transformer a current to 1A can be sent

11.3   The control pulse for the antiparallel SCR's need to be applied between the gate and cathode. Without galvanic separation (with the aid of the pulse transformers) both control circuits with their common ground would be connected to the cathode of their SCR. Both cathodes would then be connected so that the antiparallel circuit of both SCRs would be short circuited

11.4   The IN4001 and BZX83C20 allow magnetic energy to escape via the pulse transformer. These diodes limit the over voltage to the BD 675 to about 21 volt (zener diode breaks down at about 20 V and there is about 1V across the IN4001). The voltage is then in series with the 15V supply, so that there is about 36V across the BD675

11.5
- 1N4001: at switch off of the pulse transformer there is a negative pulse on the secondary and this diode prevents the voltage peak appearing at the gate of the SCR
- 2.2 kΩ
  - advantages:
    - raises $(du/dt)_{crit}$ of the SCR
    - reduces the turn off time t_{off} of the SCR
  - disadvantage:
    - increases the holding current and latching current of the SCR

11.6   By closing S, pin 6 is set low and the pulses to pins 14 and 15 are blocked

11.7   This is a snubber network

11.8 $i = \hat{i} \cdot \sin\omega t$

1) $P = 25\text{W} \longrightarrow I = \dfrac{25}{230} = 0.1087\text{A} \longrightarrow \hat{i} = 0.1537\text{A}$

$24.10^{-3} = 0.1537 \cdot \sin 2 \cdot \pi \cdot 50 \cdot t_1 \longrightarrow t_1 = 499\mu\text{s}$

2) $P = 1000\text{W} \longrightarrow \hat{i} = 6.148\text{A} \longrightarrow t_2 = 12.4\mu\text{s}$

11.9 $V_0 \cdot t$ = time voltage integral, this is the product of the pulse height and the pulse width (at half height)

V_p = RMS test voltage allowed for 1 minute between primary and secondary

CHAPTER 12

12.1   PWM = pulse width modulation

PFM = pulse frequency modulation (= PRM = pulse rate modulation)

12.2   a)   This diode allows the stored energy to be released (free) after the switch (diode, thyristor, transistor) opens

b) Fig. 12-16

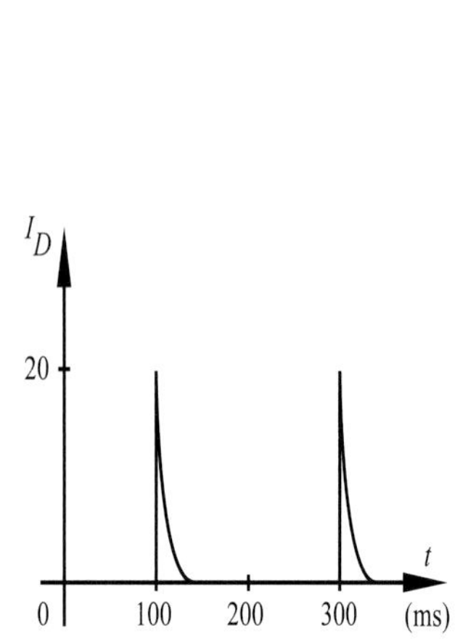

Fig. 12-16

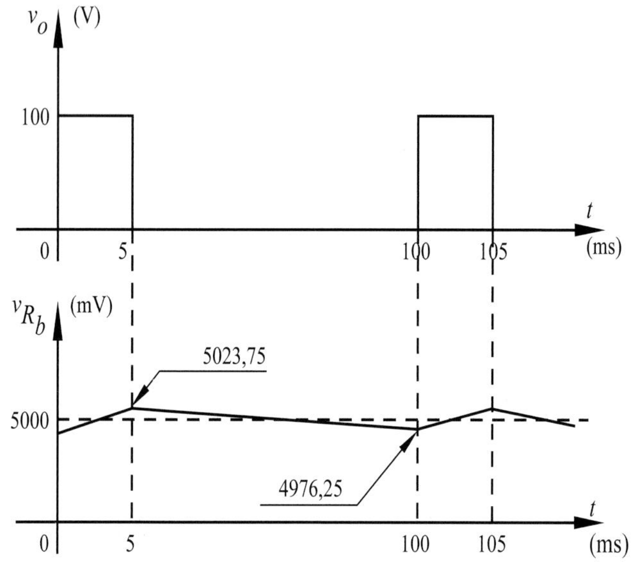

Fig. 12-17

12.3 $I_{average} = \dfrac{V}{R_b} \cdot \delta = 100/10 \times 0.05 = 0.5\text{A}$

$\Delta i = \dfrac{V}{R_b} \cdot \dfrac{T}{\tau}\,(\delta - \delta^2) = \dfrac{100}{10} \cdot \dfrac{10^{-4}}{10^{-2}} \times (0.05 - 0.0025) = 4.75\text{mA}$

$V_{R_{b(average)}} = 0.5 \times 10 = 5\text{V}\;;\; \Delta v_o = 47.5\text{mV}$ (see fig. 12-17)

12.4   • Δi : (12-6): 0/0.016/0.025/0.016/0

• $I_{average}$: (12-7): 0/2/5/8/10

12.5   a)   (12-10) $\rightarrow\rightarrow f = \dfrac{1.18}{236 \times 0.1} = 0.05\text{kHz} = 50\text{Hz}$

b) $V_g = \dfrac{5}{2} = 2.5\text{V}$. In a similar fashion to fig. 12-15 we see: $t_{on} = \dfrac{2.5 - 1}{2.5} \times T = 0.6 \times T$

$\delta = \dfrac{t_{on}}{T} \rightarrow = 60\%$

12.6 $\tau = \dfrac{L_b}{R_b} = \dfrac{1.5}{3} = 0.5\text{s}\;;\; \delta = 60\%\;;\; V = 24\text{V}\;;\; T = \dfrac{1}{50} = 0.02 = 20\text{ms}\;;$

$\dfrac{\delta}{T} = 25 \gg 1 \rightarrow\rightarrow$ (12-6) and (12-7) can be applied:

(12-7): $I_{average} = \dfrac{24}{3} \times 0.6 = 4.8\text{A}$

(12-6): $\Delta i = \dfrac{24}{3} \cdot \dfrac{0.02}{0.5} \cdot (0.6 - 0.36) = 76.8\text{mA}$

12.7   It is a DC-DC converter

12.8   Brake resistor of a frequency drive, brake resistor in a traction drive…

12.9   By connecting a choke in series

12.10   a) Comparator: compares two voltage levels with each other

b) NOR gate: logical gate with the expression : $\overline{X} = A + B + C$

CHAPTER 13

13.1	Double supply voltage: flyback and forward with single transistor
	Supply voltage across one transistor in the case with two transistors: forward
13.2	This is discontinuous current in the inductance of a non isolated converter
13.3	Boost and buck converter
13.4	Flyback: energy transfer to load with opened switch
	Forward: energy transfer with closed switch
13.5	See p. 13-25
13.6	FRD = fast recovery diode
13.7	Low losses
13.8	Direct rectification from the supply grid
13.9	The output voltage would be extremely high
13.10	Up to about 30W. 30 to 50%
13.11	a) electromagnetic interference
	b) supply transformer, smoothing (filter) capacitor
13.12	In fig. 13-13 t_1 can also be written as

$$t_1 = (1 - \delta - \frac{t_d}{T}) \cdot T \, .$$ Here t_d is the duration of discontinuous current. (= dead time)

Expression (4) on p. 13-17 then becomes:

$$V_o = \frac{\eta \cdot V_i \cdot \delta \cdot T}{n \cdot t_1} = \frac{\eta \cdot V_i \cdot \delta \cdot T}{n \cdot (1 - \delta - t_d/T) \cdot T} = \eta \cdot V_{in} \cdot \frac{\delta}{(1 - \delta - t_d/T)}$$

This is expression (13-19), but now we have the turns ratio n and the efficiency η in the formula

On the other hand if we replace (3) in (4) on p. 13.17 we arrive at expression (13-21) so that (13-21) and (13-19) are equivalent

CHAPTER 14

14.1   Power supplies for computers, measurement and control applications…

14.2   When the supply voltage is a constant DC voltage

14.3   Refers to the theoretical maximum conduction time of every switch in the inverter,

14.4   IGBT's: up to hundreds of kW's. SCR's and IGCTs for power levels >1MW

14.5   The time difference between opening one switch and closing another switch in the different bridge halves

14.6   Pulse width modulation

14.7   1) battery
2) rail + overhead wire in traction systems
3) from a rectifier connected to an AC network

14.8   180°: three-phase induction motor ; 120°: three-phase brushless DC-motor

14.9   To have less harmonics in the output of the converter

14.10 Smoothing capacitor

14.11 If all pulses of the PWM signal are next to each other resulting in the output line voltage being a square wave form

14.12 Pulse frequency control

14.13 a) amplitude $= \dfrac{2 \cdot \sqrt{3}}{\pi} \cdot V$; RMS value $= \dfrac{\sqrt{6}}{\pi} \cdot V$

b) $V_{L_1 L_2} = \sqrt{\delta} \cdot V = \sqrt{\dfrac{2}{3}} \cdot V$

c) $m = \dfrac{\dfrac{\sqrt{6}}{\pi} \cdot V}{\sqrt{\dfrac{2}{3}} \cdot V} = 0.995 = 95.5\%$

14.14

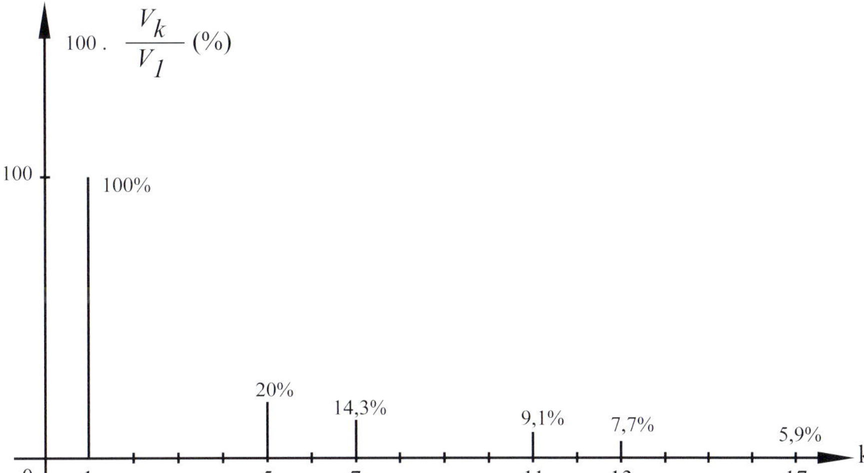

Fig. 14-27

14.15

a) ground harmonic: amplitude: $\dfrac{2 \cdot V}{\pi}$; RMS value $= \dfrac{\sqrt{2} \cdot V}{\pi}$

b) $V_R = \sqrt{\dfrac{1}{3} \cdot \left(\dfrac{2}{3} \cdot V\right)^2 + \dfrac{2}{3} \cdot \left(\dfrac{1}{3} \cdot V\right)^2} = \dfrac{\sqrt{2}}{3} \cdot V$

c) $\dfrac{\dfrac{\sqrt{2} \cdot V}{\pi}}{\dfrac{\sqrt{2}}{3} \cdot V} = 0.955 = 95.5\%$

d)

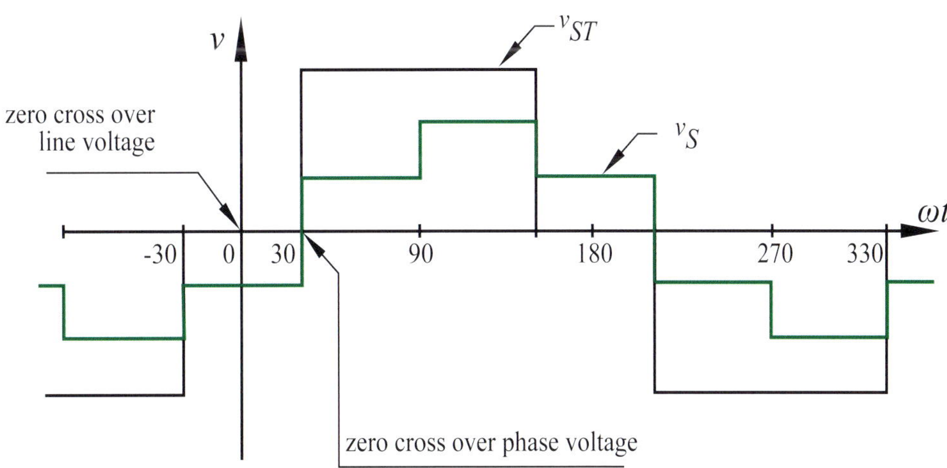

Fig. 14-28

14.16

a) $\dfrac{\dfrac{\sqrt{6} \cdot V}{\pi}}{\dfrac{\sqrt{2} \cdot V}{\pi}} = \sqrt{3}$;

b) $\dfrac{\sqrt{\dfrac{2}{3}} \cdot V}{\dfrac{\sqrt{2}}{3} \cdot V} = \sqrt{3}$

14.17

a) $B_k = \dfrac{4 \cdot V}{\pi \cdot k}$ (sin k x 5.46 + sin k x 16.18 + sin k x 38.2 + sin k x 53.499 +

sin k x 60 + sin k x 76.18 + sin k x 81.8 − sin k x 6.501 − sin k x 20.2

− sin k x 46.608 − sin k x 54.54 − sin k x 73.392 − sin k x 80.2)

$B_1 = 0.8651 . V_t$; $B_3 = 0$; $B_5 = 0.00425 . V_t$; $B_7 = 0.003861 . V_t$; $B_9 = 0$;
$B_{11} = -0.01542 . V_t$; $B_{13} = 0.275 . V_t$; $B_{15} = 0$; $B_{17} = -0.2765 . V_t$;
$B_{19} = 0.0143 . V_t$; $B_{21} = 0$; $B_{23} = 0.00304 . V_t$; $B_{25} = -0.006 . V_t$; $B_{27} = 0$;
$B_{29} = 0.1577 . V_t$; $B_{31} = 0.1561 . V_t$

b) $$V_{Leff} = \frac{V_t}{\sqrt{2}} . \sqrt{B_1^2 + \ldots\ldots + B_{31}^2} = 224V$$

c) $V_{1eff} = 198.8V$; $V_{1eff} = 0.612 . V_t$

14.18 The output voltage of a unipolar bridge inverter contains less harmonics than the output of a bipolar inverter

14.19 The order of the harmonics is given by:

N ; $N \pm 2$; $N \pm 4$; $2.N \pm 1$; $2.N \pm 3$; ...; $3.N$; $3.N \pm 2$; .....

so that with $N = 15$ there are harmonics of 11, 13, 15, 17, 19, 27, 33,...times the frequency of the fundamental harmonic

14.20 $V_{o35} = 212.13 \times 0.033 = 7V$ at 2100Hz

ENGLISH	DUTCH	GERMAN	SPANISH
abscissa	abscis	Abszisse	abscisa
AC controller	wisselstroominsteller	Wechselstromsteller	Controlador AC
acceleration torque	versnellingskoppel	Beschleunigungsmoment	torque de aceleración
AC-controller	AC controller	AC-Kontroller	Controlador AC
active component	actieve componente	aktive Komponente	componente activo
active power	aktief vermogen	Wirkleistung	potencia activa
aerodynamic	aërodynamisch	aerodynamisch	aerodinámico
air gap	luchtspleet	Luftspalte	entrehierro
algorithm	algoritme	Algorithmus	algoritmo
alternating current (AC)	wisselstroom	Wechselstrom	corriente alterna
ambient temperature	omgevingstemperatuur	Umgebungstemperatur	temperatura ambiente
amplifier	versterker	Verstärker	amplificador
analogons	analogons	analoges Modell	modelo analógico
analogue computer	analoge computer	Analogrechner	ordenador analógico
analogue	analoog	analog	análogo
angular displacement sensor	hoekstandopnemer	Winkelencoder	sensor de posición angular
angular position	hoekpositie	Winkelposition	posición angular
anode	anode	Anode	ánodo
anodizing	geanodiseerd	eloxiert	anodizado
antistatic packaging	antistatische verpakking	antistatische Verpackung	Empaque antiestatico
apparent power	schijnbaar vermogen	Scheinleistung	potencia aparente
appliance motor	huishoudelijke motor	Haushaltmotor	motor doméstico
applications	toepassingen	Anwendungen	aplicaciones
arc furnace	boogoven	Lichtbogenofen	horno de arco
armature reaction	ankerreactie	Ankerrückwirkung	reacción de armadura
armature	anker	Anker	armadura
asynchronous motor	asynchrone motor	Asynchronmaschine	motor asincrono
auxiliary pole	hulppool	Hilfspol	polo auxiliar
auxiliary thyristor	doofthyristor	Löschthyristor	tiristor auxiliar
auxiliary winding	hulpwikkeling	Hilfswicklung	arrollamiento auxiliar
avalanche diode	zenerdiode	Zenerdiode	diodo zener
average resistivity	soortelijke weerstand	spezifischer Widerstand	resistividad media
average value	gemiddelde waarde	(arithmetischer) Mittelwert	valor medio
back-to-back (SCRs)	antiparallel (SCRs)	antiparallele Ventilen	SCRs en antiparalelos
ballast	ballast	Vorschaltgerät	balast
band gap (energy gap)	bandafstand	Energielücke (Bandabstand)	banda prohibida
bandwidth	bandbreedte	Bandbreite	anchura de banda
bank of accumulators	accubatterij	Akkumulatorenbatterie	bateria de acumuladores
barrier layer	sperlaag	Sperrschicht	capa de barrera
base voltage	basisspanning	Basisspannung	tensión de base
base	basis	Basis	base
battery charger	batterijlader	Batterieladegerät	cargador de batería
biasing	voormagnetisatie	Vormagnetisierung	polarización magnética
bidirectional	bidirectioneel	bidirektional	bidireccional
bipolar transistor	bipolaire junctietransistor	bipolarer Transistor	transistor bipolar
bistable	bistabiel	bistabil	biestable
block diagram	blokschema	Blockschaltbild	diagrama duncional
block orientated	blokgeoriënteerd	blockorientiert	bloque orientación

blocking state SCR	sperren SCR	löschen Thyristor	estado de bloque directo
blocking voltage	blokkeerspanning	Blockierspannung	tensión de bloquear
blocking	blokkering	Sperrung	bloqueo
blow up	opblazen	instabilisieren	desbordamiento
brake (chopper)	rem (chopper)	Bremse (Zehrhacker)	freno (recortador)
breakaway torque	lostrekkoppel	Losbrechdrehmoment	par de despegue
break-over voltage	doorslagspanning	Kippspannung	tensión de voltear
breakover	doorslag	Durchbruch	(tensión de) disparo
break-over	kippen	kippen	voltear
bridge	brug	Brücke	puente
brushless motor	borstelloze motor	Bürstenlose Motor	motor sin escobillas
burst firing	periodesturing	Schwingungspaketsteuerung	Control de periodo
cadmiumsulfide	cadmiumsulfide	Kadmiumsulfid	sulfuro de cadmio
capacitor (motor)	condensator (motor)	Kondensator (motor)	(motor con)condensador
capacity	capaciteit	Kapazität	capacidad
case (package)	behuizing	Behausung	caja
cathode	kathode	Kathode	cátodo
cell density	celdichtheid	Zelledichte	Densidad de celula
centrifugal switch	centrifugaal schakelaar	zentrifugaler Schalter	contacto centrifugada
channel	kanaal	Kanal	canal
characteristics	karakteristieken	Kennlinien	características
charge (to)	opladen	aufladen	carga
charging current	laadstroom	Ladestrom	corriente de carga
choke	smoorspoel	Glättungsdrossel	bobina de choque
chopper	hakker	Zehrhacker /Gleichstromsteller	recortador
circuit	kring	Kreis	circuito
circuit	schakeling	Stromkreis	circuito
circular frequency	pulsatie (hoekfrequentie)	Kreisfrequenz	pulsacion
circulating current-free	kringstroomvrij	kreisstromfrei	sin corriente circular
clamp diode	klemdiode	Klemmdiode	diodo de libre reccorido
code disc	codeerschijf	Kodierungscheibe	disco de codigo
coercitivy	coërcitieve kracht	Koerzitivkraft	coercitividad
coil	spoel	Drossel / Spule	bobina de inductancia
collector	collector	Kollektor	colector
colour code	kleurencode	Farbekode	código de colores
commissioning rules	instelregels	Optimierungsvorschriften	reglas de ajuste
commutation	commutatie	Kommutierung	conmutación
commutatormachine	commutator machine	Kommutatormaschine	máquina de comutación
comparator	comparator	Vergleicher (Komparator)	comparador
compatibility	uitwisselbaarheid	Kompatibilität	compatibilidad
compensation coil	compensatiewikkeling	Kompensationswicklung	arrollamiento de compensación
compliance	compliantie	Nadelnachgiebigkeit	cumplimiento
compressor	compressor	Kompressor	compresor
conducting pattern	geleidingspatroon	Leitungsvorlage	modelo de conducción
conduction angle	geleidingshoek (vloeihoek)	Flusswinkel	ángulo de conducción
conductor	geleider	Leiter	conductor
connection wire	aansluitdraad	Anschlussdraht	conductor del conexión
connection	aansluiting	Anschluss	conexión
connection	schakeling	Schaltung	conexión

constant current (source)	constante (stroombron)	Konstantstromquelle	fuente de alimantación de corriente constante
construction (structure)	opbouw	Aufbau	estructura
consumer	verbruiker	Verbraucher	consumidor
contactor	contactor	Magnetschalter	contactor
continuous (duty)	continu (bedrijf)	Dauer (betrieb)	servicio continuo
continuous current	continue stroom	nichtlückender Betrieb	corriente permanente
continuous service	leemtevrij bedrijf	nichtlückender Betrieb	corriente continua
control (thyristor)	controle (thyristor)	Steuersätzer (Stromrichter)	control de tiristor
control characteristic	stuurkarakteristiek	Steuerkennlinie	caracteristic de control
control circuit	stuurcircuit	Steuerungskreis	circuito de excitación
control circuitry	triggerketen	Steuersatz	circuito de control
control IC	stuur-IC	integrierte Zündimpulsstufe	IC de control
control loop	regelkring	Regelkreis	bucle de control
control signal	controlesignaal	Steuersignal	control señal
control	besturing	Steuerung	control
controller	regelaar	Regler	regulador
convergence	convergentie	Konvergenz	convergencia
converter	convertor	Stromrichter	convertidor
cooking plate	kookplaat	Kochplatte	hornillo
cooling fin	koelvin	Kühlrippe	aleta de refrigeración
cooling	koeling	Kühlung	refrigeración
coordinate system	assenkruis	Systemkoordinatensystem	coordenadas
copper losses	koperverliezen	Kupferverluste	pérdidas de cobre
core	kern	Kern	núcleo
counter emf	tegen-emk	gegen emk	fuerza contra electromotriz
criterion (equal surface)	criterium (gelijke opp.)	Kriterium	criterio
current	stroom	Strom	corriente
current (rate of rise)	stroom(steilheid)	Strom (Anstiegsteilheit)	corriente (pendiente)
current circuit	stroomkring	Stromkreis	circuito de corriente
current control	stroomregeling	Stromregulierung	control de corriente
current density	stroomdichtheid	Stromdichte	densidad de corriente
current direction	stroomrichting	Stromrichtung	sentido de corriente
current flow	stroomsterkte	Stromstärke	ensidad de corriente
current form	stroomvorm	Ausbildung der Ströme	forma de la corriente
current gain factor	stroomversterkingsfactor	Stromverstärkung	factor del amplificación de la corriente
current limit	stroombegrenzing	Stromgrenzwert	límite de corriente
current propagation	stroomverloop	Stromverlauf	propagación de corriente
current source (controlled)	gestuurde stroombron	gesteuerte Stromquelle	fuente de corriente controlado
current source inverter	stroombroninvertor	Umrichter mit Zwischenkreis	invertor de fuente de corriente
current surge	stroomstoot	Stromstoss	golpe de corriente
curve	curve	Kennlinie	curva
cut off	gesperd	Sperrzustand	estado de bloqueo
cycloconvertor	cycloconverter	Direktumrichter	convertidor ciclón
dam (model)	dam-mode	Damm-Modell	dama (modelo)
damping	demping	Dämpfung	amortiguamiento
data (sheet)	gegevens (-blad)	Daten (-blatt)	datos (- página)

DC motor	gelijkstroommotor	Gleichstrommotor	motor corriente continua
DC voltage link	spanningstussenkring	Spannungszwischenkreis	circuito intermedio de tensión
DC-controller	mutator	Stromrichter	mutador
DC-link converter	tussenkringomzetter	Zwischenkreisumrichter	regulador de frecuencia con circuito intermedio
dead time	dode tijd	Totzeit	tiempo muerto
delay angle	ontsteekhoek	Zündwinkel	el ángulo de encendido
delay dissipation time	ontsteekuitbreidingstijd	Zündausbreitungszeit	(delay dissipation time)
delay of firing	ontsteekvertraging	Zündverzögerung	duración de precebado
delay time	looptijd	Laufzeit	retardo
delay time	vertragingstijd	Einschaltverzugszeit	tiempo de retardo
delay	vertraging	Verzögerung	retardo
depletion region	verarmingslaag	Übergangsgebiet	región de transición
derivative	afgeleide	Differentialquotient	derivada
detection	detectie	Demodulation	detección
detent torque	restkoppel	Rastmoment	par de detencion
diac	diac	Zweiwegschaltdiode	diac
diagram	diagram	Diagramm	diagrama
diagram	schema	Schaltbild	diagrama
dielectric	diëlectricum	Dielektrikum	dieléctrico
differential amplifier	verschilversterker	Differenzverstärker	amplificador diferencial
differential equation	differentiaalvergelijking	Differentialgleichung	ecuación diferencial
digital computer	digitale computer	digitale Rechenanlage	digital computador
digital	digitaal	digital	digital
direct current (DC)	gelijkstroom	Gleichstrom	corriente continua
discharge current	ontladingsstroom	Entladestrom	corriente de descarga
discharge lamp	ontladingslamp	Entladunglampe	lámpara de descarga
discontinuous current	stroomleemte	Lückgrenze	corriente discontinua
displacement factor	verschuivingsfactor	Phasenverschiebungswinkel	factor de desviación de fase
displacement rule	verschuivingsregel	Verschiebungsvorschrift	(displacement rule)
distortion factor	distorsiefactor	Verzerrungsfaktor	distorsión factor
distortion	vervorming	Verzerrung	distorsión
disturbance	storing	Störung	perturbación
doping	doteren	dotieren	dopado
double pulse	dubbelpuls	Doppelimpuls	impulso doble
double star connection	dubbele sterschakeling	Doppelsternschaltung	conexión en doble estrella
drain (current)	drain (stroom)	Drain (strom)	(corriente de) drenaje
drain	afvoerelektrode	Abfluss (drain)	drenador
drift velocity	driftsnelheid	Driftgeschwindigkeit	velocidad de deriva
drive system	aandrijfsystemen	Antriebe	sistema de propulsión
drive system	aandrijvingen	Antriebe	accionamiento
driver stage	stuurtrap	Treiberstufe	excitador
driving the base	uitsturen basis	steuern Basis	control de base
driving voltage	drijvende spanning	treibende Spannung	(driving voltage)
duty cycle	werkverhouding	Arbeitszyklus	ciclo de trabajo
duty	bedrijf	Betrieb	servicio
dynamic behavior	dynamisch (gedrag)	dynamisches (Verhalten)	conducta dinámica

dynamo	dynamo	Gleichstromgenerator	dinamo
eddy (currents)	wervelstroom (verliezen)	Wirbelstrom (verluste)	corriente Foucault (pérdidas)
efficiency	rendement	Wirkungsgrad	rendimiento
elasticity	elasticiteit	Elastizität	elasticidad
electric charge	lading	Ladung	carga
electric equivalent	elektrisch equivalent	Ersatzschaltbild	eléctrica equivalente
electric field	elektrisch veld	elektrisches Feld	campo eléctrico
electric machine	elektrische machine	elektrische Maschine	máquina eléctrica
electric motor	motor	Elektromotor	motor électro
electricity	elektriciteitsleer	Elektrizität	electricidad
electrochemical	elektro-chemisch	elektrochemisch	electroquímico
electrolysis	elektrolyse	Elektrolyse	electrólisis
electromagnetic	elektromagnetische	elektromagnetisch	elecromagnética
electromechanical	elektromechanisch	elektromechanisch	electromecánico
electronic motor control	motorcontrole	elektronische Motorkontrolle	control electrónica de motores
electrostatic (shield)	elektrostatisch (scherm)	elektrostatische Abschirmung	(pantalla) electrostático
elektrode	electrode	Elektrode	electrodo
elementary charge (electron)	elementaire lading	Elementarladung	carga elementaria
emergency supply	noodvoeding	Notspeisung	grid de reserva
emf (counter-)	emk (tegenwerkende-)	emk (Gegen-)	f.e.m. (contra)
emitter	emitter	Emitter	emisor
enamelled winding wire	wikkeldraad (geëmailleerd)	Lackdraht	hilo esmaltado para bobina
encoder	encoder	Kodierer	codificador
encremental	incrementeel	inkremental	incremental
energy (renewable)	energie (hernieuwbaar)	Energie (erneute)	energía (renovada)
energy converter	energie-omzetter	Umrichter elektrischer Energie	energía inversor
enhancement (FET)	verrijkingstype (FET)	Anreicherungstyp (FET)	(FET) incrementado
enhancement type	verrijkingstype	Anreicherungstyp	tipo de enriquecimiento
equal surface area	gelijke oppervlakte	gleichbleibende Oberfläche	superficie igual
equation	uitdrukking (vergelijking)	Beziehung (Relation/Gleichung)	ecuación
equivalent (circuit)	vervangings(schema)	Ersatz(schaltbild)	(diagrama) equivalente
equivalent resistance	vervangingsweerstand	Ersatzwiderstand	resistencia equivalente
excerpts from databook	uittreksels uit databoek	Auszug aus Datenbuch	extractos de datos
excit. from coupled exciter	afzonderlijk bekrachtigd	Eigenerregung	excitación separada
excitation	excitatie	Erregung	excitación
excite	bekrachtigen	erregen	excitación
external P-layer	uitwendige P-laag	äussere P-Schicht	(external P-layer)
extinction angle	doofhoek	Löschwinkel	ángulo de extinción
fall time	daaltijd (afvaltijd)	Abfallzeit	tiempo de caída
fast recovery diode	snelle hersteldiode	Schnelle Sperrdiode	(fast recovery diode)
fault current	foutstroom	Teilfehlerstrom	corriente de falta
feedback	terugkoppeling	Rückkopplung	realimentación
feeder clamp	voedingsklem	Stromklemme	grifa de alimentación
feeding cable	voedingsleiding	Speiseleitung	arteria alimentadora
ferrite core	ferrietkern	Ferritkern	núcleo de ferrita
ferromagnetic	ferromagnetisch	ferromagnetisch	ferromagnético
field control	veldregeling	Feldregelung	control de campo
field coordinates	veldcoördinaten	Feldkoordinate	coordenadas de campo

field intensity	veldsterkte	Feldstärke	intensidad de campo
field line	veldlijn	Feldlinie	linea de campo
field weakening	veldverzwakking	Feldschwächung	disminución de campo
filter	filter	Filter	filtro
final value theorem	eindwaardetheorema	Endwerttheorem	teorema del valor final
finit	eindig	endlich	finito
firing	ontsteken	zünden	encendido
first order system	eerste-orde systeem	Verzögerungsgliede 1.Ordnung	sistema del primer orden
flashing light	flikkerlicht	Funkelfeuer	luz centelleante
floating net	zwevend net	erdfreies Netz	red flotante
fluorescent lamp	fluorescentielamp	Fluoreszenzlampe	lámpara fluorescente
flux	flux	Fluss	flujo
forced commutation	gedwongen commutatie	gezwungene Kommutierung	conmutación forzado
form factor	vormfactor	Formfaktor	factor de forma
forward (current)	doorlaatrichting (stroom)	Durchlass(strom)	dirección de paso (corriente)
forward characteristic	doorlaatkarakteristiek	Durchlasskennlinie	característica directa
forward	voorwaarts	Durchlass(Vorwärtsrichtung)	dirección de paso
Fourier components	Fouriercomponenten	Fourier-Komponenten	Fourier componentes
Fourier series	Fourierreeks	Fourier-Reihe	Fourier serie
Fourier series	reeks van Fourier	Fourier-Reihe	serie de Fourier
freewheel diode	vrijloopdiode	Freilaufdiode (Löschdiode)	diodo de libre reccorido
frequency (converter)	frequentie(regelaar)	Frequenz(umrichter)	(regulador de) frecuencia
frequency spectrum	frequentiespectrum	Frequenzspektrum	espectro de frecuencias
friction (coefficient)	wrijving (coëfficiënt)	Friktion (koeffizient)	fricción (coeficiente)
full controlled	volgestuurd	vollgesteuert	(full controlled)
full wave (rectifier)	dubbelzijdig (gelijkrichter)	Zweiweg (-gleichrichter)	(rectificador de) onda completa
function	functie	Funktion	función
fundamental wave	grondgolf	Grundschwingung	onda fundamental
fuse	smeltveiligheid	Sicherung	fisible
gallium arsenide	galliumarsenide	Galliumarsenid	arseniuro de galio
gate (terminal)	poort(aansluiting)	Gate(-Anschluss)	(borne de) puerta
gate current	poortstroom	Gatestrom	corriente de puerta
gate power	poortvermogen	Gateleistung	potencia de puerta
gate turn off thyristor	uitschakelbare thyristor	Abschaltthyristor (GTO)	tiristor bloqueable
gate voltage	poortspanning	Steuerspannung	tensión de puerta
gate	poort	Gate / Gatter	puerta
generator mode	generatorbedrijf	Generatorbetrieb	funcionando como generador
gondola	gondel	Gondel	góndola
grid / mains / supply	net	Netz	red
ground wave	grondgolf	Grundschwingung	onda fundamental
half wave	halve netperiode	Netzhalbwelle	red medio ciclo
half-wave, three-phase	enkelzijdig, driefasen	Einweg, Dreiphasen	media onda, trifásica
Hall-effect	Hall effect	Hall-Effekt	efecto Hall
halogenlamp	halogeen lamp	Jodlamp	lámpara halógena
harmonics (order)	harmonischen (orde)	harmonische (Ordnungszahl)	armonicas (orden)
heat dissipation	warmte-afvoer	Wärmeableitung	disipación de calor

heatsink	koellichaam/koelplaat	Kühlkörper	disipador
heavy saturation	harde verzadiging	harte Sättigung	saturación fuerte
high voltage	hoogspanning	Hochspannung	alta tensión
hoisting apparatus	hefwerktuig	Hebemaschine	utensilio de leventados
holding current	houdstroom	Haltestrom	corriente de cierre
holding torque	houdkoppel	Haltemoment	torque de mantenimiento
holes and electrons	gaten en elektronen	Löcher und Elektronen	huecos y electrones
hot spots	warme punten	heisse Punkte	(hot spots)
human eye curve	oogbalcurve	spektrale Hellempfindlichkeit	sensibilidad espectro del ojo
hydraulical (labo)	hydraulisch laboratorium	hydraulisches Laboratorium	(labo) hydraulicamente
hyperbole	hyperbool	Hyperbel	hipérbola
hysteresis loss	hysteresisverliezen	Hysteresisverlust	pérdidas de histéresis
I^2-t integral	lastintegraal	Lastintegral	I^2-t integral
impedance	impedantie	Impedanz	impedancia
Impulse	impuls	Impuls	impulso
impurity	verontreiniging	Verunreinigung	impureza
induced emf	inductiespanning	Induktionsspannung	f.e.m. de inducción
inductance	inductantie	Selbstinduktion	inductancia
induction motor	inductiemotor	Induktionsmotor	motor de inducción
inductive heating	inductieve verwarming	Induktionsheizung	calentamiento inductivo
inductive load	inductieve belasting	induktiver Last	carga inductiva
industry standard	industriestandaard	Industriestandard	estándar industrial
inertia force	inertiekracht	Trägheitskraft	fuerza inercia
inertia resistance	traagheidskracht	Trägheitswiderstand	Fuerza de inercia
infra-red	infrarood	ultrarot	infrarrojo
initial value theorem	beginwaardetheorema	Anfangwerttheorem	teorema del valor inicial
input	ingang	Eingang	entrada
instant	tijdstip	Zeitpunkt	instante
instantaneous value	momentele waarde	Momentanwert	valor instantáneo
instantaneous value	ogenblikkelijke waarde	Momentanwert	inmediatamente
instrument transformer	meettransformator	Messwandler	transformador de medida
insulator	isolator	Isolator	aislador
integrated circuit	geïntegreerde schakeling	integrierte Schaltung	microcircuito
integration	integreren	Integration	integrar
interference suppression	ontstoring	Entstörung	supresión de interferencias
interference	interferentie	Interferenz	interferencia
intersection	snijpunt	Schnittpunkt	punto de interseccion
inversion layer	inversielaag	Inversionsschicht	capa de inversión
inverter	invertor	Wechselrichter	inversor
inverter	wisselrichter	Umkehrstromrichter	ondulador
iron losses	ijzerverliezen	Eisenverluste	pérdidas en el hierro
iteration	iteratie	Iteration	iteración
Joule losses	Joule-verliezen	Joule-Verluste	perdidas de Joule
Joule-effect	Joule effect	Joule-Effekt	efecto de Joule
junction	junctie	Sperrschicht (PN-Übergänge)	Unión PN
junction diode	junctiediode	Flächendiode	diodo de Union PN
kinetic (energy)	kinetische (energie)	kinetische (Energie)	cinétic (energia)

knee voltage	knik of knie	Kniespannung	tensión de rodillo
laminated core	blikpakket	Blechpaket	nucleo de chapas
Laplacetransform	Laplace-transformatie	Laplace-Transformation	transformada de Laplace
latching current	vergrendelstroom	Einraststrom	corriente de retención
leading screw transmission	schroefspiloverbrenging	Leitspindelübertragung	transmissión por pivote de tornillo
leadsulphite	loodsulfide	Bleisulfid	sulfito de plomo
leakage current	lekstroom	Leckstrom	corriente de fuga
leakage inductance	spreidingsinductantie	Streuinduktivität	inductancia de fuga
light (guide)	licht (geleider)	Licht (führung)	(guía de) luz
light pulse	lichtpuls	Lichtimpuls	impulso de luz
lightbeam	lichtbundel	Strahlenbündel	haz luminoso
lighting	verlichting	Beleuchtung	iluminación
limitation resistance	begrenzingsweerstand	Begrenzungswiderstand	Resistencia de limitación
line(current)	lijn(stroom)	Leiter(strom)	(corriente de) linea
linear	lineair	linear	lineal
load impedance	belastingsimpedantie	Lastimpedanz	impedancia de carga
load line	belastingslijn	Arbeitskennlinie	linea de carga
load voltage	klemspanning	Klemmenspannung	tensión en carga
load	belasting	Last	carga
low voltage	laagspanning	Niederspannung	baja tensión
luminous intensity	lichtsterkte	Lichtstärke	intensidad de luz
machine	machine	Maschine	máquina
magnetic (flux)	magnetische (flux)	Magnet (fluss)	(flujo) magnético
magnetic charging	magnetisch opladen	magnetisches Aufladen	cargar magnetico amente
magnetic inductance	magnetiseringsinductantie	magnetische Selbstinduktion	inductancia de magnetización
magnetomotive force	magnetomotorische kracht	magnetomotorische Kraft	Fuerza magnetomotriz
main pole	hoofdpool	Hauptpol	polo principal
majority- carriers	meerderheidsladingsdragers	Majoritätsträger	portadores mayoritarios
mass inertia	massatraagheidsmoment	Massenträgheitsmoment	momento de inercia de masa
matrix	matrix	Matrix	matriz
maximum current	maximale stroom	Spitzenstrom	corriente máxima
maximum reverse current	maximale sperstroom	überhöhter Sperrstrom	corriente inversa máxima
maximum value	amplitude	Scheitelwert	amplitud
measurement equipment	meetapparatuur	Messgerät	equipo de medida
measurement equipment	meetopstelling	Messaufbau	Equipo de medición
mechatronics	mechatronica	Mechatronik	mecatrónica
melting energy	smeltenergie	Schmelzeenergie	energia de fundición
memory scope	geheugenscope	Speicheroszilloskop	memoria osciloscopio
metallization	metallisatie	Metallizierung	metalización
microcontroller	microcontroller	Mikrokontroller	microcontrolador
microprocessor	microprocessor	Mikroprozessor	microprocesador
minority carriers	minderheidsladingsdragers	Minoritätsträger	portadores minoritarios
mixing	mengen	mischen	mezcla
modulation depth	modulatiediepte	Modulationsgrad	modulación medio
moment of torque	werkkoppel	Drehmoment	torque de trabajo
monopolar	enkelpolig	einpolig	unipolar

Mos field effect transistor	Mos-veldeffecttransistor	Mos-Feldeffekttransistor	transistor de efecto de campo Mos
multimeter	multimeter	Vielfachmessgerät	Multimetro
multi-pulse circuits	meerpulsige schakelingen	mehrpulsige Schaltungen	Circuito multi impulsos
natural commutation	natuurlijke commutatie	natürliche Kommutierung	conmutación natural
negative feedback	tegenkoppeling	Gegenkopplung	enlace contrario
network transformer	transformator (Net)	Netztransformator	transformador de red
network	netwerk	Netzwerk	red
neutral point (transformer)	sterpunt	Sternpunkt	punto neutro
N-material	N-materiaal	N-leitendes Material	N-material
noise (electrical)	ruis (elektrisch)	Rauschen (elektrisches)	ruido (eléctrico)
no load (current)	nullast(stroom)	Leerlauf(strom)	(corriente) en vacio
no load speed	leegloop toerental	Leerlaufdrehzahl vacio	velocidad de marcha en
nominal current	nominale stroom	Nennstrom	corriente nominal
nomograph	nomogram	Nomogramm	nomograma
notes	notaties	Notationen	apuntes
number indicator	cijferindicator	Zifferindikator	indicado de número
numeric example	cijfervoorbeeld	Rechenbeispiel	ejemplo cifrado
numerical analysis	numerieke technieken	numerische Mathematik	analísis numérico
ohmic load	ohms belast	mit Ohmischer Last	carga óhmica
ON	AAN	EIN-Zustand	(ON)
on-characteristic	aan-karakteristiek	Durchlasskennlinie	(on-characteristic)
opamp	opamp	Operationsverstärker	amplificador operacional
open circuit test	nullastproef	Leerlaufversuch	ensayo en circuito abierto
open loop control	sturing	Steuerung	control de buclo abierto
operation	werking	Arbeitsweise	operación
optimum amount	bedragoptimum	Betragsoptimum	óptimo de comportamiento
optocoupler	optokoppeling	Optoelektronisches	(optocoupler)
ordinate	ordinaat	Ordinate	ordenada
oscillation	oscillatie	Schwingung	oscilación
OUT	UIT	AUS-Zustand	(OUT)
output line	uitgangslijn	Ausgangsleitung	linea de salida
output stage	eindtrap	Endstufe	etapa de salida
output	uitgang	Ausgang	salida
oven (furnace)	oven	Ofen	horno
overlap angle	overlappingshoek	Überlappungswinkel	ángulo de solape
overshoot	doorschot	Überschwingweite	sobre exceso
overvoltage	overspanning	Überspannung	sobretensión
partial fraction	partieelbreuk	partieller Bruch	(partial fraction)
peak value	piekwaarde	Spitzenwert	valor de cresta
peak value	topwaarde	Scheitelwert	valor de cresta
period	periode	Periode	periodo
permanent magnet	permanente magneet	Dauermagnet	imán permanente
permeability	permeabiliteit	Permeabilität	permeabilidad
phase conductor	fasegeleider	Phasenleiter	conductor de fase
phase control	fase-aansnijding (controle)	Phasenanschnitt(steuerung)	control de fase
phase displacement	faseverschuiving	Phasenverschiebung	desfase
phase drop off	fase-afsnijding	Abschnittsteuerung	(phase drop off)

phase	fase	Phase	fase
phase-sequence	fasevolgorde	Phasenfolge	orden de fases
photomask	fotomasker	Photomaske	fotomáscara
photosensitive	fotogevoelig	photoempfindlich	fotosensible
photovoltaic cell	zonnecel	Photovoltaikzelle	célula fotovoltaica
photovoltaic	fotovoltaïsch	photovoltaik	fotovoltaico
pilot signal lamp	signaallamp	Signallampe	lámpara de señal
PNPN diode	vierlagendiode	Vierschichtdiode	diodo PNPN
polarized	gepolariseerd	polarisiert	polarizado
pole pairs (number of)	poolparen (aantal)	Polpaare (Zahl)	pares de polos (número de)
pole	pool	Pol	polo
position (measurement)	positie(meting)	Position(-smessung)	(medidade la) posición
potential energy	potentiële energie	potentielle Energie	energia potencial
potentiometer	potentiometer	Potentiometer	potenciómetro
power (dirty)	netvervuiling	Fehlerstrom	contaminación del red ?
power circuit	vermogenkring	Hauptkreis	circuito de potencia
power factor	arbeidsfactor	Leistungsfaktor	factor de potencia
power losses	vermogenverlies	Leistungsverluste	perdidas de potencia
power mosfet	vermogenmosfet	Leistungsmosfet	MOSFET de potencia
power supply	stroomvoorziening	Stromversorgung	sistema de alimentación
power supply	voeding	Netzteil	alimentación
primary	primaire zijde	Primärseite	circuito primario
principle configuration	principe-opstelling	Prinzipschaltbild	conexión de principio
pulley	riemschijf	Riemscheibe	polea
pulse train	impulstrein	Pulszug / Impulsfolge	tren de pulsos
pulse-shaping circuit	pulsvormer	Impulsformer	modelador de impulsos
pump	pomp	Pumpe	bomba
push-button	duwknop	Druckknopf	botón pulsador
PWM wave	PBM-golf	PBM-Wellenform	onda con modulación de anchura por impulso
quadrant	kwadrant	Quadrant	cuadrante
quelle (Source)	source	Quelle	fuente
radian	radiaal	radial	tradián
radiator	radiator	Radiator(Heizkörper)	radiador
radio frequency	Hoogfrequent	Hochfrequenz	alta frecuencia
radio receiver	radio-ontvanger	Rundfunkgerät	receptor de radio
rate of rise	steilheid	Steilheit	pendiente
ratings	grenswaarden	Grenzdaten	especificaciones máximas
reactive component	reactieve componente	Blindkomponente	componente reactivo
reactive moment	tegenwerkend koppel	Widerstandsmoment	torque de carga
reactive power	blind vermogen (wattloos)	Blindleistung	energía reactiva
receiver	ontvanger	Empfänger	receptor
recovery time	herstel (tijd)	Erholungszeit	tiempo de restitución
rectangular wave	rechthoeksgolf	Rechteckschwingung	onda rectangular
rectifier (B_2) (B_6)	mutator (B_2) (B_6)	Brückenschaltung (B_2) (B_6)	rectificador regulado
rectifier	gelijkrichter	Gleichrichter	rectificador
redundancy	redundantie	Redundanz	redundancia
regenerative (braking)	recuperatie (remmen)	rekuperatives Bremsen	frenado regenerativo
regulated process	proces(geregeld)	Regelstrecke	campo de regulación
regulation	regeling	Regelung	regulación
reluctance motor	reluctantiemotor	Reluktanzmotor	motor de reluctancia

residual voltage	restspanning	Restspannung	tensión residual
resistance heated furnace	weerstandsoven	Widerstandsofen	horno de resistencia
resistance	weerstand	Widerstand	resistencia
resistive	resistief	Ohms	resistivo
resonance	resonantie	Resonanz	resonancia
response	responsie	Ansprechverhalten	respuesta
retardation test	uitloopproef	Auslaufversuch	ensayo de desaceleración
retarding torque	vertragingskoppel	Verzögerungsmoment	torque de desaceleracion
reverse (current)	inverse (sperstroom)	negativer Sperrstrom	corriente inversa
reverse	invers	Rückwärtsrichtung	inverso
ring core	ringkern	ringförmiger Kern	núcleo anular
ripple	rimpel	Welligkeit	ondulación
rise time	stijgtijd	Anstiegzeit	tiempo de subida
rise time	doorschakeltijd	Durchschaltzeit	tiempo de ascenso
RMS	kwadratisch gemiddelde	Effektivwert	valor eficaz
root-mean-square value	effectieve waarde	Effektivwert	valor eficaz
rotating field	draaiveld	Drehfeld	campo giratorio
rotation	rotatie	Rotation	rotacional
rotor (frequency)	rotor (frequentie)	Läufer (frequenz)	(frecuencia del) rotor
safety facto	veiligheitsfactor	Sicherheitsfaktor	factor de seguridad
safety transformer	veiligheidstransformator	Sicherheitstransformator	transformador de seguridad
salvo	salvo	Salve	salva
saturation (hard)	verzadiging (hard)	Sättigung (hart)	saturación (duro)
saturation (quasi)	verzadiging (quasi)	Sättigung (quasi)	saturación (casi)
saw tooth (voltage)	zaagtand (spanning)	Sägezahn (spannung)	dente de sierra (tensión)
scale model	schaalmodel	Modell wirkliche Grösse	escala modelo
Scott transformer	Scott-transformator	Transfo in Scott-Schaltung	transformador de Scott
SCR	eénrichtingsthyristor	Einwegthyristor (SCR)	tiristor (SCR)
SCR(controlled rectifier)	gestuurde gelijkrichter	netzgeführter Stromrichter	rectificador controlado
screen grid vacuum tube	schermroosterbuis	Schirmgitterröhre	pentodo
screw tap	draadtap	Schraubenbohrer	(screw tap)
secondary	secondaire	Sekundärseite	secundario (transfo)
selector switch	omschakelaar	Umschalter	conmutador selector
self-excitation	zelfbekrachtiging	Selbsterregung	auto-excitación
self-inductance	zelfinductie	Selbstinduktivität	inductancia propia
semiconductor	halfgeleider	Halbleiter	semiconductor
sensor	opnemer	Messfühler	captador
sensor	voeler	Fühler, Sensor	captor, sensor
series motor	seriemotor	Reihenschlussmotor	motor serie
series resistor	voorschakelweerstand	Vorschaltwiderstand	resistencia adicional
Servomechanism	servomechanisme	Stellantrieb	servomecanismo
set value	ingestelde waarde	Sollwertführungsgrösse	valor especificado
settling time	inslingertijd	Ausregelzeit	tiempo de corrección
shaded pole motor	spleetpoolmotor	Spaltpolmotor	motor de anillos de desfase
short circuit to earth	aardsluiting	Erdschluss	cortocircuito a tierra
short circuit voltage	kortsluitspanning	Kurzschluss-Spannung	tensión de cortocircuito
signal line	signaallijn	Signalleitungen	linea de señal
simulation	simulatie	Simulation	simulación
sine	sinus	Sinus	seno
sine wave	sinusgolf	sinusförmige Welle	onda sinusoidal

single-phase, half-wave	éénfase, enkelzijdig	Einphase, Einweg	monofásico, media onda
six pulse configuration	zespulsig	sechspulsige Schaltung	circuito con seis impulsos
skin effect	skineffect	Skineffekt	efecto pelicular
slew rate	flanksteilheid	Anstiegsgeschwindigkeit	ritmo limitado
slip ring (armature)	sleepring (anker)	Schleifring (läufer)	anillo de fricción (rotor)
slotted armature	trommelanker	genuteter Anker	inducido dentado
smoothed	afgevlakt	geglättet	nivelada
smoothing capacitor	afvlakcondensator	Glättungskondensator	capaidad de filtrado
smoothing coil	afvlakspoel	Glättungsdrossel	bobina de filtrado
snubber (RCD)	RCD-snubber	RCD-snubber	(snubber)
snubber circuit	snubber	Snubber Netzwerk	circuito (snubber)
solar cell array	zonnepaneel	Solarzellenanordnung	campo fotovoltaico
soldering iron	soldeerbout	Lötkolben	soldador
solid state relay	solid state relay	Halbleiterrelais (statisches Relais)	(solid state relay)
source	bron	Quelle	fuente
space charge	ruimtelading	Raumladung	carga de espacio
space vector	ruimtevector	Raumvektor	vector espacial
specification	specificatie	Anforderung (Spezifikation)	especificaciones
spectral response	spectrale gevoeligheid	Spektralempfindlichkeit	sensibilidad espectral
speed profile	snelheidsprofiel	Geschwindigkeitsprofil	perfil de velocidad
speed (rated)	toerental (nominaal)	(Nenn)drehzahl	velocidad (nominal)
speed control	snelheidsregeling	Drehzahlregelung	control de velocidad
square wave	blokgolf	Rechteckwelle	onda cuadrada
squirrel cage induct. Motor	kooiankermotor	Käfigläufer-Induktionsmotor	motor de inducción de jaula
star (connection)	sterschakeling	Stern(schaltung)	conectado en estrella
start	aanlopen	anfahren / anlaufen	arranque
starting (point in) time	aanlooptijdstip	Anfahraugenblick	momento de arranque
starting torque	startkoppel	Durchdrehmoment	par de arranque
start-stop	start-stop	Start-Stopp	(start-stop)
static	statisch	statisch	estático
stator (number of poles)	stator (aantal poolparen)	Ständer (Polpaarzahl)	estator (pares de polos)
steam turbine	stoomturbine	Dampfturbine	turbina de vapor
steaty state (current)	regime (stroom)	Ausgleich (strom)	(corriente) de régimen
step accurateness	stapnauwkeurigheid	Schrittpräzision	escalón exactitud
step function	sprongfunctie	Sprungfunktion	función escalón
step function	stapfunctie	Sprungfunktion	función escalón
step response	stapresponsie	Sprungantwort	repuesta en escalón
stepper motor	stappenmotor	Schrittmotor	motor paso a paso
storage time	opslagtijd	Speicherzeit	tiempo de almacenamiento
strong inductive load	sterk inductief belast	stark induktiv belastet	carga inductiva fuerte
submarine cable	onderzeekabel	Unterseekabel	cable submarino
subscripts	voetletter	Indexzeichen	tipificación
substrate	grondlaag	Grundkörper	sustrato
substrate	substraat (grondlaag)	Substrat (Grundkörper)	sustrato
summing amplifier	sommeerversterker	Summierverstärker	amplificador sumador
supply network	distributienet	Versorgungsnetz	red de distribución
supply voltage	voedingsspanning	Speisespannung	tensión de alimentación
surface	oppervlak	Oberfläche	superficie
surge current	éénmalige piekstroom	Einschaltstromspitze	(surge current)

sustained short-circuit test	kortsluitproef	Dauerkurzschlussversuch	ensayo en cortocircuito
switch (to)	schakelen	schalten	conmutación
switch mode supply	geschakelde voeding	Schaltnetzteil	alimentación conmutado
switched regulator	geschakelde regelaar	geschalteter Regler	regulador controlado
switching (cycle)	schakel (cyclus)	Schalt (zyklus)	(ciclo de) conmutación
symbol	symbool	Symbol (Schaltzeichen)	simbolo
symmetrical optimum	symmetrisch optimum	symmetrisches Optimum	óptimo simetrico
synchron motor	synchrone motor	Drehstrom Synchronmotor	motor sincrón
synchron velocity	synchroon toerental	Synchrondrehzahl	velocidad sincrona
system	systeem	System	sistema
tail current	staartstroom	Schwanzstrom	(tail current)
teletransmission	telecommunicatie	Nachrichtentechnik	teletransmisión
television receiver	TV-ontvanger	Fernsehempfänger	receptor de televisión
temperature (control)	temperatuur(regeling)	Temperatur(regelung)	termorregulado
terminal board	klemmenbord	Klemmbrett	tablero de bornes
test	test	Test	ensayo
theorem	theorema	Theorem	teorema
thermal resistance	thermische weerstand	Wärmewiderstand	resistencia térmica
thermal resistance	warmteweerstand	thermischer Widerstand	resistividad térmica
thermistor	thermistor	Thermistor (Heissleiter)	termistor
three layer diode	drielagendiode	Dreischichtdiode	(three layer diode)
three term controller	PID-regelaar	PID-Regler	regulador de acción propor., integr. y deriv.
three-phase (bridge)	driefasen (brug)	Drehstrom (brücke)	(puente) trifásico
three-phase current	draaistroom	Drehstrom	corriente trifásica
three-phase grid	driefasennet	Drehstromnetz	red trifásico
threshold voltage	drempelspanning	Schwellenspannung	tensión umbral
thyristor (auxiliary)	hulpthyristor	Löschthyristor	auxiliar tiristor
thyristor (disc-)	thyristor (schijf-)	Thyristor(tablette)	Tiristor (disco-)
time interval	tijdsinterval	Zeitlücke	intervalo de tiempo
time ratio control	tijdsverhouding (sturen)	Steuerung mit stellbarem Zeitverhältnis	(time ratio control)
time-constant	tijdconstante	Zeitkonstante	constante de tiempo
to degauss	ontmagnetiseren	entmagnetisieren	desmagnetizar
TO3-package	TO3-omhulling	TO3-Gehäuse	TO3-encapsulado
tooth gear case	tandwielkast	Zahnradkasten	caja de engranaje
topology	topologie	Topologie	topologia
torque control	koppelregeling	Momentregelung	regulación del torque
torque	koppel	(Dreh-)Moment	par
traction motor	tractiemotor	Fahrmotor	motor de tracción
transconductance	transconductantie	Steilheit	transconductancia
transducer	meet(waarde)-omvormer	Messwertumformer	convertidor de medida
transfer function	transferfunctie	Übertragungsfunktion	función de transferencia
transformer (isolating)	scheidingstransformator	Trenntransformator	transformador de separación
transformer (core)	transformator(kern)	Transformator(kern)	núcleo magnético
transformer (electronic)	elektronische transfo	elektronischer Transformator	transf. electrónico
transformer (one coil)	spaartransformator	Autotransformator	transformador de ahorro
transformer (supply)	(voedings)transformator	(Versorgungs-)Transformator	transformador (de distribución)

transformer coupling	transformatorkoppeling	Transformatorkopplung	acoplamiento por transformador
transformer ratio	transformatieverhouding	Übersetzungsverhältnis	relación de transformación
transient(term)	overgangs(term)	Ausgleich(term)	(término de) transición
transmitter	zender	Sender	transmisor
trapezium converter	trapeziumconvertor	Trapezumrichter	trapecio convertidor
triac	triac	Zweiwegthyristor	tiristor triac
triangular wave	driehoeksgolf	Dreieckspannung	onda triángula
trigger current (SCR, triac)	triggerstroom	Zündstrom	corriente de control
turn-off losses	uitschakelverliezen	Abschaltverluste	perdidas de cortar
turn-off behaviour	uitschakelgedrag	Ausschaltverhalten	conducta de desconexión
turn-off characteristics	turn-off eigenschappen	Abschalteigenschaften	(turn-off characteristics)
turn-off circuit	uitschakelkring	Abschaltkreis	circuito de desconexión
turn-off gain	uitschakelversterking	Abschaltverstärkung	(turn-off gain)
turn-off time	uitschakeltijd	Verzögerungszeit	tiempo de cortar
turn-off	uitschakelen	abschalten	desconexión
turn-on losses	aanschakelverliezen	Anschaltverluste	perdidas de ncendido
turn-on time	aanschakeltijd	Anspreichzeit	tiempo de encendido
turn-on time	inschakeltijd	Einschaltzeit	tiempo de encendido
turns ratio	windingsverhouding	Windungszahlenverhältnis	relación de las vueltas de las bobinas
type indication	type-aanduiding	Typkennzeichen	indice de tipificación
uninterruptable (UPS)	onderbrekingsvrij (UPS)	unterbrechungslos(e) (UPS)	(UPS)
unipolar transistor	unipolaire transistor	Unipolartransistor	transistor unipolar
universal motor	universele motor	Universalmotor	motor universal
useable emf	meewerkende emk	mithelfende Emk	emf de cooperación
vacuum tube	elektronenbuis	Elektronenröhre	tubo electronico
valence band	valentieband	Valenzband	banda de valencia
vector control	vectorregeling	Vektorregelung	control vectorial
very inductive	zeer inductief	sehr induktiv	fuerte inductivo
voltage drop	spanningsval	Spannungsabfall	caida de tensión
voltage drop	spanningsverlies	Spannungsabfall	bajada de tensión
voltage peak (line transient)	spanningspiek	Spannungsspitze	cresta de tensión
voltage	spanning	Spannung	tensión
water heater (boiler)	waterverwarmer	Heisswasserspeicher (Boiler)	calentador
wave guide	golfgeleider	Hohlleiter	guiaondas
wave	golf	Welle	onda
waveform	golfvorm	Wellenform	forma de onda
welding transformer	lastransformator	Schweissung (transformator)	transformador de soldadura
wind park	windpark	Windpark	parque de viento
wind turbine	windturbine	Windturbine	turbina eólica
winding	wikkeling	Wicklung	devanado
work coil	werkspoel	Heizinduktor	inductor de calentamiento
working point	instelpunt	Arbeitspunkt	punto defuncionamiento
working quadrant	werkingskwadrant	Arbeitsquadrant	cuadrante de trabajo
Zener knee voltage	Zenerspanning	Zenerspannung	tension de Zener
zero crossing	nulpunt (nuldoorgang)	Nulldurchgang	cruce de cero
zerocross reference	referentie-nulpunt	Referenz-Nulldurchgang	(zerocross reference)
zigzag connection	zigzagschakeling	Zickzackschaltung	conexión en zigzag

INDEX